TREATISE ON ANALYSIS

Volume III

This is Volume 10-III in
PURE AND APPLIED MATHEMATICS
A series of Monographs and Textbooks
Editors: PAUL A. SMITH AND SAMUEL EILENBERG
A complete list of titles in this series appears at the end of this volume.

Volume 10

TREATISE ON ANALYSIS

10-I. Chapters I–XI (Foundations of Modern Analysis, enlarged and corrected printing, 1969).

10-II. Chapters XII–XV, 1970

10-III. Chapters XVI–XVII, 1972

TREATISE ON

ANALYSIS

J. DIEUDONNÉ
Université de Nice
Faculté des Sciences
Parc Valrose, Nice, France

Volume III

Translated by

I. G. Macdonald
University of Manchester
Manchester, England

ACADEMIC PRESS New York and London 1972

ACADEMIC PRESS, INC.
111 Fifth Avenue, New York, New York 10003

United Kingdom Edition published by
ACADEMIC PRESS, INC. (LONDON) LTD.
24/28 Oval Road, London NW1

LIBRARY OF CONGRESS CATALOG CARD NUMBER: 69-12275

AMS(MOS) 1970 Subject Classifications: 58-01, 58A05, 58A10, 58A30

PRINTED IN THE UNITED STATES OF AMERICA

"Treatise on Analysis," Volume III

First published in the French Language under the title "Elements d'Analyse," tome 3 and copyrighted in 1970 by Gauthier-Villars, Éditeur, Paris, France.

SCHEMATIC PLAN OF THE WORK

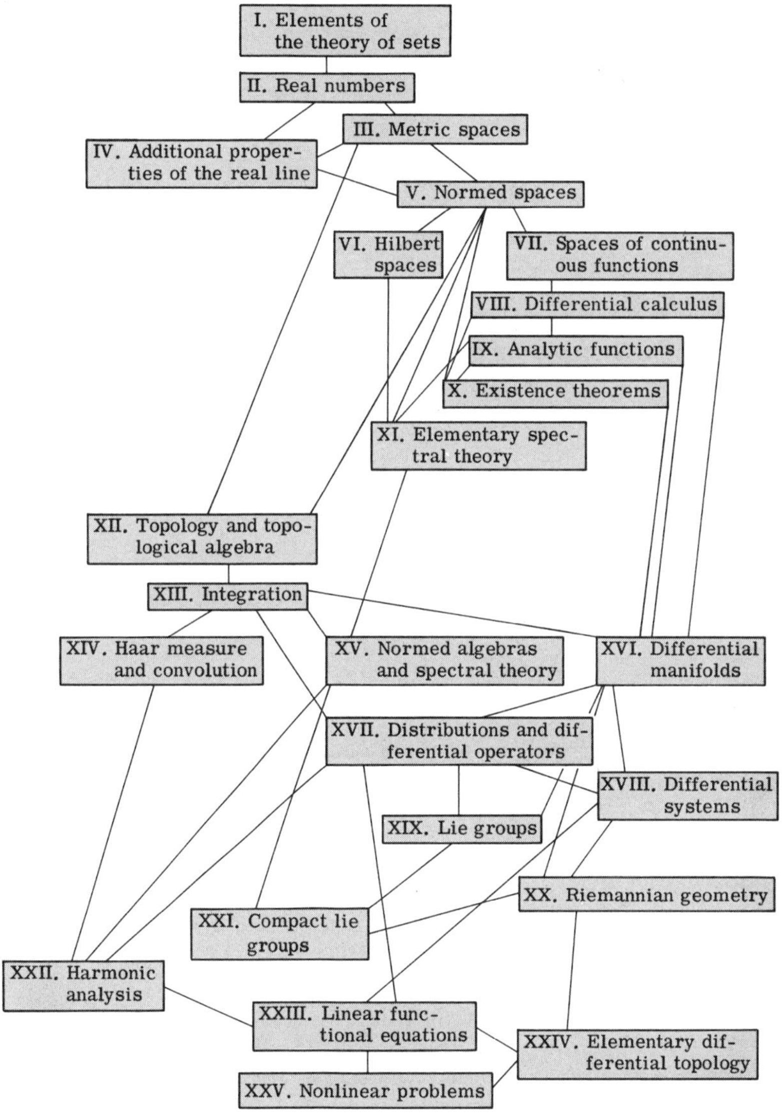

CONTENTS

Notation ix

Chapter XVI

DIFFERENTIAL MANIFOLDS 1

1. Charts, atlases, manifolds. 2. Examples of differential manifolds. Diffeomorphisms. 3. Differentiable mappings. 4. Differentiable partitions of unity. 5. Tangent spaces, tangent linear mappings, rank. 6. Products of manifolds. 7. Immersions, submersions, subimmersions. 8. Submanifolds. 9. Lie groups. 10. Orbit spaces and homogeneous spaces. 11. Examples: unitary groups, Stiefel manifolds, Grassmannians, projective spaces. 12. Fibrations. 13. Definition of fibrations by means of charts. 14. Principal fiber bundles. 15. Vector bundles. 16. Operations on vector bundles. 17. Exact sequences, subbundles, and quotient bundles. 18. Canonical morphisms of vector bundles. 19. Inverse image of a vector bundle. 20. Differential forms. 21. Orientable manifolds and orientations. 22. Change of variables in multiple integrals. Lebesgue measures. 23. Sard's theorem. 24. Integral of a differential n-form over an oriented pure manifold of dimension n. 25. Embedding and approximation theorems. Tubular neighborhoods. 26. Differentiable homotopies and isotopies. 27. The fundamental group of a connected manifold. 28. Covering spaces and the fundamental group. 29. The universal covering of a differential manifold. 30. Covering spaces of a Lie group.

Chapter XVII

DIFFERENTIAL CALCULUS ON A DIFFERENTIAL MANIFOLD
I. DISTRIBUTIONS AND DIFFERENTIAL OPERATORS. 230

1. The spaces $\mathscr{E}^{(r)}(\mathrm{U})$ (U open in $\mathbf{R}^n$). 2. Spaces of C^∞ (resp. C^r) sections of vector

bundles. 3. Currents and distributions. 4. Local definition of a current. Support of a current. 5. Currents on an oriented manifold. Distributions on $\mathbf{R}^n$. 6. Real distributions. Positive distributions. 7. Distributions with compact support. Point-distributions. 8. The weak topology on spaces of distributions. 9. Example: finite parts of divergent integrals. 10. Tensor products of distributions. 11. Convolution of distributions on a Lie group. 12. Regularization of distributions. 13. Differential operators and fields of point-distributions. 14. Vector fields as differential operators. 15. The exterior differential of a differential p-form. 16. Connections in a vector bundle. 17. Differential operators associated with a connection. 18. Connections on a differential manifold. 19. The covariant exterior differential. 20. Curvature and torsion of a connection.

Appendix
MULTILINEAR ALGEBRA 347

8. Modules. Free modules. 9. Duality for free modules. 10. Tensor product of free modules. 11. Tensors. 12. Symmetric and antisymmetric tensors. 13. The exterior algebra. 14. Duality in the exterior algebra. 15. Interior products. 16. Nondegenerate alternating bilinear forms. Symplectic groups. 17. The symmetric algebra. 18. Derivations and antiderivations of graded algebras. 19. Lie algebras.

References 378

Index 381

NOTATION

In the following definitions, the first number is the number of the chapter and the second number is the number of the section within that chapter.

$\dim_x(X)$ (resp. $\dim_{\mathbf{C},x}(X)$)	dimension of a differential (resp. complex-analytic) manifold X at a point x: 16.1
$\dim(X)$, $\dim_{\mathbf{C}}(X)$	dimension of a pure differential (resp. complex-analytic) manifold X: 16.1
$c\|V$	restriction of a chart c to an open set V: 16.1
$u(c)$	image of a chart c under a homeomorphism u: 16.2
$T_x(X)$	tangent space at a point x of a differential manifold X: 16.5
$\boldsymbol{\theta}_{c,x}$, $\boldsymbol{\theta}_c$	mapping of $T_x(X)$ onto $\mathbf{R}^n$ induced by a chart $c = (U, \varphi, n)$ at the point x: 16.5
$\boldsymbol{\tau}_x$	canonical mapping of $T_x(E)$ onto E, where E is a finite-dimensional affine space: 16.5
$T_x(f)$	tangent linear mapping to f at the point x: 16.5
$\mathrm{rk}_x f$	rank of f at the point x: 16.5

$d_x\mathbf{f}$ (resp. $d_x f$)	differential at the point x of the mapping $\mathbf{f}$ (resp. f) of X into a vector space (resp. into $\mathbf{R}$): 16.5
$\mathrm{Hess}_x(f)$	Hessian of f at the point x: 16.5
$X_{\mid\mathbf{R}}$	differential manifold underlying a complex-analytic manifold X: 16.5
$\boldsymbol{\theta}_{c\mid\mathbf{R}}$	mapping of $T_x(X_{\mid\mathbf{R}})$ onto $\mathbf{R}^{2n}$, where X is a complex-analytic manifold of dimension n: 16.5
J_x	$\mathbf{R}$-linear automorphism of the tangent space at x to a complex-analytic manifold, defined by multiplication by i: 16.5
$\overline{T_x(X)^*}$	space of antilinear forms on the tangent space at x to a complex-analytic manifold X: 16.5
$d'_x f$, $d''_x f$	$\mathbf{C}$-linear and $\mathbf{C}$-antilinear parts of $d_x f$, where f is a complex function on a complex-analytic manifold: 16.5
$\mathrm{grad}\, f(x)$	gradient of f at the point x: 16.5, Problem 7
$J_x^k(X, Y)_y$, $J_x^k(X, Y)$, $J^k(X, Y)_y$, $J^k(X, Y)$	sets of jets of order k from X to Y: 16.5, Problem 9
$J_x^k(f)$	jet of order k of the mapping f at the point x: 16.5, Problem 9
$X \times_Z Y$	fiber-product of two manifolds X, Y over Z: 16.8, Problem 10
$\mathbf{GL}(n, \mathbf{R})$, $\mathbf{GL}(n, \mathbf{C})$, $\mathbf{GL}(n, \mathbf{H})$	general linear groups in n variables over $\mathbf{R}$, $\mathbf{C}$, $\mathbf{H}$: 16.9
$s \cdot \mathbf{h}_x$, $\mathbf{h}_x \cdot s^{-1}$, $s \cdot \mathbf{h}_x \cdot t^{-1}$	actions of elements s, t of a Lie group G on a tangent vector $\mathbf{h}_x$ at a point $x \in G$: 16.9
$\mathbf{SL}(n, \mathbf{R})$, $\mathbf{SL}(n, \mathbf{C})$	special linear groups: 16.9
$s \cdot \mathbf{k}_x$, $\mathbf{h}_s \cdot x$	s an element of a Lie group G, x a point of a manifold X on which G acts, $\mathbf{h}_s$ a tangent vector to G at s, $\mathbf{k}_x$ a tangent vector to X at x: 16.10
$\mathscr{H}_{p,q}(E)$, $\mathscr{H}(E)$	spaces of quadratic (resp. hermitian) forms over a real (resp. complex or quaternionic) vector space E: 16.11
$\mathbf{O}(\Phi)$, $\mathbf{O}(n, \mathbf{R})$, $\mathbf{O}(n)$	orthogonal groups: 16.11
$\mathbf{SO}(n, \mathbf{R})$, $\mathbf{SO}(n)$	rotation groups: 16.11
$\mathbf{U}(\Phi)$, $\mathbf{U}(n, \mathbf{C})$, $\mathbf{U}(n, \mathbf{H})$, $\mathbf{U}$	unitary groups: 16.11

$\mathbf{SU}(n)$	special unitary group: 16.11
$\mathbf{S}_{n,p}(\mathbf{R})$, $\mathbf{S}_{n,p}$, $\mathbf{S}_{n,p}(\mathbf{C})$, $\mathbf{S}_{n,p}(\mathbf{H})$	Stiefel manifolds: 16.11
$\mathbf{G}_{n,p}(\mathbf{R})$, $\mathbf{G}_{n,p}$, $\mathbf{G}_{n,p}(\mathbf{C})$, $\mathbf{G}_{n,p}(\mathbf{H})$	Grassmannians: 16.11
$\mathbf{P}_n(\mathbf{R})$, $\mathbf{P}_n(\mathbf{C})$, $\mathbf{P}_n(\mathbf{H})$	projective spaces: 16.11
$\mathbf{SO}(\Phi)$	rotation group: 16.11, Problem 4
$\mathbf{Sp}(\Phi)$	symplectic group: 16.11, Problem 6
$B' \times_B X, f^*(\lambda)$	inverse image of a fiber bundle: 16.12
$f^*(s)$	inverse image of a section: 16.12
$\Gamma(B', X)$	set of C^∞-sections of a bundle X over a submanifold B′ of the base: 16.12
$X \times_B X'$	fiber-product of two bundles over B: 16.12
$X \times^G F$	bundle associated with a principal bundle X (with group G), with fiber-type F: 16.14
$x \cdot y$	element of $X \times^G F$ corresponding to $x \in X$ and $y \in F$: 16.14
$\mathbf{h}_x \cdot y, x \cdot \mathbf{k}_y$	x a point of X, y a point of F, $\mathbf{h}_x$ a tangent vector at x to X, $\mathbf{k}_y$ a tangent vector at y to F: 16.14
$\mathbf{0}, \mathbf{0}_E$	zero section of a vector bundle E: 16.15
$\mathbf{0}_b$	zero element of a fiber E_b of a vector bundle E: 16.15
$\mathrm{rk}_b(E)$	rank of a vector bundle E at a point b of the base manifold: 16.15
$T(M)$	tangent bundle of a differential manifold M: 16.15
o_M	canonical projection of the tangent bundle T(M): 16.15
$\mathscr{E}(B; \mathbf{R})$ (resp. $\mathscr{E}(B; \mathbf{C})$), $\mathscr{E}(B)$	algebra of real- (resp. complex-) valued C^∞-functions on B: 16.15
$\mathrm{Mor}(E, E')$	vector space of morphisms of E into E′, where E and E′ are vector bundles over the same base B: 16.15
$\Gamma(B, E), \Gamma(E)$	vector space of C^∞-sections of a vector bundle E over B: 16.15
$E' \oplus E''$	direct sum (Whitney sum) of two vector bundles E′, E″ over B: 16.16
$\mathbf{s}' \oplus \mathbf{s}''$	direct sum of a section $\mathbf{s}'$ of E′ and a section $\mathbf{s}''$ of E″: 16.16
$E' \otimes E''$	tensor product of two vector bundles E′, E″ over B: 16.16

$\mathbf{s}' \otimes \mathbf{s}''$	tensor product of a section $\mathbf{s}'$ of E′ and a section $\mathbf{s}''$ of E″: 16.16
$m\mathrm{E}$, $\mathrm{E}^{\otimes m}$	direct sum (tensor product) of m copies of a vector bundle E: 16.16
$\bigwedge^m \mathrm{E}$	mth exterior power of a vector bundle E: 16.16
$\mathbf{s}_1 \wedge \mathbf{s}_2 \wedge \cdots \wedge \mathbf{s}_m$	exterior product of m sections of E: 16.16
$\bigwedge^m u$	mth exterior power of a homomorphism of vector bundles: 16.16
$\mathrm{E}^{\otimes 0}$, $\bigwedge^0 \mathrm{E}$, I	trivial line-bundle: 16.16
Hom(E′, E″)	vector bundle of homomorphisms of a vector bundle E′ into a vector bundle E″: 16.16
E*	dual of a vector bundle E: 16.16
$\langle \mathbf{s}, \mathbf{s}^* \rangle$, $\langle \mathbf{s}^*, \mathbf{s} \rangle$	$\mathbf{s}$ a section of E, $\mathbf{s}^*$ a section of E*: 16.16
${}^t u$	transpose of a homomorphism u of vector bundles: 16.16
Hom(u', u'')	homomorphism of Hom(E′, E″) into Hom(F′, F″) corresponding to two vector bundle homomorphisms $u' : \mathrm{F}' \to \mathrm{E}'$, $u'' : \mathrm{E}'' \to \mathrm{F}''$: 16.16
$\mathrm{U}_{n,p}(\mathbf{R})$, $\mathrm{U}_{n,p}(\mathbf{C})$	canonical vector bundle over $\mathbf{G}_{n,p}(\mathbf{R})$, (resp. $\mathbf{G}_{n,p}(\mathbf{C})$): 16.16, Problem 1
$\mathbf{S}_m(\mathrm{E})$, $\mathbf{A}_m(\mathrm{E})$	bundle of symmetric (antisymmetric) tensors of order m over a vector bundle E: 16.17
$\mathbf{S}_m(\mathrm{E}_b)$, $\mathbf{A}_m(\mathrm{E}_b)$	space of symmetric (antisymmetric) tensors of order m over the fiber E_b: 16.17
c_k^j	contraction of the contravariant index j and the covariant index k: 16.18
Tr(u)	trace of an endomorphism u of a vector bundle: 16.18
$\boldsymbol{a}$	antisymmetrization operator: 16.18
$\mathbf{s} \wedge \mathbf{t}$	exterior product of a section of $\bigwedge^p \mathrm{E}$ and a section of $\bigwedge^q \mathrm{E}$: 16.18
$\bigwedge \mathrm{E}$	exterior algebra bundle of a vector bundle E: 16.18
$\bigwedge^m \mathrm{E}^*$	mth exterior power of the dual (= dual of the mth exterior power) of E: 16.18
$i(\mathbf{s})\mathbf{z}^*$, $i_{\mathbf{s}} \cdot \mathbf{z}^*$	interior product of a section $\mathbf{s}$ of E and

	a section $\mathbf{z}^*$ of $\bigwedge^p E^*$: 16.18
$E_{(\mathbf{C})}$	complex vector bundle obtained by extension of scalars from a real vector bundle E: 16.18
$f^*(E)$	inverse image of a vector bundle E: 16.19
$\mathbf{T}_q^p(M)$	tensor bundle of type (p, q) over a manifold M: 16.20
$\mathscr{T}_q^p(M)$	vector space of C^∞ tensor fields of type (p, q) on M: 16.20
$\mathscr{E}_{p,\mathbf{R}}(M)$, $\mathscr{E}_p(M)$	vector space of C^∞ real differential p-forms on M: 16.20
df	differential of a function f of class C^1: 16.20
κ_X	canonical differential 1-form on the cotangent bundle $T(X)^*$: 16.20
${}^tf(\alpha)$, ${}^tf(\mathbf{Z})$	inverse image under f of a p-form α and of a covariant tensor field $\mathbf{Z}$: 16.20
$\boldsymbol{\omega}' \wedge_B \boldsymbol{\omega}''$, $\boldsymbol{\omega}' \wedge \boldsymbol{\omega}''$, $B(\boldsymbol{\omega}, \boldsymbol{\omega})$	exterior products of vector-valued differential forms relative to a bilinear mapping B: 16.20
$\upsilon(x) > 0$	υ a differential n-form on an oriented manifold X: 16.21
υ_x/ζ_y, υ/ζ_y, $\upsilon/{}^tf(\zeta_y)$	f a submersion of X onto Y (oriented manifolds of dimensions n and m, respectively), ζ_y an m-covector at the point $y = f(x)$, υ_x (resp. υ) an n-covector at the point x (resp. an n-form): 16.21
$\int \upsilon$, $\int_X \upsilon$, $\int_X \upsilon(x)$	integral of an n-form υ over an oriented manifold X of dimension n: 16.24
$-X$	X an oriented manifold: 16.24
$\int_Y \sigma$	Y an oriented submanifold of dimension p of the manifold X, σ a p-form on X: 16.24
Ω_n	surface measure of the sphere $\mathbf{S}_{n-1}$: 16.24
$\alpha^\flat$	integral along the fibers of a differential p-form α: 16.24, Problem 11
$X_1 \# X_2$	connected sum of two connected manifolds of the same dimension: 16.26, Problem 15
$\pi_1(X)$, $\pi_1(X, a)$	fundamental group: 16.27

$\pi_n(X)$, $\pi_n(X, a)$	nth homotopy group ($n \geqq 2$): 16.30, Problem 3
$\pi_n(X, A, a)$	nth relative homotopy group: 16.30, Problem 4
$\mathscr{E}(U)$, $\mathscr{E}_{\mathbf{C}}(U)$	space of complex-valued C^∞-functions on the open set U: 17.1
$\mathscr{E}^{(r)}(U)$, $\mathscr{E}^{(r)}_{\mathbf{C}}(U)$	space of complex-valued C^r-functions on the open set U: 17.1
$\Gamma^{(r)}(U, E)$	space of C^r-sections of a vector bundle E over an open set U: 17.2
$\mathscr{E}_p(X)$, $\mathscr{E}^{(r)}_p(X)$	space of C^∞ (C^r) complex-valued differential p-forms on X: 17.3
$\mathscr{D}_p(X; K)$, $\mathscr{D}^{(r)}_p(X; K)$	the subspace of $\mathscr{E}_p(X)$ ($\mathscr{E}^{(r)}_p(X)$) consisting of p-forms with support contained in the compact set K: 17.3
$\mathscr{D}_p(X)$, $\mathscr{D}^{(r)}_p(X)$	the union of the $\mathscr{D}_p(X; K)$ ($\mathscr{D}^{(r)}_p(X; K)$) for all compact subsets K of X: 17.3
$\mathscr{D}(X; K)$, $\mathscr{D}^{(r)}(X; K)$, $\mathscr{D}(X)$, $\mathscr{D}^{(r)}(X)$	particular cases of $\mathscr{D}_p(X; K)$, $\mathscr{D}^{(r)}_p(X; K)$, $\mathscr{D}_p(X)$, $\mathscr{D}^{(r)}_p(X)$ for $p = 0$: 17.3
$\varepsilon_{\mathbf{z}_x}$	Dirac p-current defined by the tangent p-vector $\mathbf{z}_x$: 17.3
$T \wedge \omega$	T a p-current, ω a differential q-form ($q \leqq p$): 17.3
${}^t i_Y \cdot T$	T a p-current, Y a vector field: 17.3
$u(T)$	T a current, u a proper mapping: 17.3
$\mathscr{D}'_p(X)$, $\mathscr{D}'^{(r)}_p(X)$	space of p-currents (p-currents of order $\leqq r$) on X: 17.3
$\mathscr{D}'(X)$, $\mathscr{D}'^{(r)}(X)$	space of distributions (distributions of order $\leqq r$) on X: 17.3
$\langle T, \alpha \rangle$, $\langle \alpha, T \rangle$	T a p-current, α a differential p-form with compact support: 17.3
$\langle T, f \rangle$, $\langle f, T \rangle$, $\int f(x)\, dT(x)$	T a distribution, f a function: 17.3
Supp(T)	support of a current T: 17.4
${}^t\pi(T)$	inverse image of a current under a local diffeomorphism π: 17.4
$\mathscr{E}_{n-p,\mathrm{loc}}(X)$	space of locally integrable $(n-p)$-forms on X: 17.5
T_β	p-current defined by an $(n-p)$-form β on an oriented manifold: 17.5
T_f	n-current defined by a scalar function f: 17.5

$\mathrm{T}_{\|\upsilon_0}$	n-current defined by a distribution T, relative to an n-form υ_0: 17.5
$\mathrm{T}_f\ (=\mathrm{T}_{f\upsilon_0})$	distribution defined by a function f (relative to an n-form υ_0): 17.5
$\mathrm{D}^\nu\mathrm{T}$	derivative of order ν (ν = multi-index) of a distribution $\mathrm{T} \in \mathscr{D}'(\mathrm{U})$, where U is an open set in $\mathbf{R}^n$: 17.5
Y, Y_1	Heaviside's function: 17.5
Supp sing (T)	singular support of a distribution T: 17.5
${}^t\pi(\mathrm{T})$	inverse image of a current under a submersion: 17.5, Problem 8
$\mathrm{tr}(f)$	value of the trace measure tr on a function f: 17.5, Problem 10
$\mathscr{E}^{(r)}_{p,\mathbf{R}}(\mathrm{X})$	space of real-valued C^r differential p-forms on X: 17.6
$\mathscr{E}'(\mathrm{U})$	space of distributions with compact support: 17.7
$\pi(\mathrm{T})$	image of a compactly supported current T of order $\leqq r-1$ under a C^r mapping π: 17.7
$\langle \mathrm{T}, \alpha \rangle$	T a p-current with compact support, α a differential p-form: 17.8
Δ	Laplacian: 17.9
$\mathrm{Pf}(r^\zeta)$	finite part of r^ζ: 17.9
$\mathrm{Pf}(x_+^\zeta)$	finite part of x_+^ζ: 17.9
$\mathrm{Y}_\zeta = \dfrac{1}{\Gamma(\zeta)}\,\mathrm{Pf}(x_+^{\zeta-1})$	distribution on $\mathbf{R}$: 17.9
$\square$	d'Alembertian: 17.9
Z_ζ	distribution on $\mathbf{R}^n$: 17.9
$\mathrm{P.V.}\left(\dfrac{1}{x}\right)$, $\mathrm{P.V.}(g(x))$	Cauchy principal value: 17.9, Problem 1
$\mathrm{S} \otimes \mathrm{T}$	tensor product of two distributions: 17.10
$\iint f(x, y)\, d\mathrm{S}(x)\, d\mathrm{T}(y)$	value of $\mathrm{S} \otimes \mathrm{T}$ at f: 17.10
$\mathrm{T}_1 * \mathrm{T}_2 * \cdots * \mathrm{T}_n$	convolution of n distributions T_j on a Lie group: 17.11
$\check{\mathrm{T}}$	T a distribution on a Lie group: 17.11
Diff(X)	algebra of differential operators on X: 17.13
${}^tP \cdot \mathrm{T}$	P a differential operator, T a distribution: 17.13
θ_X	Lie derivative relative to a vector field X:

	17.14
$[X, Y]$	Lie bracket of two vector fields X, Y: 17.14
$d\alpha$	exterior differential of a p-form α: 17.15
$b\mathrm{T}$	boundary of a p-current T: 17.15
$d\boldsymbol{\alpha}$	exterior differential of a vector-valued p-form $\boldsymbol{\alpha}$: 17.15
$d'\alpha, d''\alpha$	**C**-linear and **C**-antilinear parts of $d\alpha$ on a complex manifold: 17.15
$\mathbf{C}$	linear connection in a vector bundle: 17.16
$\mathrm{rel}_{\mathbf{C}}$	horizontal lifting of a vector field relative to a connection $\mathbf{C}$: 17.16
$\nabla_{\mathbf{h}_z} \cdot \boldsymbol{G}$	covariant derivative of a mapping $\boldsymbol{G}$ of a manifold N into a vector bundle E, in the direction of a tangent vector $\mathbf{h}_z$ to N, relative to a linear connection in E: 17.17
$\nabla\boldsymbol{U}$	covariant differential of a tensor field $\boldsymbol{U}$ on M, relative to a linear connection on M: 17.18
$\mathbf{d}\boldsymbol{\omega}$	covariant exterior differential (relative to a linear connection on E) of a differential p-form with values in E: 17.19
$\boldsymbol{r}_f$	curvature morphism relative to the mapping f: 17.20
$\boldsymbol{r}$	curvature morphism (or curvature) of a connection in a vector bundle E: 17.20
$\mathbf{r}$	curvature tensor of a connection in E: 17.20
$\boldsymbol{t}$	torsion morphism of a connection on M: 17.20
$\mathbf{t}$	torsion tensor of a connection on M: 17.20
$\langle x, x^* \rangle, \langle x^*, x \rangle$	x an element of a module, x^* an element of its dual: A.9.1
c_{E}	canonical mapping of a module E into its bidual: A.9.1
${}^t u$	transpose of a linear mapping: A.9.3
${}^t u^{-1}$	contragradient of an isomorphism of one module onto another: A.9.3
${}^t X$	transpose of a matrix X: A.9.4

$x_1^* \otimes \cdots \otimes x_n^*$	tensor product of linear mappings: A.10
$E_1 \otimes_A E_2 \otimes_A \cdots \otimes_A E_n$	tensor product of finitely-generated free A-modules: A.10.3
$E^{\otimes n}$, $\mathbf{T}^n(E)$, $\mathbf{T}_0^n(E)$	nth tensor power of a finitely-generated free module: A.11.1
$\mathbf{T}_q^p(E)$ $(p \geqq 0, q \geqq 0)$	module of tensors of type (p, q): A.11.1
c_j^i	contraction of the contravariant index i and the covariant index j: A.11.3
$\mathrm{Tr}(u)$	trace of an endomorphism: A.11.3
$\sigma \cdot z$	transform of a contravariant tensor z by a permutation σ: A.12.1
$\boldsymbol{s} \cdot z$, $\boldsymbol{a} \cdot z$	symmetrization and antisymmetrization of a tensor: A.12.2
$u \wedge v$	exterior product of two antisymmetric tensors: A.13.2
$\bigwedge^m E$	mth exterior power of a finitely-generated free module: A.13.3
e_H	basis elements of $\bigwedge^m E$: A.13.3
$\bigwedge^m u$	mth exterior power of a linear mapping: A.13.4
$\bigwedge E$	exterior algebra of a finitely-generated free A-module E: A.13.5
$z_q \lrcorner u_{p+q}^*$	interior product of a q-vector and a $(p+q)$-form: A.15.1
$i(x)$	interior product by a vector x: A.15.4
$\mathbf{Sp}(E, B)$, $\mathbf{Sp}(n, K)$	symplectic groups: A.16.4
$z_p z_q$	symmetric product of two symmetric tensors: A.17.1
$\mathbf{S}_n(E)$	nth symmetric power of a finitely-generated free A-module E: A.17
$\mathbf{S}(E)$	symmetric algebra of a finitely-generated free A-module E: A.17
e^α	basis element of a symmetric power: A.17
$\mathrm{ad}(a)$	inner derivation in an associative algebra: A.18.2
$\mathrm{ad}_{\mathfrak{g}}(x)$, $\mathrm{ad}(x)$	inner derivation in a Lie algebra: A.19.4
$\mathrm{Der}(\mathfrak{g})$	algebra of derivations of a Lie algebra: A.19.4

CHAPTER XVI

DIFFERENTIAL MANIFOLDS

One of the dominant themes of modern mathematics may be described as "analysis on differential manifolds." Of course, the word "analysis" is to be understood here in its widest sense; in it are inextricably blended a range and variety of concepts whose latent fertility would be stunted by confining them within the old and artificial divisions of algebra, geometry, and analysis in the classical sense of these words.

The traditional domain of differential geometry, namely the study of curves and surfaces in three-dimensional space, was soon realized to be inadequate, particularly under the influence of mechanics. A solid body depends on six parameters, and a system of n points depends on $3n$ parameters, in general connected by certain relations. It is therefore natural to represent such systems by points in a space $\mathbf{R}^N$ (where N is any positive integer) restricted to lie on a "submanifold" of this space, defined by certain equations; and it is important to have available on such a "manifold" algorithms generalizing those of differential and integral calculus on open sets in $\mathbf{R}^N$, which were developed in Chapters VIII, X, and XIII (see [37], Introduction). This and the two following chapters are devoted to the development of such algorithms.

The classical viewpoint in the differential study of surfaces consists in regarding them as embedded in the ambient space $\mathbf{R}^3$. This point of view can be generalized, and in fact there is no loss of generality in considering only differential manifolds embedded in $\mathbf{R}^N$ (Section 16.25, Problem 2). In certain problems, such an embedding can be a useful device (Sections 16.25 and 16.26). Nevertheless, this conception is patently artificial (for example, a sphere remains the same surface whether we embed it in $\mathbf{R}^3$ or in $\mathbf{R}^4$). Moreover, the necessity of embedding every manifold in a space $\mathbf{R}^N$ would be an intolerable constraint in relation to many operations which arise naturally on manifolds (for example, the formation of "orbit manifolds" (16.10.3)).

The first of the two fundamental notions which are developed in this Chapter is therefore the concept (which goes back to Gauss) of a differential manifold as an *intrinsic* object, independent of any adventitious embedding in a Euclidean space. The essential idea is that *locally* a differential manifold of dimension n can be identified with an open subset of $\mathbf{R}^n$ by means of a *chart* (16.1), and the structure of the manifold resides in the manner in which the charts are patched together.

The property, for two mappings of $\mathbf{R}$ into $\mathbf{R}^n$, of being *tangent* at a point $t_0 \in \mathbf{R}$ (8.1) is invariant under a differentiable change of variables in $\mathbf{R}$ or in $\mathbf{R}^n$. This fact makes it possible to define, for two differentiable mappings f, g of $\mathbf{R}$ into a differential manifold M of dimension n, the relation "f and g are tangent at a point $t_0 \in \mathbf{R}$" (which implies that $f(t_0) = g(t_0) = x_0 \in \mathrm{M}$). The equivalence classes for this relation (with t_0 and x_0 fixed) are no longer called "derivatives" but *tangent vectors to* M *at the point* x_0. They form in a natural way a real vector space $\mathrm{T}_{x_0}(\mathrm{M})$ of dimension n, called the *tangent vector space to* M *at the point* x_0 (16.5). The fundamental difference, compared with analysis in vector spaces, is that the space $\mathrm{T}_x(\mathrm{M})$ *varies with* x (whereas all the values of the derivative of a function with values in $\mathbf{R}^n$ are considered as belonging to the same vector space). This variation of the tangent space may seem familiar, from the example of surfaces in $\mathbf{R}^3$; unfortunately, geometric intuition here is misleading, because it suggests that the "tangent plane" also is embedded in $\mathbf{R}^3$. To see that one cannot arrive in this way at a correct conception of tangent vectors, it is enough to remark that a tangent vector at a point x of a surface S depends firstly on the two parameters which determine x, and then for each x on two more parameters which fix the vector in the tangent plane $\mathrm{T}_x(\mathrm{S})$. The tangent vectors to S must therefore be considered as forming a *four*-dimensional manifold, which clearly cannot be embedded in $\mathbf{R}^3$. The notion which is appropriate here, and which enables us to "pull" the tangent vectors out of the ambient space, is the second fundamental idea in this chapter, that of a *fiber bundle*. In its various forms it dominates nowadays not only differential geometry, but all of topology (see [43] and [49]).

This chapter is very long, and the greater part of it consists essentially of *transcriptions*: either in order to give intrinsic expression to properties (usually of a *local* nature) of mappings of one differential manifold into another, by reducing to the case of open sets in $\mathbf{R}^n$ by means of charts; or in order to transpose to the context of vector bundles the elementary notions and results of linear and multilinear algebra. Apart from the theorem on the existence of orbit-manifolds (16.10.3) and its consequences, there is no substantial theorem before Section 16.21. The only aid to digestion of this accumulation of definitions and trivialities that I have been able to devise is to make the chapter yet longer, by inserting in the text and in the problems as many and various examples as possible, in order to show the full richness of the ideas introduced. The most important of these examples are connected with the notion of Lie group (which will be studied in greater depth in Chapters XIX and XXI) and the closely related notion of homogeneous spaces. It is a fact of experience that the most important manifolds in applications are homogeneous space (for example, the upper half-plane $\mathscr{I}z > 0$ is a homogeneous space of the unimodular group $\mathbf{SL}(2, \mathbf{R})$). From the fact that such a space can be put in the form G/H, one has algorithms available arising from the group structure of G; the precept of "lifting everything up to the group" has shown its validity in all studies of homogeneous spaces, and the reader will have many opportunities to see it in action.

From Section 16.21 onwards, we can at last begin to study some elementary *global* questions on differential manifolds: orientation (16.21), integration on a manifold (16.22–16.24), elementary properties of approximation and homotopy (16.25 and 16.26), and finally the theory of covering spaces, limited to differential manifolds (16.27–16.30).

1. CHARTS, ATLASES, MANIFOLDS

Let X be a topological space. A *chart* of X is a triplet $c = (\mathrm{U}, \varphi, n)$, where U is an open set in X, n is an integer $\geqq 0$, and φ is a homeomorphism of U onto an open set in $\mathbf{R}^n$. The integer n is called the *dimension* of the chart c, and the open set U is its *domain of definition.* If V is an open set contained in U, it is clear that the restriction $\varphi | \mathrm{V}$ is a homeomorphism of V onto an open set in $\mathbf{R}^n$, and therefore $c | \mathrm{V} = (\mathrm{V}, \varphi | \mathrm{V}, n)$ is a chart, called the *restriction* of c to V.

Consider two charts $c = (\mathrm{U}, \varphi, n)$ and $c' = (\mathrm{U}, \varphi', n')$ of X with the same domain of definition U. They are said to be *compatible* if the two homeomorphisms

$$\varphi' \circ \varphi^{-1} : \varphi(\mathrm{U}) \to \varphi'(\mathrm{U}),$$
$$\varphi \circ \varphi'^{-1} : \varphi'(\mathrm{U}) \to \varphi(\mathrm{U})$$

(called the *transition homeomorphisms*) are *indefinitely differentiable* (8.12). This implies that, for each $x \in \varphi(\mathrm{U})$, the derivative $\mathrm{D}(\varphi' \circ \varphi^{-1})(x)$ is a *bijective* linear mapping of $\mathbf{R}^n$ onto $\mathbf{R}^{n'}$, hence $n = n'$.

(16.1.1) *For two charts* (U, φ, n) *and* $(\mathrm{U}, \varphi', n)$ *with the same domain of definition to be compatible, it is necessary and sufficient that* $\varphi' \circ \varphi^{-1}$ *should be an indefinitely differentiable bijection of* $\varphi(\mathrm{U})$ *onto* $\varphi'(\mathrm{U})$ *whose derivative* $\mathrm{D}(\varphi' \circ \varphi^{-1})$ *has rank n at every point of* $\varphi(\mathrm{U})$.

The condition is clearly necessary, and the sufficiency follows from (10.2.5).

Two arbitrary charts $c = (\mathrm{U}, \varphi, n)$ and $c' = (\mathrm{U}', \varphi', n')$ of X are said to be *compatible* if either $\mathrm{U} \cap \mathrm{U}' = \varnothing$ or the restrictions $(\mathrm{U} \cap \mathrm{U}', \varphi|(\mathrm{U} \cap \mathrm{U}'), n)$ and $(\mathrm{U} \cap \mathrm{U}', \varphi'|(\mathrm{U} \cap \mathrm{U}'), n')$ of c and c' to $\mathrm{U} \cap \mathrm{U}'$ are compatible. If c and c' are compatible, then for each pair of open sets $\mathrm{V} \subset \mathrm{U}$ and $\mathrm{V}' \subset \mathrm{U}'$, the restrictions $c|\mathrm{V}$ and $c'|\mathrm{V}'$ are compatible.

An *atlas* of X is a set $\mathfrak{A}$ of charts of X, each pair of which are compatible and whose domains of definition cover X. Two atlases $\mathfrak{A}$, $\mathfrak{B}$ of X are said to be *compatible* if $\mathfrak{A} \cup \mathfrak{B}$ is an atlas of X, or equivalently if each chart in $\mathfrak{A}$ is compatible with each chart in $\mathfrak{B}$.

(16.1.2) *On the set of atlases of* X, *the relation* R: "$\mathfrak{A}$ *and* $\mathfrak{B}$ *are compatible*" *is an equivalence relation.*

Reflexivity and symmetry are obvious; we have to prove that the relation is transitive. Let $\mathfrak{A}_1$, $\mathfrak{A}_2$, $\mathfrak{A}_3$ be three atlases of X such that $\mathfrak{A}_1$ and $\mathfrak{A}_2$ are compatible, and $\mathfrak{A}_2$ and $\mathfrak{A}_3$ compatible. We shall show that if $c_1 = (\mathrm{U}_1, \varphi_1, n_1)$ and $c_3 = (\mathrm{U}_3, \varphi_3, n_3)$ are charts belonging to $\mathfrak{A}_1$ and $\mathfrak{A}_3$, respectively, then they are compatible. We may assume that $\mathrm{U}_1 \cap \mathrm{U}_3 \neq \varnothing$, or there is nothing to prove. If f_1, f_3 are the restrictions of φ_1, φ_3 to $\mathrm{U}_1 \cap \mathrm{U}_3$, it is clear that $f_3 \circ f_1^{-1}$ is a bijection of $\varphi_1(\mathrm{U}_1 \cap \mathrm{U}_3)$ onto $\varphi_3(\mathrm{U}_1 \cap \mathrm{U}_3)$. Moreover, for each $x \in \mathrm{U}_1 \cap \mathrm{U}_3$, there is a chart $c_2 = (\mathrm{U}_2, \varphi_2, n_2)$ in $\mathfrak{A}_2$ such that $x \in \mathrm{U}_2$. If g_1, g_2, g_3 are the restrictions of φ_1, φ_2, φ_3 to $\mathrm{U}_1 \cap \mathrm{U}_2 \cap \mathrm{U}_3$, the hypothesis that $\mathfrak{A}_1$ and $\mathfrak{A}_2$ (resp. $\mathfrak{A}_2$ and $\mathfrak{A}_3$) are compatible implies that $n_1 = n_2$ (resp. $n_2 = n_3$), and (denoting the common value of n_1, n_2, n_3 by n) that $g_2 \circ g_1^{-1}$ (resp. $g_3 \circ g_2^{-1}$) is indefinitely differentiable with derivative of rank n at every point of $\varphi_1(\mathrm{U}_1 \cap \mathrm{U}_2 \cap \mathrm{U}_3)$ (resp. $\varphi_2(\mathrm{U}_1 \cap \mathrm{U}_2 \cap \mathrm{U}_3)$). It follows that $g_3 \circ g_1^{-1} = (g_3 \circ g_2^{-1}) \circ (g_2 \circ g_1^{-1})$ is indefinitely differentiable with derivative of rank n at every point of $\varphi_1(\mathrm{U}_1 \cap \mathrm{U}_2 \cap \mathrm{U}_3)$ ((8.12.10) and (8.2.1)). This shows that $f_3 \circ f_1^{-1}$ is indefinitely differentiable and has derivative of rank n at every point of $\varphi_1(\mathrm{U}_1 \cap \mathrm{U}_3)$. Hence the result, by (16.1.1).

(16.1.3) It follows from (16.1.2) that the union of all the atlases of a given equivalence class is the *largest* atlas in this class. Such an atlas is said to be *saturated.* A chart which is compatible with all the charts in an atlas $\mathfrak{A}$ belongs to the saturated atlas of the equivalence class of $\mathfrak{A}$.

A *differential manifold* is by definition a *separable metrizable* topological space X on which is given an equivalence class of atlases (with respect to the relation R) or, equivalently, a saturated atlas. The topological space X is called the *underlying* topological space of the differential manifold defined be a saturated atlas on X.

If X is a differential manifold, an *atlas of* X is any atlas in the equivalence class defining X, and a *chart of* X is any chart belonging to one of these atlases (or, equivalently, any chart belonging to the saturated atlas of X). If $c = (\mathrm{U}, \varphi, n)$ is a chart on X, we say that c is a *chart of* X *at* a for each point $a \in \mathrm{U}$. The real-valued functions $\varphi^i = \mathrm{pr}_i \circ \varphi : \mathrm{U} \to \mathbf{R}$ $(1 \leqq i \leqq n)$ are called *coordinates* in U (for the chart c). For each point $a \in \mathrm{U}$, we say that $(\varphi^i)_{1 \leqq i \leqq n}$ is a *system of local coordinates at* a, and the numbers $\varphi^i(a)$ are the *local coordinates of* a for the chart c.

Since all translations in $\mathbf{R}^n$ are indefinitely differentiable, there always exists a chart (U, φ, n) of X at a point $a \in \mathrm{X}$ such that $\varphi(a) = 0$.

An atlas $\mathfrak{A}$ on a separable metrizable space X defines a structure of differential manifold on X, namely that defined by the equivalence class of $\mathfrak{A}$.

(16.1.4) (i) *A differential manifold is locally compact and locally connected, and every point has a neighborhood homeomorphic to a complete metric space. The set of (open) connected components of a differential manifold is at most denumerable.*

(ii) *For each open covering* $(\mathrm{V}_\alpha)_{\alpha \in \mathrm{I}}$ *of a differential manifold* X, *there exists a locally finite denumerable open covering* (U_n) *which is finer than* (V_α) *and consists of relatively compact connected sets which are domains of definition of charts on* X.

Assertion (i) is an immediate consequence of the definitions (cf. (3.18.1), (3.19.1), and (3.20.16)) and of the fact that a differential manifold is separable. Since clearly there exists an open covering which is finer than (V_α) and consists of connected domains of definition of charts on X, assertion (ii) follows from (12.6.1) and from the fact that the restriction of a chart on X to an open set contained in its domain of definition is again a chart on X.

When the differential manifold X is compact, the number of connected components is finite (because they form an open covering of X) and in (16.1.4, (ii)) we may take the covering (U_n) to be *finite*.

(16.1.5) Let X be a differential manifold, x a point of X. Then for all the charts (U, φ, n) on X such that $x \in \mathrm{U}$, the integer n is the same; it is called the *dimension of* X *at the point* x and is written $\dim_x(\mathrm{X})$. Since $\dim_y(\mathrm{X}) = \dim_x(\mathrm{X})$ for all $y \in \mathrm{U}$, the function $x \mapsto \dim_x(\mathrm{X})$ is a continuous mapping of X into the discrete space $\mathbf{N}$; hence it is constant on each connected component of X (3.19.7). When $x \mapsto \dim_x(X)$ is constant over X, the manifold X is said to be *pure*. If X is pure and not empty, the common value of the numbers $\dim_x(X)$ is called the *dimension* of X, written dim(X). A pure differential manifold of dimension 1 (resp. 2) is often called a *curve* (resp. *surface*); but these terms can lead to confusion, since they have several commonly accepted meanings.

Remarks

(16.1.6) (i) A separable metrizable space X is said to be a *topological manifold* if these exists a family of charts on X whose domains of definition cover X. The topological space underlying a differential manifold is therefore a topological manifold, but examples are known of compact topological manifolds which are not the underlying topological spaces of *any* differential manifold.

(ii) If in the definition of compatible charts given above we replace "indefinitely differentiable" by "analytic" (9.3), we have the notion of *real-analytically compatible* charts. A *real-analytic atlas* on a topological space X is a set of charts each pair of which are real-analytically compatible and whose domains of definition cover X. The notion of "real-analytic compatibility" for two such atlases is defined as above, and (16.1.2) extends immediately to this new definition, by virtue of (9.3.2). A *real-analytic manifold* is then a separable metrizable space endowed with an equivalence class of real-analytic atlases. Since a real-analytic atlas is an atlas and since analytically compatible atlases are compatible, it follows that the analytic atlases of a real-analytic manifold X define on X a structure of differentiable manifold, called the differential manifold *underlying* the analytic manifold X.

(iii) We can also replace $\mathbf{R}^n$ by $\mathbf{C}^n$ in all the definitions: in this way we define a *complex-analytic atlas* and a *complex-analytic manifold.* A complex-analytic atlas of a complex-analytic manifold X is also a real-analytic atlas, hence defines on X a structure of a real-analytic manifold, called the real-analytic manifold *underlying* the complex-analytic manifold X.

If $(\varphi^j)_{1 \leqq j \leqq n}$ is a system of (complex) local coordinates at a point a of a complex-analytic manifold X, the $2n$ real-valued functions $\mathscr{R}\varphi^j$ and $\mathscr{I}\varphi^j$ $(1 \leqq j \leqq n)$ form a system of local coordinates at the point a for the real-analytic manifold underlying X. If at a point $x \in X$ the dimension of the *complex*-analytic manifold X (written $\dim_{\mathbf{C},x}(X)$) is n, then the dimension at x of the underlying real-analytic manifold is $2n$. If X is a pure manifold, the common value of the numbers $\dim_{\mathbf{C},x}(X)$ is written $\dim_{\mathbf{C}}(X)$ and called the (complex) *dimension* of X.

A *Riemann surface* is a pure complex-analytic manifold of dimension 1 (so that the underlying real analytic manifold has dimension 2).

In this book we shall not study the general theory of analytic manifolds; we shall merely mention, from time to time, the existence of analytic structures which arise naturally on certain differential manifolds (see [7], [17], [18], [19], [39], [41], [42], [45]).

PROBLEMS

1. Let X be the subspace of $\mathbf{R}^3$ consisting of the points (x_1, x_2, x_3) satisfying $x_3^2 = x_1^2 + x_2^2$ (a cone of revolution). Show that X is not a topological manifold. (Consider the connected components of $V - \{x\}$, where V is an open neighborhood of a point $x \in X$.)

2. The differential manifolds defined in the text are also called "C^∞-manifolds," and topological manifolds (16.1.6) are also called "C^0-manifolds." Define in the same way, for each integer $r \geqq 1$, *C^r-manifolds*, by replacing in the definition of differential manifolds the phrase "indefinitely differentiable" by "r times continuously differentiable." Examine the validity for C^r-manifolds of the properties proved in the text.

2. EXAMPLES OF DIFFERENTIAL MANIFOLDS: DIFFEOMORPHISMS

(16.2.1) On any (at most denumerable) *discrete* space X there is a unique structure of a pure differential manifold of dimension 0. If (x_n) is the sequence of points of X, the charts are the triplets $(\{x_n\}, \varphi_n, 0)$ where φ_n is the unique mapping of $\{x_n\}$ onto $\mathbf{R}^0 = \{0\}$.

(16.2.2) Let E be a real vector space of finite dimension n, endowed with the unique Hausdorff topology compatible with its vector space structure (12.13.2). Let $\varphi : E \to \mathbf{R}^n$ be a bijective linear mapping. The triplet $c = (E, \varphi, n)$ is a chart and $\mathfrak{A} = \{c\}$ is an atlas. Moreover, if φ' is another bijective linear mapping of E onto $\mathbf{R}^n$, then $\varphi' \circ \varphi^{-1}$ is a linear bijection of $\mathbf{R}^n$ onto itself, hence is indefinitely differentiable (indeed analytic). The equivalence class of the atlas $\mathfrak{A}$ is therefore independent of the choice of the linear bijection φ. In future, whenever we consider a finite-dimensional vector space E as a differential or analytic manifold, it is always the structure (called *canonical*) defined by this equivalence class that is to be understood, unless the contrary is expressly stated. E is a pure manifold of dimension n.

(16.2.3) Let $(\mathbf{e}_i)_{0 \leqq i \leqq n}$ be the canonical basis of the space $\mathbf{R}^{n+1}$, and let us identify $\mathbf{R}^n$ with the hyperplane spanned by $\mathbf{e}_1, \ldots, \mathbf{e}_n$. Let $(\mathbf{x}|\mathbf{y})$ be the usual scalar product on $\mathbf{R}^{n+1}$, such that $(\mathbf{e}_i|\mathbf{e}_j)=\delta_{ij}$ (Kronecker delta), and $\|\mathbf{x}\|$ the corresponding norm (6.2). The sphere whose equation is $\|\mathbf{x}\|=1$ relative to this norm is denoted by $\mathbf{S}_n$ and is called the "*Euclidean unit sphere of dimension n*," considered as a subspace of $\mathbf{R}^{n+1}$. We shall define on $\mathbf{S}_n$ a structure of a pure real-analytic manifold of dimension n. To do this, we associate with each point $\mathbf{x}\neq\mathbf{e}_0$ of $\mathbf{S}_n$ with coordinates ξ^i $(0\leqq i\leqq n)$ the point $\mathbf{y}$ where the line through $\mathbf{e}_0$ and $\mathbf{x}$ meets the hyperplane $\mathbf{R}^n$. A simple calculation gives

(16.2.3.1)
$$\begin{cases}\mathbf{y}=(1-\xi^0)^{-1}(\mathbf{x}-\xi^0\mathbf{e}_0),\\ \mathbf{x}=\dfrac{\|\mathbf{y}\|^2-1}{\|\mathbf{y}\|^2+1}\mathbf{e}_0+\dfrac{2}{\|\mathbf{y}\|^2+1}\mathbf{y},\end{cases}$$

and consequently these formulas define a homeomorphism

$$\varphi_1:\quad \mathbf{S}_n-\{\mathbf{e}_0\}\to\mathbf{R}^n$$

called *sterographic projection* with *pole* $\mathbf{e}_0$. In the same way we define the stereographic projection

$$\varphi_2:\quad \mathbf{S}_n-\{-\mathbf{e}_0\}\to\mathbf{R}^n$$

with pole $-\mathbf{e}_0$, such that for $\mathbf{x}\neq-\mathbf{e}_0$

(16.2.3.2)
$$\varphi_2(\mathbf{x})=(1+\xi^0)^{-1}(\mathbf{x}-\xi^0\mathbf{e}_0).$$

We have thus defined two charts

$$c_1=(\mathbf{S}_n-\{\mathbf{e}_0\},\varphi_1,n),\qquad c_2=(\mathbf{S}_n-\{-\mathbf{e}_0\},\varphi_2,n).$$

Let us show that they are *analytically compatible*. For each $\mathbf{y}\neq 0$ in $\mathbf{R}^n$ we have, by (16.2.3.1) and (16.2.3.2),

(16.2.3.3)
$$(\varphi_2\circ\varphi_1^{-1})(\mathbf{y})=\frac{\mathbf{y}}{\|\mathbf{y}\|^2},$$

and since

$$\begin{aligned}\varphi_1((\mathbf{S}_n-\{\mathbf{e}_0\})\cap(\mathbf{S}_n-\{-\mathbf{e}_0\}))&=\varphi_2((\mathbf{S}_n-\{\mathbf{e}_0\})\cap(\mathbf{S}_n-\{-\mathbf{e}_0\}))\\&=\mathbf{R}^n-\{0\},\end{aligned}$$

this proves our assertion, since $\|\mathbf{y}\|^2=\sum_{i=1}^{n}(\eta^i)^2$ is a polynomial. Since $\mathbf{S}_n-\{\mathbf{e}_0\}$ and $\mathbf{S}_n-\{-\mathbf{e}_0\}$ cover $\mathbf{S}_n$, we have defined an analytic atlas $\mathfrak{A}=\{c_1,c_2\}$ on $\mathbf{S}_n$. In future, whenever we speak of $\mathbf{S}_n$ as a manifold (analytic or differential), it is always the structure defined by the equivalence class of the atlas $\mathfrak{A}$ that is meant (cf. (16.2.7)), unless the contrary is expressly stated. The differential manifold $\mathbf{S}_n$ so defined is *compact*, since it is a bounded closed subset of $\mathbf{R}^{n+1}$ ((3.15.1), (3.17.3), (3.17.6), and (3.20.16)) and *connected* if $n \geqq 1$ by virtue of (3.19.2). We remark that, since a nonempty open set in $\mathbf{R}^n$ cannot be compact if $n \geqq 1$, an atlas of a nondiscrete compact manifold contains at least two charts (cf. Problem 5).

(16.2.4) Let X be a differential manifold, $\mathfrak{A}$ an atlas of X, and Y an open set in X. We have seen (16.1) that the restrictions to Y of any two charts c, c' belonging to $\mathfrak{A}$ are compatible. Hence, as c runs through $\mathfrak{A}$, the restrictions $c|Y$ form an atlas on Y, called the *restriction* of $\mathfrak{A}$ to Y, and written $\mathfrak{A}|Y$. Moreover, the equivalence class of $\mathfrak{A}|Y$ depends only on that of $\mathfrak{A}$, and therefore defines on Y a structure of a differential manifold depending only on that of X. This structure on Y is said to be *induced* by that on X. Here again, whenever we consider an open subset Y of a differential manifold as a differential manifold, it is always the induced structure that is meant. If Y is an open set in $\mathbf{R}^n$, then $(Y, 1_Y, n)$ is a chart on Y, called the *canonical* chart.

(16.2.5) Let X be a separable metrizable space and $(X_\alpha)_{\alpha \in I}$ an open covering of X. Suppose we are given on each X_α a structure of a differential manifold in such a way that for each pair (α, β) the structures of differential manifold induced on the open set $X_\alpha \cap X_\beta$ by those on X_α and X_β (16.2.4) are the *same*. It then follows immediately from the definitions that if $\mathfrak{A}_\alpha$ is an atlas of X_α and $\mathfrak{A}$ the union of the $\mathfrak{A}_\alpha$, then $\mathfrak{A}$ is an atlas of X, whose equivalence class depends only on those of the $\mathfrak{A}_\alpha$. The structure of differential manifold on X defined by $\mathfrak{A}$ is said to be obtained by *patching together* the differential manifolds X_α; it is clear that it induces on each X_α the given structure of differential manifold.

(16.2.6) Let X be a differential manifold, X′ a topological space, $u: X \to X'$ a homeomorphism of X onto X′. For each chart $c = (U, \varphi, n)$ of X, the triplet $(u(U), \varphi \circ u^{-1}, n)$ is a chart on X′, which we denote by $u(c)$. If c and $c' = (U', \varphi', n')$ are compatible, then so are $u(c)$ and $u(c')$, because $(\varphi' \circ u^{-1}) \circ (\varphi \circ u^{-1})^{-1} = \varphi' \circ \varphi^{-1}$. As c runs through the saturated atlas of charts on X, the $u(c)$ from a saturated atlas of X′, defining a structure of differential manifold, which is said to be obtained by *transporting* the structure on X by means of the homeomorphism u.

If X, Y are two differential manifolds, a mapping $u: X \to Y$ is a *diffeomorphism* (or an *isomorphism of differential manifolds*) if u is a homeomorphism and if the structure of differential manifold on Y is the same as that obtained by transporting the structure on X by means of u. Two differential manifolds X, Y are said to be *diffeomorphic* if there exists a diffeomorphism of X onto Y.

Remarks

(16.2.7) Consider the real line $\mathbf{R}$, endowed with its canonical structure of differential manifold (16.2.2), and let u be the real-valued function such that $u(t) = t$ for $t \leqq 0$, and $u(t) = 2t$ for $t \geqq 0$. It is clear that u is a homeomorphism of $\mathbf{R}$ onto itself (4.2.2), and we may therefore endow $\mathbf{R}$ with the structure of differential manifold defined by the single chart $u(c)$, where $c = (\mathbf{R}, 1_{\mathbf{R}}, 1)$ is the single chart defining the canonical structure. Since u is not differentiable at the point $t = 0$, the charts c and $u(c)$ are *not* compatible. If X_1 and X_2 are the differential manifolds defined on the underlying space $\mathbf{R}$ by c and $u(c)$, respectively, then u is a diffeomorphism of X_1 onto X_2, and we have therefore defined on $\mathbf{R}$ two *distinct* (but isomorphic) structures of differential manifold. In other words, the identity mapping $1_{\mathbf{R}}$ is *not* a diffeomorphism of X_1 onto X_2.

It can be shown that, for certain values of $n \geqq 7$, there exist on the topological space $\mathbf{S}_n$ several *nonisomorphic* structures of differential manifold, having the same underlying topology.

It can also be shown that the only connected differential manifolds of dimension 1 are (up to diffeomorphism) $\mathbf{R}$ and $\mathbf{S}_1$ (Problem 6).

(16.2.8) Let X be a differential manifold, X′ a set, $u: X \to X'$ a *bijection* of X onto X′. We can begin by *transporting* the topology of X to X′ by means of u, by defining the open sets in X′ to be the images under u of the open sets in X. Since u then becomes a homeomorphism of X onto X′ we can transport to X′ (again by means of u) the structure of differential manifold on X, as explained in (16.2.6).

PROBLEMS

1. Show that the space $\mathbf{T}^n = \mathbf{R}^n/\mathbf{Z}^n$ (12.11) is endowed with a structure of real-analytic manifold for which there exists an atlas of $n+1$ charts whose images are translations in $\mathbf{R}^n$ of the open cube I^n, where $\mathrm{I} = \,]0, 1[$ (cf. (16.10.6)).

2. (a) Let K be a compact subset of $\mathbf{R}^n$ and B a closed ball whose interior contains K. Show that for each $\varepsilon > 0$ there exists a homeomorphism f of $\mathbf{R}^n$ onto itself, such that $f(x) = x$ for all $x \notin \mathrm{B}$ and such that the diameter of $f(\mathrm{K})$ is $\leqq \varepsilon$. (We may assume that 0 is the center of B; take f to be of the form $f(x) = x\varphi(\|x\|)$, where φ is a suitably chosen real-valued function.)

(b) Let W be an open set in $\mathbf{R}^n$ and a a point of W. Show that, for each open ball B with center a contained in W, there exists a homeomorphism of W onto an open neighborhood of a contained in B, which coincides with the identity on a neighborhood of a (same method).

3. Let X be a metrizable space and A a compact subset of X such that there exists a fundamental system (V_k) of relatively compact open neighborhoods of A which are all homeomorphic to $\mathbf{R}^n$. In these conditions, the space X/A (Section 12.5, Problem 10) is homeomorphic to X; moreover, for each relatively compact neighborhood U of A, there exists a homeomorphism h of X/A onto X such that, if $\pi : X \to X/A$ is the canonical mapping, $h \circ \pi$ coincides with 1_X on $X - U$. (We may restrict ourselves to the case where $U = V_1$, and $\bar{V}_{k+1} \subset V_k$. Using Problem 2(a) and (3.16.5), show that there exists a sequence (g_k) of homeomorphisms of X onto itself with the following properties: (i) $g_1 = 1_X$; (ii) g_{k+1} agrees with g_k on a neighborhood of $X - V_k$; (iii) the diameter of $g_k(V_k)$ is $\leqq 1/k$. Deduce that the sequence (g_k) converges uniformly to a continuous mapping g of X into itself such that $g(A)$ is a single point, and show that g factorizes into $h \circ \pi$; h is the required mapping.)

4. Let A be a compact subset of $\mathbf{R}^n$, and suppose that there exists a homeomorphism h of $\mathbf{R}^n/A$ onto an open set W in $\mathbf{R}^n$ such that, if $\pi : \mathbf{R}^n \to \mathbf{R}^n/A$ is the canonical mapping, we have $h(\pi(A)) = \{a\}$. Show that there exists a fundamental system (V_k) of relatively compact open neighborhoods of A which are homeomorphic to $\mathbf{R}^n$. (Using Problem 2(b), show that there exists a fundamental sequence (U_k) of relatively compact open neighborhoods of a in $\mathbf{R}^n$, and for each k a homeomorphism f_k of W onto U_k, which coincides with $1_{\mathbf{R}^n}$ on a neighborhood of a. Take $V_k = \pi^{-1}(h^{-1}(U_k))$, and show that there exists a homeomorphism g_k of $\mathbf{R}^n$ onto V_k which coincides with $1_{\mathbf{R}^n}$ on a neighborhood of A and is such that $h \circ \pi \circ g_k = f_k \circ h \circ \pi$.)

5. Let M be a compact connected metrizable space which has an open covering consisting of two subspaces X, Y each homeomorphic to $\mathbf{R}^n$. Then M is homeomorphic to $\mathbf{S}_n$ (*Morton Brown's theorem*). (If $A = M - Y \subset X$, observe that M/A is homeomorphic to $\mathbf{S}_n$, hence X/A is homeomorphic to an open set in $\mathbf{R}^n$; then apply Problems 3 and 4.)

6. Let X be a connected differential manifold of dimension 1. Then there exists an atlas $((U_k, \varphi_k, 1))$, where the $\varphi_k(U_k)$ are open intervals in $\mathbf{R}$, such that (U_k) is an at most denumerable locally finite covering of X by relatively compact open sets, and such that φ_k extends to a homeomorphism of an open neighborhood of $\bar{U}_k$ onto an open interval in $\mathbf{R}$.

(a) Suppose that $U_h \cap U_k \neq \varnothing$ but that neither of U_h, U_k is contained in the other. Show that there are just two possibilities:

(α) $\varphi_h(U_h \cap U_k)$ is an interval, one of whose endpoints is also an endpoint of $\varphi_h(U_h)$, and $\varphi_k(U_h \cap U_k)$ is an interval, one of whose endpoints is also an endpoint of $\varphi_k(U_k)$;

(β) $\varphi_h(U_h \cap U_k)$ is the union of two disjoint intervals, each of which has an endpoint which is also an endpoint of $\varphi_h(U_h)$, and likewise for $\varphi_k(U_h \cap U_k)$ and $\varphi_k(U_k)$. In this case (β), show that every other U_j is contained in $U_h \cup U_k$ and that X is diffeomorphic to $\mathbf{S}_1$.

(b) Deduce from (a) that X is diffeomorphic to either **R** or $\mathbf{S}_1$. (Assume that X is not diffeomorphic to $\mathbf{S}_1$, that each U_{k+1} intersects $U_1 \cup \cdots \cup U_k$ and that neither of these two sets is contained in the other. Using (a) and induction, construct a diffeomorphism f_k of $U_1 \cup \cdots \cup U_k$ onto an open set in **R**, such that f_{k+1} extends f_k.)

3. DIFFERENTIABLE MAPPINGS

Let X, Y be two differential manifolds. For each integer $p \geqq 0$, a mapping $f: X \to Y$ is said to be *p times continuously differentiable* (resp. *indefinitely differentiable*) if f is continuous on X and satisfies the following condition: for each pair of charts (U, φ, n) and (V, ψ, m) of X and Y, respectively, such that $f(U) \subset V$, the mapping

$$F = \psi \circ (f \mid U) \circ \varphi^{-1} : \varphi(U) \to \psi(V)$$

(which is called the *local expression* of f for the charts under consideration) is p times continuously differentiable (resp. indefinitely differentiable) (8.12). (For $p = 0$ we make the convention that the derivative of order 0 of F is F itself.)

(16.3.1) *For a mapping $f: X \to Y$ to be p times continuously differentiable* (resp. *indefinitely differentiable*) *it is necessary and sufficient that for each $x_0 \in X$ there exists a chart* (U, φ, n) *of* X, *a chart* (V, ψ, m) *of* Y *and a mapping* $F: \varphi(U) \to \psi(V)$ *which is p times continuously differentiable* (resp. *indefinitely differentiable*), *such that $x_0 \in U$, $f(x_0) \in V$ and such that $f \mid U = \psi^{-1} \circ F \circ \varphi$.*

The condition is clearly necessary. Conversely, suppose that it is satisfied. Clearly it implies that f is continuous on X. Let (U', φ', n') and (V', ψ', m') be charts of X and Y, respectively, such that $f(U') \subset V'$. We have to show that $\psi' \circ (f \mid U') \circ \varphi'^{-1}$ satisfies the appropriate differentiability condition. For each $x_0 \in U'$, let (U, φ, n), (V, ψ, m), and F be two charts and a mapping satisfying the conditions of the proposition. Then first of all we have $n = n'$ and $m = m'$, because $x_0 \in U \cap U'$ and $f(x_0) \in V \cap V'$. Replacing U and U′ by $U \cap U'$, and V and V′ by $V \cap V'$, we may assume that $U = U'$ and $V = V'$. However, then we have

$$\psi' \circ (f \mid U) \circ \varphi'^{-1} = (\psi' \circ \psi^{-1}) \circ F \circ (\varphi \circ \varphi'^{-1})$$

and the result follows from the definition of compatible charts (16.1) and from (8.12.10).

In the notation of (16.3.1), if $z = (\zeta^i)_{1 \leqq i \leqq n} \in \varphi(U)$, the local expression F of f is of the form

$$F(z) = F(\zeta^1, \ldots, \zeta^n) = (F^1(\zeta^1, \ldots, \zeta^n), \ldots, F^m(\zeta^1, \ldots, \zeta^n))$$

where the F^j $(1 \leqq j \leqq m)$ are scalar functions defined on $\varphi(U)$; to say that F is p times continuously differentiable (resp. indefinitely differentiable) means that the F^j have this property (8.12.6). If (φ^i), (ψ^j) are coordinates in U, V, respectively (16.1), we have

(16.3.1.1) $$\psi^j(f(x)) = F^j(\varphi^1(x), \ldots, \varphi_n(x)) \qquad (1 \leqq j \leqq m)$$

for all $x \in U$. The F^j $(1 \leqq j \leqq m)$ are said to constitute the *local expression* of f for the given charts.

A p times continuously differentiable (resp. indefinitely differentiable) mapping is also called a *mapping of class* C^p (resp. a *mapping of class* C^∞, or a *morphism of differential manifolds*). A mapping of class C^p (resp. C^∞) into **R** is also called (if there is no risk of confusion) a *function of class* C^p (resp. *of class* C^∞) defined on X.

It is clear that a mapping of class C^p (where p is an integer or ∞) is also of class C^q for all $q < p$.

(16.3.2) *The sum and product of two functions of class* C^p *on* X *are functions of class* C^p. *If* f *is a function of class* C^p *such that* $f(x) \neq 0$ *for all* $x \in X$, *then* $1/f$ *is a function of class* C^p.

This follows from (8.12.9), (8.12.10), and (8.12.11).

(16.3.3) (i) *Let* X, Y, Z *be three differential manifolds and* $f: X \to Y$, $g: Y \to Z$ *two mappings. If* f *and* g *are of class* C^p (p *an integer or* ∞), *then so is* $g \circ f$.

(ii) *For a mapping* $f: X \to Y$ *to be a diffeomorphism of* X *onto* Y *it is necessary and sufficient that* f *be bijective and that* f *and* f^{-1} *be of class* C^∞.

(i) Let $x \in X$ and let (U, α, a) and (V, β, b) be charts on X and Y, respectively, such that $x \in U$, $f(x) \in V$ and $f|U = \beta^{-1} \circ f_1 \circ \alpha$, where f_1 is of class C^p. Likewise, let (V', β', b') and (W, γ, c) be charts on Y and Z, respectively, such that $f(x) \in V'$, $g(f(x)) \in W$ and $g|V' = \gamma^{-1} \circ g_1 \circ \beta'$, where g_1 is of class C^p. Replacing V and V' by $V \cap V'$, and U by $f^{-1}(V \cap V')$, we may assume that $V' = V$, from which it follows that $b' = b$. We have then

$$(g \circ f)|U = \gamma^{-1} \circ (g_1 \circ (\beta' \circ \beta^{-1}) \circ f_1) \circ \alpha,$$

and the result now follows from the definition of compatible charts and from (8.12.10).

(ii) The necessity of the condition is an immediate consequence of the definitions. To prove the sufficiency it is enough to show, for each chart $c = (U, \varphi, n)$ on X, if we put $f_1 = f|U$, that $f(c) = (f(U), \varphi \circ f_1^{-1}, n)$ is a chart on Y. Since f is a homeomorphism, it is clear first of all that $f(c)$ is a chart of the *topological space* Y, and it is enough to show that it is compatible with every chart $c' = (V, \psi, m)$ of the *manifold* Y (16.1). We may assume that $V = f(U)$, and then it follows from the definition of morphisms and from the hypotheses that $\psi \circ (\varphi \circ f_1^{-1})^{-1} = \psi \circ f_1 \circ \varphi^{-1}$ and $(\varphi \circ f_1^{-1}) \circ \psi^{-1}$ are indefinitely differentiable. Hence the result.

Examples

(16.3.4) (i) When X and Y are open subsets of finite-dimensional real vector spaces, the definition of a mapping of class C^p (p an integer or ∞) agrees with that of (8.12), by virtue of (16.2.2). If X is any differential manifold and (U, φ, n) is any chart on X, then φ is a diffeomorphism of U onto the open set $\varphi(U)$ in $\mathbf{R}^n$. Conversely, every diffeomorphism φ of an open set U in X onto an open set $\varphi(U)$ in $\mathbf{R}^n$ defines a chart (U, φ, n) on X.

If Y is an open subset of a differential manifold X, the canonical injection of Y into X is a mapping of class C^∞.

(ii) The mapping

(16.3.4.1) $$f : \mathbf{x} \mapsto \frac{2\mathbf{x}}{1 - \|\mathbf{x}\|^2}$$

is a diffeomorphism of the open ball B: $\|\mathbf{x}\| < 1$ in $\mathbf{R}^n$ ($\|\mathbf{x}\|$ being the Euclidean norm (16.2.3)) onto $\mathbf{R}^n$. The inverse diffeomorphism is

(16.3.4.2) $$f^{-1} : \mathbf{y} \mapsto \frac{\mathbf{y}}{1 + (1 + \|\mathbf{y}\|^2)^{1/2}}.$$

The mapping

(16.3.4.3) $$g : \mathbf{x} \mapsto \frac{\mathbf{x}}{\|\mathbf{x}\|^2}$$

is a diffeomorphism of the exterior $\|\mathbf{x}\| > 1$ of B onto the complement of $\{0\}$ in B. The composition $f \circ g$ is therefore a diffeomorphism of the exterior of B onto $\mathbf{R}^n - \{0\}$.

(iii) For the two manifolds X_1, X_2 defined in (16.2.7), both of which have $\mathbf{R}$ as underlying space, the identity map $1_{\mathbf{R}}$ is not of class C^1, whether considered as a mapping from X_1 to X_2 or from X_2 to X_1. If v is the real-valued function $t \mapsto t^3$ (which is a homeomorphism of $\mathbf{R}$ onto itself) and if we endow $\mathbf{R}$ with the structure of differential manifold defined by the single chart $v(c)$, we obtain a differential manifold X_3 again having $\mathbf{R}$ as underlying space and distinct from both X_1 and X_2. This time, the mapping $1_{\mathbf{R}}$, considered as a mapping from X_1 to X_3, is of class C^∞ but is not a diffeomorphism, since the inverse mapping is not even of class C^1.

Remark

(16.3.5) If in the definition of a mapping of class C^∞ we replace differential manifolds by real-analytic (resp. complex-analytic) manifolds, and indefinitely differentiable mappings of open sets of $\mathbf{R}^n$ (resp. $\mathbf{C}^n$) into $\mathbf{R}^m$ (resp. $\mathbf{C}^m$) by analytic mappings (9.3), we arrive at the definition of an *analytic mapping* of one real-analytic (resp. complex-analytic) manifold into another. In the complex case, such mappings are also called *holomorphic*. We leave to the reader the task of formulating for such mappings the analogues of the propositions of this section.

PROBLEMS

1. Let X be a pure differential manifold (resp. a real-analytic manifold, resp. a complex-analytic manifold) of dimension n. For each open set $U \subset X$, let $\mathscr{F}(U)$ be the set of C^∞-mappings of U into $\mathbf{R}$ (resp. real-analytic mappings of U into $\mathbf{R}$, resp. complex-analytic mappings of U into $\mathbf{C}$). Show that the sets $\mathscr{F}(U)$ have the following properties:

   (a) For each open set $V \subset U$, the restrictions to V of the functions $f \in \mathscr{F}(U)$ belong to $\mathscr{F}(V)$.
   (b) For each open set $U \subset X$ and each covering (U_α) of U by open sets contained in U, if a function f defined on U is such that $f | U_\alpha \in \mathscr{F}(U_\alpha)$ for each α, then $f \in \mathscr{F}(U)$.
   (c) For each point $x \in X$, there exists a homeomorphism u of an open neighborhood U of x onto an open set in $\mathbf{R}^n$ (resp. $\mathbf{R}^n$, resp. $\mathbf{C}^n$) such that, for each open set $V \subset U$, $\mathscr{F}(V)$ is the set of all functions of the form $g \circ u$ where g runs through the set of C^∞-mappings of $u(U)$ into $\mathbf{R}$ (resp. real-analytic mappings of $u(U)$ into $\mathbf{R}$, resp. complex-analytic mappings of $u(U)$ into $\mathbf{C}$).

   Conversely, let X be a separable metrizable space and suppose we are given, for each open set U in X, a set $\mathscr{F}(U)$ with the above properties. Show that there exists a unique structure of differential manifold (resp. real-analytic manifold, resp. complex-analytic manifold) on X for which $\mathscr{F}(U)$ is the set of C^∞-mappings of U into $\mathbf{R}$ (resp. real-analytic mappings of U into $\mathbf{R}$, resp. complex-analytic mappings of u into $\mathbf{C}$) for each open set U in x. (Observe that, if $u = (u^1, \dots, u^j)$, the functions u^j belong to $\mathscr{F}(U)$.)

2. In $\mathbf{C}$, considered as a complex-analytic manifold, every nonempty simply connected open set other than $\mathbf{C}$ is isomorphic to the unit disk $|z| < 1$, and the latter is not isomorphic to $\mathbf{C}$. Hence there are two classes of simply connected nonempty open sets with respect to the relation of isomorphism (Section 10.3, Problem 4). Deduce that in the plane $\mathbf{R}^2$ any two simply connected nonempty open sets are diffeomorphic. Give an example of two nonisomorphic complex-analytic manifolds having the same underlying structure of differential manifold.

3. (a) Let X, Y be two connected real-analytic (resp. complex-analytic) manifolds and f, g two analytic mappings of X into Y. Show that if there exists a nonempty open set $U \subset X$ on which f and g agree, then $f = g$ (9.4.2).

(b) Let X be a connected complex analytic manifold and let f be a holomorphic complex-valued function on X, not identically zero. Show that the set of points $x \in X$ such that $f(x) \neq 0$ is a *connected* dense open set. (If a, b are two points of X, show that there exists a sequence $(c_i)_{0 \leq i \leq n}$ of points of X such that $c_0 = a$, $c_n = b$ and such that for each $i = 0, 1, \ldots, n-1$, the points c_i and c_{i+1} both belong to the domain of definition U of a chart (U, φ, n), where $\varphi(U)$ is the polydisk $|z_j| < 1$ $(1 \leq j \leq n)$ in $\mathbf{C}^n$.)

4. Give an example of two distinct structures of real-analytic manifold on **R** (resp. two distinct structures of complex-analytic manifold on **C**) which are isomorphic and have as underlying structure of differential manifold the canonical structure (16.2.2).

5. Let X be a compact metrizable space and let B be a Banach subalgebra of $\mathscr{C}_{\mathbf{C}}(X)$ which is a *Dirichlet algebra* (Section 15.3, Problem 9(c)). Let χ_0 be a character of B and let μ be the unique representative measure of χ_0. Let $P \subset \mathbf{X}(B)$ be the Gleason part of χ_0 (Section 15.3, Problem 18). For each character $\lambda \in P$, the unique representative measure of λ can be written $\psi_\lambda \cdot \mu$, where ψ_λ and $1/\psi_\lambda$ are bounded in measure (relative to μ) (Section 15.3, Problem 19). For each function $f \in \mathscr{H}^2(\mu)$ (Section 15.3, Problem 15) put $f(\lambda) = \int f\psi_\lambda \, d\mu$. If $f, g \in \mathscr{H}^2(\mu)$, we have $f(\lambda)g(\lambda) = \int fg\psi_\lambda \, d\mu$ (Section 15.3, Problem 13(f)). If also $fg \in \mathscr{L}^2(\mu)$, then $fg \in \mathscr{H}^2(\mu)$ (Section 15.3, Problem 15).

(a) Suppose that P does not consist of the single point χ_0, and let $\chi_1 \in P$ be distinct from χ_0. The set $C \subset \mathscr{H}^2(\mu)$ of functions f such that $f(\chi_1) = 0$ is of the form $q\mathscr{H}^2(\mu)$ where $q \in C$ is such that $|q| = 1$ on X (Section 15.3, Problem 15(c)). Show that $|q(\lambda)| < 1$ for all $\lambda \in P$. (If not, q would be almost everywhere equal to a constant, which would contradict the relation $q(\chi_1) = 0$.)

(b) Put $U \cdot f = q^{-1}(f - f(\chi_1))$ for $f \in \mathscr{H}^2(\mu)$. Then U is a continuous linear mapping of $\mathscr{H}^2(\mu)$ into itself. Show that there exists a constant β such that $\|U^n\| \leq \beta$ for all integers $n \geq 1$. (Calculate the norm of U considered as an operator on $\mathscr{L}^2(\mu_1)$, where $\mu_1 = \psi_{\chi_1} \cdot \mu$ is the representative measure of χ_1, and show that this norm is ≤ 1.) Deduce that, for each $f \in \mathscr{H}^2(\mu)$ and each integer $n \geq 1$, we have

$$|(U^n \cdot f)(\lambda)| \leq \beta \cdot N_2(f) N_2(\psi_\lambda)$$

for all $\lambda \in P$. Deduce that the function

$$\hat{f}(z) = \sum_{n=0}^{\infty} (U^n \cdot f)(\chi_1) z^n$$

is holomorphic in the disk $|z| < 1$ and that

$$f(\lambda) = \hat{f}(q(\lambda))$$

for all $\lambda \in P$. (Observe that $f = \sum_{k=0}^{n-1} (U^k \cdot f)(\chi_1) q^k + (U^n \cdot f) q^n$.)

(c) If f, g, and fg are all in $\mathscr{H}^2(\mu)$, show that for all integers $n \geq 1$ we have

$$(U^n \cdot (fg))(\chi_1) = \sum_{k=0}^{n} (U^k \cdot f)(\chi_1)(U^{n-k} \cdot g)(\chi_1).$$

(Multiply together the expressions for f and g in terms of the $U^k \cdot f$ and $U^k \cdot g$ used in (b).) Deduce that $\hat{f}\hat{g} = (fg)^\wedge$.

(d) Show that the mapping $\lambda \mapsto q(\lambda)$ is a bijection of P onto the unit disk $|z| < 1$. (Use (a) to show that $q(\lambda_1) = q(\lambda_2)$ implies $\lambda_1 = \lambda_2$. By virtue of (c), for each z_0 in the unit disk, $f \mapsto \hat{f}(z_0)$ is a character $\lambda_0 \in \mathbf{X}(\mathrm{B})$. To show that $\lambda_0 \in \mathrm{P}$, observe that if z_1, z_2 are two points of the disk $|z| < 1$, there exists a constant $c < 2$ such that $|\mathrm{F}(z_1) - \mathrm{F}(z_2)| \leqq c$ for every holomorphic function F such that $|\mathrm{F}(z)| \leqq 1$ at all points of the disk. Finally, by passing to the limit in $\mathscr{L}^2(\mu)$ show that we have $f(\lambda_0) = \hat{f}(z_0)$ for all $f \in \mathscr{H}^2(\mu)$, and in particular for $f = q$.)

(e) Let φ be the inverse of the mapping $\lambda \mapsto q(\lambda)$. Show that φ is continuous on the disk $|z| < 1$. (Argue by contradiction: observe, by using the compactness of $\mathbf{X}(\mathrm{B})$, that if φ were not continuous at a point z_0 there would exist a character $\chi \in \mathbf{X}(\mathrm{B})$ distinct from $\varphi(z_0)$ and such that $f(\chi) = f(\varphi(z_0))$ for all $f \in \mathrm{B}$, by remarking that $f \circ \varphi$ is continuous.)

The set P is therefore endowed with a structure of complex-analytic manifold.

4. DIFFERENTIABLE PARTITIONS OF UNITY

The following proposition is a sharpening of (12.6.3) for differential manifolds:

(16.4.1) *If (A_n) is an at most denumerable locally finite open covering of a differential manifold X, there exists a partition of unity (f_n) on X subordinate to (A_n) and consisting of functions of class C^∞.*

We shall apply the remark (12.6.5) to the set $\mathscr{F}$ of functions of class C^∞ on X; since the properties (2) and (3) in this remark follow from (16.3.2), we have only to establish property (1). The proof of this consists of several steps.

(16.4.1.1) *For each integer $n \geqq 0$, we have* $\lim_{t \to +\infty} t^{-n} e^t = +\infty$.

For it follows from the power-series expansion of e^t that $e^t \geqq t^{n+1}/(n+1)!$ for $t \geqq 0$.

(16.4.1.2) *The function $h : \mathbf{R} \to \mathbf{R}$ defined by*

(16.4.1.3)
$$h(t) = \begin{cases} 0 & \text{for } t \leqq 0, \\ \exp(-t^2) & \text{for } t > 0 \end{cases}$$

is indefinitely differentiable.

For it is easily shown (8.8) by induction on n that for $t > 0$ we have

$$D^n h(t) = P_n(t^{-1}) \exp(-t^2)$$

where P_n is a polynomial; hence $\lim\limits_{t \to 0, t > 0} t^{-1} D^n h(t) = 0$ by (16.4.1.1). This proves (16.4.1.2) by induction on n.

(16.4.1.4) *Let* I *be the interval* $[-1, +1]$ *in* **R**. *There exists a function* g *of class* C^∞ *on* $\mathbf{R}^n$ *which is* >0 *in the interior of* $K = I^n$, *zero on the exterior of* K, *and such that*

$$\int \cdots \int g(t_1, \ldots, t_n) \, dt_1 \cdots dt_n = 1.$$

Put $h_0(t) = h(1 + t)h(1 - t)$, where h is the function defined by (16.4.1.3). Then we may take $g(t_1, \ldots, t_n) = c h_0(t_1) \cdots h_0(t_n)$ with a suitable constant c.

(16.4.1.5) *End of the proof* Let M be a compact subset of X, and N a closed subset of X such that $M \cap N = \varnothing$. For each $x \in M$, there exists a chart (U_x, φ_x, n_x) such that $x \in U_x$, $U_x \cap N = \varnothing$, $\varphi_x(U_x) \supset I^n$ and $\varphi_x(x) = 0$. The real-valued function f_x which is equal to $g \circ \varphi_x$ on U_x and 0 on the complement of U_x is of class C^∞ and is >0 on $V_x = \varphi^{-1}(\mathring{I}^n)$, which is an open neighborhood of x. We can cover M by a finite number of such neighborhoods V_{x_i}; the function $\sum_i f_{x_i}$ is of class C^∞, vanishes everywhere on N and is >0 everywhere on M. If $\alpha = \inf\limits_{x \in M} \sum_i f_{x_i}(x)$, we have $\alpha > 0$ (3.17.10), and the function $f = d^{-1} \sum_i f_{x_i}$ satisfies condition (1) of (12.6.5).

(16.4.2) *Let* X *be a differential manifold,* K *a compact subset of* X, *and* $(A_k)_{1 \leqq k \leqq m}$ *a finite covering of* K *by open subsets of* X. *Then there exist* m *functions* f_k *of class* C^∞ *on* X *with values in the interval* $[0, 1]$ *such that* $\operatorname{Supp}(f_k) \subset A_k$ *for* $1 \leqq k \leqq m$, $\sum_k f_k(x) = 1$ *for all* $x \in K$ *and* $\sum_k f_k(x) \leqq 1$ *for all* $x \in X$.

This follows from the preceding result and from (12.6.5) and (12.6.4).

(16.4.3) *Let* X *be a differential manifold,* F *a closed subset of* X, *and* g *a mapping of* F *into* **R**. *Suppose that, for each* $x \in F$, *there exists an open neighborhood* V_x *of* x *in* X *and a function* f_x *of class* C^r (r *an integer or* $+\infty$) *on* V_x *which is equal to* g *on* $V_x \cap F$. *Then for each open neighborhood* U *of* F *there exists a function* f *of class* C^r *which is zero on the complement of* U *and equal to* g *on* F.

For each $x \in \complement F$, let V_x be an open neighborhood of x which does not intersect F. For each $x \in F$, on the other hand, we may assume that $V_x \subset U$ (by replacing V_x by $V_x \cap U$). Let (A_n) be a denumerable open covering of X which is locally finite and finer than the covering $(V_x)_{x \in X}$ (12.6.1). For each n, choose an x such that $A_n \subset V_x$. If $x \notin F$, let f_n denote the zero function on A_n, and if $x \in F$, let f_n denote the restriction of f_x to A_n. If (h_n) is a C^∞ partition of unity on X subordinate to the covering (A_n) (16.4.1), then the function g_n which is equal to $h_n f_n$ on A_n and is zero on the complement of A_n is of class C^r on X, and the function $f = \sum_n g_n$ satisfies the required conditions.

(16.4.4) For brevity we shall say that a function g with the property stated in (16.4.3) is of *class* C^r *on* F (although in general F is not a differential manifold); equivalently, g is the restriction to F of a function of class C^r on X (cf. Problem 6).

PROBLEMS

1. Let K_0, K_1 be disjoint closed subsets of the sphere S_n. Show that there exists a C^∞-function f on $\mathbf{R}^{n+1} - \{0\}$ which is equal to 0 on K_0 and to 1 on K_1, satisfies $f(tx) = f(x)$ for all real numbers $t > 0$ and is such that for each multi-index α, $\|x\|^{|\alpha|} D^\alpha f(x)$ remains bounded as $x \to 0$ ($\|x\|$ denotes the Euclidean norm). For each C^∞-function g on $\mathbf{R}^{n+1}$ such that $D^\alpha g(0) = 0$ for each multi-index α, the function gf extends to a C^∞-function on $\mathbf{R}^{n+1}$.

2. Let $(x_k)_{k \geq 1}$ be a sequence of distinct points of $\mathbf{R}^n$ tending to 0. For each k let α_k be a multi-index such that $|\alpha_k| \to +\infty$, and let $(c_{k,\nu})_{\nu \in \mathbf{N}^n}$ be a multiple sequence of numbers such that $c_{k,\nu} = 0$ for $|\nu| \leq |\alpha_k|$ and $\nu \neq \alpha_k$. Show that there exists a C^∞-function f on $\mathbf{R}^n$ with the following properties:
(a) $D^\nu f(x_k) = c_{k,\nu}$ for all $\nu \in \mathbf{N}^n$;
(b) $D^\nu f(0) = 0$ for all ν.
(Use the method of Problem 4 of Section 8.14 to construct by induction a sequence of C^∞-functions f_k, whose supports are pairwise disjoint and do not contain 0, such that (i) $D^\nu f_k(x_k) = c_{k,\nu}$ for each multi-index ν, and (ii) $\|D^\nu f_k\| \leq 2^{-k}$ for all ν such that $|\nu| < |\alpha_k|$. Then take $f = \sum_k f_k$.)

3. Let X be a differential manifold. Show that for each $x \in X$ there exists a chart (U, φ, n) at the point x such that φ is the restriction to U of a C^∞-mapping of X into $\mathbf{R}^n$.

4. Let F be a closed subset of $\mathbf{R}^n$ and U its complement. Let h, k, η be three numbers in the interval $]0, 1[$. Show that there exists a denumerable covering of U by open Euclidean balls $B(a_i, r_i)$ with the following properties:

(a) the balls $B(a_i, kr_i)$ cover U;
(b) $r_i = hd(a_i, F)$ for each i;
(c) there exists an integer $N(h, k, \eta)$ depending only on h, k, η such that for each $x \in U$ the closed ball with center x and radius $\eta d(x, F)$ meets at most N closed balls $B'(a_i, r_i)$.
(Let ε be a real number >0. For each $m \in \mathbf{Z}$, let F_m be the set of points $x \in U$ such that $d(x, F) = (1+\varepsilon)^m$, and let T_m be an at most denumerable subset of F_m consisting of points whose mutual distances are $\geqq \varepsilon(1+\varepsilon)^m$, and such that the open balls with centers at these points and radii equal to $\varepsilon(1+\varepsilon)^m$ cover F_m. Let (a_i) be the sequence consisting of the points of $\bigcup_{m \in \mathbf{Z}} T_m$, arranged in any order. Show that if we take $r_i = hd(a_i, F)$, the required conditions are satisfied provided that $\varepsilon < \frac{1}{2}hk$. If $x \in U$ and $\delta = d(x, F)$, observe that there exists $m \in \mathbf{Z}$ such that $(1+\varepsilon)^m \leqq \delta < (1+\varepsilon)^{m+1}$, and deduce that $d(x, T_m) \leqq 2\varepsilon(1+\varepsilon)^m$. Then show that there exist two constants $c > 0$ and $C > 0$, depending only on h and η, and such that (i) if $B'(a_i, r_i)$ meets the ball with center x and radius $\eta\delta$, then $d(x, a_i) \leqq C\delta$, and (ii) if $j \neq i$ is another index with the same property, then $d(a_i, a_j) \geqq c\delta$.)

5. With the notation of Problem 4, put $B_i = B(a_i, r_i)$. Show that there exists a C^∞-partition of unity (U_i) subordinate to the covering (B_i), and for each $\alpha \in \mathbf{N}^n$ a constant C_α such that $\|D^\alpha u_i(x)\| \leqq C_\alpha(d(x, F))^{-|\alpha|}$ for all $x \in \mathbf{R}^n$, i and α.

6. Let F be a closed subset of $\mathbf{R}^n$. Generalizing the definition of a mapping of class C^r, a mapping $f: F \to \mathbf{R}^m$ is said to be of *class* C^r (resp. C^∞) if for each multi-index $\alpha \in \mathbf{N}^n$ such that $|\alpha| \leqq r$ (resp. for each multi-index α) there exists a mapping $f_\alpha : F \to \mathbf{R}^m$ with $f_0 = f$, such that the following conditions are satisfied: if for each integer $s \leqq r$ (resp. each integer $s \geqq 0$) we write

$$f_\alpha(x) = \sum_{|\alpha+\beta| \leqq s} f_{\alpha+\beta}(z) \cdot \frac{(x-z)^\beta}{\beta!} + R_{\alpha,s}(x, z)$$

where $x \in F$, $z \in F$, and $|\alpha| \leqq s$, then for each $x_0 \in F$, each $\varepsilon > 0$, and each pair (α, s) with $|\alpha| \leqq s$, there exists $\rho > 0$ such that $\|R_{\alpha,s}(x, z)\| \leqq \varepsilon \|x - z\|^{s-|\alpha|}$ for all $x, z \in F$ such that $\|x - x_0\| < \rho$ and $\|z - x_0\| < \rho$.
These conditions imply that the f_α are continuous on F. When $F = \mathbf{R}^n$, this definition is equivalent to the previous definition of functions of class C^r (resp. C^∞).

(a) Show that if the mapping $f: F \to \mathbf{R}^m$ is of class C^r then f can be extended to a mapping $h: \mathbf{R}^n \to \mathbf{R}^m$ of class C^r on $\mathbf{R}^n$ and of class C^∞ on $U = \complement F$, and hence the definition above agrees with (16.4.4). (The *Taylor polynomial* of order $s \leqq r$ of f at the point $z \in F$ is the polynomial in $x^1, \dots, x^n$

$$T_z^s f(x) = \sum_{|\alpha| \leqq s} f_\alpha(z) \cdot \frac{(x-z)^\alpha}{\alpha!}$$

With the notation of Problem 5, show that the function h defined by

$$h(x) = \begin{cases} f(x) & \text{if } x \in F, \\ \sum_i u_i(x) T_{b_i}^r f(x) & \text{if } x \in U, \end{cases}$$

where $b_i \in F$ is such that $d(a_i, b_i) = d(a_i, F)$, satisfies the required conditions. For this purpose, show that if $|\alpha| \leqq r$, then $D^\alpha h(x) - D^\alpha T_{x_0}^r f(x) \to 0$ as $x \to a \in \operatorname{Fr}(F)$, where

$x_0 \in F$ is such that $d(x, x_0) = d(x, F)$. Using Leibniz's formula, this reduces to majorizing the norm $\|D^\beta T^r_{b_i} f(x) - D^\beta T^r_{x_0} f(x)\|$ for $\beta \leq \alpha$, by using the results of Problems 4 and 5.)

(b) Show that if f is of class C^∞ there exists an extension h of f to $\mathbf{R}^n$ which is of class C^∞ (*Whitney's extension theorem*). As a consequence, the definition above agrees with that of (16.4.4). (For each integer r, show that there exists a number d_r such that the relations $z \in F$, $x \in \mathbf{R}^n$, $\|x - z\| \leq d_r$, $s \leq r$, and $|\alpha| \leq s$ imply

$$\|D^\alpha T^r_z f(x) - D^\alpha T^s_z f(x)\| \leq \|x - z\|^{s - |\alpha|}.$$

If V_r is the neighborhood of F consisting of the points x such that $d(x, F) \leq \frac{1}{2} d_r$, let r_i denote the largest r such that $a_i \in V_r$ (we can always suppose that the sequence (d_r) tends to 0). Then define

$$h(x) = \begin{cases} f(x) & \text{if } x \in F, \\ \sum_i u_i(x) T^{r_i}_{b_i} f(x) & \text{if } x \in V, \end{cases}$$

the points b_i being defined as in (a). Show that h has the required properties by arguing as in (a).)

7. With the notation of Problems 4 and 5, suppose that F is compact. Show that for each $\rho > 0$ one can define a function $v_\rho \geq 0$ of class C^∞ on $\mathbf{R}^n$, such that $v_\rho(x) = 1$ whenever $d(x, F) \leq \rho$, $v_\rho(x) = 0$ whenever $d(x, F) \geq 2\rho$, and such that for every function $f : \mathbf{R}^n \to \mathbf{R}$ of class C^r which vanishes on F together with all its partial derivatives $D^\alpha f$ of order $|\alpha| \leq r$, the functions $v_\rho f$, and their partial derivatives $D^\alpha(v_\rho f)$ of order $|\alpha| \leq r$ tend uniformly to 0 on $\mathbf{R}^n$ as $\rho \to 0$. (Take v_ρ to be a sum of certain of the functions u_i defined in Problem 5.)

8. Let F be a closed subset of $\mathbf{R}^n$ and let f be a real-valued function of class C^r on F, and g a real-valued function of class C^r on the open set $\complement F$. For each $z \in \mathbf{R}^n$ let P_z be the polynomial in $x^1, \ldots, x^n$ which is equal to $T^r_z f$ if $z \in F$, and is equal to $T^r_z g$ if $z \notin F$, in the notation of Problem 6. Show that there exists a function h of class C^r on $\mathbf{R}^n$ such that $P_z = T^r_z h$ for all $z \in \mathbf{R}^n$ if and only if the coefficients of P_z are continuous functions of z. (Reduce to the case $r = 1$ by induction, then to the case $n = 1$ by using (8.9). Then we have $P_z(x) = a(z) + (x - \alpha)b(z)$ where a and b are continuous functions of $z \in \mathbf{R}$. Reduce to the case $b = 0$ and then show that the function $a(z)$ has zero derivative at each $z \in \mathbf{R}$.)

9. (a) Let f be a real-valued function ≥ 0 of class C^2 on a neighborhood of 0 in $\mathbf{R}^N$. Suppose that f and its derivatives of order ≤ 2 all vanish at 0, and that there exist positive real numbers c, M such that $|D_i D_j f(x)| \leq M$ for all pairs of indices (i, j) and all $x \in \mathbf{R}^N$ such that $|x^j| \leq 2c$ $(1 \leq j \leq N)$. Show that if

$$|x^1| + |x^2| + \cdots + |x^N| \leq c,$$

we have

$$|D_j f(x)|^2 \leq 2M f(x) \qquad (1 \leq j \leq N). \tag{1}$$

(Observe first that the relation $\sum_j |x^j| \leq c$ implies that $|D_j f(x)| \leq Mc$ for all j. Then argue by contradiction, by supposing that at some point x satisfying $\sum_j |x^j| \leq c$ the inequality (1) is false for some index j, with $f(x) > 0$; use Taylor's formula to conclude that $f(y) < 0$ for some point y such that $|y^j| \leq 2c$ $(1 \leq j \leq N)$.)

(b) Let f be a real-valued function ≥ 0 of class C^2 on an open set U in $\mathbf{R}^n$, such that all the derivatives of order $\leqq 2$ of f vanish at the zeros of f. Show that $f^{1/2}$ is of class C^1 on U (use (a)).

(c) For each $\varepsilon > 0$, let f_ε be the function defined by $f_\varepsilon(x) = (x^2 + \varepsilon^2)u(x)$, where u is a function of class C^∞ on **R** such that $u(x) = 1$ for $|x| \leqq \frac{1}{2}$ and $u(x) = 0$ for $|x| \geqq \frac{3}{4}$. Let (α_n) be a sequence of strictly positive numbers such that the series

$$s = 1 + 2(\alpha_1 + \alpha_2 + \cdots + \alpha_n + \cdots)$$

converges, and let (β_n) be another sequence of strictly positive numbers which tend to 0 sufficiently rapidly so that $\beta_n/\alpha_n^k \to 0$ for each integer $k > 0$. Finally let (ε_n) be a sequence of strictly positive numbers tending to 0. Put

$$s_n = 1 + 2(\alpha_1 + \cdots + \alpha_{n-1}) + \alpha_n,$$

$$g_n(x) = \beta_n f_{\varepsilon_n}\left(\frac{x - s_n}{\alpha_n}\right).$$

Show that the function $g(x) = \sum_n g_n(x)$ is $\geqq 0$ and of class C^∞ on **R**, and that all its derivatives vanish at the zeros of g. For each $\mu \in]0, \frac{1}{2}[$, show that it is possible to choose the sequence (ε_n) so that the function g^μ is not of class C^1; also that it is possible to choose the sequence (ε_n) so that the function $g^{1/2}$ (which is of class C^1 by virtue of (b) above) is not of class C^2.

10. Let E be a finite-dimensional real vector space and let M_j $(1 \leqq j \leqq r)$ be vector subspaces of E. Show that the following conditions are equivalent:

(a) If $m_j = \operatorname{codim} M_j$, then for each subset H of $[1, r]$ in **N** the codimension of $\bigcap_{j \in H} M_j$ is $\sum_{j \in H} m_j$.

(b) The sum of the annihilators $M_j^0 \subset E^*$ of the M_j in the dual E^* of E is direct.

(c) There exists a direct sum decomposition of E of the form $P \oplus N_1 \oplus \cdots \oplus N_r$ such that each M_j is the direct sum of P and the N_k with $k \neq j$.

(d) If $P = \bigcap_{1 \leqq j \leqq r} M_j$, then $\operatorname{codim} P = \sum_{j=1}^{r} m_j$.

(To show that (a) implies (c), observe that if $P = \bigcap_{1 \leqq j \leqq r} M_j$ and $Q_j = \bigcap_{k \neq j} M_k$, then P has codimension m_j in Q_j.)

A family of vector subspaces M_j satisfying these conditions is said to be *in general position* in E.

Let V be the union of the M_j and let f be a real-valued function on V such that the restriction of f to M_j is of class C^k for $1 \leqq j \leqq r$. Show that f is the restriction to V of a function of class C^k on E. (Proceed by induction on r.)

11. Give an example of a C^∞-mapping $f\colon \mathbf{R} \to \mathbf{R}^2$ such that $f(\mathbf{R})$ is the square $\sup(|x^1|, |x^2|) = 1$.

5. TANGENT SPACES, TANGENT LINEAR MAPPINGS, RANK

(16.5.1) Let X, Y be two differential manifolds, x a point of X. Let f_1, f_2 be two C^1-functions, each defined on an open neighborhood of x, with values in Y. The functions f_1, f_2 are said to be *tangent* at the point x if

$f_1(x) = f_2(x)$ and if the following condition is satisfied: If (U, φ, n) is a chart of X at x such that U is contained in the domains of definition of f_1 and f_2 and if (V, ψ, m) is a chart of Y at the point $f_1(x)$ such that $f_1(U)$ and $f_2(U)$ are contained in V, then the functions $\psi \circ (f_1 | U) \circ \varphi^{-1}$ and $\psi \circ (f_2 | U) \circ \varphi^{-1}$ (that is to say, the local expressions of f_1 and f_2) are *tangent* at the point $\varphi(x)$ (8.1), i.e., they have the same derivative at this point. If this condition is satisfied for one choice of the charts (U, φ, n) and (V, ψ, m), then it is satisfied for any other pair of charts (U', φ', n) and (V', ψ', m) satisfying the same conditions. For we may assume without loss of generality that $U = U'$ and $V = V'$, and then we have

$$\psi' \circ (f_i | U) \circ \varphi'^{-1} = (\psi' \circ \psi^{-1}) \circ (\psi \circ (f_i | U) \circ \varphi^{-1}) \circ (\varphi' \circ \varphi^{-1})^{-1}$$

for $i = 1, 2$, and the assertion therefore follows from (8.2.1). Furthermore, it follows immediately from this definition that the relation "f_1 and f_2 are tangent at the point x" is an equivalence relation.

(16.5.1.1) Consider in particular the real line $\mathbf{R}$, a differential manifold X, a point $x \in X$, and the relation

"f_1 and f_2 are tangent at the point 0"

between two functions f_1 and f_2 of class C^1, defined on an open neighborhood of 0 in $\mathbf{R}$, with values in X and such that $f_1(0) = f_2(0) = x$. The equivalence classes for this relation are called the *tangent vectors to* X *at the point* x, and the set of them is denoted by $T_x(X)$. Let $c = (U, \varphi, n)$ be a chart on X at the point x. Then the definition just given shows that we obtain a bijection $\boldsymbol{\theta}_c : T_x(X) \to \mathbf{R}^n$ (also denoted by $\boldsymbol{\theta}_{c,x}$) by mapping the equivalence class of a mapping $f : V \to X$ (where V is an open neighborhood of 0 in $\mathbf{R}$) of class C^1 and such that $f(0) = x$, to the vector $(D(\varphi \circ f))(0)$. The inverse of this bijection maps a vector $\mathbf{h} \in \mathbf{R}^n$ to the tangent vector, belonging to $T_x(X)$, which is the equivalence class of the mapping $\xi \mapsto \varphi^{-1}(\varphi(x) + \xi\mathbf{h})$, where ξ belongs to a sufficiently small neighborhood of 0 in $\mathbf{R}$. If $c' = (U, \varphi', n)$ is another chart of X at x (we may assume that c' and c have the same domain of definition), then the mapping $\boldsymbol{\theta}_{c'} \circ \boldsymbol{\theta}_c^{-1}$ is the bijective *linear* mapping

(16.5.1.2) $$\boldsymbol{\theta}_{c'} \circ \boldsymbol{\theta}_c^{-1} : \mathbf{h} \mapsto (D(\varphi' \circ \varphi^{-1}))(\varphi(x)) \cdot \mathbf{h}.$$

It follows that we can define a structure of a real vector space of dimension n on $T_x(X)$ by transporting by means of $\boldsymbol{\theta}_c^{-1}$ the vector space structure of $\mathbf{R}^n$, that is to say by defining

$$\boldsymbol{\theta}_c^{-1}(\mathbf{h}) + \boldsymbol{\theta}_c^{-1}(\mathbf{h}') = \boldsymbol{\theta}_c^{-1}(\mathbf{h} + \mathbf{h}') \qquad \text{and} \qquad \lambda \cdot \boldsymbol{\theta}_c^{-1}(\mathbf{h}) = \boldsymbol{\theta}_c^{-1}(\lambda \mathbf{h})$$

for $\lambda \in \mathbf{R}$; moreover, this vector space structure is independent of the choice of the chart c because the mapping (16.5.1.2) is linear.

The set $T_x(X)$, endowed with this vector space structure, is called the *tangent vector space*, or simply the *tangent space*, to the differential manifold X at the point x. If $(\mathbf{e}_i)_{1 \leqq i \leqq n}$ is the canonical basis of $\mathbf{R}^n$, the tangent vectors $\boldsymbol{\theta}_{c,x}^{-1}(\mathbf{e}_i)$ $(1 \leqq i \leqq n)$ form a basis of the tangent space $T_x(X)$. This basis is said to be *associated* with the chart c. The reader should beware of confusing the notions of tangent vector and tangent space defined here with the elementary notions of "tangent vector" or "tangent plane" defined for ordinary "surfaces" in $\mathbf{R}^3$. The relationship between these notions will be made clear in (16.8.6).

For each tangent vector $\mathbf{h}_x \in T_x(X)$,

$$\boldsymbol{\theta}_c(\mathbf{h}_x) = \sum_{j=1}^{n} \xi^j \mathbf{e}_j$$

is called the *local expression* of $\mathbf{h}_x$, relative to the chart c.

Example

(16.5.2) Let E be a real vector space of dimension n, endowed with its canonical structure of differential manifold (16.2.2). For each linear bijection $\varphi : E \to \mathbf{R}^n$ and each $x \in E$, the triplet $c(\varphi, x) = (E, \varphi, n)$ is a chart on E, hence defines a linear bijection $\boldsymbol{\theta}_{c(\varphi, x)} : T_x(E) \to \mathbf{R}^n$, and therefore by composition a linear bijection

(16.5.2.1) $$\boldsymbol{\tau}_x = \varphi^{-1} \circ \boldsymbol{\theta}_{c(\varphi, x)} : \ T_x(E) \to E$$

which is independent of the linear bijection φ, by virtue of (16.5.1) and the relation $D(\varphi' \circ \varphi^{-1})(\varphi(x)) = \varphi' \circ \varphi^{-1}$ for two linear bijections φ, φ' of E onto $\mathbf{R}^n$ (8.1.3). The bijection $\boldsymbol{\tau}_x$ is called *canonical.*

(16.5.3) Now let X, Y be two differential manifolds, $f : X \to Y$ a mapping of class C^1, x a point of X, and $y = f(x)$. Let $c = (U, \varphi, n)$ and $c' = (V, \psi, m)$ be charts of X and Y at x, y, respectively, such that $f(U) \subset V$, and consider the local expression $F = \psi \circ (f | U) \circ \varphi^{-1}$ of f relative to c and c'. This local expression is a C^1-mapping of $\varphi(U)$ into $\psi(V)$, and its derivative $F'(\varphi(x))$ (8.1) is therefore a linear mapping of $\mathbf{R}^n$ into $\mathbf{R}^m$. We shall show that the linear mapping

(16.5.3.1) $$T_x(f) = \boldsymbol{\theta}_{c'}^{-1} \circ F'(\varphi(x)) \circ \boldsymbol{\theta}_c : \ T_x(X) \to T_x(Y)$$

is independent of the choice of charts c, c' at x, y. For if we replace c and c' by two other charts $c_1 = (U_1, \varphi_1, n)$ and $c_1' = (V_1, \psi_1, m)$ at x and y, respectively, we may assume that $U = U_1$ and $V = V_1$, by replacing U and U_1 by $U \cap U_1$, and V and V_1 by $V \cap V_1$; then the local expression of f relative to the charts c_1 and c_1' is $(\psi_1 \circ \psi^{-1}) \circ F \circ (\varphi_1 \circ \varphi^{-1})^{-1}$, and the assertion follows from (8.2.1) and (16.5.1.2).

The mapping $T_x(f)$ is called the *tangent linear mapping to f at the point x*. For a point $z = (\zeta^i) \in \mathbf{R}^n$, put $F(z) = (F^1(\zeta^1, \ldots, \zeta^n), \ldots, F^m(\zeta^1, \ldots, \zeta^n))$. Since the matrix of $F'(\varphi(x))$ with respect to the canonical bases is the Jacobian matrix $(D_j F^i(\varphi^1(x), \ldots, \varphi^n(x)))$ of type (m, n) (8.10), this Jacobian matrix is also the matrix of $T_x(f)$ relative to the bases $(\boldsymbol{\theta}_c^{-1}(\mathbf{e}_i))_{1 \leq i \leq n}$ and $(\boldsymbol{\theta}_{c'}^{-1}(\mathbf{e}_j))_{1 \leq j \leq m}$. The mapping $F'(\varphi(x))$, or its matrix relative to these bases, is called the *local expression* of $T_x(f)$ relative to the charts c and c'.

To say that f and g are *tangent* at a point $x \in X$ therefore means that $f(x) = g(x)$ and $T_x(f) = T_x(g)$.

The rank of the linear mapping $T_x(f)$ is called the *rank of f at the point x* and is denoted by $\text{rk}_x(f)$. The mapping $x \mapsto \text{rk}_x(f)$ of X into the discrete space $\mathbf{N} \subset \mathbf{R}$ is lower semicontinuous on X ((10.3) and (12.7)). We have

$$\text{rk}_x(f) \leq \inf(\dim_x(X), \dim_{f(x)}(Y)).$$

(16.5.4) *Let X, Y, Z be three differential manifolds, and $f: X \to Y$, $g: Y \to Z$ two mappings of class C^1. For each $x \in X$, we have*

(16.5.4.1) $$T_x(g \circ f) = T_{f(x)}(g) \circ T_x(f).$$

This follows immediately from the definitions and from (8.2.1).

(16.5.5) *Let X be a differential manifold, Y a differential manifold, and $f: X \to Y$ a mapping of class C^1. For f to be locally constant on X it is necessary and sufficient that $T_x(f) = 0$ for all $x \in X$* (or equivalently, that $\text{rk}_x(f) = 0$ for all $x \in X$).

It is clear that the condition is necessary. Conversely, since each point $x \in X$ has a connected neighborhood contained in the domain of definition of a chart at x, the relation $T_x(f) = 0$ for all x implies that f is locally constant (8.6.1).

If X is connected, the condition $T_x(f) = 0$ for all $x \in X$ therefore forces f to be *constant* on X, because if $x_0 \in X$ the set of points $x \in X$ such that $f(x) = f(x_0)$ is both open and closed (3.15.1).

(16.5.6) *Let X, Y be two differential manifolds, $f: X \to Y$ a mapping of class C^r (r an integer >0, or ∞), x a point of X. Then the following conditions are equivalent:*

(a) $T_x(f)$ *is a bijective linear mapping;*
(b) $\mathrm{rk}_x(f) = \dim_x(X) = \dim_{f(x)}(Y)$;
(c) *There exists an open neighborhood* U *of* x *in* X *such that* $f|U$ *is a homeomorphism of* U *onto an open neighborhood* V *of* $f(x)$, *and the inverse homeomorphism is of class* C^r.

The equivalence of (a) and (b) is linear algebra (A.4.18). For the equivalence of (a) and (c) we reduce immediately, by using charts, to the situation where $X = \mathbf{R}^n$ and $Y = \mathbf{R}^m$, and then the result follows from (10.2.5).

When the conditions of (16.5.6) are satisfied with $r = \infty$, the mapping f is said to be a *local diffeomorphism* at x, or *étale* at x, and X is said to be *étale over* Y *at the point* x (relative to f). If X is an open subset of Y, endowed with the induced structure of differential manifold (16.2.4), then the canonical injection of X into Y is étale.

Remark

(16.5.6.1) A bijective local diffeomorphism is clearly a diffeomorphism, but a mapping $f: X \to Y$ can be a local diffeomorphism at each point of X without being injective, even if X is connected. An example is the analytic mapping $z \mapsto z^2$ of $\mathbf{C} - \{0\}$ onto itself (cf. (16.12.4)).

(16.5.7) Now let X be a differential manifold, E a finite-dimensional real vector space, $\mathbf{f}: X \to E$ a mapping of class C^1, and x a point of X. Then the linear mapping (cf. (16.5.2))

(16.5.7.1) $$\tau_{\mathbf{f}(x)} \circ T_x(\mathbf{f}) : \ T_x(X) \to E$$

is an element of $\mathrm{Hom}(T_x(X), E)$, called the *differential of* $\mathbf{f}$ *at the point* x, and denoted by $d_x \mathbf{f}$. In the particular case where X is also a finite-dimensional real vector space G, it is immediate that the mapping $\tau_{\mathbf{f}(x)} \circ T_x(\mathbf{f}) \circ \tau_x^{-1}$ is precisely the derivative $D\mathbf{f}(x)$ defined in (8.1) (an element of $\mathrm{Hom}(G, E)$). Hence in this case we have, if $\mathbf{h}_x \in T_x(X)$,

(16.5.7.2) $$d_x \mathbf{f} \cdot \mathbf{h}_x = D\mathbf{f}(x) \cdot \tau_x(\mathbf{h}_x).$$

If u is any linear mapping of E into another finite-dimensional real vector space F, it follows immediately from the above definition that

(16.5.7.3) $$d_x(u \circ \mathbf{f}) = u \circ d_x \mathbf{f}.$$

In particular, if we take a basis $(\mathbf{b}_j)_{1 \leqq j \leqq m}$ of E, so that

$$\mathbf{f} = f^1 \mathbf{b}_1 + \cdots + f^m \mathbf{b}_m,$$

where the f^j are real-valued functions of class C^1 on X, then by taking for u in (16.5.7.3) the coordinate functions on E we obtain

(16.5.7.4) $$d_x \mathbf{f} \cdot \mathbf{h}_x = \sum_{j=1}^{m} \langle d_x f^j, \mathbf{h}_x \rangle \mathbf{b}_j$$

for $\mathbf{h}_x \in T_x(X)$. This is also written in the abbreviated form

(16.5.7.5) $$d_x \mathbf{f} = \sum_j (d_x f^j) \mathbf{b}_j$$

(instead of $\sum_j (d_x f^j) \otimes \mathbf{b}_j$, which is the correct form when $T_x(X)^* \otimes E$ is identified with $\mathrm{Hom}(T_x(X), E)$). The differentials $d_x f^j$ belong to the dual $T_x(X)^*$ of $T_x(X)$. The elements of this dual space are called *tangent covectors to* X *at* x (or simply *covectors at* x).

(16.5.8) Let $c = (U, \varphi, n)$ be a chart on X at x. Then the bijection $\boldsymbol{\theta}_c$ introduced earlier is given by

(16.5.8.1) $$\boldsymbol{\theta}_c = d_x \varphi,$$

for if we take the chart $c' = (\mathbf{R}^n, 1_{\mathbf{R}^n}, n)$ on $\mathbf{R}^n$, the definition (16.5.3.1) shows that $T_x(\varphi) = \boldsymbol{\theta}_{c'}^{-1} \circ \boldsymbol{\theta}_c$, and our assertion follows from (16.5.2.1) and the definition of the differential (16.5.7.1). This shows that the covectors $d_x \varphi^i$ form the basis dual to the basis $(\boldsymbol{\theta}_c^{-1}(\mathbf{e}_i))$ of $T_x(X)$. This dual basis is likewise said to be *associated* with the chart c.

From this result and the definition (16.5.7.1) we see that if $\mathbf{f}$ is a C^1-mapping of X into a finite-dimensional real vector space E, and if

$$\mathbf{F} = (\mathbf{f} \mid U) \circ \varphi^{-1} \colon \mathbf{R}^n \to E,$$

then

(16.5.8.2) $$d_x \mathbf{f} \cdot \mathbf{h}_x = \sum_{i=1}^{n} D_i \mathbf{F}(\varphi(x)) \langle d_x \varphi^i, \mathbf{h}_x \rangle$$

for $\mathbf{h}_x \in T_x(X)$, where $D_i \mathbf{F}(\varphi(x))$, the partial derivative of $\mathbf{F}$ at the point $\varphi(x) \in \mathbf{R}^n$, is identified with a vector in E (8.4). If we identify $\mathrm{Hom}(T_x(X), E)$ canonically with $E \otimes (T_x(X))^*$, then the formula (16.5.8.2) takes the form

(16.5.8.3) $$d_x\mathbf{f} = \sum_{i=1}^{n} \mathrm{D}_i\mathbf{F}(\varphi(x)) \otimes d_x\varphi^i,$$

and in particular, when $\mathrm{E} = \mathbf{R}$ (so that the $\mathrm{D}_i\mathrm{F}(\varphi(x))$ are scalars)

(16.5.8.4) $$d_x f = \sum_{i=1}^{n} \mathrm{D}_i\mathrm{F}(\varphi(x))\, d_x\varphi^i.$$

These are the *local expressions* of $d_x\mathbf{f}$ and $d_x f$ relative to the chart c.

Finally, consider a mapping $\pi : \mathrm{Y} \to \mathrm{X}$ of class C^1. For each C^1-mapping $\mathbf{f} : \mathrm{X} \to \mathrm{E}$, where as above E is a finite-dimensional real vector space, we have for each $y \in \mathrm{Y}$

(16.5.8.5) $$d_y(\mathbf{f} \circ \pi) = d_{\pi(y)}\mathbf{f} \circ \mathrm{T}_y(\pi)$$

and in particular, when $\mathrm{E} = \mathbf{R}$,

(16.5.8.6) $$d_y(f \circ \pi) = {}^t\mathrm{T}_y(\pi) \circ d_{\pi(y)} f$$

by the definition of the transpose of a linear mapping.

(16.5.9) *Let* X *be a differential manifold, let* $f^1, \dots, f_n$ *be* n *functions of class* C^∞ *defined on an open neighborhood* V *of a point* $x \in \mathrm{X}$, *and let* $\mathbf{f}$ *denote the mapping* $(f^i)_{1 \leq i \leq n}$ *of* V *into* $\mathbf{R}^n$. *Then the following conditions are equivalent:*

(a) *There exists an open neighborhood* $\mathrm{U} \subset \mathrm{V}$ *of* x *such that* $(\mathrm{U}, \mathbf{f}\,|\,\mathrm{U}, n)$ *is a chart on* X *at the point* x;

(b) *The differentials* $d_x f^i$ $(1 \leq i \leq n)$ *form a basis of* $(\mathrm{T}_x(\mathrm{X}))^*$.

For if (W, φ, n) is a chart on X at x, and we put $\mathrm{F}^i = f^i \circ \varphi^{-1}$, we have

$$d_x f^i = \sum_{j=1}^{n} \mathrm{D}_j\mathrm{F}^i(\varphi(x))\, d_x\varphi^j,$$

and condition (b) signifies that the Jacobian matrix $(\mathrm{D}_j\mathrm{F}^i(\varphi(x)))$ is invertible. The result therefore follows from (16.5.6).

(16.5.10) *Let* f *be a real-valued function of class* C^1 *on a differential manifold* X. *If* f *attains a relative minimum* (resp. *a relative maximum*) *at a point* $x_0 \in \mathrm{X}$, *that is to say if* $f(x) \geq f(x_0)$ (resp. $f(x) \leq f(x_0)$) *for all points* x *in some neighborhood of* x_0, *then* $d_{x_0} f = 0$.

We reduce immediately to the case $\mathrm{X} = \mathbf{R}^n$, and then it is enough to prove that the partial derivatives $\mathrm{D}_i f(x_0)$ are all zero, and so we reduce to the case $n = 1$. However, then $f'(x_0)$ is the limit at the point 0 of the function $h \mapsto (f(x_0 + h) - f(x_0))/h$, which is defined for all sufficiently small $h \neq 0$, and is $\geqq 0$ for $h > 0$ and $\leqq 0$ for $h < 0$. Hence the result, by (3.15.4).

(16.5.11) The converse of the proposition (16.5.10) is false, as is already shown by the example of the function $t \mapsto t^3$ at the point $t = 0$. At the points $x \in X$ such that $d_x f = 0$ we say that f is *stationary*, or that x is a *critical point* of f; the number $f(x)$ is called a *critical value* of f. To see whether, at such a point, f has a relative minimum or maximum or neither, we introduce (assuming that f is of class C^2) a quadratic form on the vector space $T_x(X)$, as follows. Consider a C^2-mapping $u : V \to X$, where V is a neighborhood of 0 in $\mathbf{R}$ and $u(0) = x$. We shall show that the hypothesis that f is stationary at the point x implies that, for the real-valued function $v = f \circ u$ of class C^2, *the value* $v''(0)$ *depends only on the tangent vector* $\mathbf{h}_x$ which is the class of the function u. To see this, let $c = (U, \varphi, n)$ be a chart of X at the point x, and let $F = f \circ \varphi^{-1}$ be the corresponding local expression of f; we may write $v = F \circ w$, where $w = \varphi \circ u$ is a C^2-mapping of V into $\mathbf{R}^n$. Then we have, by (8.1.4) and (8.12.1),

$$v'(t) = DF(w(t)) \cdot w'(t),$$

$$v''(t) = D^2F(w(t)) \cdot (w'(t), w'(t)) + DF(w(t)) \cdot w''(t);$$

but by hypothesis $DF(\varphi(x)) = 0$, so that

(16.5.11.1)
$$v''(0) = D^2F(\varphi(x)) \cdot (w'(0), w'(0)) = D^2F(\varphi(x)) \cdot (\boldsymbol{\theta}_c(\mathbf{h}_x), \boldsymbol{\theta}_c(\mathbf{h}_x)).$$

If $c_1 = (U, \varphi_1, n)$ is another chart on X at x and $\psi = \varphi \circ \varphi_1^{-1}$ the transition homeomorphism, and if we put $F_1 = f \circ \varphi_1^{-1}$, $w_1 = \varphi_1 \circ u$, then we have $F_1 = F \circ \psi$, so that for all $y \in U$ and $\mathbf{t} \in \mathbf{R}^n$

$$DF_1(\varphi_1(y)) \cdot \mathbf{t} = DF(\varphi(y)) \cdot (D\psi(\varphi_1(y)) \cdot \mathbf{t}).$$

Differentiating again, putting $y = x$, and remembering that $DF(\varphi(x)) = 0$, we shall obtain

$$D^2F_1(\varphi_1(x)) \cdot (\mathbf{s}, \mathbf{t}) = D^2F(\varphi(x)) \cdot (D\psi(\varphi_1(x)) \cdot \mathbf{s}, D\psi(\varphi_1(x)) \cdot \mathbf{t}).$$

Since on the other hand $w = \psi \circ w_1$, we have $\boldsymbol{\theta}_c(\mathbf{h}_x) = D\psi(\varphi_1(x)) \cdot \boldsymbol{\theta}_{c_1}(\mathbf{h}_x)$ and this shows that $v''(0)$ depends only on $\mathbf{h}_x$. The formula (16.5.11.1) shows moreover that there is a symmetric bilinear form on $T_x(x)$, called the *Hessian of f at the point x* and denoted by $\mathrm{Hess}_x(f)$, such that

$$v''(0) = \mathrm{Hess}_x(f) \cdot (\mathbf{h}_x, \mathbf{h}_x).$$

The symmetric bilinear form $D^2F(\varphi(x))$ on $\mathbf{R}^n$ is the *local expression* of the Hessian of f at the critical point x relative to the chart c; its matrix with respect to the canonical basis of $\mathbf{R}^n$ is therefore the symmetric matrix $(D_i D_j F(\varphi(x)))$, called the *Hessian matrix* of F at the point $\varphi(x)$ (8.12.3).

We have now the following *sufficient* criterion for a C^2-function to have a relative minimum or maximum at a point of X:

(16.5.12) *Let f be a function of class C^2 on a differential manifold X. If at a point $x \in X$ we have $d_x f = 0$ and if* $\text{Hess}_x(f)$ *is positive definite* (resp. *negative definitive*), *then f attains a relative minimum* (resp. *relative maximum*), *at the point x.*

We reduce immediately to the case $X = \mathbf{R}^n$. Suppose that the Hessian is positive definite. Then as $\mathbf{h}$ runs over the sphere S_{n-1}, the continuous function $\mathbf{h} \mapsto D^2 f(x) \cdot (\mathbf{h}, \mathbf{h})$ is always >0; hence its greatest lower bound α is >0 (3.17.10). Since the function $(y, \mathbf{h}) \mapsto D^2 f(y) \cdot (\mathbf{h}, \mathbf{h})$ is continuous on $X \times S_{n-1}$, there exists $\rho > 0$ such that $D^2 f(y) \cdot (\mathbf{h}, \mathbf{h}) \geqq \frac{1}{2}\alpha$ for all y such that $\|y - x\| < \rho$ and all $\mathbf{h} \in S_{n-1}$. Now Taylor's formula (8.14.2) gives, for $\xi \in \mathbf{R}$,

$$f(x + \xi\mathbf{h}) = f(x) + \xi^2 \int_0^1 (1 - t) D^2 f(x + t\xi\mathbf{h}) \cdot (\mathbf{h}, \mathbf{h})\, dt$$

$$\geqq f(x) + \frac{\alpha}{4}\xi^2,$$

and the result follows.

Remark

(16.5.13) Let X be a complex-analytic manifold and let $X_{|\mathbf{R}}$ denote the underlying differential manifold (16.1.6). As at the beginning of this section we can define the notion of holomorphic mappings f_1, f_2 of X into a complex-analytic manifold Y which are tangent at a point. In particular, the *tangent vectors* to X at a point x will be the equivalence classes of holomorphic functions defined on a neighborhood of 0 in $\mathbf{C}$, with values in X. A chart $c = (U, \varphi, n)$ on X at x defines a bijection $\boldsymbol{\theta}_c : T_x(X) \to \mathbf{C}^n$ as before, and we deduce that $T_x(X)$ is endowed intrinsically with the structure of a *complex* vector space of dimension n. However, since c is also a chart of $X_{|\mathbf{R}}$, there is also a bijection $\boldsymbol{\theta}_{c|\mathbf{R}} : T_x(X_{|\mathbf{R}}) \to \mathbf{R}^{2n}$, and therefore, by identifying canonically $\mathbf{C}^n$ with $\mathbf{R}^{2n}$, a bijection $\boldsymbol{\theta}_{c|\mathbf{R}}^{-1} \circ \boldsymbol{\theta}_c : T_x(X) \to T_x(X_{|\mathbf{R}})$ which is $\mathbf{R}$-linear and does not depend on the choice of the chart c. Hence, by means of this canonical bijection, we may *identify* $T_x(X_{|\mathbf{R}})$ with the real vector space obtained by restricting the scalars to $\mathbf{R}$ in $T_x(X)$. Multiplication by $i = \sqrt{-1}$ is an $\mathbf{R}$-automorphism $J_x : \mathbf{h}_x \mapsto i\mathbf{h}_x$ of the real vector space $T_x(X_{|\mathbf{R}})$ such that $J_x^2 = -I_x$, where I_x is the identity automorphism.

The notion of a *differential* is defined just as above for holomorphic mappings of X into a *complex* vector space E of finite dimension. The elements of the dual $T_x(X)^*$ of the *complex* vector space $T_x(X)$ are called *covectors* at x.

The dual $T_x(X)^* = \mathrm{Hom}_{\mathbf{C}}(T_x(X), \mathbf{C})$ can be embedded canonically in

$$\mathrm{Hom}_{\mathbf{R}}(T_x(X_{|\mathbf{R}}), \mathbf{C}) = T_x(X_{|\mathbf{R}})^* \oplus iT_x(X_{|\mathbf{R}})^* = (T_x(X_{|\mathbf{R}})^*)_{(\mathbf{C})}.$$

To be precise, we have by transposition an automorphism ${}^tJ_x$ of $T_x(X_{|\mathbf{R}})^*$, which extends canonically to a $\mathbf{C}$-automorphism (also denoted by ${}^tJ_x$) of $(T_x(X_{|\mathbf{R}})^*)_{(\mathbf{C})}$:

$${}^tJ_x \cdot (\mathbf{h}_x^* \otimes \xi) = ({}^tJ_x \cdot \mathbf{h}_x^*) \otimes \xi$$

for $\xi \in \mathbf{C}$. The $\mathbf{C}$-endomorphisms

$$p_x' = \tfrac{1}{2}(I_x - i\,{}^tJ_x), \qquad p_x'' = \tfrac{1}{2}(I_x + i\,{}^tJ_x)$$

of $(T_x(X_{|\mathbf{R}})^*)_{(\mathbf{C})}$, are *projectors* on this space, such that $p_x' + p_x'' = I_x$, and their respective images are $T_x(X)^*$ (the space of $\mathbf{C}$-*linear* forms on $T_x(X)$) and $\overline{T_x(X)^*}$ (the space of $\mathbf{C}$-*antilinear* forms on $T_x(X)$, or equivalently of complex conjugates of $\mathbf{C}$-linear forms), so that we have

$$(T_x(X_{|\mathbf{R}})^*)_{(\mathbf{C})} = T_x(X)^* \oplus \overline{T_x(X)^*}.$$

If $c = (U, \varphi, n)$ is a chart of the *complex-analytic* manifold X at the point x, then the forms $d_x\varphi^j$ form a basis of $T_x(X)^*$ over $\mathbf{C}$, and their complex conjugates $\overline{d_x\varphi^j}$ a basis of $\overline{T_x(X)^*}$ over $\mathbf{C}$.

If now f is a C^1-mapping of the *differential* manifold $X_{|\mathbf{R}}$ into $\mathbf{C}$, so that $d_x f \in T_x(X_{|\mathbf{R}})^* \oplus iT_x(X_{|\mathbf{R}})^*$, we put

(16.5.13.1) $$d_x' f = p_x'(d_x f), \qquad d_x'' f = p_x''(d_x f).$$

If we consider $c = (U, \varphi, n)$ as a chart on $X_{|\mathbf{R}}$, the corresponding local coordinates are $\mathscr{R}\varphi^j$ and $\mathscr{I}\varphi^j$ (16.1.6). If $F(\xi^1, \eta^1, \dots, \xi^n, \eta^n)$ is the local expression of f relative to this chart, then by (16.5.7.4) we have

$$d_x\varphi^j = d_x(\mathscr{R}\varphi^j) + i\,d_x(\mathscr{I}\varphi^j), \qquad \overline{d_x\varphi^j} = d_x(\mathscr{R}\varphi^j) - i\,d_x(\mathscr{I}\varphi^j)$$

and consequently

$$d_x' f = \sum_{j=1}^{n} \frac{1}{2}\left(\frac{\partial F}{\partial \xi^j} - i\frac{\partial F}{\partial \eta^j}\right) d_x\varphi^j, \qquad d_x'' f = \sum_{j=1}^{n} \frac{1}{2}\left(\frac{\partial F}{\partial \xi^j} + i\frac{\partial F}{\partial \eta^j}\right) \overline{d_x\varphi^j}$$

(the derivatives of F being taken at the point $\varphi(x)$). It follows that f is *holomorphic* if and only if $d''_x f = 0$ for all $x \in X$ (9.10.2). In the same way we define $d'_x \mathbf{f}$ and $d''_x \mathbf{f}$, where $\mathbf{f}$ is a C^1-mapping of $X_{|\mathbf{R}}$ into a finite-dimensional complex vector space.

PROBLEMS

1. Let f be a real-valued function of class C^2 on a differential manifold X.

   (a) If f attains a relative minimum at a point $x \in X$, show that the symmetric bilinear form $\text{Hess}_x(f)$ is positive (definite or semidefinite).
   (b) For the functions $f_1(\xi, \eta) = \xi^2 + \eta^4$ and $f_2(\xi, \eta) = \xi^2 - \eta^4$, defined on the plane $\mathbf{R}^2$, the point $(0, 0)$ is a critical point at which the Hessian is positive but not positive definite. For f_1, show that this point is a relative minimum, but not for f_2.

2. Let f be a real-valued function of class C^2 on a differential manifold X, and let x be a critical point of f. For each pair of tangent vectors $\mathbf{h}_x, \mathbf{k}_x$ at x, there exists a C^∞-mapping w of a neighborhood $V \subset \mathbf{R}^2$ of $(0, 0)$ into X such that $w(0, 0) = x$, $T_0(w) \cdot \mathbf{e}_1 = \mathbf{h}_x$, $T_0(w) \cdot \mathbf{e}_2 = \mathbf{k}_x$. If $F = f \circ w$, show that $D_1 D_2 F(0) = \text{Hess}_x(f) \cdot (\mathbf{h}_x, \mathbf{k}_x)$.

3. Let f be a real-valued function of class C^2 on a pure differential manifold of dimension n. At a critical point x of f, the *Morse index* of f at x is defined to be the number of coefficients < 0 in any reduction of $\text{Hess}_x(f)$ to diagonal form (or equivalently the maximum dimension of a subspace of $T_x(X)$ on which the Hessian is negative definite). Suppose that X is an open set in $\mathbf{R}^n$. Let K be a compact subset of X, and suppose that at every critical point x of f belonging to K the Morse index of f is $\geqq k$. Show that there exists $\varepsilon > 0$ such that, if g is any C^2-function on X satisfying the conditions

$$|D_i(g - f)(z)| \leqq \varepsilon, \qquad |D_i D_j(g - f)(z)| \leqq \varepsilon$$

   for all i, j and all $z \in K$, then at each critical point of g belonging to K the Morse index of g is $\geqq k$. (For each $z \in X$ put

$$c_f(z) = \sum_{i=1}^{n} |D_i f(z)|$$

   and let

$$v_f^1(z) \leqq v_f^2(z) \leqq \cdots \leqq v_f^n(z)$$

   be the sequence of (real) eigenvalues of the matrix $(D_i D_j f(z))_{1 \leqq i, j \leqq n}$, each counted according to its multiplicity. Observe that the hypothesis on f implies that the number $m_f(z) = \sup(c_f(z), -v_f^k(z))$ is strictly positive at each point $z \in K$; then show that the function $z \mapsto m_f(z)$ is continuous on X, by using (9.17.4) or Problem 8 of Section 11.5; finally compare m_g and m_f.)

4. Let f be a function of class C^2 on a pure differential manifold X of dimension n. A critical point x of f is said to be *nondegenerate* if the symmetric bilinear form $\text{Hess}_x(f)$ is nondegenerate.

(a) If x is a nondegenerate critical point of f, show that there exists a local coordinate system at x for which the local expression of f is

$$F(\zeta^1, \ldots, \zeta^n) = f(x) - (\zeta^1)^2 - \cdots - (\zeta^k)^2 + (\zeta^{k+1})^2 + \cdots + (\zeta^n)^2$$

(cf. Section **8.14**, Problem 7).
(b) Deduce from (a) that the nondegenerate critical points of f are *isolated.*

5. Let G be a finite group of diffeomorphisms of a differential manifold X, and let X^G be the set of points of X which are fixed by G.

(a) If $x \in X^G$, show that there exists a chart on X at x such that the local expressions of the diffeomorphisms $s \in G$ are *linear* mappings. (Reduce to the case where X is a neighborhood of 0 in $\mathbf{R}^n$. If f is a positive definite quadratic form on $\mathbf{R}^n$, consider the function $g(x) = \sum_{s \in G} f(s \cdot x)$ and use Section **8.14**, Problem 7.) (Cf. Section **19.1**, Problem 6.)
(b) Deduce that X^G is a closed submanifold of X (**16.8.3**).
(c) Suppose that X is connected. If $s \in G$ is such that there exists a point $x_0 \in X^G$ such that the tangent linear mapping $T_{x_0}(s)$ is the identity, show that s is the identity mapping. (Use (a) to show that the set of points $x \in X$ such that $s(x) = x$ is both open and closed.)

6. If we fix an origin in a finite-dimensional real affine space E, the canonical topology (**12.13.2**) of the vector space so obtained does not depend on the choice of origin, and is called the *canonical topology* on E. The *dimension* of a convex set in E is the dimension of the affine-linear variety generated by the set. A *convex body* in E is by definition a closed convex set in E, of dimension equal to the dimension of E; equivalently, it is a closed convex set in E whose interior is not empty (Section **12.14**, Problem 11(d)).
A *convex polyhedron* in E is the intersection of a finite number of closed half-spaces. Hence the intersection of two convex polyhedra is a convex polyhedron. Show that the frontier of a convex polyhedron P of dimension n is the union of a finite number of convex polyhedra of dimension $n - 1$ which are intersections of P with hyperplanes of support (Section **5.8**, Problem 3) of P. These are well determined by this condition and are called the *faces* of P.

7. Let E be a real affine space of dimension n and let $f: E \to \mathbf{R}$ be a C^2-function, bounded below. Suppose that for each $x \in E$ the symmetric bilinear form $(\mathbf{h}, \mathbf{k}) \mapsto D^2 f(x) \cdot (\mathbf{h}, \mathbf{k})$ is positive definite. Show that f is strictly convex, and that for $\alpha > \inf_{x \in E} f(x)$ the set $A_\alpha = \{x \in E : f(x) \leq \alpha\}$ is a closed strictly convex set of dimension n, whose frontier is the set $F_\alpha = \{x \in E : f(x) = \alpha\}$. Through each point $x \in F_\alpha$ there passes a unique hyperplane of support, whose equation is $\langle Df(x), y - x \rangle = 0$.

Suppose that $E = \mathbf{R}^n$, and let $(\mathbf{x} | \mathbf{y})$ be the Euclidean scalar product on $\mathbf{R}^n$. Suppose that A_α is compact and contains 0 in its interior. For each $x \in E$, let grad $f(x)$ be the vector defined by

$$(\operatorname{grad} f(x) | \mathbf{u}) = \langle Df(x), \mathbf{u} \rangle$$

for all $\mathbf{u} \in E$. If $\alpha > \inf_{x \in E} f(x)$, then $\operatorname{grad} f(x) \neq 0$ for all $x \in F_\alpha$. Put $g(x) = (\operatorname{grad} f(x)) / \|\operatorname{grad} f(x)\|$ and show that g is a homeomorphism of F_α onto S_{n-1} and that both g and the inverse homeomorphism h_0 are of class C^1. For each $\mathbf{z} = t\mathbf{u}$ in $\mathbf{R}^n$, with $t \geqq 0$ and $\|\mathbf{u}\| = 1$, let $H(\mathbf{z}) = (\mathbf{z} | h_0(\mathbf{u}))$, which is a C^1-function on $\mathbf{R}^n - \{0\}$. We have

$$H(\mathbf{z}) = \sup_{\mathbf{y} \in A_\alpha} (\mathbf{y} | \mathbf{z})$$

(the *function of support* of A_α). The function H is convex and positively homogeneous.

8. Let A be a compact convex body in $\mathbf{R}^n$, having 0 as an interior point.

(a) Prove that for each $\varepsilon > 0$ there exists a convex polyhedron P such that $A \subset P \subset (1 + \varepsilon)A$. (Separate each point of the frontier of $(1 + \varepsilon)A$ from A by a hyperplane (Section 12.15, Problem 4(d)).)

(b) Let P be a compact convex polyhedron of dimension n in $\mathbf{R}^n$, having 0 as an interior point. We may suppose that P is defined by m inequalities $g_j(x) \leqq 1$, where each g_j is a nonzero linear form on $\mathbf{R}^n$. Let $N > 0$ and put

$$f(x) = m^{-1} \sum_{j=1}^{m} \exp(N(g_j(x) - 1)).$$

Show that the real-analytic function f satisfies the conditions of Problem 7, and that the convex set $B = \{x \in \mathbf{R}^n : f(x) \leqq 1\}$ satisfies $P \subset B \subset (1 + N^{-1} \log m) P$.

(c) Deduce from (a) and (b) that for each $\varepsilon > 0$ there exists a real-analytic function f on $\mathbf{R}^n$ satisfying the conditions of Problem 7 and such that if $B = \{x \in \mathbf{R}^n : f(x) \leqq 1\}$, we have $A \subset B \subset (1 + \varepsilon)A$.

9. Let X, Y be two differential manifolds and f, g two mappings of class $C^r (r \geqq 1)$ defined on an open neighborhood of a point $x \in X$, with values in Y. If k is an integer such that $0 \leqq k \leqq r$, the functions f and g have *contact of order* $\geqq k$ *at the point* $x \in X$ if $f(x) = g(x)$ and if, for each chart (U, φ, n) on X at x and each chart (V, ψ, m) on Y at the point $f(x) = g(x)$, the local expressions F, G of f, g are such that $\|F(t) - G(t)\| / \|z - t\|^k$ tends to 0 as $t \in \mathbf{R}^n$ tends to $z = \varphi(x)$; or, equivalently, if $D^pF(z) = D^pG(z)$ for $1 \leqq p \leqq k$. If this condition is satisfied for one pair of charts, then it is satisfied for all pairs. If f and g have contact of order $\geqq k$ for all k, they are said to have *contact of infinite order at* x. The relation "f and g have contact of order $\geqq k$ at the point x" is an equivalence relation between C^k-mappings defined on a neighborhood of x with values in Y. An equivalence class for this relation is called a *jet of order k from* X *to* Y, *with source x and target y* (the common value of the mappings in the equivalence class). The equivalence class of f is denoted by $J_x^k(f)$ and is called the *jet of order k of f at the point x*. The set of jets of order k with source x and target y is written $J_x^k(X, Y)_y$; the set of jets of order k with source x (resp. with target y) is written $J_x^k(X, Y)$ (resp. $J^k(X, Y)_y$). The union of the sets $J_x^k(X, Y)_y$ for all $x \in X$ and all $y \in Y$ is written $J^k(X, Y)$. If $Y = \mathbf{R}$, we write $P_x^k(X)$ in place of $J_x^k(X, Y)$, and $P_x^k(f)$ in place of $J_x^k(f)$. The set $P_x^k(X)$ has a natural $\mathbf{R}$-algebra structure, and we have

$$P_x^k(f + g) = P_x^k(f) + P_x^k(g); \quad P_x^k(\alpha f) = \alpha P_x^k(f) \quad \text{for } \alpha \in \mathbf{R}; \quad P_x^k(fg) = P_x^k(f) P_x^k(g).$$

Then $P_x^k(X)_0 = \mathfrak{m}$ is the unique maximal ideal of this algebra; we have $\mathfrak{m}^{k+1} = 0$, and $\mathfrak{m}/\mathfrak{m}^2$ is canonically isomorphic to the vector space $T_x(X)^*$ of covectors at the point x.

The set $J_0^k(\mathbf{R}^n, \mathbf{R}^m)_0$ of jets of order k from $\mathbf{R}^n$ to $\mathbf{R}^m$ with source and target at the origins of these spaces is denoted by $L_{n,m}^k$. This set carries a natural structure of a real vector space of dimension $m\binom{k+n}{n} - m$, and the jets of the monomials $x \mapsto x^\alpha \cdot \mathbf{e}_j$ $(1 \leqq j \leqq m, 0 < |\alpha| \leqq k)$ form a canonical basis. Every set of jets $J_x^k(X, Y)_y$ is in one–

one correspondence with $L^k_{n,m}$ by means of charts at the points x, y (where $\dim_x(X) = n$ and $\dim_y(Y) = m$), but if $k \geqq 2$ the vector space structure on $J^k_x(X, Y)_y$ obtained by transporting that of $L^k_{n,m}$ depends on the choice of charts. When $k = 1$, $J^1_0(\mathbf{R}, X)_x$ is the tangent space $T_x(X)$.

6. PRODUCTS OF MANIFOLDS

All the definitions and all the results of the next three sections (16.6)–(16.8) (with the single exception of (16.8.9)) can be transposed to the contexts of real- or complex-analytic manifolds, simply by replacing C^∞-mappings by analytic mappings in the statements and the proofs. We shall therefore make use of them for real- and complex-analytic manifolds without further comment.

Let X_1, X_2 be two topological spaces. If

$$c_1 = (U_1, \varphi_1, n_1) \qquad \text{and} \qquad c_2 = (U_2, \varphi_2, n_2)$$

are charts of X_1, X_2, respectively, the triple $(U_1 \times U_2, \varphi_1 \times \varphi_2, n_1 + n_2)$ is a chart of $X_1 \times X_2$ (3.20.15 and 12.5); it is denoted by $c_1 \times c_2$. If c'_1, c'_2 are two other charts on X_1, X_2, respectively, and if c_i and c'_i are compatible for $i = 1, 2$, then $c_1 \times c_2$ and $c'_1 \times c'_2$ are compatible, by (8.12.6). If $\mathfrak{A}_1$ is an atlas of X_1 and $\mathfrak{A}_2$ is an atlas of X_2, the set $\mathfrak{A}$ of charts $c_1 \times c_2$, where $c_1 \in \mathfrak{A}_1$ and $c_2 \in \mathfrak{A}_2$, is therefore an atlas of $X_1 \times X_2$, and is denoted (by abuse of notation) by $\mathfrak{A}_1 \times \mathfrak{A}_2$. Moreover, if $\mathfrak{A}_i$ and $\mathfrak{A}'_i$ are compatible atlases of X_i ($i = 1, 2$), then the atlases $\mathfrak{A}_1 \times \mathfrak{A}_2$ and $\mathfrak{A}'_1 \times \mathfrak{A}'_2$ are compatible. If X_1 and X_2 are differential manifolds, the product space $X = X_1 \times X_2$ is separable and metrizable (3.20.16), and the atlases $\mathfrak{A}_1 \times \mathfrak{A}_2$, where $\mathfrak{A}_1$ (resp. $\mathfrak{A}_2$) runs through the equivalence class of atlases defining the structure of differential manifold on X_1 (resp. X_2), are all equivalent. Hence their equivalence class defines on X a structure of differential manifold which depends only on the structures of X_1 and X_2. The space X endowed with this structure is called the *product* of the differential manifolds X_1 and X_2. It should be noted that even if $\mathfrak{A}_1$ and $\mathfrak{A}_2$ are saturated atlases, $\mathfrak{A}_1 \times \mathfrak{A}_2$ will in general not be saturated.

Whenever we consider $X_1 \times X_2$ as a differential manifold, it is always the product structure as defined above that is meant, unless the contrary is expressly stated.

Example

(16.6.1) If E_1, E_2 are two finite-dimensional real vector spaces, each endowed with its canonical structure of differential manifold, it follows from the definitions (16.2.2) that the product manifold $E_1 \times E_2$ is the product vector space endowed with its canonical structure of differential manifold.

(16.6.2) *Let* X_1, X_2 *be two differential manifolds*, $X = X_1 \times X_2$ *their product. The projections* $\mathrm{pr}_1 : X \to X_1$, $\mathrm{pr}_2 : X \to X_2$ *are morphisms* (16.3). *For each point* $(x_1, x_2) \in X$, *the mapping*

$$(T_{(x_1, x_2)}(\mathrm{pr}_1), T_{(x_1, x_2)}(\mathrm{pr}_2)) : T_{(x_1, x_2)}(X_1 \times X_2) \to T_{x_1}(X_1) \times T_{x_2}(X_2)$$

is an isomorphism of vector spaces.

In view of the definition of the product manifold structure, we reduce immediately to the situation where X_1 and X_2 are open sets in $\mathbf{R}^{n_1}$ and $\mathbf{R}^{n_2}$ respectively. The first assertion is then a trivial consequence of (8.12.10), and the second follows from (8.1.5) applied to a C^1-mapping of a neighborhood of 0 in $\mathbf{R}$, with values in $\mathbf{R}^{n_1} \times \mathbf{R}^{n_2}$.

We shall identify canonically $T_{(x_1, x_2)}(X_1 \times X_2)$ with the product $T_{x_1}(X_1) \times T_{x_2}(X_2)$ by means of the isomorphism defined in (16.6.2). The canonical injection $T_{x_1}(X_1) \to T_{(x_1, x_2)}(X_1 \times X_2)$ resulting from this identification is just the tangent linear mapping at the point x_1 to the injection $y_1 \mapsto (y_1, x_2)$, which is a morphism of X_1 into X. Likewise for the canonical injection $T_{x_2}(X_2) \to T_{(x_1, x_2)}(X_1 \times X_2)$. It is clear that

(16.6.3) $$\dim_{(x_1, x_2)}(X_1 \times X_2) = \dim_{x_1}(X_1) + \dim_{x_2}(X_2).$$

(16.6.4) *Let* Y, X_1, X_2 *be three differential manifolds and* $f_1 : Y \to X_1$, $f_2 : Y \to X_2$ *two mappings. Then the mapping* $f = (f_1, f_2) : Y \to X_1 \times X_2$ *is of class* C^r (r *an integer* >0, *or* ∞) *if and only if* f_1 *and* f_2 *are of class* C^r. *Moreover, for all* $y \in Y$, *we have*

$$T_y((f_1, f_2)) = (T_y(f_1), T_y(f_2))$$

with the identification (16.6.2).

Once again we reduce to the case in which X_1 and X_2 are open sets in $\mathbf{R}^{n_1}$ and $\mathbf{R}^{n_2}$, and the result then follows from (8.12.6).

(16.6.5) *Let* X_1, X_2, Y_1, Y_2 *be differential manifolds, and* $f_i : Y_i \to X_i$ ($i = 1, 2$) *mappings of class* C^r. *Then* $f_1 \times f_2 : Y_1 \times Y_2 \to X_1 \times X_2$ *is a mapping of class* C^r, *and we have*

$$T_{(y_1, y_2)}(f_1 \times f_2) = T_{y_1}(f_1) \times T_{y_2}(f_2),$$
$$\mathrm{rk}_{(y_1, y_2)}(f_1 \times f_2) = \mathrm{rk}_{y_1}(f_1) + \mathrm{rk}_{y_2}(f_2).$$

The second formula is a trivial consequence of the first, and that follows from (16.6.4) and (16.6.2), since $f_1 \times f_2 = (f_1 \circ \mathrm{pr}_1, f_2 \circ \mathrm{pr}_2)$.

(16.6.6) *Let* X_1, X_2, Z *be three differential manifolds,* $f: X_1 \times X_2 \to Z$ *a mapping of class* C^r *(r an integer* >0*, or* ∞*), and* (a_1, a_2) *a point of* $X_1 \times X_2$. *Let* $f(a_1, \cdot)$ (resp. $f(\cdot, a_2)$) *denote the partial mapping* $x_2 \mapsto f(a_1, x_2)$ (resp. $x_1 \mapsto f(x_1, a_2)$). *Then we have*

$$T_{(a_1, a_2)}(f) = T_{a_1}(f(\cdot, a_2)) \circ p_1 + T_{a_2}(f(a_1, \cdot)) \circ p_2,$$

where

$$p_1 = T_{(a_1, a_2)}(\mathrm{pr}_1) : T_{(a_1, a_2)}(X_1 \times X_2) \to T_{a_1}(X_1),$$

$$p_2 = T_{(a_1, a_2)}(\mathrm{pr}_2) : T_{(a_1, a_2)}(X_1 \times X_2) \to T_{a_2}(X_2)$$

are the canonical projections (with the identification (16.6.2)).

Once more, the proof reduces to the case where X_1, X_2, Z are open sets in $\mathbf{R}^{n_1}$, $\mathbf{R}^{n_2}$, and $\mathbf{R}^m$, and then it follows from (8.9.1).

In particular, if $Z = E$ is a finite-dimensional real vector space, we have

(16.6.7) $$(d_{(a_1, a_2)}\mathbf{f}) \cdot (\mathbf{h}_1, \mathbf{h}_2) = (d_{a_1}\mathbf{f}(\cdot, a_2)) \cdot \mathbf{h}_1 + (d_{a_2}\mathbf{f}(a_1, \cdot)) \cdot \mathbf{h}_2.$$

(16.6.8) *With the hypotheses and notation of* (16.6.6), *suppose that*

$$T_{a_2} f(a_1, \cdot) : T_{a_2}(X_2) \to T_c(Z) \qquad (\textit{where} \quad c = f(a_1, a_2))$$

is bijective. Then there exists an open neighborhood U_1 *of* a_1 *in* X_1 *and an open neighborhood* U_2 *of* a_2 *in* X_2 *with the following properties: for each* $x_1 \in U_1$ *there exists a unique point* $u(x_1) \in U_2$ *such that* $f(x_1, u(x_1)) = c$*, and* u *is a* C^r*-mapping of* U_1 *into* U_2. *Furthermore, we have*

(16.6.8.1) $$T_{a_1}(u) = -(T_{a_2} f(a_1, \cdot))^{-1} \circ T_{a_1} f(\cdot, a_2)$$

("implicit function theorem").

We reduce to the case where X_1, X_2, and Z are open sets in $\mathbf{R}^{n_1}$, $\mathbf{R}^{n_2}$, and $\mathbf{R}^m$, respectively, and then the theorem is a particular case of (10.2.3).

7. IMMERSIONS, SUBMERSIONS, SUBIMMERSIONS

(16.7.1) Let X, Y be two differential manifolds, $f: X \to Y$ a mapping of class C^∞, and x a point of X. The mapping f is said to be a *subimmersion* at the point x if there exists a neighborhood U of x in X such that the function $x' \mapsto \mathrm{rk}_{x'}(f)$ is constant on U. The mapping f is said to be an *immersion* (resp. a *submersion*) at x if the linear mapping $T_x(f)$ is injective (resp. surjective). By virtue of the lower semicontinuity of the rank of f (**16.5**), this implies that $\text{rk}_{x'}(f) = \dim_{x'}(X)$ (resp. $\text{rk}_{x'}(f) = \dim_{f(x')}(Y)$) for all x' in some neighborhood of x, and hence f is a subimmersion at the point x. A mapping $f : X \to Y$ of class C^∞ is both an immersion and a submersion at the point x if and only if f is *étale* at x (**16.5.6**).

It is clear that the set U of points of X at which f is a subimmersion (resp. a submersion, resp. an immersion, resp. étale) is open in X.

The mapping f is said to be an *immersion* (resp. a *submersion*, a *subimmersion*, *étale*) if $U = X$.

For example, the projections of a product manifold $X_1 \times X_2$ onto its factors X_1, X_2 are submersions (**16.6.2**).

(**16.7.2**) *If $f : X \to Y$ and $g : Y \to Z$ are both submersions* (resp. *both immersions*), *then $g \circ f : X \to Z$ is a submersion* (resp. *an immersion*).

This follows immediately from the definitions (**16.7.1**) and from (**16.5.4**).

We remark that the composition of two subimmersions is not necessarily a subimmersion (Section **16.8**, Problem 1(b)).

(**16.7.3**) *If $f_1 : X_1 \to Y_1$ and $f_2 : X_2 \to Y_2$ are both submersions* (resp. *immersions*, resp. *subimmersions*), *then $f_1 \times f_2 : X_1 \times X_2 \to Y_1 \times Y_2$ is a submersion* (resp. *an immersion*, resp. *a subimmersion*).

This follows from the definitions (**16.7.1**) and from (**16.6.5**).

(**16.7.4**) *Let $f : X \to Y$ be a mapping of class C^∞. In order that f should be a subimmersion of rank r at a point $x \in X$, it is necessary and sufficient that there should exist a chart* (U, φ, n) *of* X, *a chart* (V, ψ, m) *of* Y *and a C^∞-mapping* $F : \varphi(U) \to \psi(V)$ *such that* $x \in U$, $f(x) \in V$, $\varphi(x) = 0$, $\psi(f(x)) = 0$, $f|U = \psi^{-1} \circ F \circ \varphi$, *and such that the local expression* F *of f is the restriction to* $\varphi(U)$ *of the mapping*

(**16.7.4.1**)
$$(\xi^1, \ldots, \xi^n) \mapsto (\xi^1, \ldots, \xi^r, 0, \ldots, 0)$$

of $\mathbf{R}^n$ *into* $\mathbf{R}^m$.

This is an immediate consequence of the rank theorem (**10.3.1**).

(**16.7.5**) *If $f : X \to Y$ is a submersion, the image under f of any open set* U *in* X *is open in* Y.

If $x \in U$, it follows from (16.7.4) that there exists an open neighborhood $W \subset U$ of x such that $f(W)$ is open in Y; now apply axiom (O_I) of topological spaces (12.1).

(16.7.6) It should be remarked that if $f: X \to Y$ is an *immersion*, $f(X)$ is not necessarily closed, *nor even locally compact*, even if f is injective (16.9.9.3).

On the other hand, a C^∞-mapping $f: X \to Y$ can be injective (resp. surjective) without being an immersion (resp. a submersion), as is shown by the example of the bijective mapping $\xi \mapsto \xi^3$ of **R** onto **R** (cf. however Section 10.3, Problem 2).

(16.7.7) (i) *Let $f: X \to Y$ be an injective immersion, and let $g: Z \to X$ be a continuous mapping* (X, Y, Z *being differential manifolds*). *For g to be of class C^r it is necessary and sufficient that $f \circ g: Z \to Y$ should be of class C^r.*

(ii) *Let $f: X \to Y$ be a surjective submersion and let $g: Y \to Z$ be a mapping* (X, Y, Z *being differential manifolds*). *For g to be of class C^r it is necessary and sufficient that $g \circ f: X \to Z$ should be of class C^r.*

Only the sufficiency of these conditions requires proof, by (16.3.3). Also the questions are local on X, Y, and Z by virtue of the continuity of g (which follows from the continuity of $g \circ f$ in (ii)). In case (i), we may therefore suppose (16.7.4) that f is the mapping $(\xi^1, \dots, \xi^n) \mapsto (\xi^1, \dots, \xi^n, 0, \dots, 0)$ of $\mathbf{R}^n$ into $\mathbf{R}^m$ (with $n \leqq m$). The assertion of (i) is then that a mapping $z \mapsto (g^1(z), \dots, g^n(z), 0, \dots, 0)$ of Z into $\mathbf{R}^m$ is of class C^r provided that the g^j are of class C^r. In case (ii) we may likewise suppose that f is the mapping $(\xi^1, \dots, \xi^n) \mapsto (\xi^1, \dots, \xi^m)$ of $\mathbf{R}^n$ onto $\mathbf{R}^m$ (with $n \geqq m$). The assertion of (ii) is then that the mapping $(\xi^1, \dots, \xi^m) \mapsto g(\xi^1, \dots, \xi^m)$ is of class C^r on $\mathbf{R}^m$ if and only if the mapping $(\xi^1, \dots, \xi^n) \mapsto g(\xi^1, \dots, \xi^m)$ is of class C^r on $\mathbf{R}^n$ (16.6.6).

Remark

(16.7.8) The preceding results can be extended immediately to the situation where C^r-mappings (r a positive integer) replace C^∞-mappings, subimmersions are replaced by mappings of locally constant rank, and submersions (resp. immersions) by mappings f such that $T_x(f)$ is surjective (resp. injective).

8. SUBMANIFOLDS

(16.8.1) *Let X be a separable metrizable space, Y a differential manifold, f a mapping of X into Y. In order that there should exist on X a structure of differential manifold for which the underlying topology is the given topology* *on* X *and such that the mapping* f *is an immersion, it is necessary and sufficient that the following condition be satisfied:*

(16.8.1.1) *For each* $a \in X$, *there exists an open neighborhood* U *of* a *in* X *and a chart* (V, ψ, m) *of* Y *such that* $f(U) \subset V$ *and such that* $\psi \circ (f|U)$ *is a homeomorphism of* U *onto the intersection of* $\psi(V)$ *with a linear subvariety of* $\mathbf{R}^m$.

When this condition is satisfied, the structure of differential manifold on X *satisfying the conditions above is unique.*

The necessity of the condition follows immediately from (16.7.4). To prove sufficiency, consider for each $a \in X$ a neighborhood U_a of a in X and a chart (V_a, ψ_a, m_a) of Y satisfying (16.8.1.1), and let E_a denote the linear subvariety of $\mathbf{R}^{m_a}$ such that $\psi_a(V_a) \cap E_a = \psi_a(f(U_a))$. ($E_a$ is unique because $E_a \cap \psi_a(V_a)$ is a nonempty open subset of E_a.) Let $n_a = \dim E_a$ and let λ_a be an affine-linear bijection of E_a onto $\mathbf{R}^{n_a}$. Finally let φ_a be the composition of λ_a and $\psi_a \circ (f|U_a)$. We shall show that the charts $c_a = (U_a, \varphi_a, n_a)$ form an *atlas* of X. Suppose therefore that $U_a \cap U_b \neq \varnothing$; then (16.8.1.1) implies that $\varphi_a(U_a \cap U_b) = W_{ab}$ and $\varphi_b(U_a \cap U_b) = W_{ba}$ are open sets in $\mathbf{R}^{n_a}$ and $\mathbf{R}^{n_b}$, respectively. If ψ_{ab} (resp. ψ_{ba}) is the restriction of ψ_a (resp. ψ_b) to $V_a \cap V_b$, and φ_{ab} (resp. φ_{ba}) the restriction of φ_a (resp. φ_b) to $U_a \cap U_b$, it is immediate that $\varphi_{ba} \circ \varphi_{ab}^{-1} = \lambda_b \circ (\psi_{ba} \circ \psi_{ab}^{-1}) \circ ((\lambda_a^{-1})|W_{ab})$. Hence to show that $\varphi_{ba} \circ \varphi_{ab}^{-1}$ is indefinitely differentiable, it is enough to observe that the restriction of an indefinitely differentiable mapping

$$\psi_{ba} \circ \psi_{ab}^{-1} : V_a \cap V_b \to \mathbf{R}^m \qquad (\text{where } m = m_a = m_b)$$

to the intersection of the open set $V_a \cap V_b$ in $\mathbf{R}^m$ with a linear subvariety E_a of $\mathbf{R}^m$ is indefinitely differentiable on this intersection, which is obvious by (8.12.8).

The uniqueness of the structure of differential manifold on X follows from (16.7.7(i)). For by replacing X by an open neighborhood of a point of X, we reduce to the case where f is injective, and apply (16.7.7(i)) to $g = 1_X$ (considering two structures of differential manifold on X satisfying the conditions of (16.8.1)).

When the condition (16.8.1.1) is satisfied, the unique structure of differential manifold defined in (16.8.1) is called the *inverse image* under f of the structure of differential manifold on Y.

In particular:

(16.8.2) *Let* X *be a separable metrizable space,* Y *a differential manifold,* f *a mapping of* X *into* Y *with the property that for each* $x \in X$ *there exists an open neighborhood* U *of* x *such that* $f|\mathrm{U}$ *is a homeomorphism of* U *onto an open set in* Y. *Then there exists a unique structure of differential manifold on* X *for which* f *is an immersion, and* f *is then in fact étale* (16.5.6).

(16.8.3) Let Y be a differential manifold, X a *subspace* of Y. If the canonical injection $f : \mathrm{X} \to \mathrm{Y}$ satisfies the condition (16.8.1.1), then the space X, endowed with the structure of differential manifold which is the inverse image under f of that of Y, is said to be a *submanifold* of Y. We also say that X *is* a submanifold of Y; this abuse of language is justified by the property of uniqueness in (16.8.2). The condition (16.8.1.1), in the present situation, is that *for each* $x \in \mathrm{X}$ *there exists a chart* (V, ψ, m) *of* Y *such that* $x \in \mathrm{V}$, $\psi(x) = 0$ *and such that* $\psi(\mathrm{V} \cap \mathrm{X})$ *is the intersection of the open set* $\psi(\mathrm{V})$ *of* $\mathbf{R}^m$ *and the vector subspace of* $\mathbf{R}^m$ *given by the equations* $\xi^{n+1} = 0, \dots, \xi^m = 0$; and then $(\mathrm{V} \cap \mathrm{X}, \psi|(\mathrm{V} \cap \mathrm{X}), n)$ is a *chart* of the submanifold X. It follows that $\mathrm{V} \cap \mathrm{X}$ is *closed* in V, and hence that X is *locally closed* in Y (12.2.3). Moreover, there is an open neighborhood $\mathrm{W} \subset \mathrm{V}$ of x in Y which is diffeomorphic to $(\mathrm{W} \cap \mathrm{X}) \times \mathrm{Z}$, where Z is a *submanifold of dimension* $n - m$ of Y, containing x. To see this we reduce to the case where $\mathrm{Y} = \psi(\mathrm{V}) \subset \mathbf{R}^m$ and $\mathrm{X} = \psi(\mathrm{V}) \cap \mathbf{R}^n$, and then it is obvious.

(16.8.3.1) In view of (16.7.4), the condition for X to be a submanifold of Y can be expressed as follows: *for each* $x \in \mathrm{X}$ *there exists an open neighborhood* U *of* x *in* Y *and a submersion* $g : \mathrm{U} \to \mathbf{R}^{n-m}$ *such that* $\mathrm{X} \cap \mathrm{U}$ *is the set of points* $z \in \mathrm{U}$ *such that* $g(z) = 0$.

(16.8.3.2) In particular, when $\mathrm{Y} = \mathbf{R}^n$, we may always assume (by translation if necessary) that $0 \in \mathrm{X}$, and then, by permuting the coordinates, that if $g = (g^1, \dots, g^{n-m})$, the determinant of the matrix formed by the first $n - m$ columns of the Jacobian matrix of g is $\neq 0$ at the point 0. If we identify $\mathbf{R}^n$ with $\mathbf{R}^m \times \mathbf{R}^{n-m}$, it follows from the implicit function theorem (10.2.2) that there exists an open neighborhood V of 0 in $\mathbf{R}^m$ such that $\mathrm{U} \cap (\mathrm{V} \times \mathbf{R}^{n-m})$ is the graph of a C^∞-mapping $f = (f^1, \dots, f^{n-m})$ of V into $\mathbf{R}^{n-m}$, or in other words this submanifold is the set of points $x = (\xi^1, \dots, \xi^n)$ such that $\xi^j \in \mathrm{V}$ for $1 \leq j \leq m$ and $\xi^{m+k} - f^k(\xi^1, \dots, \xi^m) = 0$ for $1 \leq k \leq n - m$.

(16.8.3.3) Every open subset of a manifold Y, endowed with the induced structure (16.2.4) is a submanifold of Y. Conversely, a submanifold X of Y such that $\dim_x(\mathrm{X}) = \dim_x(\mathrm{Y})$ for all $x \in \mathrm{X}$ is open in Y.

Every discrete subspace (necessarily at most denumerable) of Y is a submanifold of dimension 0. In a pure manifold Y of dimension n, a pure submanifold of dimension $n - 1$ is called a *hypersurface*.

(16.8.3.4) Let X be a submanifold of Y, X′ a submanifold of Y′, and $j : \mathrm{X} \to \mathrm{Y}$, $j' : \mathrm{X}' \to \mathrm{Y}'$ the canonical injections. Let $g : \mathrm{Y} \to \mathrm{Y}'$ be a C^r-mapping such that $g(\mathrm{X}) \subset \mathrm{X}'$. Then we can write $g \circ j = j' \circ f$, where f is a C^r-mapping of X into X′. This follows from **(16.7.7(i))**.

(16.8.4) *Let* X, Y *be two differential manifolds,* $f : \mathrm{X} \to \mathrm{Y}$ *an immersion. If f is a homeomorphism of* X *into the subspace* $f(\mathrm{X})$ *of* Y, *then* $f(\mathrm{X})$ *is a submanifold of* Y, *and* $f : \mathrm{X} \to f(\mathrm{X})$ *is a diffeomorphism.*

For each $x \in \mathrm{X}$ we apply **(16.7.4)** (keeping the notation used there) with $r = n = \dim_x(\mathrm{X})$ and $m = \dim_{f(x)}(\mathrm{Y})$. Since f is a homeomorphism of X onto the subspace $f(\mathrm{X})$, it follows that $\mathrm{F}(\varphi(\mathrm{U}))$ is an open neighborhood of $\mathrm{F}(\varphi(x)) = \psi(f(x))$ in $\psi(\mathrm{V}) \cap \mathbf{R}^n$, and hence there exists an open neighborhood $\mathrm{T} \subset \psi(\mathrm{V})$ of $\psi(f(x))$ in $\mathbf{R}^m$ such that $\mathrm{T} \cap \mathbf{R}^n = \mathrm{F}(\varphi(\mathrm{U}))$. Putting $\mathrm{W} = \psi^{-1}(\mathrm{T})$, the chart $(\mathrm{W}, \psi \,|\, \mathrm{W}, m)$ on Y satisfies the condition of **(16.8.3)** relative to the subspace $f(\mathrm{X})$. Furthermore, it follows from **(16.7.4)** that F is a diffeomorphism of $\varphi(\mathrm{U})$ onto the open set $\mathrm{F}(\varphi(\mathrm{U}))$, and the second assertion of **(16.8.4)** follows.

An immersion f which satisfies the hypotheses of **(16.8.4)** is called an *embedding* of X in Y.

Remark

(16.8.5) It can happen that an immersion $f : \mathrm{X} \to \mathrm{Y}$ is injective and that $f(\mathrm{X})$ is closed in Y but that f is not an embedding. For example, take X to be the open interval $]-\infty, 1[$ in $\mathbf{R}$, $\mathrm{Y} = \mathbf{R}^2$, and f to be the immersion

$$t \mapsto \left(\frac{t^2 - 1}{t^2 + 1}, \frac{t(t^2 - 1)}{t^2 + 1} \right).$$

This immersion is not an embedding, because $f(-1) = \lim_{t \to 1} f(t)$ (cf. Problem 2).

(16.8.6) If X is a submanifold of Y and $j : \mathrm{X} \to \mathrm{Y}$ is the canonical injection, it follows from the definition of an immersion **(16.7.1)** that for each $x \in \mathrm{X}$ the linear mapping $\mathrm{T}_x(j) : \mathrm{T}_x(\mathrm{X}) \to \mathrm{T}_x(\mathrm{Y})$ is an injection, by means of which we shall *identify* canonically $\mathrm{T}_x(\mathrm{X})$ with a vector subspace of $\mathrm{T}_x(\mathrm{Y})$.

In the particular case where Y is a vector space E of finite dimension n, we recall **(16.5.2)** that there is a canonical linear bijection $\boldsymbol{\tau}_x : \mathrm{T}_x(\mathrm{E}) \to \mathrm{E}$. The image of $\boldsymbol{\tau}_x(\mathrm{T}_x(\mathrm{X}))$ under the translation $\mathbf{h} \mapsto \mathbf{h} + x$ is an affine-linear variety in E, passing through the point x, of dimension $m = \dim_x(\mathrm{X})$. This is called the *tangent affine-linear variety* to X at the point x (or the *tangent* to X at x if $m = 1$, the *tangent plane* if $m = 2$, the *tangent hyperplane* if $m = n - 1$). It is the set of points $x + \tau_x(\mathbf{h}_x)$ in E, as $\mathbf{h}_x$ runs through $T_x(X)$. It should be observed that the possibility of defining such a "tangent linear variety" as a sub*manifold* of E depends essentially on the group-structure of E, and that there is no analogous definition when E is replaced by an arbitrary differential manifold Y.

Remark

(16.8.6.1) If X is a differential manifold, a a point of X, and E a vector subspace of $T_a(X)$, then there exists a submanifold Z of X containing a and such that $T_a(Z) = E$. To see this, it is enough to consider the case where X is an open set in $\mathbf{R}^n$ and $a = 0$, and then we may take Z to be the intersection of X with a vector subspace of $\mathbf{R}^n$.

(16.8.7) (i) *Let* Z *be a differential manifold,* Y *a submanifold of* Z, X *a subspace of* Y. *Then* X *is a submanifold of* Z *if and only if* X *is a submanifold of* Y.

(ii) *If* X_1 (resp. X_2) *is a submanifold of* Y_1 (resp. Y_2), *then* $X_1 \times X_2$ *is a submanifold of* $Y_1 \times Y_2$.

Assertion (ii) follows immediately from the definitions of a submanifold (16.8.3) and a product manifold (16.6). As to (i), suppose first that X is a submanifold of Y. Using local charts, we reduce to the case where $Z = \mathbf{R}^m$, Y is an open subset of $\mathbf{R}^n$ (where $n < m$), and there exists a submersion $f : Y \to \mathbf{R}^{n-p}$ such that X is the set of points $y \in Y$ satisfying $f(y) = 0$. We then extend f to a submersion $g : Y \times \mathbf{R}^{m-n} \to \mathbf{R}^{m-p}$ by defining $g(y, t) = (f(y), t)$ for $y \in Y$ and $t \in \mathbf{R}^{m-n}$. Then X is the set of points (y, t) satisfying $g(y, t) = 0$, hence is a submanifold of Z. Conversely, suppose that X is a submanifold of Z. This time we may suppose that $Z = \mathbf{R}^m$ and that X is an open set in $\mathbf{R}^p$ containing the origin. The tangent space $T_0(Y)$ can be identified with a vector subspace E of $\mathbf{R}^m$ containing $\mathbf{R}^p$. Let F be a supplement of $\mathbf{R}^p$ in E, let G be a supplement of E in $\mathbf{R}^m$, and let π be the projection of $\mathbf{R}^m$ onto F parallel to $G + \mathbf{R}^p$. The restriction h of π to Y is a submersion, because by definition the rank of $T_0(h)$ is $\dim(E) - p = \dim(F)$. The set X′ of points $y \in Y$ such that $h(y) = 0$ is therefore a submanifold of Y, hence of Z, and of the same dimension as X. Since X′ contains X and the canonical injection j of X into X′ is of class C^∞ (16.7.7(i)), it follows that j is a local diffeomorphism (16.5.6); hence X is open in X′ and therefore is a submanifold of Y.

(16.8.8) *Let* $f : X \to Y$ *be a subimmersion* (16.7.1), *a a point of* X, *and* $b = f(a)$.

(i) *The subspace* $f^{-1}(b)$ *is a closed submanifold of* X. *The tangent space* $T_a(f^{-1}(b))$ *to* $f^{-1}(b)$ *is the kernel of* $T_a(f)$, *and hence we have an exact sequence*

$$0 \to T_a(f^{-1}(b)) \to T_a(X) \xrightarrow{T_a(f)} T_b(Y).$$

(ii) *There exists an open neighborhood* U *of* a *in* X *such that* $f(U)$ *is a submanifold of* Y, *and we have*

(16.8.8.1) $$\dim_a(X) = \dim_b(f(U)) + \dim_a(f^{-1}(b)).$$

Furthermore, if E *is any supplement of* $T_a(f^{-1}(b))$ *in* $T_a(X)$, *there exists a submanifold of* V *of* U *whose tangent space at the point* a *is* E. *For each submanifold* V *of* U *having this property, there exists an open neighborhood* W *of* b *in* Y *such that the restriction of* f *to* $V' = V \cap f^{-1}(W)$ *is an isomorphism of* V′ *onto* $W \cap f(U)$, *and such that* $T_x(V)$ *is a supplement of* $T_x(f^{-1}(f(x)))$ *in* $T_x(X)$ *for all* $x \in V'$.

(iii) *If* $T_a(f)$ *is not surjective,* V *may also be chosen so that* $f(U)$ *is nowhere dense* (12.16) *in* Y.

(iv) *If* f *is an injective subimmersion, then* f *is an immersion. If* f *is a bijective submersion, then* f *is a diffeomorphism* (cf. Problem 3).

We can apply (16.7.4) to the point $a \in X$, and then it is clear that

$$U \cap f^{-1}(b) = \varphi^{-1}(\varphi(U) \cap F^{-1}(0)).$$

Hence we reduce to the case where X and Y are open sets in finite-dimensional real vector spaces and f is the restriction to X of a linear mapping. Parts (i)–(iii) of the proposition are now obvious ((16.5.2) and (12.16)). As to (iv), it follows from (i) that if f is injective, then $T_a(f)$ is injective, for each $a \in X$, hence f is an immersion. If in addition f is a submersion, then f is a local diffeomorphism. Finally, if f is also bijective, then f is a diffeomorphism.

(16.8.9) *Let* Y *be a differential manifold and let* $(f_i)_{1 \leq i \leq r}$ *be a sequence of real-valued* C^∞*-functions on* Y. *Let* X *be the set of* $x \in Y$ *such that* $f_i(x) = 0$ *for* $1 \leq i \leq r$. *Suppose that, for each* $x \in X$, *the differentials* $d_x f_i$ $(1 \leq i \leq r)$ *are linearly independent covectors in* $T_x(Y)^*$. *Then*:

(i) X *is a closed submanifold of* Y, *and for each* $x \in X$ *the tangent space* $T_x(X)$ *is the annihilator in* $T_x(Y)$ *of the subspace of* $T_x(Y)^*$ *spanned by the differentials* $d_x f_i$, *and consequently is of dimension* $\dim_x(Y) - r$.

(ii) *Let* F *be a* C^∞*-function on* Y *which vanishes at all points of* X. *Then for each* $x_0 \in X$ *there exists an open neighborhood* U *of* x_0 *in* Y *and* r *functions* F_j $(1 \leq j \leq r)$ *of class* C^∞ *on* U *such that*

$$F(y) = \sum_{j=1}^{r} F_j(y) f_j(y)$$

for all $y \in U$. *If* Y *is a real-* (resp. *complex-*) *analytic manifold and the functions* F *and* f_j *are analytic, then the functions* F_j *may be chosen to be analytic.*

(i) Since the differentials $d_x f_i$ are linearly independent, there exists an open neighborhood $V(x)$ of x in Y such that the differentials $d_{x'} f_i$ are linearly independent for all $x' \in V(x)$ (16.5.8.4). Replacing Y by the open set which is the union of the neighborhoods $V(x)$ as x runs through X, we may therefore assume that the $d_y f_i$ are linearly independent for all $y \in Y$. However, then the mapping $g : y \mapsto (f_i(y))_{1 \leqq i \leqq r}$ of Y into $\mathbf{R}^r$ is a submersion, by virtue of (16.5.7.2) and the definition (16.7.1). Since $X = g^{-1}(0)$, we can now apply (16.8.8).

(ii) By virtue of (16.7.4) we may limit ourselves to the case where Y is an open set in $\mathbf{R}^n$ and $f_j(x) = x^j$, the jth coordinate of x ($1 \leqq j \leqq r$), so that $X = U \cap \mathbf{R}^{n-r}$ (where $\mathbf{R}^{n-r}$ is identified with the subspace spanned by the last $n - r$ vectors of the canonical basis of $\mathbf{R}^n$). Moreover we may take x_0 to be the origin. Then the assertion (for C^∞-functions) is a consequence of the following lemma:

(16.8.9.1) *Let* F *be a real-valued function of class* C^∞ *on an open cube* $I^n \subset \mathbf{R}^n$ (*where* I *is an open interval in* $\mathbf{R}$). *Then in* I^n *we can write*

(16.8.9.2)
$$F(x^1, \dots, x^n) = F(0, \dots, 0) + x^1 F_1(x^1, \dots, x^n) + x^2 F_2(x^2, \dots, x^n) + \cdots + x^{n-1} F_{n-1}(x^{n-1}, x^n) + x^n F_n(x^n),$$

where $F_1, \dots, F_n$ *are* C^∞*-functions on* I^n.

Assuming this lemma, since $F(x^1, \dots, x^n) = 0$ whenever $x^1 = \cdots = x^r = 0$ we obtain successively that $F_n, F_{n-1}, \dots, F_{r+1}$ are identically zero: first we put all the x^i except x^n equal to zero, then all the x^i except x^{n-1} and x^n, and so on.

To prove (16.8.9.1) we write

$$F(x^1, \dots, x^n) = (F(x^1, x^2, \dots, x^n) - F(0, x^2, \dots, x^n)) + F(0, x^2, \dots, x^n)$$

so that by induction on n we are reduced to proving that the function

(16.8.9.3) $$G(x^1, \dots, x^n) = (x^1)^{-1}(F(x^1, \dots, x^n) - F(0, x^2, \dots, x^n))$$

which is defined whenever $x^1 \neq 0$, tends to a finite limit as $x^1 \to 0$, and that the function so extended is of class C^∞ on I^n. Using Taylor's formula (8.14.2) we have, for $x^1 \neq 0$ and any integer $p \geqq 1$,

(16.8.9.4)

$$G(x^1, \ldots, x^n) = D_1F(0, x^2, \ldots, x^n) + \frac{x^1}{2!} D_1^2F(0, x^2, \ldots, x^n) + \cdots$$
$$+ \frac{(x^1)^{p-1}}{p!} D_1^pF(0, x^2, \ldots, x^n) + (x^1)^{-1}H(x^1, \ldots, x^n),$$

where

$$\text{(16.8.9.5)} \quad H(x^1, \ldots, x^n) = \int_0^{x^1} \frac{(x^1 - t)^p}{p!} D_1^{p+1}F(t, x^2, \ldots, x^n)\, dt$$

is of class C^∞ on I^n by hypothesis. Differentiating under the integral sign (8.11.2) and replacing F by some derivative of the form $D_2^{\nu_2}D_3^{\nu_3} \cdots D_n^{\nu_n}F$, we are reduced to showing that, as $x^1 \to 0$, the derivatives

$$\text{(16.8.9.6)} \quad D_1^k((x^1)^{-1}H) = \sum_{j=0}^{k} (-1)^j \binom{k}{j} j!(x^1)^{-j-1}D_1^{k-j}H$$

$(0 \leqq k \leqq p - 1)$, calculated by Leibniz's rule (8.13.2), tend to zero uniformly in $(x^2, \ldots, x^n)$ on a neighborhood V of 0 in $\mathbf{R}^{n-1}$; but by (8.11.2) we have

$$\text{(16.8.9.7)} \quad D_1^{k-j}H(x^1, \ldots, x^n) = \int_0^{x^1} \frac{(x^1 - t)^{p-k+j}}{(p - k + j)!} D_1^{p+1}F(t, x^2, \ldots, x^n)\, dt$$

and hence by the mean value theorem

$$|D_1^{k-j}H(x^1, \ldots, x^n)| \leqq C|x^1|^{p-k+j+1}$$

where C is a constant, for all $(x^2, \ldots, x^n) \in V$. Using this inequality in (16.8.9.6) now completes the proof of the lemma.

In the case where F is (real- or complex-) analytic, the proof is much simpler, by considering the Taylor expansion of F at the origin, and it follows from (9.1.4) that the F_j are analytic in a neighborhood of 0.

We remark that (16.8.3.2) shows conversely that for *each* submanifold X of Y and each point $a \in X$, there exists an open neighborhood V of a such that $V \cap X$ is defined by equations satisfying the conditions of (16.8.9).

Examples

(16.8.10) In (16.8.9) let us take $Y = \mathbf{R}^{n+1} - \{0\}$ and $r = 1$, and the sequence (f_i) to consist of the single function $f: \mathbf{y} \mapsto \|\mathbf{y}\|$, the Euclidean norm on Y.

Then $X = f^{-1}(1)$, as a topological space, is just the unit sphere $\mathbf{S}_n$; since Df is of rank 1, it follows that $\mathbf{S}_n$ is endowed with a structure of a submanifold of Y. Let us show that this structure of differential manifold on $\mathbf{S}_n$ is the same as that defined in (16.2.3). For this it is enough to observe, in view of (16.8.3), that the formulas (16.2.3.1) define a *diffeomorphism* of the submanifold $\mathbf{S}_n - \{\mathbf{e}_0\}$ of $\mathbf{R}^{n+1}$ onto $\mathbf{R}^n$.

Now consider the mapping $g : \mathbf{y} \mapsto (\mathbf{y}/\|\mathbf{y}\|, \|\mathbf{y}\|)$ of Y into $\mathbf{S}_n \times \mathbf{R}_+^*$. This mapping is a bijection, whose inverse is $(\mathbf{z}, \zeta) \mapsto \zeta\mathbf{z}$; also g is a submersion, because $Df \neq 0$ and the restriction of the mapping $\mathbf{y} \mapsto \mathbf{y}/\|\mathbf{y}\|$ to a sphere $\lambda\mathbf{S}_n$ (where $\lambda > 0$) is a homothety of this sphere onto $\mathbf{S}_n$, hence a diffeomorphism. It follows that g is a diffeomorphism, by virtue of (16.6.4) and (16.8.8(iv)).

(16.8.11) Suppose that the conditions of (16.6.8) are satisfied, so that $f : X_1 \times X_2 \to Z$ is a submersion at the point (a_1, a_2). Then there exists an open neighborhood W of (a_1, a_2) in $X_1 \times X_2$ such that, if Y is the set of points (x_1, x_2) satisfying $f(x_1, x_2) = c$, then $Y \cap W$ is a submanifold of $X_1 \times X_2$, and the restriction of pr_1 to $Y \cap W$ is an isomorphism of this submanifold onto an open subset of X_1. If $X_1 = X_2 = Z = \mathbf{C}$ and if f is holomorphic, the set Y_0 of points of Y where $D_2 f(x_1, x_2) \neq 0$ is an open subset of Y; this complex-analytic submanifold of $\mathbf{C}^2$ is called the *Riemann surface* (relative to the second coordinate) *defined by the holomorphic function f*. For example, if

$$f(x_1, x_2) = x_1 - e^{x_2},$$

we have $Y = Y_0$, and Y is an analytic subgroup of the complex-analytic group $\mathbf{C}^* \times \mathbf{C}$ (16.9.10). This surface Y is called the *Riemann surface of the logarithmic function* (cf. Problem 12). The restriction of pr_1 to Y is an étale morphism of Y onto $\mathbf{C}^* = \mathbf{C} - \{0\}$. For each $t \in Y$ we write $\log(t) = \mathrm{pr}_2 t$, so that we have $\log(tt') = \log(t) + \log(t')$ for t, t' in Y (the law of composition in Y being $(x, y)(x', y') = (xx', y + y')$), and the mapping $t \mapsto \log t$ is a holomorphic mapping (16.3.5) of Y into $\mathbf{C}$.

(16.8.12) *Let $f : X \to Y$ be a submersion, Z any submanifold of Y. Then $f^{-1}(Z)$ is a submanifold of X, and the restriction $f^{-1}(Z) \to Z$ of f is a submersion* (cf. Problem 17).

Using a chart satisfying (16.7.4), we may assume that Y is an open set in $\mathbf{R}^n$ and X an open set in $\mathbf{R}^m$ $(n \leqq m)$, and that f is the restriction to X of the canonical projection $(\xi^1, \ldots, \xi^m) \mapsto (\xi^1, \ldots, \xi^n)$. Let $x \in f^{-1}(Z)$ and let $y \in Z$ be its projection. Then by hypothesis there exists a chart (U, ψ, n) on Y at the point y such that $\psi(U \cap Z) = \psi(U) \cap \mathbf{R}^p$, where $p \leqq n$. If we denote by φ the restriction to $f^{-1}(\mathrm{U})$ of $\psi \times 1_{\mathbf{R}^{m-n}}: \mathrm{U} \times \mathbf{R}^{m-n} \to \mathbf{R}^m$, then $(f^{-1}(\mathrm{U}), \varphi, m)$ is a chart on X at the point x, such that

$$\varphi(f^{-1}(\mathrm{U}) \cap f^{-1}(\mathrm{Z})) = \varphi(f^{-1}(\mathrm{U})) \cap \mathbf{R}^{p+m-n}.$$

Hence the result.

(16.8.13) *Let* X, Y *be two differential manifolds and* $f: \mathrm{X} \to \mathrm{Y}$ *a mapping of class* C^∞. *Then the graph* Γ_f *of* f *in* $\mathrm{X} \times \mathrm{Y}$ *is a closed submanifold of* $\mathrm{X} \times \mathrm{Y}$, *the mapping* $g: x \mapsto (x, f(x))$ *is an embedding* (16.8.4) *and the vector subspace* $\mathrm{T}_{(x, f(x))}(\Gamma_f)$ *of* $\mathrm{T}_{(x, f(x))}(\mathrm{X} \times \mathrm{Y})$ *is the graph of the linear mapping* $\mathrm{T}_x(f)$.

We know that g is a homeomorphism of X onto Γ_f (the inverse of g being the restriction to Γ_f of the projection pr_1), and that Γ_f, being the set of points $z \in \mathrm{X} \times \mathrm{Y}$ such that $\mathrm{pr}_2(z) = f(\mathrm{pr}_1(z))$, is closed in $\mathrm{X} \times \mathrm{Y}$. Hence (16.8.4) it is enough to prove that g is an immersion, but since $\mathrm{T}_x(g) = (\mathrm{T}_x(1_\mathrm{X}), \mathrm{T}_x(f))$ by virtue of (16.6.4), it is clear that $\mathrm{T}_x(g)$ is a linear mapping of rank equal to $\dim_x(\mathrm{X})$. Hence the result.

In particular, when $\mathrm{Y} = \mathrm{X}$, the *diagonal* Δ of $\mathrm{X} \times \mathrm{X}$ is a closed submanifold of $\mathrm{X} \times \mathrm{X}$, and the diagonal mapping $x \mapsto (x, x)$ is a diffeomorphism of X onto Δ. In $\mathbf{R} \times \mathbf{R}$, the set of pairs (ξ, η) such that $\xi \neq 0$ and $\eta = \sin(1/\xi)$ is an (analytic) submanifold whose closure is not locally connected.

Remark

(16.8.14) Let Z be a submanifold of $\mathrm{X} \times \mathrm{Y}$ such that at a point $(a, b) \in \mathrm{Z}$ the restriction to $\mathrm{T}_{(a,b)}(\mathrm{Z})$ of the projection $\mathrm{T}_a(\mathrm{X}) \times \mathrm{T}_b(\mathrm{Y}) \to \mathrm{T}_a(\mathrm{X})$ is a bijection onto $\mathrm{T}_a(\mathrm{X})$. Since this restriction is equal to $\mathrm{T}_{(a,b)}(p)$, where $p: \mathrm{Z} \to \mathrm{X}$ is the restriction of $\mathrm{pr}_1: \mathrm{X} \times \mathrm{Y} \to \mathrm{X}$, it follows (16.5.6) that there exists an open neighborhood U of (a, b) such that $p | \mathrm{U}$ is a diffeomorphism onto an open neighborhood V of a in X. Since $p(x, y) = x$, the inverse diffeomorphism $g: \mathrm{V} \to \mathrm{U}$ is of the form $x \mapsto (x, f(x))$, where $f: \mathrm{V} \to \mathrm{Y}$ is of class C^∞, and U is therefore the *graph* of f in $\mathrm{V} \times \mathrm{Y}$.

PROBLEMS

1. (a) Let $f: \mathrm{X} \to \mathrm{Y}$ be a submersion and $g: \mathrm{Y} \to \mathrm{Z}$ an immersion. Show that $g \circ f: \mathrm{X} \to \mathrm{Z}$ is a subimmersion.

   (*b*) The mapping $f: t \mapsto (t, t^2, t^3)$ of $\mathbf{R}$ into $\mathbf{R}^3$ is an immersion, and the projection $g: (x, y, z) \mapsto (y, z)$ of $\mathbf{R}^3$ into $\mathbf{R}^2$ is a submersion, but $g \circ f$ is not a subimmersion of $\mathbf{R}$ into $\mathbf{R}^2$, although it is injective.

2. Let $f : X \to Y$ be an injective immersion which is *proper* (Section 12.7, Problem 2). Show that f is an embedding (observe that the image of a closed subset of X is closed in Y). Give an example of an embedding of **R** into $\mathbf{R}^2$ which is not proper (consider a "spiral").

3. (a) In $\mathbf{R}^3$, the union of the line $z = 1, y = 0$ and the complement in the plane $Y : z = 0$ of the line $z = 0, y = 0$ is a nonconnected manifold X. Show that the restriction to X of the projection $(x, y, z) \mapsto (x, y)$ of $\mathbf{R}^3$ onto Y is a bijective immersion of X onto Y which is not a diffeomorphism.
(b) Let X be a *connected* differential manifold, Y a differential manifold, and $f : X \to Y$ a bijective subimmersion of X onto Y. Show that f is a diffeomorphism. (Observe that the set of points $x \in X$ at which f is a submersion is both open and closed in X; to show that this set is nonempty, use (16.8.8(iii)), (12.6.1), and Baire's theorem (12.16.1).)

4. Give the analogs of the results of Sections 16.3–16.8 for manifolds of class C^r $(r \geqq 1)$, which were defined in Section 16.1, Problem 2. Show that the analog of (16.8.9(ii)) is false.

5. Let E be a finite-dimensional real vector space, F a closed subset of E, and $\|x\|$ a norm defining the topology of E (12.13.2). If a is a nonisolated point of F, the *contingent* of F at a is the union of the rays through 0 whose direction vectors of norm 1 are limits of sequences of the form $((x_n - a)/\|x_n - a\|)$, where (x_n) is a sequence of points of F, distinct from a and with a as limit. The *paratingent* of F at a is the union of the lines through 0 whose direction vectors are limits of sequences of the form

$$((x_n - y_n)/\|x_n - y_n\|),$$

where (x_n) and (y_n) are two sequences of points of F, distinct from a and with a as limit, and such that $x_n \neq y_n$ for all n.

Show that F is a submanifold of class C^1 of E if and only if, for each nonisolated point $a \in F$,

(i) the paratingent of F at a is a vector subspace P of E, and
(ii) if N is a supplement of P in E and $p : E \to P$ is the projection parallel to N, then the image under p of any neighborhood of a in F is a neighborhood of $p(a)$ in P.

(Show first that there exists a compact neighborhood U of a in F such that $p | U$ is a bijection onto a compact neighborhood V of $p(a)$, by contradiction. If we identify E with $P \times N$, then U is the graph of a continuous mapping $f : V \to N$. Show that f is of class C^1 by using condition (i) above, Problem 3 of Section 8.6, and arguing by contradiction.)

Give an example in which the contingent and the paratingent of F at a are each equal to the whole of E but the condition (ii) above is not satisfied. (Take $E = \mathbf{R}^2$.)

6. In a differential manifold X, let $Y_1, \dots, Y_r$ be submanifolds with a common point a. Show that the union of the Y_i cannot be a submanifold of X unless it has the same dimension at a as one of the Y_i. (Use Problem 5.)

7. Show that a complex-analytic submanifold of $\mathbf{C}^n$ which is compact and connected consists of a single point (see Section 16.3, Problem 3).

8. Let X, Y be two differential manifolds, $f : X \to Y$ a mapping of class C^∞, and U a connected open subset of X. If r is the least upper bound of $\mathrm{rk}_x(f)$ as x runs over U, show that r is finite and that the set of points $x \in U$ at which $\mathrm{rk}_x(f) = r$ is open. Deduce that the set of points at which f is a subimmersion is a dense open subset of X (argue by contradiction). If f is an open mapping, the set of points at which f is a submersion is dense in X.

9. Let X, Y be two differential manifolds, $f : X \to Y$ a mapping of class C^1. If Z is a submanifold of Y, the mapping f is said to be *transversal over* Z *at* $x \in f^{-1}(Z)$ if the tangent space $T_{f(x)}(Y)$ is the sum of $T_{f(x)}(Z)$ and $T_x(f)(T_x(X))$, and f is said to be *transversal over* Z if this condition is satisfied for all $x \in f^{-1}(Z)$. If so, then $f^{-1}(Z)$ is a submanifold of X, and for each $x \in f^{-1}(Z)$ the tangent space $T_x(f^{-1}(Z))$ is the inverse image under $T_x(f)$ of $T_{f(x)}(Z)$. (Since the question is a local one, we may take Z to be a submanifold given by an equation $g(y) = 0$, where $g : Y \to \mathbf{R}^p$ is a submersion; consider the composite mapping $g \circ f$.)

In particular, if X and Z are submanifolds of Y, we say that X and Z are *transversal at a point* $x \in X \cap Z$ if the canonical injection of X into Y is transversal over Z at x, or equivalently if $T_x(Y) = T_x(X) + T_x(Z)$, which is symmetrical in X and Z. The submanifolds X and Z are said to be *transversal* if they are transversal at all points $x \in X \cap Z$; in that case, $X \cap Z$ is a submanifold of Y.

10. Let $f : X \to Z$ and $g : Y \to Z$ be two mappings of class C^∞, and consider their product $f \times g : X \times Y \to Z \times Z$, which is also of class C^∞. Show that $f \times g$ is transversal over the diagonal Δ of $Z \times Z$ if and only if, for each pair $(x, y) \in X \times Y$ such that $f(x) = g(y)$, we have

$$(*) \qquad T_z(Z) = T_x(f)(T_x(X)) + T_y(g)(T_y(Y)),$$

where $z = f(x) = g(y)$. This condition is always satisfied if either f or g is a submersion.

When condition $(*)$ is satisfied, the set of points $(x, y) \in X \times Y$ such that $f(x) = g(y)$ is a submanifold of $X \times Y$, which is called the *fiber product of* X *and* Y *over* Z and is written $X \times_Z Y$. The tangent space at the point $(x, y) \in X \times_Z Y$ to the fiber product is the subspace of $T_x(X) \times T_y(Y)$ consisting of the pairs $(\mathbf{h}, \mathbf{k})$ such that

$$T_x(f) \cdot \mathbf{h} = T_y(g) \cdot \mathbf{k}.$$

In this situation, f and g are said to be *transversal* mappings into Z. Show that if f is a submersion (resp. an immersion, resp. a subimmersion), then so is the restriction $X \times_Z Y \to Y$ of pr_2.

11. Let Y be a differential (resp. real-analytic, resp. complex-analytic) manifold, X a Hausdorff topological space, and $p : X \to Y$ a mapping with the following property: For each $x \in X$ there exists an open neighborhood V of x such that $p|V$ is a homeomorphism of the subspace V onto a submanifold of Y.

(a) Show that X is locally connected, that each point of x has a closed neighborhood which is homeomorphic to a closed ball in $\mathbf{R}^n$, and that for each $y \in Y$ the fiber $p^{-1}(y)$ is a discrete subspace of X.

(b) Let $\mathfrak{B}$ be a denumerable basis for the topology of Y. A pair (W, U) is said to be *distinguished* if $U \in \mathfrak{B}$ and if W is a connected component of $p^{-1}(U)$ such that $\overline{W}$ is compact and metrizable and $p|\overline{W}$ is a homeomorphism of $\overline{W}$ onto a subspace of Y.

Show that for each $x \in X$ there exists a distinguished pair (W, U) such that $x \in W$. Show also that if (W, U) is a distinguished pair, the set of distinguished pairs (W', U') such that $W \cap W' \neq \varnothing$ is denumerable (use the fact that W is separable).
(c) Deduce from (b) that each connected component X_0 of X is metrizable and separable. (Consider the following relation between two points x, x' in X: There exists a finite sequence of distinguished pairs (W_i, U_i) $(1 \leqq i \leqq r)$ such that $x \in W_1$, $x' \in W_r$, and $W_i \cap W_{i+1} \neq \varnothing$ for $1 \leqq i \leqq r-1$. Then apply (12.4.7).)
(d) Show that there exists on X_0 a unique structure of differential (resp. real-analytic, resp. complex-analytic) manifold such that $p|X_0$ is an immersion of X_0 into Y (*Poincaré–Volterra theorem*).

12. Let L be the set of all pairs $\lambda = (P_\lambda, f_\lambda)$, where P_λ is a nonempty open polydisk in $\mathbf{C}^n$ (resp. $\mathbf{R}^n$) and f_λ is a complex (resp. real) analytic function on P_λ. For each pair of elements λ, μ in L, define a set $A_{\lambda\mu}$ as follows: $A_{\lambda\mu} = \varnothing$ if $P_\lambda \cap P_\mu = \varnothing$ or if the restrictions of f_λ and f_μ to $P_\lambda \cap P_\mu$ are distinct; $A_{\lambda\mu} = P_\lambda \cap P_\mu$ if the restrictions of f_λ and f_μ to $P_\lambda \cap P_\mu$ are equal. Let $h_{\mu\lambda}$ be the identity mapping of $A_{\lambda\mu}$ onto itself. Show that the mappings $h_{\mu\lambda}$ satisfy the patching condition (12.2.4.1), and hence that we obtain a topological space X by patching together the P_λ along the $A_{\lambda\mu}$ by means of the $h_{\lambda\mu}$. Let $\pi_\lambda : P_\lambda \to X$ be the canonical mapping and let $X_\lambda = \pi_\lambda(P_\lambda)$ be its image, which is an open subset of X. If $j_\lambda : P_\lambda \to \mathbf{C}^n$ (resp. $j_\lambda : P_\lambda \to \mathbf{R}^n$) is the canonical injection, show that there exists a unique mapping $p : X \to \mathbf{C}^n$ (resp. $p : X \to \mathbf{R}^n$) such that $p \circ \pi_\lambda = j_\lambda$ for all λ. The restriction $p|X_\lambda$ is a homeomorphism of X_λ onto p_λ, and π_λ is the inverse homeomorphism. Show that X is Hausdorff (use (9.4.1)). Deduce that the results of Problem 11 apply to X and p: If (Y_α) is the family of connected components of X, then there exists on each Y_α a unique structure of a complex (resp. real) analytic manifold such that $p|Y_\alpha$ is a local isomorphism of analytic manifolds. For each index α, there exists a complex (resp. real) analytic function F_α on Y_α such that, for each index λ for which $X_\lambda \subset Y_\alpha$, the restriction of F_α to X_λ is equal to $f_\lambda \circ (p|X_\lambda)$; for each such index λ, Y_α is said to be the *analytic manifold defined by* f_λ (the *Riemann surface of* f_λ in the complex case when $n = 1$), and F_α is the *natural continuation of* f_λ.

13. (a) Let X, Y be two pure differential manifolds of the same dimension, and let $f : X \to Y$ be a mapping of class C^∞. Let $S \subset X$ be the closed set of points at which f is not a local diffeomorphism, and suppose that the set of nonisolated points of S is *discrete*. Show that, for each point $x_0 \in X$, the image under f of any neighborhood of x_0 is a neighborhood of $f(x_0)$, and hence that f is an open mapping. (Reduce to the case $X = Y = \mathbf{R}^n$, and show that it is impossible that on each sphere $\|x - x_0\| = \rho$ in $\mathbf{R}^n$ there should exist a point x such that $f(x) = f(x_0)$, by a compactness argument. Let $\|x - x_0\| = \rho$ be a sphere on which $f(x) \neq f(x_0)$, so that $\|f(x) - f(x_0)\| \geqq \alpha > 0$; let D be the open ball $\|x - x_0\| < \rho$ and D′ the open ball $\|y - f(x_0)\| < \alpha$; and let $G = D \cap \complement S$ and $H = D' \cap \complement\overline{f(S)}$. Show that $f(\mathrm{Fr}(G))$ does not intersect H; deduce that $f(G) \supset H$ and hence that $f(D)$ is a neighborhood of $f(x_0)$.)
(b) Deduce from (a) that $\mathrm{Fr}(f(U)) \subset f(\mathrm{Fr}(U))$ for every relatively compact open set U in X. Deduce that if in addition f is proper (Section 12.7, Problem 2) and X, Y are connected, then $f(X) = Y$.
(c) If $Y = \mathbf{R}^n$, show that for each $a \in \mathbf{R}^n$, $\inf_{x \in X} \|f(x) - a\|$ is not attained at a point $x_0 \in X$ unless $f(x_0) = a$.
(d) Suppose that $Y = \mathbf{R}^2$ and that X is an open neighborhood of the disk $D : |z| \leqq 1$ in $\mathbf{R}^2$. Show that if f satisfies the conditions of (a), then $f(S_1)$ cannot be a Bernoulli

lemniscate B (with equation $(\xi_1^2+\xi_2^2)^2=\xi_1^2-\xi_2^2$). (Observe that the image of the interior of D cannot contain any point of the unbounded connected component of the complement of B.)

14. Let X be a connected *complex*-analytic manifold of dimension n, and let Y_j $(1 \leqq j \leqq r)$ be a finite number of closed submanifolds of X, of dimensions $\leqq n-1$. Show that the complement in X of the union of the Y_j is a *connected* dense open set (use (12.6.1) and Section 16.3, Problem 3(c)).

15. Let X be a real-analytic manifold and Y_0 a differential manifold whose underlying set is contained in X. Suppose that, for each point $x \in Y_0$, there exists a chart $c = (U, \varphi, n)$ of X at x and a neighborhood $V \subset U$ of x in Y_0, such that $\varphi | V$ maps V onto $\varphi(U) \cap \mathbf{R}^m$ and $(V, \varphi | V, m)$ is a chart of Y_0. In these conditions show that there exists a unique real-analytic submanifold Y of X whose underlying differential manifold is Y_0.

16. Let X_0 be a differential manifold, Y a real-analytic manifold, and suppose that the differential manifold Y_0 underlying Y is a submanifold of X_0. Show that for each $y \in Y$ there exists a chart (U, φ, n) of X_0 and an open neighborhood V of y in Y contained in U, such that $(V, \varphi | V, m)$ is a chart of Y for which $\varphi(V) = \varphi(U) \cap \mathbf{R}^m$.

17. If $j : \mathbf{R} \to \mathbf{R}^2$ is the canonical injection, give examples of submanifolds $Y \subset \mathbf{R}^2$ such that $j^{-1}(Y)$ is not a submanifold of $\mathbf{R}$.

9. LIE GROUPS

(16.9.1) Let G be a set endowed with a group structure and a structure of a differential manifold. These two structures are said to be *compatible* if the mappings $(x, y) \mapsto xy$ of $G \times G$ into G and $x \mapsto x^{-1}$ of G into G are of class C^∞.

It comes to the same thing to require that the mapping $(x, y) \mapsto xy^{-1}$ (or $(x, y) \mapsto x^{-1}y$) should be of class C^∞, by virtue of the relations $x^{-1} = ex^{-1}$ and $xy = x(y^{-1})^{-1}$. A group endowed with a structure of differential manifold which is compatible with its group structure is called a *Lie group* (or a *real Lie group*). It is clear that the topology of a Lie group G is compatible with the group structure, and the topological group so defined is *metrizable*, *separable*, *locally compact*, and *locally connected* (16.1.3), and the set of its connected components is therefore *at most denumerable*. Moreover, this metrizable group is *complete* (12.9.5).

An *isomorphism* of a Lie group G onto a Lie group G′ is by definition an isomorphism of the group G onto the group G′ which is also a diffeomorphism (16.2.6). If G = G′, we say *automorphism* in place of isomorphism.

For each $a \in G$, the left and right translations $\gamma(a) : x \mapsto ax$ and $\delta(a^{-1}) : x \mapsto xa$

are diffeomorphisms of G onto itself (16.6) In particular, it follows that a Lie group is a *pure* differential manifold (16.1.3). For each $a \in G$, the inner automorphism $\text{Int}(a) : x \mapsto axa^{-1}$ is a Lie group automorphism of G.

Examples

(16.9.2) If E is a finite-dimensional real vector space, the canonical structure of differential manifold on E (16.2.2) is compatible with the additive group structure of E. Hence E is endowed with a canonical structure of (commutative) Lie group.

(16.9.3) Let A be an **R**-algebra of finite dimension with unit element. Then A is normable (15.1.8) by virtue of the continuity of polynomial functions on $\mathbf{R}^n$ and hence (because $\mathbf{R}^n$ is complete) the multiplicative group A^* of invertible elements of A is a nonempty open set in A (15.2.4). The structure of differential manifold induced on A^* by the canonical structure on the vector space A (16.2.2) is compatible with the group structure of A^*, by the argument of (8.12.11), which applies to any Banach algebra, not merely to $\mathscr{L}(E; E)$.

In particular, if E is a real vector space of dimension n, the algebra $A = \mathscr{L}(E; E) = \text{End}(E)$ may be identified, together with its canonical structure of differential manifold, with the vector space $\mathbf{R}^{n^2}$, and therefore the linear group $\mathbf{GL}(E) = A^*$ is a Lie group of dimension n^2. Likewise, if E is a vector space of dimension n over the field of complex numbers **C** (resp. the division ring of quaternions **H**†), then **GL**(E) is a (real) Lie group of dimension $2n^2$ (resp. $4n^2$). When $E = \mathbf{R}^n$ (resp. $E = \mathbf{C}^n$, resp. $E = \mathbf{H}^n$ (the *left* vector space)), we write $\mathbf{GL}(n, \mathbf{R})$ (resp. $\mathbf{GL}(n, \mathbf{C})$, resp. $\mathbf{GL}(n, \mathbf{H})$) instead of **GL**(E).

(16.9.4) If $G_1, G_2, \ldots, G_n$ are Lie groups, $G = G_1 \times G_2 \times \cdots \times G_n$ is a Lie group when endowed with the product group structure and the product manifold structure; this follows immediately from (16.6.5). The Lie group G so defined is called the *product* of the Lie groups G_i.

(16.9.5) If G is a Lie group, then the set G endowed with the same structure of differential manifold and the *opposite* group structure is again a Lie group, by virtue of (16.6.5); it is called the *opposite* of the Lie group G, and is denoted by G^0.

† For the elementary algebraic properties of quaternions, see the author's book "Linear Algebra and Geometry," Houghton, Boston, Massachusetts, 1969.

(16.9.6) Let G be a Lie group and H a subgroup of G which is a submanifold of G (16.8.3). Then the group structure and manifold structure of H are compatible. For H × H is a submanifold of G × G, and if

$$j : \mathrm{H} \times \mathrm{H} \to \mathrm{G} \times \mathrm{G} \qquad \text{and} \qquad j' : \mathrm{H} \to \mathrm{G}$$

are the canonical injections, the mapping $g : (x, y) \mapsto xy^{-1}$ of H × H into H is such that $j' \circ g = f \circ j$, where f is the mapping $(x, y) \mapsto xy^{-1}$ of G × G into G; our assertion now follows from (16.8.3.4). The set H, endowed with its structures of group and differential manifold, is called a *Lie subgroup* of G. It is *closed* in G by (12.9.6).

Every open subgroup of a Lie group G is a Lie subgroup. In particular, the *neutral component* G_0 of G is a Lie subgroup, because G is locally connected (12.8.7). Every *discrete* subgroup of G is a Lie subgroup. In a product $\mathrm{G}_1 \times \cdots \times \mathrm{G}_m$ of Lie groups, if H_i is a Lie subgroup of G_i for $1 \leqq i \leqq m$, then $\mathrm{H}_1 \times \cdots \times \mathrm{H}_m$ is a Lie subgroup of $\mathrm{G}_1 \times \cdots \times \mathrm{G}_m$. For each $n \geqq 1$, the Lie group $\mathbf{GL}(n, \mathbf{R})$ is a Lie subgroup of $\mathbf{GL}(n, \mathbf{C})$, which in turn is a Lie subgroup of $\mathbf{GL}(n, \mathbf{H})$ (if we identify $\mathbf{C}$ with the subfield of $\mathbf{H}$ generated by 1 and i).

For each pair of integers $p \geqq 1$, $q \geqq 1$, the product group

$$\mathbf{GL}(p, \mathbf{R}) \times \mathbf{GL}(q, \mathbf{R})$$

may be canonically identified with the Lie subgroup of $\mathbf{GL}(p + q, \mathbf{R})$ consisting of matrices of the form $\begin{pmatrix} S & 0 \\ 0 & T \end{pmatrix}$, where $S \in \mathbf{GL}(p, \mathbf{R})$ and $T \in \mathbf{GL}(q, \mathbf{R})$; likewise when $\mathbf{R}$ is replaced by $\mathbf{C}$ or $\mathbf{H}$.

Remark

(16.9.6.1) To verify that a subgroup H of a Lie group G is a Lie subgroup of G, it is enough to verify that at *one* point $x_0 \in \mathrm{H}$ there exists a chart (V, ψ, m) on G such that $x_0 \in \mathrm{H}$, $\psi(x_0) = 0$ and $\psi(\mathrm{V} \cap \mathrm{H})$ is the intersection of the open set $\psi(\mathrm{V})$ of $\mathbf{R}^m$ with a vector subspace of $\mathbf{R}^m$. For if x is any other point of H, we shall obtain a chart having analogous properties by taking $((xx_0^{-1})\mathrm{V}, \psi \circ \gamma(x_0 x^{-1}), m)$, because $(x_0 x^{-1}) \cdot ((xx_0^{-1})\mathrm{V} \cap \mathrm{H}) = \mathrm{V} \cap \mathrm{H}$.

(16.9.7) Let G, G′ be two Lie groups. A mapping $u : \mathrm{G} \to \mathrm{G}'$ is said to be a *Lie group homomorphism* (or simply a *homomorphism*) if u is a homomorphism of groups and a morphism of differential manifolds. In order that a group homomorphism $u : \mathrm{G} \to \mathrm{G}'$ should be a Lie group homomorphism, it is necessary and sufficient that there should exist an open neighborhood U of e in G such that $u | \mathrm{U}$ is of class C^∞; for then u is of class C^∞ on $a\mathrm{U}$ for all $a \in \mathrm{G}$, because $u(x) = u(a)u(a^{-1}x)$ for all $x \in a\mathrm{U}$.

A Lie group homomorphism of a Lie group G into a linear group $\mathbf{GL}(E)$ (where E is a finite-dimensional real vector space) is called a *linear representation of* G *on* E.

(16.9.8) Let G be a Lie group and s an element of G. We have already seen that the translations

$$\gamma(s) : x \mapsto sx \qquad \delta(s) = x \mapsto xs^{-1}$$

are diffeomorphisms of G onto itself. Hence for each $x \in G$ we have tangent linear mappings

$$T_x(\gamma(s)) : T_x(G) \to T_{sx}(G), \qquad T_x(\delta(s)) : T_x(G) \to T_{xs^{-1}}(G)$$

which are *bijections* (16.5.6). The image of a vector $\mathbf{h}_x \in T_x(G)$ under $T_x(\gamma(s))$ (resp. under $T_x(\delta(s))$) is denoted by $s \cdot \mathbf{h}_x$ (resp. $\mathbf{h}_x \cdot s^{-1}$) when there is no risk of confusion. If $s, t \in G$ and $\mathbf{h}_x \in T_x(G)$, it is clear that $s \cdot (\mathbf{h}_x \cdot t^{-1}) = (s \cdot \mathbf{h}_x) \cdot t^{-1}$; this is an element of $T_{sxt^{-1}}(G)$, which we denote by $s \cdot \mathbf{h}_x \cdot t^{-1}$.

If $s, t \in G$ and $\mathbf{h}_x \in T_x(G)$, we have

$$(st) \cdot \mathbf{h}_x = s \cdot (t \cdot \mathbf{h}_x),$$

and in particular $s \cdot (s^{-1} \cdot \mathbf{h}_x) = e \cdot \mathbf{h}_x = \mathbf{h}_x$. This follows from (16.5.4).

(16.9.9) (i) *Let $i : G \to G$ be the mapping $x \mapsto x^{-1}$ on a Lie group* G. *Then for all $x \in G$ and $\mathbf{h}_x \in T_x(G)$ we have*

$$T_x(i) \cdot \mathbf{h}_x = -x^{-1} \cdot \mathbf{h}_x \cdot x^{-1}.$$

(ii) *The C^∞ mapping $m : (x, y) \mapsto xy$ of $G \times G$ into G is a submersion, and with the identification* (16.6.2) *we have*

$$T_{(x, y)}(m) \cdot (\mathbf{h}_x, \mathbf{h}_y) = x \cdot \mathbf{h}_y + \mathbf{h}_x \cdot y.$$

(iii) *If $u : G \to G'$ is a Lie group homomorphism, then u is a subimmersion of constant rank*; Ker(u) *is a normal Lie subgroup of* G, *and for all $x \in G$, we have*

$$T_x(u) \cdot \mathbf{h}_x = u(x) \cdot (T_e(u) \cdot (x^{-1} \cdot \mathbf{h}_x)),$$

where $\mathbf{h}_x \in T_x(G)$ and e is the identity element of G.

(iv) *If $u : G \to G'$ is a surjective Lie group homomorphism, then u is a submersion and we have*

(16.9.9.1) $$\dim(G) - \dim(G') = \dim(\mathrm{Ker}(u)).$$

In particular, a bijective homomorphism is an isomorphism.

At a point $(x_0, y_0) \in G \times G$, the partial mappings $m(\cdot, y_0)$ and $m(x_0, \cdot)$ are just the translations $\delta(y_0^{-1})$ and $\gamma(x_0)$, so that (ii) follows from (16.6.6). Next, we have $m(x, i(x)) = e$ for all $x \in G$, so that from (ii) and (16.5.4) we deduce that

$$\mathbf{h}_x \cdot x^{-1} + x \cdot (T_x(i) \cdot \mathbf{h}_x) = 0$$

for all $\mathbf{h}_x \in T_x(G)$; this establishes (i). Assertion (iii) follows from the relation $u \circ \gamma(x^{-1}) = \gamma(u(x^{-1})) \circ u$, which by (16.5.4) leads to

$$T_e(u) \cdot (x^{-1} \cdot \mathbf{h}_x) = u(x)^{-1} \cdot (T_x(u) \cdot \mathbf{h}_x)$$

for all $\mathbf{h}_x \in T_x(G)$. It follows that $T_x(u)$ and $T_e(u)$ have the same rank, hence u is a subimmersion. The assertion about $\operatorname{Ker}(u)$ then follows from (16.8.8).

To prove (iv) we argue by contradiction. If u is not a submersion, then there exists a point $x_0 \in G$ and a compact neighborhood $V(x_0)$ of x_0 in G such that $u(V(x_0))$ is nowhere dense in G′ (16.8.8). If follows that for all $x \in G$ the set $u((xx_0^{-1})V(x_0)) = u(xx_0^{-1})u(V(x_0))$ is nowhere dense in G′. Since there exists a denumerable open covering (A_n) of G which is finer than the covering formed by the sets $(xx_0^{-1})V(x_0)$ (12.6.1), we conclude that G′ is a denumerable union of nowhere dense subsets, which is absurd (12.16.1).

Examples

(16.9.9.2) The mapping $X \mapsto \det(X)$ is a homomorphism of the Lie group $\mathbf{GL}(n, \mathbf{R})$ (resp. $\mathbf{GL}(n, \mathbf{C})$) onto the Lie group $\mathbf{R}^*$ (resp. $\mathbf{C}^*$). The kernel, which is denoted by $\mathbf{SL}(n, \mathbf{R})$ (resp. $\mathbf{SL}(n, \mathbf{C})$), is a real Lie group of dimension $n^2 - 1$ (resp. $2(n^2 - 1)$), called the *unimodular group* (or *special linear group*) in n variables.

Remarks

(16.9.9.3) The image $u(G)$ is not necessarily a *Lie* subgroup of G′. This is shown by the example where $G = \mathbf{Z} \times \mathbf{Z}$ and $G' = \mathbf{R}$, the homomorphism u being defined by $u(m, n) = m + n\theta$, where θ is a fixed irrational number (12.8.2.1).

(16.9.9.4) Let G, G′ be two topological groups. A *local homomorphism from* G *to* G′ is by definition a continuous mapping h of an open neighborhood U of the identity element e of G with values in G′, such that $h(xy) = h(x)h(y)$ whenever x, y and xy all lie in U. Since there exists a symmetric open neighborhood V of e such that $V^2 \subset U$ (12.8.3), these conditions are satisfied whenever $x \in V$ and $y \in V$. A *local isomorphism* from G to G′ is defined to be a local homomorphism h which is a *homeomorphism* of U onto a neighborhood U′ of the identity element e' of G′. If we put $V' = h(V)$, then V′ is an open neighborhood of e' in G′. Putting $x' = h(x)$ and $y' = h(y)$ with $x, y \in V$, we have $x'y' = h(x)h(y) = h(xy)$, so that $x'y' \in U'$ and

$$h^{-1}(x')h^{-1}(y') = h^{-1}(x'y').$$

This shows that $h^{-1} \mid V'$ is a local isomorphism from G′ to G. Two topological groups G, G′ are said to be *locally isomorphic* if there exists a local isomorphism from G to G′. This relation is an equivalence relation; for we have just shown that it is symmetric, and it is clearly reflexive and transitive. For example, if H is a *discrete* normal subgroup of G, then G and G/H are locally isomorphic (12.11.2(iii)).

If G and G′ are Lie groups and h is a local homomorphism of class C^∞ from G to G′, the argument of (16.9.9(iii)) shows that h is a subimmersion at all points of V, and the same calculation gives its tangent linear mapping $T_x(h)$ at all $x \in V$.

(16.9.10) *Let* G, G′ *be two Lie groups,* $u : G \to G'$ *a homomorphism of (abstract) groups. In order that* u *should be a Lie group homomorphism, it is necessary and sufficient that the graph* Γ_u *should be a Lie subgroup of* $G \times G'$.

The necessity of the condition follows from (16.8.13). Conversely, if the condition is satisfied, let h be the restriction of the projection pr_1 to the submanifold Γ_u. Then clearly h is a bijective Lie group homomorphism, hence an isomorphism (16.9.9(iv)).

Remarks

(16.9.11) Everything in this section remains valid, *mutatis mutandis*, when we replace differential manifolds by real- or complex-analytic manifolds (16.1.4) and C^∞-mappings by analytic mappings (16.3.5). This leads us to the notions of a *real-* (resp. *complex-*) *analytic group*. Such a group has an underlying structure of (real) Lie group (resp. real-analytic group). We shall prove later (in Chapter XXI) that every *real* Lie group can be endowed with a structure of a real-analytic group, such that the given Lie group structure is the underlying structure. On the other hand, a real Lie group cannot necessarily be endowed with a structure of a *complex*-analytic group for which the given structure is the underlying one, even when the group is of even dimension.

(16.9.12) We shall prove in Chapter XIX (19.10.1) that every *closed* subgroup of a (real) Lie group is a Lie subgroup. The example of **C** and its subgroup **R** shows that the corresponding statement for complex-analytic groups is false.

PROBLEMS

1. Let X, Y, Z be three differential manifolds and $f: X \to Y$, $g: Y \to Z$ two mappings of class C^k ($k \geq 1$); let x be a point of X, and put $y = f(x)$. Then the jet $J_x^k(g \circ f)$ (Section 16.5, Problem 9) depends only on $J_x^k(f)$ and $J_y^k(g)$; it is called the *composition* of these two jets and is written $J_y^k(g) \circ J_x^k(f)$. A jet u of order k from X to Y is said to be *invertible* if there exists a jet v of order k of Y into X such that $v \circ u$ and $u \circ v$ are defined and equal to the jets of the identity mappings 1_X and 1_Y, respectively. A jet is invertible if and only if it is the jet of a local diffeomorphism of X into Y.

 The set $G^k(n) \subset L_{n,n}^k$ of invertible jets from $\mathbf{R}^n$ to $\mathbf{R}^n$ with source and target at the origin is a group with respect to composition of jets, and is an open subset of the vector space $L_{n,n}^k$ endowed with its canonical topology. Show that the structure of analytic manifold induced on $G^k(n)$ by that of $L_{n,n}^k$ is compatible with the group structure, and that $G^k(n)$ acts analytically on the left on $L_{n,m}^k$, and that $G^k(m)$ acts analytically on the right on $L_{n,m}^k$.

 If $r \leq s$, the jet $J_x^r(f)$ of a C^s-mapping $f: X \to Y$ depends only on the jet $J_x^s(f)$ of order s, so that we have canonical surjections $J_x^s(X, Y)_y \to J_x^r(X, Y)_y$ and $P_x^s(X) \to P_x^r(X)$, the latter being an **R**-algebra homomorphism. In particular, we have a surjective mapping $L_{n,m}^s \to L_{n,m}^r$, which is linear and whose kernel is the set of jets of order s from $\mathbf{R}^n$ to $\mathbf{R}^m$ with source and target at the respective origins of these spaces and which have contact of order $\geq r$ with the zero mapping. In particular, by restricting the canonical mapping $L_{n,n}^s \to L_{n,n}^r$ to $G^s(n)$, we obtain a surjective Lie group homomorphism $G^s(n) \to G^r(n)$. If $s = r + 1$ and $r \geq 1$, show that the kernel of this homomorphism is an additive group $\mathbf{R}^N$, and calculate N; show also that the group $G^1(n)$ is isomorphic to $\mathbf{GL}(n, \mathbf{R})$.

2. Let G be a Lie group, e its identity element, and let A be a commutative Lie group, written additively. Suppose that there exists a C^∞-mapping $B: G \times G \to A$ satisfying the relations

$$B(x, e) = B(e, x) = 0 \qquad \text{for all} \quad x \in G;$$

$$B(x, y) + B(xy, z) = B(x, yz) + B(y, z) \qquad \text{for all} \quad x, y, z \in G.$$

Show that on the product manifold $G \times A$ the law of composition

$$(x, u)(y, v) = (xy, u + v + B(x, y))$$

defines a Lie group structure. The identity element is $(e, 0)$, and the set $\{(e, u) : u \in A\}$ is a Lie subgroup isomorphic to A. The center of the group is $N \times A$, where N is the set of all elements n belonging to the center of G such that $B(n, x) = B(x, n)$ for all $x \in G$. The quotient group $(G \times A)/A$ is isomorphic to G.

 When G and A are real vector spaces (G now being written in additive notation) we may take B to be a bilinear function on $G \times G$ with values in A. This generalizes to the situation where G is an arbitrary Lie group and B is a C^∞-mapping such that the mappings $x \mapsto B(x, y)$ and $x \mapsto B(y, x)$ are homomorphisms of G into A for each $y \in G$.

3. (a) Let M be a differential manifold, e a point of M, U an open neighborhood of e, and $m: U \times U \to M$ a mapping of class C^∞, satisfying the following conditions:

(1) $m(e, x) = m(x, e) = x$ for all $x \in \mathrm{U}$; (2) there exists an open neighborhood V of e contained in U such that $m(\mathrm{V} \times \mathrm{V}) \subset \mathrm{U}$ and such that $m(m(x, y), z) = m(x, m(y, z))$ for all x, y, z in V. Show that there exists an open neighborhood W of e contained in V and a diffeomorphism $\theta : \mathrm{W} \to \mathrm{W}$ such that $\theta(e) = e$, $\theta(\theta(x)) = x$ and

$$m(x, \theta(x)) = m(\theta(x), x) = e$$

for all $x \in \mathrm{W}$. (Apply the implicit function theorem (16.6.8) to the equations $m(x, y) = e$ and $m(y, x) = e$.)

(b) Let M be a differential manifold and $m : \mathrm{M} \times \mathrm{M} \to \mathrm{M}$ a law of composition of class C^∞ on M, which is associative and admits an identity element. Let G be the set of elements of M which are invertible with respect to m. Show that G is open in M and is a Lie group for the structure of differential manifold and law of composition induced from M. (Remark that if $s \in \mathrm{G}$, then $x \mapsto m(s, x)$ is a diffeomorphism of M onto M, and use (a) above.)

10. ORBIT SPACES AND HOMOGENEOUS SPACES

(16.10.1) Let G be a Lie group, X a differential manifold. We say that G *acts differentiably* on the left on X if we are given a left action $(s, x) \mapsto s \cdot x$ of G on X (12.10) which is a C^∞*-mapping* of $\mathrm{G} \times \mathrm{X}$ into X. Usually we shall omit the word "differentiably" when there is no risk of ambiguity. Similarly we define a (differentiable) right action of G on X.

For example, if $\rho : \mathrm{G} \to \mathbf{GL}(\mathrm{E})$ is a *linear representation* of G on a finite-dimensional real vector space (16.9.7), then G acts differentiably on E by $(s, \mathbf{x}) \mapsto \rho(s) \cdot \mathbf{x}$. Conversely, if G acts differentiably on E in such a way that, for each $s \in \mathrm{G}$, the mapping $\rho(s) : \mathbf{x} \mapsto s \cdot \mathbf{x}$ is *linear*, then if $(\mathbf{a}_j)_{1 \leq j \leq m}$ is a basis of E, the mapping $s \mapsto \rho(s) \cdot \mathbf{a}_j$ is a C^∞-mapping of G into E, and hence the entries in the matrix of $\rho(s)$ relative to the basis $(\mathbf{a}_j)$ are C^∞-functions on G. This shows that $s \mapsto \rho(s)$ is a Lie group homomorphism of G into $\mathbf{GL}(\mathrm{E})$.

Suppose that G acts on the left on X, and put $m(s, x) = s \cdot x$. We denote the tangent linear mappings

$$\mathrm{T}_x(m(s, \cdot)) : \mathrm{T}_x(\mathrm{X}) \to \mathrm{T}_{s \cdot x}(\mathrm{X}) \qquad \text{and} \qquad \mathrm{T}_s(m(\cdot, x)) : \mathrm{T}_s(\mathrm{G}) \to \mathrm{T}_{s \cdot x}(\mathrm{X}),$$

respectively, by

$$\mathbf{k}_x \mapsto s \cdot \mathbf{k}_x, \qquad \mathbf{h}_s \mapsto \mathbf{h}_s \cdot x.$$

Since $m(s, \cdot) : x \mapsto s \cdot x$ is a *diffeomorphism* of X onto itself for each $s \in \mathrm{G}$ (12.10.2) it follows that $\mathbf{k}_x \mapsto s \cdot \mathbf{k}_x$ is a linear bijection, and by (16.5.4) we have the formulas

(16.10.1.1) $\quad (st) \cdot \mathbf{k}_x = s \cdot (t \cdot \mathbf{k}_x), \qquad t \cdot (\mathbf{h}_s \cdot x) = (t \cdot \mathbf{h}_s) \cdot x,$

$$\mathbf{h}_s \cdot (t \cdot x) = (\mathbf{h}_s \cdot t) \cdot x$$

for all s, t in G and x and X: the notation $t \cdot \mathbf{h}_s$ and $\mathbf{h}_s \cdot t$ was introduced in (16.9.8). Further, it follows from (16.6.6) that the tangent linear mapping to m is given by

(16.10.1.2) $$T_{(s,x)}(m) \cdot (\mathbf{h}_s, \mathbf{k}_x) = s \cdot \mathbf{k}_x + \mathbf{h}_s \cdot x$$

which proves that m is a *submersion.*

(16.10.2) *For each $x \in X$, the mapping $s \mapsto s \cdot x$ of G into X is a subimmersion of constant rank. The stabilizer S_x of x is a Lie subgroup of* G.

The first assertion follows from the facts that $u : \mathbf{h}_s \mapsto (ts^{-1}) \cdot \mathbf{h}_s$ is a bijection of $T_s(G)$ onto $T_t(G)$ (16.9.8), $v : \mathbf{k}_{s \cdot x} \mapsto (ts^{-1}) \cdot \mathbf{k}_{s \cdot x}$ a bijection of $T_{s \cdot x}(X)$ onto $T_{t \cdot x}(X)$, and the diagram

$$\begin{array}{ccc} T_s(G) & \xrightarrow{u} & T_t(G) \\ {\scriptstyle f}\downarrow & & \downarrow{\scriptstyle g} \\ T_{s \cdot x}(X) & \xrightarrow[v]{} & T_{t \cdot x}(X) \end{array}$$

(in which $f : \mathbf{h}_s \mapsto \mathbf{h}_s \cdot x$ and $g : \mathbf{h}_t \mapsto \mathbf{h}_t \cdot x$ are the tangent linear mappings defined above) is commutative. The second assertion follows from the first and from (16.8.8).

(16.10.3) *In order that there should exist on the orbit space* X/G (12.10) *a structure of differential manifold for which the underlying topological space is the topological space* X/G *and for which the canonical mapping $\pi : X \to X/G$ is a submersion, it is necessary and sufficient that the set* R *of pairs (x, y) belonging to the same orbit* (*that is to say, the graph* (1.3) *of the equivalence relation* "there exists $s \in G$ such that $y = s \cdot x$") *should be a closed submanifold of the product manifold* X × X. *The structure of differential manifold on* X/G *satisfying these requirements is then unique.*

If π is a submersion, then so also is $\pi \times \pi : X \times X \to (X/G) \times (X/G)$, and we have $R = (\pi \times \pi)^{-1}(\Delta)$, where Δ is the diagonal of $(X/G) \times (X/G)$, which is a closed submanifold of this product (16.8.13). Hence R is a closed submanifold of X × X by (16.8.12).

Conversely, suppose that R is a closed submanifold of X × X. Then (12.10.7) the space X/G is Hausdorff. We shall prove:

(16.10.3.1) *For each $x \in X$, there exists a chart* (U, φ, n) *on* X *at the point* x, *such that $\varphi(x) = 0$ and $\varphi(U) = V \times W$, where* V, W *are open subsets of* $\mathbf{R}^m$, $\mathbf{R}^{n-m}$, *respectively, and such that the relation $(z, z') \in R \cap (U \times U)$ is equivalent to* $\mathrm{pr}_2(\varphi(z)) = \mathrm{pr}_2(\varphi(z'))$.

This result is a consequence of the following:

(16.10.3.2) *For each* $x \in X$, *there exists an open neighborhood* U *of* x *in* X, *a submanifold* S *of* U *containing* x *and a submersion* $s : U \to S$ *such that for each* $z \in U$, $s(z)$ *is the only point of* S *for which* $(z, s(z)) \in R$.

Let us assume **(16.10.3.2)** for the moment. Replacing U by a smaller open neighborhood of x if necessary, we can assume that there exists a chart (U, φ, n) on X at x for which the submersion s is of the form indicated in **(16.7.4)**, and it is then clear that this chart (again restricted, if necessary, to the inverse image of a product of open sets in $\mathbf{R}^m$ and $\mathbf{R}^{n-m}$) satisfies the conditions of **(16.10.3.1)**.

For the proof of **(16.10.3.2)**, let $n = \dim_x(X)$; observe that since the submanifold R contains the diagonal of $X \times X$, the dimension of R at the point (x, x) is of the form $m + n$ where $0 \leqq m \leqq n$, and $T_{(x,x)}(R)$ contains the diagonal Δ of $T_{(x,x)}(X \times X) = T_x(X) \times T_x(X)$. Hence $T_{(x,x)}(R)$ is the direct sum of Δ and a subspace $\{0\} \times E = T_{(x,x)}(R) \cap (\{0\} \times T_x(X))$, of dimension m. By **(16.8.3)** there exists an open neighborhood U_0 of x in X and a submersion $f : U_0 \times U_0 \to \mathbf{R}^{n-m}$, such that $R \cap (U_0 \times U_0)$ is the set of pairs $(z, z') \in U_0 \times U_0$ satisfying $f(z, z') = 0$. Furthermore (**(16.8.7)** and **(16.8.8)**), the intersection $R \cap (\{x\} \times U_0)$ is a submanifold of $\{x\} \times X$, of dimension m at the point (x, x), and consisting of the points $(x, z) \in \{x\} \times U_0$ such that $f(x, z) = 0$; the tangent space to this submanifold at the point (x, x) is $\{0\} \times E$, the mapping $z \mapsto f(x, z)$ of U_0 into $\mathbf{R}^{n-m}$ being a *submersion* at the point x, because Δ is contained in the kernel of $T_{(x,x)}(f)$. Observe that the mapping $z \mapsto f(x, z)$ is also a *submersion* of U_0 into $\mathbf{R}^{n-m}$ at the point x; for since the relation R is symmetric, the kernel of $T_{(x,x)}(f)$, which is the tangent space to R at (x, x), is invariant under the mapping $(\mathbf{u}, \mathbf{v}) \mapsto (\mathbf{v}, \mathbf{u})$ of $T_x(X) \times T_x(X)$ onto itself, and hence its intersection with $T_x(X) \times \{0\}$ is of dimension m, which proves the assertion. Replacing U_0 if necessary by a smaller neighborhood of x, we can assume **(16.8.8)** that there exists a *submersion* $g : U_0 \to \mathbf{R}^m$ such that, if N is the submanifold of U_0 given by the equation $g(z) = 0$, then $x \in N$ and the tangent space at (x, x) to $\{x\} \times N$ is a supplement $\{0\} \times F$ of $\{0\} \times E$ in $\{0\} \times T_x(X)$. This being so, consider the mapping $u : (z, z') \mapsto (f(z, z'), g(z'))$ of $U_0 \times U_0$ into $\mathbf{R}^n = \mathbf{R}^{n-m} \times \mathbf{R}^m$. The choices of f and g show that the partial mapping $z' \mapsto (f(x, z'), g(z'))$ of U_0 into $\mathbf{R}^n$ has a *bijective* tangent linear mapping at the point x. For if a vector $\mathbf{h} \in \{0\} \times T_x(X)$ runs through $\{0\} \times E$, then its image under $T_x(f(x, .))$ is zero and its image under $T_x(g)$ runs through $\mathbf{R}^m$; and if, on the other hand, $\mathbf{h}$ runs through $\{0\} \times F$, its image under $T_x(g)$ is zero and its image under $T_x(f(x, \cdot))$ runs through $\mathbf{R}^{n-m}$. By the implicit function theorem **(16.6.8)** there exist two open neighborhoods U_1, U_2 of x in X and a C^∞-mapping

$v : U_1 \to U_2$ such that for each $z \in U_1$, the *only* solution $z' \in U_2$ of the system of equations

$$f(z, z') = 0, \qquad g(z') = 0$$

is $z' = v(z)$. Moreover, since $z \mapsto f(z, x)$ is a submersion of U_0 into $\mathbf{R}^{n-m}$ at the point x, it follows that $T_x(f(\cdot, x))$ is of rank $n - m$, hence (16.6.8.1) $T_x(v)$ is of rank $n - m$, or in other words v is a submersion, at the point x, of U_1 into the submanifold N of U_0. Replacing U_0 by a smaller neighborhood we can therefore assume that v is a submersion of U_1 into N, so that

$$v(U_1) \subset U_2 \cap N$$

is open in N. If $S = v(U_1) \cap U_1$ and $U = v^{-1}(S) \cap U_1$, then U, S and $s = v \mid U$ satisfy the conditions of (16.10.3.2).

Now that (16.10.3.1) is established, we shall return to the proof of the sufficiency of the condition in (16.10.3). For each $x \in X$, let (U_x, φ_x, n_x) be a chart of X at x satisfying the condition of (16.10.3.1), where V, W, m are replaced by V_x, W_x, m_x. Observe first that if K_x is the image under φ_x^{-1} of a compact neighborhood of 0 in $\{0\} \times W_x$, then $\pi(K_x)$ is a neighborhood of $\pi(x)$ in X/G and the restriction of π to K_x is injective. Hence (12.10.9) X/G is metrizable, locally compact, and separable. Moreover, the open sets in X/G contained in $\pi(U_x)$ are the sets of the form $\pi(\varphi_x^{-1}(V_x \times T))$, where T runs through the open sets in W_x (12.10.5); in other words, there is a homeomorphism $\omega_x : \pi(U_x) \to W_x$ such that $\omega_x^{-1}(w) = \pi(\varphi_x^{-1}(0, w))$. We have to show next that the charts $(\pi(U_x), \omega_x, n_x - m_x)$ are *mutually compatible*. So let (U_x, φ_x, n_x) and $(U_{x'}, \varphi_{x'}, n_{x'})$ be two charts of X of the family considered above; put

$$S = \pi^{-1}(\pi(U_x) \cap \pi(U_{x'})) \cap U_x, \qquad S' = \pi^{-1}(\pi(U_x) \cap \pi(U_{x'})) \cap U_{x'}$$

which are open sets, and put

$$Q = \varphi_x(S) \subset V_x \times W_x, \qquad Q' = \varphi_{x'}(S') \subset V_{x'} \times W_{x'}.$$

The projections

$$P = \mathrm{pr}_2(Q) \subset W_x, \qquad P' = \mathrm{pr}_2(Q') \subset W_{x'}$$

are then open sets such that $Q = V_x \times P$ and $Q' = V_{x'} \times P'$. Since $\pi(S) = \pi(S') = \pi(V_x) \cap \pi(V_{x'})$ by definition, for each $p \in P$, there exists a unique point $p' = f(p) \in P'$ such that

$$\pi(\varphi_x^{-1}(V_x \times \{p\})) = \pi(\varphi_{x'}^{-1}(V_{x'} \times \{p'\})),$$

and what has to be shown is that the bijection $f : P \to P'$ so defined is *of class* C^∞ in a neighborhood of each point $p \in P$. Now, let $q \in Q$ (resp. $q' \in Q'$)

be such that $\mathrm{pr}_2(q) = p$ (resp. $\mathrm{pr}_2(q') = p'$); if $z = \varphi_x^{-1}(q)$ and $z' = \varphi_{x'}^{-1}(q')$, there exists an element $s \in G$ such that $z' = s \cdot z$. Consider the diffeomorphism $g : u \mapsto s \cdot u$ of X onto itself; since it maps z to z', there exists an open neighborhood $T \subset U_x$ of z such that $g(T) \subset U_{x'}$, and the composite mapping $u \mapsto \varphi_{x'}(g(u))$ defined on T furnishes a *chart* on X at the point z, with domain T, which is therefore compatible with that defined by φ_x. The mapping $r \mapsto \varphi_{x'}(g(\varphi_x^{-1}(r)))$, defined on the neighborhood $\varphi_x(T)$ of q, is therefore of class C^∞. However, by its definition it is of the form $r \mapsto (\theta(r), f(\mathrm{pr}_2(r)))$, which shows that f is of class C^∞ in a neighborhood of $\mathrm{pr}_2(q) = p$.

Finally, the uniqueness of the structure of differential manifold on X/G in (16.10.3) is a consequence of (16.7.7(ii)) (take $f = \pi$ and $g = 1_{X/G}$ there).

Q.E.D.

When the condition of (16.10.3) is satisfied, the space X/G endowed with the structure of differential manifold defined in (16.10.3) is called the *orbit manifold* of the action of G on X.

If $\pi : X \to X/G$ is the canonical submersion, then we have $\pi(s \cdot x) = \pi(x)$ for all $s \in G$ and all $x \in X$; taking tangent mappings and using the notation introduced earlier, we obtain

(16.10.3.3) $$T_{s \cdot x}(\pi) \cdot (s \cdot \mathbf{h}_x) = T_x(\pi) \cdot \mathbf{h}_x$$

for all $\mathbf{h}_x \in T_x(X)$.

(16.10.3.4) It should be noted that the condition in (16.10.3) is not always satisfied, even when G is a *finite* group. Consider for example the case where $X = \mathbf{R}$ and G is the multiplicative subgroup $\{1, -1\}$ of $\mathbf{R}^*$, the action of G on X being multiplication (cf. Problem 1).

(16.10.4) *Suppose that the orbit manifold* X/G *exists. Then a mapping* $\Phi : X/G \to Y$, *where* Y *is a differential manifold, is of class* C^r (resp. *a subimmersion*, resp. *a submersion*) *if and only if the composite mapping* $\Phi \circ \pi : X \to Y$ *has the same property.*

The assertion relative to class C^r is a particular case of (16.7.7(ii)). The other assertions follow from the relation $\mathrm{rk}_x(\Phi \circ \pi) = \mathrm{rk}_{\pi(x)}(\Phi)$, which comes from the fact that π is a submersion.

(16.10.5) *Let* G (resp. G′) *be a Lie group acting differentiably on a differential manifold* X (resp. X′). *Then* $G \times G'$ *acts differentiably on* $X \times X'$.

If the orbit manifolds X/G *and* X'/G' *exist, then so does the orbit manifold* $(X \times X')/(G \times G')$, *and the canonical mapping*

$$(X \times X')/(G \times G') \to (X/G) \times (X'/G')$$

is a diffeomorphism.

The first two assertions follow from (16.6.5), (16.10.3), and (16.8.7(ii)). The third is a consequence of (16.10.4) and (16.5.6).

(16.10.6) *Let* H *be a Lie subgroup of a Lie group* G, *and consider* H *as acting on* G *on the right by translation. Then the orbit manifold* G/H *exists,* G *acts differentiably on the left on* G/H, *and we have*

$$\dim(G/H) = \dim(G) - \dim(H).$$

If H *is normal in* G, *the manifold structure of* G/H *is compatible with its group structure.*

To verify that the condition of (16.10.3) is satisfied in the present situation, we observe that the set $R \subset G \times G$ is here the set of pairs (x, y) such that $x^{-1}y \in H$. Now the mapping $(x, y) \mapsto x^{-1}y$ of $G \times G$ into G is a submersion (16.9.9), and H is a submanifold of G, hence (16.8.12) R is a submanifold of $G \times G$. To show that G acts differentiably on G/H, let p denote the mapping $(x, y) \mapsto xy$ of $G \times G$ into G, $\dot{p}$ the mapping $(x, \dot{y}) \mapsto x \cdot \dot{y}$ of $G \times (G/H)$ into G/H, and $\pi : G \to G/H$ the canonical mapping. Then we have a commutative diagram

$$\begin{array}{ccc} G \times G & \xrightarrow{\ p\ } & G \\ {\scriptstyle 1_G \times \pi}\downarrow & & \downarrow{\scriptstyle \pi} \\ G \times (G/H) & \xrightarrow[\ \dot{p}\ ]{} & G/H \end{array}$$

and $G \times (G/H)$ may be identified with the orbit manifold $(G \times G)/(\{e\} \times H)$ (16.10.5). The fact that $\dot{p}$ is of class C^{∞} now follows from (16.10.4), since p and π are C^{∞}-mappings. When H is normal in G, let m denote the mapping $(x, y) \mapsto xy^{-1}$ of $G \times G$ into G, and $\dot{m}$ the mapping $(\dot{x}, \dot{y}) \mapsto \dot{x}\dot{y}^{-1}$ of $(G/H) \times (G/H)$ into G/H. Then we have a commutative diagram

$$\begin{array}{ccc} G \times G & \xrightarrow{\ m\ } & G \\ {\scriptstyle \pi \times \pi}\downarrow & & \downarrow{\scriptstyle \pi} \\ (G/H) \times (G/H) & \xrightarrow[\ \dot{m}\ ]{} & G/H \end{array}$$

Identifying $(G/H) \times (G/H)$ with $(G \times G)/(H \times H)$, it follows as above that $\dot{m}$ is of class C^∞. Finally, the dimension formula follows immediately from (16.10.3.1).

Examples

(16.10.6.1) If H is a *discrete* normal subgroup of G, the canonical mapping $\pi : G \to G/H$ is a *local diffeomorphism* (16.5.6). As an example, consider the *n-dimensional torus* $\mathbf{T}^n = \mathbf{R}^n/\mathbf{Z}^n$, which is a compact connected commutative Lie group, being the canonical image of the cube $[0, 1]^n$ in $\mathbf{R}^n$.

(16.10.7) *Let* G *be a Lie group acting differentiably on a differential manifold* X. *If a point* $x \in X$ *is such that the orbit* $G \cdot x$ *is a locally closed subspace of* X, *then* $G \cdot x$ *is a submanifold of* X, *and the canonical mapping* $f_x : G/S_x \to G \cdot x$ (12.11.4) *is an isomorphism of differential manifolds. In particular, the above condition is satisfied for all* $x \in X$ *wherever the orbit manifold* X/G *exists.*

Every point of the subspace $G \cdot x$ of X has by hypothesis a neighborhood homeomorphic to a complete metric space (3.14.5), hence (12.16.12) f_x is a homeomorphism. Next, since the composite mapping $h_x : G \overset{\pi_x}{\to} G/S_x \overset{f_x}{\to} G \cdot x$ is a subimmersion (16.10.2), it follows from (16.10.6), (16.10.4), and (16.8.8) that f_x is an immersion; hence the first two assertions follow from (16.8.4). Finally, if the manifold X/G exists, then since $G \cdot x$ is the section R(x) of the set R defined in (16.10.3) it follows that $G \cdot x$ is closed in X (3.20.12).

The example of the group $\mathbf{Z}$ acting differentiably on $\mathbf{T}$ by the rule $n \cdot x = x + \varphi(n\theta)$, where $\varphi : \mathbf{R} \to \mathbf{T}$ is the canonical homomorphism and θ is irrational, shows that the hypothesis on $G \cdot x$ in (16.10.7) is not always satisfied.

In particular:

(16.10.8) (i) *If* G *is a Lie group which acts differentiably and transitively on a differential manifold* X, *then for each* $x \in X$ *the canonical mapping* $f_x : G/S_x \to X$ *is a diffeomorphism.*

(ii) *If* $u : G \to G'$ *is a surjective homomorphism of Lie groups with kernel* H, *then the canonical mapping* $G/H \to G'$ *is an isomorphism of Lie groups.*

The assertion (ii) follows from (i) by considering G as acting transitively on G′ by means of the mapping $(x, x') \mapsto u(x)x'$.

(16.10.9) *Let* $u : G \to G'$ *be a homomorphism of Lie groups,* H *a Lie subgroup of* G, *and* H′ *a Lie subgroup of* G′ *such that* $u(H) \subset H'$. *Then the unique mapping* $\tilde{u} : G/H \to G'/H'$ *for which the diagram*

$$\begin{array}{ccc} G & \xrightarrow{u} & G' \\ \big\downarrow{\scriptstyle \pi} & & \big\downarrow{\scriptstyle \pi'} \\ G/H & \xrightarrow[\tilde{u}]{} & G'/H' \end{array}$$

(π, π' being the canonical mappings) *is commutative, is of class* C^∞.

We have $\pi' \circ u = \tilde{u} \circ \pi$, and it is clear that $\pi' \circ u$ is of class C^∞; now apply (16.10.4) and (16.10.6).

(16.10.10) Let G (resp. G′) be a Lie group acting differentiably on a differential manifold X (resp. X′). If $\rho : G \to G'$ is a Lie group homeomorphism and $f : X \to X'$ a C^∞-mapping, then G and G′ are said to act *equivariantly* (relative to ρ and f) on X and X′ if the diagram

$$\begin{array}{ccc} G \times X & \xrightarrow{m} & X \\ {\scriptstyle \rho \times f}\big\downarrow & & \big\downarrow{\scriptstyle f} \\ G' \times X' & \xrightarrow[m']{} & X' \end{array}$$

is commutative (where m, m' define the actions of G on X and G′ on X′ respectively).

This leads, for each pair $(s, x) \in G \times X$, to a commutative diagram of tangent linear mappings (16.5.4):

$$\begin{array}{ccc} T_s(G) \times T_x(X) & \xrightarrow{T_{(s,x)}(m)} & T_x(X) \\ {\scriptstyle T_s(\rho) \times T_x(f)}\big\downarrow & & \big\downarrow{\scriptstyle T_x(f)} \\ T_{\rho(s)}(G) \times T_{f(x)}(X') & \xrightarrow[T_{(\rho(s), f(x))}(m')]{} & T_{f(x)}(X') \end{array}$$

Remarks

(16.10.11) We leave it to the reader to transpose the results of this section to the context of real- (resp. complex-) analytic groups acting analytically on real- (resp. complex-) manifolds.

(16.10.12) Let E be a set (not *a priori* equipped with a topology), G a Lie group acting transitively on E, and suppose that the stabilizer of each element of E is a Lie subgroup of G (in fact it is sufficient that this should be the case for *one* point of E). Then there exists on E a unique structure of differential manifold such that G acts differentiably on E: this follows from (16.10.6) and (16.10.8).

PROBLEMS

1. If a Lie group G acts differentiably and properly (Section 12.10, Problem 1) on a differential manifold X, the orbit-manifold X/G does not necessarily exist (cf. (16.10.3.4)). Show however that the orbit-manifold X/G does exist if, in addition to the hypotheses above, G acts *freely* on X.

2. Let G be a Lie group, H a Lie subgroup of G, and X a differential manifold on which H acts differentiably (on the left).

   (a) The Lie group H acts differentiably on the right on $G \times X$ by the rule

$$(s, x) \cdot t = (st, t^{-1} \cdot x).$$

   Show that with respect to this action the orbit manifold $Y = H\backslash(G \times X)$ always exists. If $H = G$, the orbit manifold is diffeomorphic to X. If $\pi : G \times X \to Y$ is the canonical mapping, show that G acts differentiably on the left on Y by the rule $s' \cdot \pi(s, x) = \pi(s's, x)$.
   (b) Show that the mapping $h : x \mapsto \pi(e, x)$ is a diffeomorphism of X onto a submanifold X′ of Y, that X′ is stable under the action of $H \subset G$, and that H acts equivariantly (relative to h) on X and X′. If $s \in G$ is such that $s \cdot X' \cap X' \neq \varnothing$, then $s \in H$. The stabilizer of $\pi(e, x)$ under the action of G is equal to the stabilizer of x under the action of H.
   (c) Show that the mapping $(s, y) \mapsto s \cdot y$ is a surjective submersion of $G \times X'$ onto Y (cf. 16.14.8).

3. Let G be a Lie group, H a Lie subgroup of G, and $\pi : G \to G/H$ the canonical submersion.

   (a) For each Lie subgroup G′ of G, show that $G' \cap H$ is a Lie subgroup and that $\pi | G'$ is a subimmersion of G′ into G/H. If $\pi(G')$ is closed in G/H (which will be the case if either H or G′ is compact), then $\pi(G')$ is a submanifold of G/H diffeomorphic to $G'/(G' \cap H)$. (Use (16.10.7).)
   (b) Consider the Lie group $G = \mathbf{R} \times \mathbf{T}^2$ and the Lie subgroup $H = \mathbf{R} \times \{0\}$ of G. If $\varphi : \mathbf{R} \to \mathbf{T}$ is the canonical homomorphism, let G′ be the subgroup of G which is the image of $\mathbf{R}$ under the homomorphism $x \mapsto (x, \varphi(x), \varphi(\theta x))$, where θ is a fixed irrational number. Show that G′ is a Lie subgroup of G but that $\pi(G')$ is dense and not closed in $G/H = \mathbf{T}^2$.
   (c) If $\dim(G') - \dim(G' \cap H) = \dim(G) - \dim(H)$, the restriction of π to G′ is a submersion into G/H, and hence factorizes into $G' \to G'/(G' \cap H) \xrightarrow{u} G/H$, where u is

a diffeomorphism of $G'/(G' \cap H)$ onto an open submanifold of G/H. If either G′ or H is compact and G is connected, then u is a diffeomorphism of $G'/(G' \cap H)$ onto G/H. Give an example in which the image of u is a nondense open set in G/H. (Take G = **SL**(2, **R**) and G′ to be the subgroup of upper triangular matrices $\begin{pmatrix} x & y \\ 0 & x^{-1} \end{pmatrix}$, where $x > 0$, and H the subgroup of lower unitriangular matrices $\begin{pmatrix} 1 & 0 \\ z & 1 \end{pmatrix}$.)

11. EXAMPLES: UNITARY GROUPS, STIEFEL MANIFOLDS, GRASSMANNIANS, PROJECTIVE SPACES

(16.11.1) Let E be a real vector space of dimension n, and let $\mathscr{H}(E)$ be the set of all symmetric bilinear forms on E × E, which is a real vector space of dimension $\frac{1}{2}n(n+1)$. For each pair (p, q) of integers $\geqq 0$ such that $p + q = n$, the subset $\mathscr{H}_{p,q}(E)$ of symmetric bilinear forms of signature (p, q) on E is *open* in the vector space $\mathscr{H}(E)$. To see this, let Φ_0 be a form belonging to $\mathscr{H}_{p,q}(E)$; then there exists a direct sum decomposition $E = P \oplus N$, where P and N are vector subspaces of dimensions p and q, respectively, such that $\Phi_0(\mathbf{x}, \mathbf{x}) > 0$ for $\mathbf{x} \neq 0$ in P, and $\Phi_0(\mathbf{x}, \mathbf{x}) < 0$ for $\mathbf{x} \neq 0$ in N. If $\|\mathbf{x}\|$ is a norm which defines the topology on E, then there exist two real numbers $a > 0$ and $b > 0$ such that $\Phi_0(\mathbf{x}, \mathbf{x}) \geqq a\|\mathbf{x}\|^2$ for all $\mathbf{x} \in P$ and $\Phi_0(\mathbf{x}, \mathbf{x}) \leqq -b\|\mathbf{x}\|^2$ for all $\mathbf{x} \in N$, because spheres are compact (**3.17.10**). If Ψ is a sufficiently small symmetrical bilinear form such that $|\Psi(\mathbf{x}, \mathbf{x})| \leqq \frac{1}{2}\inf(a, b)\|\mathbf{x}\|^2$ for all $\mathbf{x} \in E$, and if $\Phi = \Phi_0 + \Psi$, then we shall have $\Phi(\mathbf{x}, \mathbf{x}) \geqq \frac{1}{2}a\|\mathbf{x}\|^2$ for $\mathbf{x} \in P$ and $\Phi(\mathbf{x}, \mathbf{x}) \leqq -\frac{1}{2}b\|\mathbf{x}\|^2$ for $\mathbf{x} \in N$; this shows that Φ has signature (p, q), by virtue of the law of inertia.

(16.11.2) The group **GL**(E) acts differentiably (indeed analytically) on $\mathscr{H}(E)$ and on each of the $\mathscr{H}_{p,q}(E)$. Namely, if Φ is any symmetric bilinear form and $s \in \mathbf{GL}(E)$, then $s \cdot \Phi$ is the form

$$(\mathbf{x}, \mathbf{y}) \mapsto \Phi(s^{-1} \cdot \mathbf{x}, s^{-1} \cdot \mathbf{y}).$$

Moreover, each of the open sets $\mathscr{H}_{p,q}(E)$ is an *orbit* of this action. Hence it follows from (**16.10.2**) that, for each $\Phi \in \mathscr{H}_{p,q}(E)$, the subgroup of elements $s \in \mathbf{GL}(E)$ such that $s \cdot \Phi = \Phi$ is a *Lie subgroup* of **GL**(E) of dimension $n^2 - \frac{1}{2}n(n+1) = \frac{1}{2}n(n-1)$. This Lie group is called the *orthogonal group of the form* Φ and is denoted by $\mathbf{O}(\Phi)$. When $p = 3$ and $q = 1$ it is called the *Lorentz group*. When $p = n$ and $q = 0$, it is called simply the *orthogonal group* in n variables. All orthogonal groups $\mathbf{O}(\Phi)$ with Φ of signature $(n, 0)$ are isomorphic to the group corresponding to $E = \mathbf{R}^n$ and Φ the Euclidean scalar product, namely

$$\Phi(\mathbf{x}, \mathbf{y}) = (\mathbf{x} | \mathbf{y}) = \sum_{j=1}^{n} \xi_j \eta_j,$$

where $\mathbf{x} = (\xi_j)$, $\mathbf{y} = (\eta_j)$. This group is also denoted by $\mathbf{O}(n, \mathbf{R})$, or simply $\mathbf{O}(n)$ if there is no risk of ambiguity. It is *compact*, because the matrices $S = (\alpha_{ij})$ belonging to $\mathbf{O}(n)$ are characterized by the relation ${}^tS \cdot S = I$ and therefore in particular satisfy the relations $\sum_{j=1}^{n} \alpha_{ij}^2 = 1$ for $1 \leqq i \leqq n$; hence they form a bounded closed subset of $\mathbf{R}^{n^2}$. The kernel in $\mathbf{O}(n)$ of the homomorphism $s \mapsto \det(s)$ is a Lie subgroup of $\mathbf{O}(n)$ of index 2 (because $\det(s) = -1$ if s is a reflection in a hyperplane) called the *rotation group* or *special orthogonal group* in n variables, and denoted by $\mathbf{SO}(n, \mathbf{R})$ or $\mathbf{SO}(n)$; it is an open subgroup of $\mathbf{O}(n)$.

(16.11.3) There are analogous definitions and results when E is taken to a vector space of dimension n over the field of complex numbers $\mathbf{C}$, or a left vector space of dimension n over the division ring of quaternions $\mathbf{H}$. In either case $\mathscr{H}(\mathrm{E})$ now denotes the set of *hermitian sesquilinear forms* Φ on $\mathrm{E} \times \mathrm{E}$, that is to say forms Φ satisfying

$$\Phi(\mathbf{x} + \mathbf{x}', \mathbf{y}) = \Phi(\mathbf{x}, \mathbf{y}) + \Phi(\mathbf{x}', \mathbf{y}),$$
$$\Phi(\lambda \mathbf{x}, \mathbf{y}) = \lambda \Phi(\mathbf{x}, \mathbf{y}) \qquad (\lambda \in \mathbf{C} \quad (\text{resp. } \mathbf{H})),$$
$$\Phi(\mathbf{y}, \mathbf{x}) = \overline{\Phi(\mathbf{x}, \mathbf{y})}.$$

It follows that $\Phi(\mathbf{x}, \mathbf{x})$ is always real, and therefore $\mathscr{H}(\mathrm{E})$ is a real vector space of dimension $n + n(n-1) = n^2$ in the complex case, and of dimension $n + 2n(n-1) = 2n^2 - n$ in the quaternionic case. Just as in **(16.11.1)**, we can show that for each signature (p, q) such that $p + q = n$, the subspace $\mathscr{H}_{p,q}(\mathrm{E})$ of forms of signature (p, q) is an *open* subspace of $\mathscr{H}(\mathrm{E})$. If $\Phi \in \mathscr{H}_{p,q}(\mathrm{E})$, we see as in **(16.11.2)** that the subgroup of elements $s \in \mathbf{GL}(\mathrm{E})$ such that $s \cdot \Phi = \Phi$ is a Lie subgroup of dimension $2n^2 - n^2 = n^2$ in the complex case, and of dimension $4n^2 - (2n^2 - n) = n(2n + 1)$ in the quaternionic case. This subgroup is called the *unitary group of the form* Φ, and is denoted by $\mathbf{U}(\Phi)$. When $(p, q) = (n, 0)$, it is called simply the *unitary group* in n variables. All unitary groups $\mathbf{U}(\Phi)$ with Φ of signature $(n, 0)$ are isomorphic to the group corresponding to $\mathrm{E} = \mathbf{C}^n$ (resp. $\mathrm{E} = \mathbf{H}^n$) and

$$\Phi(\mathbf{x}, \mathbf{y}) = (\mathbf{x} \mid \mathbf{y}) = \sum_{j=1}^{n} \xi_j \bar{\eta}_j,$$

where $\mathbf{x} = (\xi_j)$, $\mathbf{y} = (\eta_j)$. This group is also denoted by $\mathbf{U}(n, \mathbf{C})$ or $\mathbf{U}(n)$ (resp. by $\mathbf{U}(n, \mathbf{H})$). It is *compact*, because the matrices $S = (\alpha_{ij})$ belonging to $\mathbf{U}(n, \mathbf{C})$ (resp. $\mathbf{U}(n, \mathbf{H})$) are characterized by the relation $S \cdot {}^t\bar{S} = I$ and therefore in particular satisfy the relations $\sum_{j=1}^{n} |\alpha_{ij}|^2 = 1$ for $1 \leqq i \leqq n$; consequently they form a bounded closed subset of $\mathbf{C}^{n^2}$ (resp. $\mathbf{H}^{n^2}$).

The homomorphism $s \mapsto \det(s)$ of $\mathbf{U}(n, \mathbf{C})$ onto $\mathbf{U}(1, \mathbf{C}) = \mathbf{U}$ (the unit circle in $\mathbf{C}$) is surjective, because if $\zeta \in \mathbf{U}$ and if $(\mathbf{e}_j)_{1 \leqq j \leqq n}$ is an orthogonal basis of $\mathbf{C}^n$, the automorphism s of $\mathbf{C}^n$ defined by $s(\mathbf{e}_1) = \zeta\mathbf{e}_1$, $s(\mathbf{e}_j) = \mathbf{e}_j$ for $2 \leqq j \leqq n$ is unitary and has determinant ζ. Hence (16.9.9) the kernel of this homomorphism is a normal Lie subgroup of $\mathbf{U}(n)$ of dimension $n^2 - 1$, called the *special unitary group* and denoted by $\mathbf{SU}(n)$.

(16.11.4) Let n, p be two integers $\geqq 1$. The space $\mathbf{R}^{np}$ of sequences $(\mathbf{x}_k)_{1 \leqq k \leqq p}$ of p vectors in $\mathbf{R}^n$ can be identified with the set of real matrices X with n rows and p columns, the kth column being the vector $\mathbf{x}_k$. The group $\mathbf{GL}(n, \mathbf{R})$ acts *differentiably* (indeed analytically) on the left on $\mathbf{R}^{np}$ as follows: the automorphism $s \in \mathbf{GL}(n, \mathbf{R})$ transforms the sequence $(\mathbf{x}_k)$ into the sequence $(s \cdot \mathbf{x}_k)$. Equivalently, if we identify s with its matrix S relative to the canonical basis of $\mathbf{R}^n$, the action of $\mathbf{GL}(n, \mathbf{R})$ on $\mathbf{R}^{np}$ is left multiplication $(S, X) \to S \cdot X$ of matrices.

Now let $p \leqq n$ and let $\mathbf{S}_{n,p}$ (or $\mathbf{S}_{n,p}(\mathbf{R})$) be the subset of $\mathbf{R}^{np}$ consisting of sequences $(\mathbf{x}_k)_{1 \leqq k \leqq p}$ which are *orthonormal* relative to the Euclidean scalar product (6.5). This set may also be described as follows: the orthogonal group $\mathbf{O}(n, \mathbf{R})$ acts differentiably on $\mathbf{R}^{np}$ by restriction of the action of $\mathbf{GL}(n, \mathbf{R})$ defined above, and $\mathbf{S}_{n,p}$ is the *orbit*, under this action of $\mathbf{O}(n, \mathbf{R})$, of the orthonormal sequence $(\mathbf{e}_k)_{1 \leqq k \leqq p}$ consisting of the first p vectors of the canonical basis $(\mathbf{e}_k)_{1 \leqq k \leqq n}$ of $\mathbf{R}^n$. Since, by virtue of (12.10.5), this orbit is *compact* and hence closed in $\mathbf{R}^{np}$, it follows from (16.10.7) that $\mathbf{S}_{n,p}$ is a *compact submanifold* of $\mathbf{R}^{np}$, called the (real) *Stiefel manifold* of orthonormal systems of p vectors (sometimes called *p-frames*) in $\mathbf{R}^n$. It is clear that the subgroup of $\mathbf{O}(n, \mathbf{R})$ which stabilizes the p-frame $(\mathbf{e}_k)_{1 \leqq k \leqq p}$ may be canonically identified with the orthogonal group $\mathbf{O}(n-p, \mathbf{R})$, by identifying $\mathbf{R}^{n-p}$ with the subspace of $\mathbf{R}^n$ spanned by $\mathbf{e}_{p+1}, \ldots, \mathbf{e}_n$ (when $n = p$, this group consists only of the identity element). Hence $\mathbf{S}_{n,p}$ is isomorphic to the homogeneous space

$$\mathbf{O}(n, \mathbf{R})/\mathbf{O}(n-p, \mathbf{R}).$$

(16.11.5) When $p = n$ the Stiefel manifold $\mathbf{S}_{n,n}(\mathbf{R})$ may be identified with $\mathbf{O}(n, \mathbf{R})$, and when $p = 1$ the manifold $\mathbf{S}_{n,1}(\mathbf{R})$ may be identified with the sphere $\mathbf{S}_{n-1}$ (16.2.3). When $1 \leqq p \leqq n-1$, $\mathbf{S}_{n,p}(\mathbf{R})$ is also the orbit of the p-frame $(\mathbf{e}_k)_{1 \leqq k \leqq p}$ under the action of the rotation group $\mathbf{SO}(n, \mathbf{R})$. Since the stabilizer of $(\mathbf{e}_k)_{1 \leqq k \leqq p}$ may be identified with $\mathbf{SO}(n-p, \mathbf{R})$, it follows that $\mathbf{S}_{n,p}(\mathbf{R})$ may be identified with the homogeneous space $\mathbf{SO}(n, \mathbf{R})/\mathbf{SO}(n-p, \mathbf{R})$ for $1 \leqq p \leqq n-1$.

(16.11.6) In the considerations of (16.11.4) the field $\mathbf{R}$ can be replaced everywhere by $\mathbf{C}$ or $\mathbf{H}$, the Euclidean scalar product being replaced by the Hermitian scalar product. In this way we define *complex* and *quaternionic Stiefel manifolds* $\mathbf{S}_{n,p}(\mathbf{C})$ and $\mathbf{S}_{n,p}(\mathbf{H})$. They are isomorphic, respectively, to the homogeneous spaces $\mathbf{U}(n, \mathbf{C})/\mathbf{U}(n-p, \mathbf{C})$ and $\mathbf{U}(n, \mathbf{H})/\mathbf{U}(n-p, \mathbf{H})$; and if

$$1 \leqq p \leqq n-1,$$

$\mathbf{S}_{n,p}(\mathbf{C})$ is also isomorphic to $\mathbf{SU}(n, \mathbf{C})/\mathbf{SU}(n-p, \mathbf{C})$. They are therefore compact manifolds. When $p = 1$, $\mathbf{S}_{n,1}(\mathbf{C})$ may be identified with the sphere $\mathbf{S}_{2n-1}$, and $\mathbf{S}_{n,1}(\mathbf{H})$ with $\mathbf{S}_{4n-1}$.

(16.11.7) *The groups* $\mathbf{SO}(n, \mathbf{R})$, $\mathbf{SU}(n, \mathbf{C})$, $\mathbf{U}(n, \mathbf{C})$, *and* $\mathbf{U}(n, \mathbf{H})$ *are connected if* $n \geqq 1$; *so also are the Stiefel manifolds* $\mathbf{S}_{n,p}(\mathbf{C})$ *and* $\mathbf{S}_{n,p}(\mathbf{H})$ *if* $1 \leqq p \leqq n-1$, *and* $\mathbf{S}_{n,p}(\mathbf{R})$ *if* $1 \leqq p \leqq n-1$ *and* $n \geqq 2$.

When $n = 1$, the groups $\mathbf{SO}(1, \mathbf{R})$ and $\mathbf{SU}(1, \mathbf{C})$ consist of the identity element alone. The group $\mathbf{U}(1, \mathbf{C})$ may be identified with the unit circle $\mathbf{U} = \mathbf{S}_1$ in $\mathbf{C}$, and $\mathbf{U}(1, \mathbf{H})$ with the multiplicative group of quaternions of norm 1, which as a topological space is the sphere $\mathbf{S}_3$; hence these two groups are connected (16.2.3). It follows also from (16.2.3) that the Stiefel manifolds $\mathbf{S}_{n,1}(\mathbf{R}) = \mathbf{S}_{n-1}$ are connected if $n \geqq 2$, and that $\mathbf{S}_{n,1}(\mathbf{C}) = \mathbf{S}_{2n-1}$ and $\mathbf{S}_{n,1}(\mathbf{H}) = \mathbf{S}_{4n-1}$ are connected if $n \geqq 1$. Consequently the homogeneous spaces

$$\mathbf{SO}(n)/\mathbf{SO}(n-1), \qquad \mathbf{SU}(n)/\mathbf{SU}(n-1),$$

$$\mathbf{U}(n, \mathbf{C})/\mathbf{U}(n-1, \mathbf{C}), \qquad \mathbf{U}(n, \mathbf{H})/\mathbf{U}(n-\mathbf{H})$$

are connected if $n \geqq 2$. The first assertion of (16.11.7) now follows by induction on n, by virtue of (12.10.12). The second is an immediate consequence, by (3.19.7).

(16.11.8) If E is a vector space over a field K, we denote by $\mathbf{G}_p(\mathrm{E})$ the set of *vector subspaces of dimension p in* E. The set $\mathbf{G}_p(\mathbf{R}^n)$ is denoted by $\mathbf{G}_{n,p}(\mathbf{R})$ or simply $\mathbf{G}_{n,p}$. It is clear that the orthogonal group $\mathbf{O}(n, \mathbf{R})$ acts transitively on $\mathbf{G}_{n,p}(\mathbf{R})$. Furthermore, if F is the subspace of $\mathbf{R}^n$ generated by the first p vectors in the canonical basis, then the stabilizer of F under this action leaves fixed (as a whole) the orthogonal supplement of F, namely the subspace spanned by the last $n - p$ vectors of the canonical basis. Hence the stabilizer of F is a Lie subgroup of $\mathbf{O}(n, \mathbf{R})$ which may be identified with the product $\mathbf{O}(p, \mathbf{R}) \times \mathbf{O}(n-p, \mathbf{R})$ ((16.9.8) and (16.8.8(i))). It follows (16.10.12) that there exists on the set $\mathbf{G}_{n,p}$ a *unique* structure of differential manifold for which $\mathbf{O}(n, \mathbf{R})$ acts differentiably on $\mathbf{G}_{n,p}$. The set $\mathbf{G}_{n,p}$ endowed with this structure is called the (*real*) *Grassmannian* with indices n, p. When $p = 1$, the Grassmannian $\mathbf{G}_{n,1}$ is also denoted by $\mathbf{P}_{n-1}(\mathbf{R})$ or simply $\mathbf{P}_{n-1}$, and is called (*real*) *projective space* of dimension $n - 1$. The differential manifold $\mathbf{G}_{n,p}(\mathbf{R})$ is diffeomorphic to the homogeneous space $\mathbf{O}(n, \mathbf{R})/(\mathbf{O}(p, \mathbf{R}) \times \mathbf{O}(n-p, \mathbf{R}))$. If $p \leqq n-1$, it is also diffeomorphic to $\mathbf{SO}(n, \mathbf{R})/\mathrm{H}_p$, where H_p is the subgroup of $\mathbf{O}(p, \mathbf{R}) \times \mathbf{O}(n-p, \mathbf{R})$ consisting of pairs (t, t') such that $\det(t) = \det(t')$.

We remark that $\mathbf{G}_{n,p}$ may also be considered as the space of spheres with center 0 and dimension $p-1$ contained in the sphere $\mathbf{S}_{n-1}$: these spheres correspond one-to-one with the vector subspaces of dimension p in $\mathbf{R}^n$.

(16.11.9) The orthogonal group $\mathbf{O}(p, \mathbf{R})$ acts differentiably on the *right* on the Stiefel manifold $\mathbf{S}_{n,p}(\mathbf{R})$ by matrix multiplication $(X, T) \mapsto X \cdot T$. For each matrix $X \in \mathbf{S}_{n,p}$ can be written as $S \cdot E$, where $S \in \mathbf{O}(n, \mathbf{R})$ and E is the matrix whose columns are the vectors $\mathbf{e}_1, \ldots, \mathbf{e}_p$; thus X consists of the first p columns of S. The columns of $E \cdot T$ are the images of $\mathbf{e}_1, \ldots, \mathbf{e}_p$ under the element of the orthogonal group $\mathbf{O}(p, \mathbf{R})$ whose matrix relative to $(\mathbf{e}_k)_{1 \leqq k \leqq p}$ is T; hence these columns form a p-frame, that is to say $E \cdot T \in \mathbf{S}_{n,p}$ and therefore also $X \cdot T = S \cdot E \cdot T \in \mathbf{S}_{n,p}$. This shows also that the set of orbits for the above action may be identified with $\mathbf{G}_{n,p}$. If we endow $\mathbf{G}_{n,p}$ with the structure of differential manifold defined in **(16.11.8)** and identify $\mathbf{S}_{n,p}$ with $\mathbf{O}(n, \mathbf{R})/\mathbf{O}(n-p, \mathbf{R})$ and $\mathbf{G}_{n,p}$ with $\mathbf{O}(n, \mathbf{R})/(\mathbf{O}(p, \mathbf{R}) \times \mathbf{O}(n-p, \mathbf{R}))$, then the canonical mapping $\pi : \mathbf{S}_{n,p} \to \mathbf{G}_{n,p}$ is a *submersion* **(16.10.4)**; consequently, for the action of $\mathbf{O}(p, \mathbf{R})$ on $\mathbf{S}_{n,p}$ defined above, the orbit-manifold exists and can be identified with the Grassmannian $\mathbf{G}_{n,p}$ **(16.10.3)**. Moreover, the orbits are each diffeomorphic to $\mathbf{O}(p, \mathbf{R})$. It follows from **(16.11.6)** and **(16.11.7)** that $\mathbf{G}_{n,p}(\mathbf{R})$ is *compact* and *connected* if $n \geqq 1$ and $1 \leqq p \leqq n$.

There are analogous definitions and results for the complex and quaternionic Grassmannians $\mathbf{G}_{n,p}(\mathbf{C})$ and $\mathbf{G}_{n,p}(\mathbf{H})$, and in particular for the complex and quaternionic projective spaces $\mathbf{P}_{n-1}(\mathbf{C})$ and $\mathbf{P}_{n-1}(\mathbf{H})$. The dimensions of the differential manifolds $\mathbf{G}_{n,p}(\mathbf{R})$, $\mathbf{G}_{n,p}(\mathbf{C})$, and $\mathbf{G}_{n,p}(\mathbf{H})$ are therefore, respectively,

$$\begin{aligned} \tfrac{1}{2}n(n-1) - \tfrac{1}{2}p(p-1) - \tfrac{1}{2}(n-p)(n-p-1) &= p(n-p), \\ n^2 - p^2 - (n-p)^2 &= 2p(n-p), \\ n(2n+1) - p(2p+1) - (n-p)(2n-2p+1) &= 4p(n-p). \end{aligned}$$

(16.11.10) There is another, equivalent, definition of the structure of differential manifold on the Grassmannian $\mathbf{G}_{n,p}(\mathbf{R})$. For the group $\mathbf{GL}(n, \mathbf{R})$ also acts transitively on $\mathbf{G}_{n,p}(\mathbf{R})$, and the stabilizer of the subspace F considered in **(16.11.8)** is the subgroup H of $\mathbf{GL}(n, \mathbf{R})$ consisting of matrices of the form $\begin{pmatrix} A & B \\ 0 & C \end{pmatrix}$, where A is a square matrix of p rows and p columns. Since H is clearly a Lie subgroup of $\mathbf{GL}(n, \mathbf{R})$ (it is diffeomorphic to the product $\mathbf{GL}(p, \mathbf{R}) \times \mathbf{GL}(n-p, \mathbf{R}) \times \mathbf{R}^{p(n-p)}$) we obtain **(16.10.12)** a structure of differential manifold on $\mathbf{G}_{n,p}$. Since the action of $\mathbf{O}(n, \mathbf{R})$ on $\mathbf{G}_{n,p}$ is obtained by restriction of the action of $\mathbf{GL}(n, \mathbf{R})$, the structure so obtained is the same as that defined in (16.11.8).

Now let $\mathrm{L}_{n,p}$ denote the subset of $\mathbf{R}^{np}$ consisting of the matrices X of rank p, that is to say matrices X whose p columns $\mathbf{x}_k$ ($1 \leqq k \leqq p$) are linearly independent. This set $\mathrm{L}_{n,p}$ is *open* in $\mathbf{R}^{np}$. More precisely, for each subset J consisting of p elements $i_1 < i_2 < \cdots < i_p$ of the set $\mathrm{I} = \{1, 2, \ldots, n\}$, let $\mathrm{T_J}$ be the set of matrices X such that the matrix X_J formed by the i_1th, ..., i_pth rows of X is invertible; then it is clear that $\mathrm{T_J}$ is open in $\mathbf{R}^{np}$ and is canonically diffeomorphic to $\mathbf{GL}(p, \mathbf{R}) \times \mathbf{R}^{p(n-p)}$, and $\mathrm{L}_{n,p}$ is the union of the $\binom{n}{p}$ sets $\mathrm{T_J}$. Note that $\mathbf{GL}(p, \mathbf{R})$ acts differentiably on the right on $\mathrm{L}_{n,p}$, by matrix multiplication $(X, T) \mapsto X \cdot T$. We assert that $\mathbf{G}_{n,p}$ can be identified with the orbit-manifold of this action. Firstly, the orbit-manifold $\mathbf{GL}(p, \mathbf{R})\backslash\mathrm{L}_{n,p}$ exists: for if R is the set of pairs (X, Y) of elements of $\mathrm{L}_{n,p}$ belonging to the same orbit, then the intersection $\mathrm{R} \cap (\mathrm{T_J} \times \mathrm{L}_{n,p})$ is the graph of the C^∞-mapping $(X, T) \mapsto (XX_\mathrm{J}^{-1}T)_{\mathrm{I-J}}$ of $\mathrm{T_J} \times \mathbf{GL}(p, \mathbf{R})$ into $\mathbf{R}^{p(n-p)}$, and the existence of the orbit-manifold now follows from (16.8.13) and (16.10.3). It is clear that there is a canonical bijection ω of $\mathbf{GL}(p, \mathbf{R})\backslash\mathrm{L}_{n,p}$ onto $\mathbf{G}_{n,p}$ such that $\omega(\pi(S \cdot X)) = S \cdot \pi'(X)$ for $S \in \mathbf{GL}(n, \mathbf{R})$ and $X \in \mathrm{L}_{n,p}$, where π and π' are the canonical mappings of $\mathrm{L}_{n,p}$ onto $\mathbf{GL}(p, \mathbf{R})\backslash\mathrm{L}_{n,p}$ and $\mathbf{G}_{n,p}$, respectively. It follows (16.10.12) that ω is a diffeomorphism, and our assertion is proved.

From these considerations we can construct a convenient *atlas* for $\mathbf{G}_{n,p}$. Let $\mathrm{V_J}$ be the subset of $\mathrm{T_J}$ consisting of the matrices $X \in \mathrm{T_J}$ such that $X_\mathrm{J} = I_p$ (the unit matrix). Then it is immediately seen that the restriction of π to $\mathrm{V_J}$ is a *bijection* of $\mathrm{V_J}$ onto $\mathrm{U_J} = \pi(\mathrm{T_J})$. If φ_J is the inverse of this bijection, we have $\varphi_\mathrm{J}(\pi(X)) = XX_\mathrm{J}^{-1}$, from which we conclude (16.10.4) that φ_J is of class C^∞; since $\pi | \mathrm{V_J}$ is also of class C^∞, φ_J is a *diffeomorphism* of $\mathrm{U_J}$ onto $\mathrm{V_J}$. As $\mathrm{V_J}$ may be identified with $\mathbf{R}^{p(n-p)}$, we have an atlas of $\mathbf{GL}(p, \mathbf{R})\backslash\mathrm{L}_{n,p}$ consisting of the charts $(\mathrm{U_J}, \varphi_\mathrm{J}, p(n-p))$.

There are of course analogous results for complex and quaternionic Grassmannians. The real and quaternionic Grassmannians are real-analytic manifolds, and the complex Grassmannians are complex-analytic manifolds.

(16.11.11) By virtue of (16.11.9), the projective space $\mathbf{P}_n(\mathbf{R})$ (resp. $\mathbf{P}_n(\mathbf{C})$, resp. $\mathbf{P}_n(\mathbf{H})$) may be identified with the orbit manifold of the group consisting of the identity and the symmetry $x \mapsto -x$ acting on $\mathbf{S}_n$ (resp. the group of rotations $x \mapsto x\zeta$ with $|\zeta| = 1$ acting on $\mathbf{S}_{2n+1} \subset \mathbf{C}^{n+1}$, resp. the group of rotations $x \mapsto xq$, where q is a quaternion of norm 1, acting on $\mathbf{S}_{4n+3} \subset \mathbf{H}^{n+1}$).

(16.11.12) *The differential manifolds* $\mathbf{P}_1(\mathbf{R})$, $\mathbf{P}_1(\mathbf{C})$, *and* $\mathbf{P}_1(\mathbf{H})$ *are diffeomorphic to* $\mathbf{S}_1$, $\mathbf{S}_2$, *and* $\mathbf{S}_4$, *respectively.*

We shall prove the assertion for $\mathbf{P}_1(\mathbf{H})$. For each pair of quaternions

$$x = x_0 + ix_1 + jx_2 + kx_3, \qquad y = y_0 + iy_1 + jy_2 + ky_3$$

such that $|x|^2 + |y|^2 = 1$ (where $|x|$ is the Euclidean norm of x in $\mathbf{R}^4$), let $z = f(x, y) = (z_0, z_1, z_2, z_3, z_4)$ be the point of $\mathbf{R}^5$ (identified with $\mathbf{H} \times \mathbf{R}$) defined by

$$2x\bar{y} = z_0 + iz_1 + jz_2 + kz_3, \qquad z_4 = |x|^2 - |y|^2.$$

Since

$$z_0^2 + z_1^2 + z_2^2 + z_3^2 = 4|x\bar{y}|^2 = 4|x|^2|y|^2,$$

it is clear that $z \in \mathbf{S}_4$. Moreover, $z_4 = 2|x|^2 - 1$ can take all values in the interval $[-1, +1]$, and for a given value of $z_4 \neq \pm 1$, the quaternion $2x\bar{y}$ can take all values on the sphere of radius $1 - |z_4|^2$. Hence f is a *surjective* C^∞-mapping of $\mathbf{S}_7$ onto $\mathbf{S}_4$. Next, the relation $f(x, y) = f(x', y')$ implies firstly that $|x'| = |x|$, so that we may write $x' = xq$, where q is a quaternion of norm 1; also it implies that $x'\bar{y}' = x\bar{y}$, which gives $y' = y\bar{q}^{-1} = yq$. Hence the mapping f factorizes as follows:

$$\mathbf{S}_7 \to \mathbf{P}_1(\mathbf{H}) \xrightarrow{g} \mathbf{S}_4,$$

where g is *bijective*. It remains to show that f is a submersion. Since the rotations $x \mapsto q'x$, $y \mapsto q''y$ (where q', q'' are quaternions of norm 1) are diffeomorphisms of $\mathbf{S}_7$ which fix z_4 and transform $x\bar{y}$ into $q'x\bar{y}\bar{q}''$, it is enough to check that f is a submersion at the points $(x, y) \in \mathbf{S}_7$ such that the quaternions x, y are *scalars* x_0, y_0. It is then immediately verified (since x_0, y_0 are not both simultaneously zero) that the Jacobian matrix of f (extended to $\mathbf{H}^2 = \mathbf{R}^8$ by the same definition) is of rank 5. This fact, together with the relation $f(tx, ty) = t^2 f(x, y)$ for all scalars t, proves that g is a diffeomorphism of $\mathbf{P}_1(\mathbf{H})$ onto $\mathbf{S}_4$ (16.8.8). The proofs for $\mathbf{P}_1(\mathbf{R})$ and $\mathbf{P}_1(\mathbf{C})$ are analogous but simpler: In the case of $\mathbf{P}_1(\mathbf{C})$, we map the point $(x, y) \in \mathbf{S}_3 \subset \mathbf{R}^4 = \mathbf{C}^2$ to the point

$$f(x, y) = (2x\bar{y}, |x|^2 - |y|^2)$$

on $\mathbf{S}_2 \subset \mathbf{C} \times \mathbf{R} = \mathbf{R}^3$, and the mapping f factorizes as

$$\mathbf{S}_3 \to \mathbf{P}_1(\mathbf{C}) \xrightarrow{g} \mathbf{S}_2,$$

where g is a diffeomorphism. We may therefore transport to $\mathbf{S}_2$ by means of g the structure of complex-analytic manifold of $\mathbf{P}_1(\mathbf{C})$. The sphere $\mathbf{S}_2$, endowed with this structure, is called the *Riemann sphere*.

PROBLEMS

1. (a) For any two points $x, y \in \mathbf{S}_{2n-1} \subset \mathbf{C}^n$, put

$$\alpha(x, y) = \text{arc}\cos(\mathscr{R}(x \mid y))$$

which is a real number between 0 and π. If s, t are any two elements of the unitary group $\mathbf{U}(n)$, put

$$d(s, t) = \sup_{x \in \mathbf{S}_{2n-1}} \alpha(s \cdot x, t \cdot x).$$

Show that d is a bi-invariant distance on $\mathbf{U}(n)$.
(b) For $s \in \mathbf{U}(n)$, let $e^{i\theta_j}$ $(1 \leq j \leq m)$ be the distinct eigenvalues of s, so that $\mathbf{C}^n$ is the Hilbert sum of the eigenspaces V_j of s $(1 \leq j \leq m)$, the restriction of s to V_j being the homothety with ratio $e^{i\theta_j}$; we may assume that $-\pi < \theta_j \leq \pi$ for each j. Show that if $\theta(s) = \sup_{1 \leq j \leq m} |\theta_j|$, then $d(e, s) = \theta(s)$. (Minorize $\mathscr{R}(x \mid s \cdot x)$ by using the decomposition of x as a sum of vectors $x_j \in \mathrm{V}_j$.)
(c) Let s, t be two elements of $\mathbf{U}(n)$ such that s and the commutator $(s, t) = sts^{-1}t^{-1}$ commute, or equivalently such that s and $u = tst^{-1}$ commute. Show that if $\theta(t) < \frac{1}{2}\pi$, then s and t commute. (With the notation of (b), observe that if V'_j is the orthogonal supplement of V_j, and if $\mathrm{W}_j = t(\mathrm{V}_j)$, then W_j is the direct sum of $\mathrm{W}_j \cap \mathrm{V}_j$ and $\mathrm{W}_j \cap \mathrm{V}'_j$, and deduce from the hypothesis on t that $\mathrm{W}_j \cap \mathrm{V}'_j = \{0\}$.)

2. When $n = 2$, the two charts φ_1, φ_2 on $\mathbf{S}_2$ defined in (16.2.3) are such that $\bar{\varphi}_1$ and φ_2 define on $\mathbf{S}_2$ the structure of complex analytic manifold defined in (16.11.12).

3. Let U be an open neighborhood of 0 in $\mathbf{R}^n$. In the real-analytic manifold $\mathrm{U} \times \mathbf{P}_{n-1}(\mathbf{R})$, let U' be the subset consisting of points (x, z) such that for some system $(z^1, \ldots, z^n)$ of homogeneous coordinates for z we have $x^j z^k - x^k z^j = 0$ for all pairs of indices j, k (in which case these relations are satisfied for all systems of homogeneous coordinates for z). Show that U' is a closed analytic submanifold of dimension n in $\mathrm{U} \times \mathbf{P}_{n-1}(\mathbf{R})$ (consider the atlas of $\mathbf{P}_{n-1}(\mathbf{R})$ defined in (16.11.10)). The restriction π_{U} of the projection pr_1 to U' is a surjection of U' onto U and is proper; $\pi_{\mathrm{U}}^{-1}(0)$ is a submanifold of U' isomorphic to $\mathbf{P}_{n-1}(\mathbf{R})$, and the restriction of π_{U} to $\mathrm{U}' - \pi_{\mathrm{U}}^{-1}(0)$ is an isomorphism of this open set onto $\mathrm{U} - \{0\}$. Let r be the inverse isomorphism of $\mathrm{U} - \{0\}$ onto $\mathrm{U}' - \pi_{\mathrm{U}}^{-1}(0)$, and let f be any C^1-function defined on an open neighborhood I of 0 in $\mathbf{R}$ and with values in U, such that $f(0) = 0$ and $\mathrm{T}_0(f) \neq 0$. Then the function $t \mapsto r(f(t))$, defined on $\mathrm{I} - \{0\}$ and with values in U', extends by continuity to a mapping $f' : \mathrm{I} \to \mathrm{U}'$, such that $f'(0)$ is the canonical image in $\mathbf{P}_{n-1}(\mathbf{R})$ of the vector $\mathrm{T}_0(f) \cdot 1$. Furthermore, if f is of class C^r, then f' is of class C^{r-1}; and if f, g are two C^r-functions defined on I, such that $f(0) = g(0) = 0$, which have contact of order $\geq k \geq 1$ at the point 0, then f' and g' have contact of order $\geq k - 1$ at the point 0 (Section 16.5, Problem 9).

If V is another open neighborhood of 0 in $\mathbf{R}^n$ and if $u : \mathrm{U} \to \mathrm{V}$ is an isomorphism of analytic manifolds (resp. a diffeomorphism), then if V' and π_{V} are defined as above, there exists a unique isomorphism $u' : \mathrm{U}' \to \mathrm{V}'$ of analytic manifolds (resp. a unique diffeomorphism) such that $\pi_{\mathrm{V}} \circ u' = u \circ \pi_{\mathrm{U}}$.

Deduce that if X is a pure differential (resp. analytic) manifold of dimension n, and x a point of X, there exists a differential (resp. analytic) manifold X' of dimension n

and C^∞- (resp. analytic) surjection $\pi_X : X' \to X$ with the following properties: (a) the restriction of π_X to $X' - \pi_X^{-1}(x)$ is an isomorphism of $X' - \pi_X^{-1}(x)$ onto $X - \{x\}$; (b) there exists a chart (W, φ, n) of X at the point x such that $\varphi(x) = 0$ and $\varphi(W) = U$ is an open neighborhood of 0 in $\mathbf{R}^n$, and a diffeomorphism (resp. an isomorphism of analytic manifolds) u of $\pi_X^{-1}(W)$ onto U' (with the notation introduced above) such that $\varphi(\pi_X(x')) = \pi_U(u(x'))$ for all $x' \in \pi_X^{-1}(W)$. Moreover, these properties determine X′ up to isomorphism, and X′ is said to be obtained from X by *blowing up the point x*. Extend this construction to complex-analytic manifolds.

4. Let p, q be two integers >0 such that $p + q = n$, and let Φ be a symmetric bilinear form of signature (p, q) on $\mathbf{R}^n$.

(a) Show that the set of vectors $\mathbf{x} \in \mathbf{R}^n$ such that $\Phi(\mathbf{x}, \mathbf{x}) = 1$ is a connected submanifold of $\mathbf{R}^n$ except in the case $p = 1$, when there are two connected components.
(b) Let $\mathbf{SO}(\Phi)$ denote the subgroup of $\mathbf{O}(\Phi)$ consisting of the elements with determinant 1. Show that $\mathbf{SO}(\Phi)$ has two connected components (and hence that $\mathbf{O}(\Phi)$ has four components). (Argue as in (16.11.7), using (a) above.)
(c) What are the corresponding results when $\mathbf{R}^n$ is replaced by $\mathbf{C}^n$ or $\mathbf{H}^n$ and Φ by a Hermitian sesquilinear form?

5. Show that the groups $\mathbf{GL}(n, \mathbf{C})$ and $\mathbf{GL}(n, \mathbf{H})$ are connected when $n \geqq 1$, and that the same is true of the groups $\mathbf{SL}(n, \mathbf{R})$ and $\mathbf{SL}(n, \mathbf{C})$ for $n \geqq 1$. On the other hand, $\mathbf{GL}(n, \mathbf{R})$ has two connected components. (Use the method of (16.11.7).)

6. Let Φ be a nondegenerate alternating bilinear form on $\mathbf{R}^{2n}$. Show that the subgroup $\mathbf{Sp}(\Phi)$ of $\mathbf{GL}(2n, \mathbf{R})$ which leaves Φ invariant is a connected Lie group, and calculate its dimension. (Use the method of (16.11.7).) Consider the same problem with $\mathbf{R}$ replaced by $\mathbf{C}$.

7. Let G be a Lie group and suppose that there exists a Lie subgroup H of G and a submanifold L of G such that the mapping $(x, y) \mapsto xy$ of $L \times H$ into G is a diffeomorphism of $L \times H$ onto G. If K is any Lie subgroup of H, show that the manifold G/K is canonically isomorphic to $L \times (H/K)$. (Use (16.10.4) and (16.8.8).) Consider in particular the case where $G = \mathbf{GL}(2n, \mathbf{R})$, $H = \mathbf{O}(2n, \mathbf{R})$; then L may be taken to be the manifold consisting of the positive definite matrices (Section 11.5, Problem 15). Deduce that $\mathbf{GL}(2n, \mathbf{R})/\mathbf{U}(n, \mathbf{C})$ is diffeomorphic to the manifold $\mathbf{R}^{n(2n+1)} \times (\mathbf{O}(2n, \mathbf{R})/\mathbf{U}(n, \mathbf{C}))$.

12. FIBRATIONS

All the definitions and all the results of Sections 16.12–16.14, with the exceptions of (16.12.11) and (16.12.12), remain valid when we replace differential manifolds and C^∞-mappings by real-analytic (resp. complex-analytic) manifolds and analytic mappings in the statements and proofs. The notions which correspond in this way to those of differential fibrations and (differential) principal bundles are called *real-analytic* (resp. *complex-analytic* or *holomorphic*) fibrations and *real-analytic* (resp. *complex-analytic* or *holomorphic*) principal bundles.

(16.12.1) A *differential fibration* (or simply a *fibration*) is by definition a triple $\lambda = (X, B, \pi)$ in which X and B are differential manifolds and π is a C^∞-mapping of X into B which is *surjective* and satisfies the following condition of *local triviality*:

(LT) *For each* $b \in B$ *there exists an open neighborhood* U *of* b *in* B, *a differential manifold* F *and a diffeomorphism*

$$\varphi : U \times F \to \pi^{-1}(U)$$

such that $\pi(\varphi(y, t)) = y$ *for all* $y \in U$ *and* $t \in F$.

The restriction of π to $\pi^{-1}(U)$ is therefore $pr_1 \circ \varphi^{-1}$, which shows that π is a *submersion*. The manifold X is called the *space* of the fibration λ, the manifold B its *base*, and the mapping π its *projection*. For each $b \in B$, the inverse image $X_b = \pi^{-1}(b)$ is a closed submanifold of X, called the *fiber* of λ over b. By the local triviality condition, there exists a neighborhood U of b such that $X_{b'}$ is diffeomorphic to X_b for all $b' \in U$. By abuse of language, instead of saying that (X, B, π) is a "fibration" we shall also say that X is a *differential fiber bundle*, or simply a *fiber bundle* with base B and projection π, and that for each $x \in X$ the submanifold $X_{\pi(x)}$ is the *fiber through the point* x. If all the fibers are diffeomorphic to the same manifold F, then X is said to be a fiber bundle *of fiber-type* F. This will always be the case when B is connected, for it follows immediately from the local triviality condition that the set of points $b \in B$ such that X_b is diffeomorphic to a given fiber X_{b_0} is both open and closed.

The tangent space at a point $x \in X$ to the fiber $X_{\pi(x)}$ will be canonically identified with a subspace of $T_x(X)$, and the tangent vectors belonging to $T_x(X_{\pi(x)})$ are called the *vertical tangent vectors* at x (or the *tangent vectors along the fiber* at x). They are the elements of the kernel of $T_x(\pi)$.

If f is any mapping of a set E into B, a mapping $f' : E \to X$ is called a *lifting* of f if $\pi(f'(z)) = f(z)$ for all $z \in E$.

Let $\lambda = (X, B, \pi)$ and $\lambda' = (X', B', \pi')$ be two fibrations. A *morphism* of λ into λ' is by definition a pair (f, g) where $f : B \to B'$ and $g : X \to X'$ are C^∞-mappings such that

(16.12.1.1) $$\pi' \circ g = f \circ \pi.$$

The composition of two morphisms (f, g) and (f', g') is defined to be $(f \circ f', g \circ g')$, which is clearly a morphism.

An *isomorphism* of λ onto λ' is a morphism (f, g) such that f and g are diffeomorphisms. In that case (f^{-1}, g^{-1}) is an isomorphism of λ' onto λ, called the *inverse* of (f, g).

When $B = B'$ and $(1_B, g)$ is a morphism (resp. an isomorphism), g is said to be a B-*morphism* of λ into λ', or, by abuse of language, of X into X' (resp. a B-*isomorphism* of λ onto λ' or of X onto X').

If $g : X \to X'$ is a B-morphism, then for each $b \in B$ the relation (16.12.1.1) shows that there exists a C^∞-mapping $g_b : X_b \to X'_b$ such that $g_b(x) = g(x)$ for all $x \in X_b$ (16.8.3.4).

(16.12.2) *Let* $\lambda = (X, B, \pi)$ *and* $\lambda' = (X', B', \pi')$ *be two fibrations and let* (f, g) *be a morphism of* λ *into* λ' *such that* f *is a diffeomorphism of* B *onto* B'. *In order that* (f, g) *should be an isomorphism it is necessary and sufficient that* $g_b : X_b \to X'_{f(b)}$ *should be an isomorphism for each* $b \in B$.

The condition is clearly necessary. To prove that it is sufficient, we remark first that it implies that g is bijective, so that it is enough to show that g is a local diffeomorphism (16.5.6). By virtue of the condition (LT), we may therefore assume that $B' = B$, $f = 1_B$, $X = B \times F$, and $X' = B \times F'$; hence we may write $g(b, z) = (b, u(b, z))$ for $(b, z) \in X$. The following lemma will then complete the proof:

(16.12.2.1) *If* $u : B \times F \to F'$ *is a mapping of class* C^∞ *such that for each* $b \in B$ *the partial mapping* $u(b, \cdot) : F \to F'$ *is a diffeomorphism* (resp. *a submersion*), *then* $g : (b, z) \mapsto (b, u(b, z))$ *is a diffeomorphism* (resp. *a submersion*).

For by (16.6.5) and (16.6.6) we have

$$T_{(b,z)}(g) \cdot (\mathbf{h}'_b, \mathbf{h}''_z) = (\mathbf{h}'_b, T_b(u(\cdot, z)) \cdot \mathbf{h}'_b + T_z(u(b, \cdot)) \cdot \mathbf{h}''_z),$$

which shows that the linear mapping $T_{(b,z)}(g)$ is bijective (resp. surjective); hence the result, by (16.5.6).

Examples

(16.12.3) If B and F are two differential manifolds, the triple $(B \times F, B, \mathrm{pr}_1)$ is a fibration called the *trivial* fibration; its fibers are canonically diffeomorphic to F. A fibration $\lambda = (X, B, \pi)$ is said to be *trivializable* if there exists a B-isomorphism of λ onto a trivial fibration $(B \times F, B, \mathrm{pr}_1)$. Such an isomorphism is called a *trivialization* of λ.

It is important to realize that when a fibration $\lambda = (X, B, \pi)$ is trivializable, there is not in general a *distinguished*, uniquely determined trivialization of λ. If g_1 and g_2 are two trivializations of λ, then we have $g_2(z) = v(g_1(z))$, where

$v : (b, x) \mapsto (b, u(b, x))$ is a B-automorphism of the trivial bundle $B \times F$; that is to say (16.12.2.1) u is a C^∞-mapping such that $u(b, \cdot)$ is a diffeomorphism of F onto itself for each $b \in B$. In other words, the distinction between a trivial fibration and a trivializable one is that for the former any two fibers are *canonically* diffeomorphic, whereas for the latter they are diffeomorphic but there exists in general no distinguished diffeomorphism of one onto the other.

(16.12.4) A fiber bundle with base B whose fibers are *discrete* is called a *covering* (or *covering space*) of B. From the definition it follows immediately that the projection $\pi : X \to B$ is a *surjective local diffeomorphism* (16.5.6). Conversely, however, if $f : X \to Y$ is a surjective local diffeomorphism, it does not necessarily follow that (X, Y, f) is a covering of Y. For example, consider a covering (X_0, B, π) of B such that π is not injective; if $x_0 \in X_0$ is such that the fiber $\pi^{-1}(\pi(x_0))$ has at least two points, consider the space $X = X_0 - \{x_0\}$ and the restriction f of π to X; it is clear that f is a surjective local diffeomorphism, but (X, Y, f) is not a covering (cf. (20.18.8)). (Cf. Problem 1.)

(16.12.4.1) To say that a triple (X, B, π) is a covering of the differential manifold B is equivalent to saying that X is a differential manifold, π a surjective C^∞-mapping, and that the following condition is satisfied:

(R) *For each* $b \in B$, *there exists an open neighborhood* U *of* b *in* B *such that* $\pi^{-1}(U)$ *is the union of a (finite or infinite) sequence* (V_n) *of pairwise disjoint open subsets of* X, *with the property that for each* n *the restriction* $\pi_n : V_n \to U$ *of* π *to* V_n *is a diffeomorphism of* V_n *onto* U.

If B is connected, the condition (R) by itself implies that π is surjective. For $\pi(X)$ is open in B, and if b is in the closure of $\pi(X)$, then there exists an open neighborhood U of b in B which satisfies (R) and meets $\pi(X)$. This implies that the sets V_n are not empty, hence that $U \subset \pi(X)$; in other words, $\pi(X)$ is both open and closed in B, hence is the whole of B because B is connected.

(16.12.4.2) For example, the Riemann surface Y of the logarithmic function (16.8.11) is a *covering* of $\mathbf{C}^* = \mathbf{C} - \{0\}$ of fiber-type $\mathbf{Z}$, the projection π being the restriction of pr_1 to Y. For each point $z_0 = r_0 e^{i\theta_0} \in \mathbf{C}^*$ (where $r_0 > 0$ and $\theta_0 \in \mathbf{R}$) has an open neighborhood U in $\mathbf{C}^*$, namely the image under the bijection $(r, \theta) \mapsto re^{i\theta}$ of the open set

$$V = \{(r, \theta) : r > 0,\ \theta_0 - \pi < \theta < \theta_0 + \pi\} \subset \mathbf{R}^2.$$

It is clear that the mapping

$$\varphi : (re^{i\theta}, k) \mapsto \log r + i\theta + 2k\pi i$$

is a diffeomorphism of $U \times \mathbf{Z}$ onto $\pi^{-1}(U)$ satisfying the local triviality condition (LT). This covering of $\mathbf{C}^*$ is *not trivializable*, because it is *connected*: two points in the same fiber belong to the image in Y of $\mathbf{R}$ under a continuous mapping of the form $t \mapsto (re^{it}, \log r + it)$ (3.19).

A covering whose fiber-type is a finite set of n points is called an *n-sheeted covering*.

(16.12.5) If $\lambda = (X, B, \pi)$ and $\lambda' = (X', B', \pi')$ are two fibrations, then it is immediate that $(X \times X', B \times B', \pi \times \pi')$ is a fibration, called the *product* of the fibrations λ and λ' and written $\lambda \times \lambda'$. For each point $(b, b') \in B \times B'$, we have $(X \times X')_{(b, b')} = X_b \times X'_{b'}$.

(16.12.6) A *section* of a fibration (X, B, π) (or a section of the *fiber bundle* X) is by definition any mapping $s : B \to X$, *not necessarily continuous*, such that $\pi \circ s = 1_B$ (in other words, it is a *lifting* of 1_B). A section is necessarily an injective mapping. A C^∞-section of X may be considered as a B-morphism of the trivial bundle $(B, B, 1_B)$, identified with B, into (X, B, π). It is clear that any trivializable fibration has at least one C^∞-section, but conversely the existence of such a section does not necessarily imply that the fibration is trivializable (Section **16.16**, Problem 1).

The sections (resp. sections of class C^r) of a trivial bundle $(B \times F, B, pr_1)$ are the mappings $b \mapsto (b, f(b))$, where f is a mapping (resp. a mapping of class C^r) of B into F; hence they are in one-to-one correspondence with such mappings.

If g is a B-morphism of (X, B, π) into (X', B, π'), then for each C^r-section s of (X, B, π) (where $0 \leqq r \leqq \infty$), the mapping $g \circ s : B \to X'$ is a C^r-section of (X', B, π').

On the other hand, it should be noted that if $(f, g) : (X, B, \pi) \to (X', B', \pi')$ is a morphism of fibrations with *different* bases, it is not in general possible to define the image of a section of (X, B, π) under such a morphism, because a point of B' may be the image of several distinct points of B. However, when f is a *diffeomorphism* of B onto B', the image of a section s of (X, B, π) under the morphism (f, g) is defined to be the section $b' \mapsto g(s(f^{-1}(b')))$ of (X', B', π').

(16.12.7) (i) *A C^∞-section of a fibration* (X, B, π) *is an embedding of* B *into* X *whose image is closed in* X.

(ii) *A continuous section of a covering* (X, B, π) *is a diffeomorphism of* B *onto an open and closed submanifold of* X.

(i) Let s be a C^∞-section. Since s is injective, the question is local with respect to B, so that we may assume the fibration to be trivial; but in this case the result is immediate, since $s(B)$ is the graph of the mapping $b \mapsto \varphi(b, s(b))$ of B into F, in the notation used at the beginning of **(16.12)**.

(ii) If U is a connected open neighborhood of a point $b \in B$ over which the covering is trivializable, then $s(U)$ must be one of the connected components V_n of $\pi^{-1}(U)$ **(16.12.4.1)**, and $s|U$ is the inverse of $\pi_n = \pi|V_n$. Since π_n is a diffeomorphism, $s|U$ is the inverse diffeomorphism. Hence the section s is a local diffeomorphism, and the result now follows from (i) and **(16.7.5)**.

(16.12.8) *Let* $\lambda = (X, B, \pi)$ *be a fibration,* B' *a differential manifold, and* $f: B' \to B$ *a mapping of class* C^∞.

(i) *The set* $B' \times_B X$ *of points* $(b', x) \in B' \times X$ *such that* $f(b') = \pi(x)$ *is a closed submanifold of* $B' \times X$.

(ii) *If* π' *is the restriction to* $B' \times_B X$ *of* pr_1, *then* $\lambda' = (B' \times_B X, B', \pi')$ *is a fibration such that for each point* $b' \in B'$ *the fiber* $(B' \times_B X)_{b'}$ *is canonically diffeomorphic to* $X_{f(b')}$. *If* f' *is the restriction of* pr_2 *to* $B' \times_B X$, *then* (f, f') *is a morphism of* λ' *into* λ.

(iii) *Let* $\mu' = (Y', B', p')$ *be a fibration with base* B' *and let* $g: Y' \to X$ *be a* C^∞*-mapping such that* (f, g) *is a morphism of* μ' *into* λ. *Then there exists a unique* B'*-morphism* $u: Y' \to B' \times_B X$ *such that* $g = f' \circ u$.

(i) Put $h(b', x) = (f(b'), \pi(x))$, so that h is a C^∞-mapping of $B' \times X$ into $B \times B$ **(16.6.5)**. If Δ is the diagonal of $B \times B$, we have $B' \times_B X = h^{-1}(\Delta)$. Since the question is local with respect to B, we may assume that there exists a submersion $\psi: B \times B \to \mathbf{R}^m$ such that $\Delta = \psi^{-1}(0)$ (**(16.8.3)** and **(16.8.13)**). Putting $\theta = \psi \circ h$, we have $B' \times_B X = \theta^{-1}(0)$, and we have only to show that θ is a submersion *at every point of* $B' \times_B X$; for there will then exist a neighborhood of $B' \times_B X$ in which θ is a submersion, and we can apply **(16.8.8)**.

So let (b', x) be a point of $B' \times_B X$, and let $b = f(b') = \pi(x)$. Since π is a submersion, the image under $T_{(b',x)}(h)$ of $T_{(b',x)}(B' \times X)$ in $T_{(b,b)}(B \times B) = T_b(B) \times T_b(B)$ contains $\{0\} \times T_b(B)$, which is a supplement of the diagonal in this product; but this diagonal, when identified with $T_{(b,b)}(\Delta)$, is the kernel of $T_{(b,b)}(\psi)$. Since ψ is a submersion, the result now follows from **(16.5.4)**.

(ii) Let b'_0 be a point of B', and let $b_0 = f(b'_0)$. By hypothesis, there exists an open neighborhood U of b_0 in B, a differential manifold F, and a diffeomorphism $\varphi: U \times F \to \pi^{-1}(U)$ such that $\pi(\varphi(b, t)) = b$ for all $b \in U$ and $t \in F$. Consider now the open neighborhood $f^{-1}(U)$ of b'_0 in B'; the mapping

$$\varphi': (b', t) \mapsto (b', \varphi(f(b'), t))$$

is a *bijection* of $f^{-1}(U) \times F$ onto $\pi'^{-1}(f^{-1}(U))$ such that $\pi'(\varphi'(b', t)) = b'$. For we have $\pi(\varphi(f(b'), t)) = f(b')$; hence $\varphi'(b', t)$ belongs to $B' \times_B X$ and

$$\pi'(\varphi'(b', t)) = b';$$

conversely, if $(b', x) \in \pi'^{-1}(f^{-1}(U))$, then $b = f(b') = \pi(x) \in U$; hence there exists a unique $t \in F$ such that $\varphi(b, t) = x$. Finally, the fact that φ' is a diffeomorphism follows from (16.12.2.1) and the fact that φ is a diffeomorphism. This completes the proof of (ii).

(iii) For each $y' \in Y'$ we have $\pi(g(y')) = f(p'(y'))$, so that the point $u(y') = (p'(y'), g(y'))$ belongs to $B' \times_B X$, and it is clear that u is the unique B-morphism with the required properties.

The fibration λ' is called the *inverse image* of λ under f, and is denoted by $f^*(\lambda)$. If λ is trivial, so is $f^*(\lambda)$. For each section $s : B \to X$ of λ, the mapping

$$s' : b' \mapsto (b', s(f(b')))$$

is a section of $f^*(\lambda)$, called the *inverse image* of s under f and denoted by $f^*(s)$. If s is continuous (resp. of class C^r), then the same is true of $f^*(s)$.

If (X_1, B, π_1) and (X_2, B, π_2) are two fibrations with base B and if g is a B-morphism of X_1 into X_2, then it is immediate that the mapping

$$g' : B' \times_B X_1 \to B' \times_B X_2$$

defined by $g'(b', x_1) = (b', g(x_1))$ is a B'-morphism; it is denoted by $f^*(g)$.

(16.12.9) Let $\lambda = (X, B, \pi)$ be a fibration, B' a submanifold of B, and $j : B' \to B$ the canonical injection. The set $B' \times_B X$ is then the image under the mapping $(x, b') \to (b', x)$ of the graph of the restriction of π to $\pi^{-1}(B')$, and the mapping $x \mapsto (\pi(x), x)$ is therefore a diffeomorphism of the submanifold $\pi^{-1}(B')$ of X (16.8.12) onto $B' \times_B X$. If we identify $B' \times_B X$ with $\pi^{-1}(B')$ by means of this diffeomorphism, then π' is identified with the restriction of π to $\pi^{-1}(B')$. In future we shall always make this identification, and we shall say that the fibration $j^*(\lambda) = (\pi^{-1}(B'), B', \pi')$ is *induced by* λ *on* B'. The inverse image $j^*(s)$ of a section s of λ is then the restriction of s to B'.

A section of the induced fibration $j^*(\lambda)$ is also called a *section of* λ *over* B'. The set of C^∞-sections of λ over B' is denoted by $\Gamma(B', X)$. More generally, for any subset A of B, a *section of* λ (or *of* X) *over* A is by definition any mapping $s : A \to X$ such that $\pi \circ s = 1_A$ (in other words, such that $s(b) \in X_b$ for all $b \in A$). The sections of λ over B are sometimes called *global sections* of λ (or of X).

The condition (LT) may be stated in the form that each point $b \in B$ has an open neighborhood U in B such that the fibration induced by λ on U is trivializable, or (as we shall sometimes say) that λ is *trivializable over* U.

(16.12.10) Let $\lambda = (X, B, \pi)$ and $\lambda' = (X', B, \pi')$ be two fibrations with the same base B. We may form the product fibration

$$\lambda \times \lambda' = (X \times X', B \times B', \pi \times \pi').$$

Let δ be the diagonal mapping $B \to B \times B$, and let λ'' be the inverse image $\delta^*(\lambda \times \lambda')$. The space of this fibration is by definition the submanifold of $B \times X \times X'$ consisting of all (b, x, x') such that $b = \pi(x) = \pi'(x')$. Now, by (16.12.8(i)), the set $X \times_B X'$ of points $(x, x') \in X \times X'$ such that $\pi(x) = \pi'(x')$ is a submanifold of $X \times X'$, and up to a canonical symmetry (of $B \times X \times X'$ onto $X \times X' \times B$) the space X'' of the fibration λ'' is therefore the graph of the restriction of $\pi \circ \mathrm{pr}_1$ (or $\pi' \circ \mathrm{pr}_2$) to the submanifold $X \times_B X'$ of $X \times X$. Hence it is canonically diffeomorphic to this submanifold (16.8.13), and we shall *identify* X'' with $X \times_B X'$. Next, if π'' is the projection of λ'', and if U is an open subset of B over which X and X' are trivializable, we have diffeomorphisms $\varphi : U \times F \to \pi^{-1}(U)$, $\varphi' : U \times F' \to \pi'^{-1}(U)$, and it is immediately verified that

$$(b, (t, t')) \mapsto (b, \varphi(b, t), \varphi'(b, t'))$$

is a diffeomorphism of $U \times (F \times F')$ onto $\pi''^{-1}(U)$. In particular, over each point $b \in B$, the fiber of λ'' is diffeomorphic to $X_b \times X'_b$. The fibration λ'' (or the fiber bundle $X \times_B X'$) is called the *fiber product* of λ and λ' (or of X and X') over B.

The following proposition is the analog, for sections of fiber bundles, of the Tietze–Urysohn theorem (4.5.1):

(16.12.11) *Let $\lambda = (X, B, \pi)$ be a fibration with fibers diffeomorphic to $\mathbf{R}^N$, let S be a closed subset of B, and let $g : S \to X$ be a section of X over S such that for each $b \in S$ there exists an open neighborhood V_b of b in B and a C^r-section $(1 \leq r \leq \infty)$ s_b of λ over V_b which agrees with g on $V_b \cap S$. Then there exists a C^r-section f of λ over B which agrees with g on S.*

For brevity we shall call g a *C^r-section of λ* (or X) *over* S.

Let (A_n) be a denumerable locally finite covering of B by connected open sets such that λ is trivializable over each A_n (12.6.1). Let (B_n) be another open covering of B such that $\bar{B}_n \subset A_n$ for each n (12.6.2). Let U_n be the union of the A_k for $1 \leq k \leq n$, and let W_n be the union of the B_k for $1 \leq k \leq n$, so that $\overline{W}_n$ is the union of the $\bar{B}_k$ for $1 \leq k \leq n$. By induction on n we shall define a C^r-section f_n of λ over $\overline{W}_n$ such that: (1) f_n agrees with g on $\overline{W}_n \cap S$, and the section of $\pi^{-1}(U_n)$ over $\overline{W}_n \cup (U_n \cap S)$ which is equal to f_n on $\overline{W}_n$ and to g on $U_n \cap S$, is of class C^r; (2) f_{n+1} agrees with f_n on $\overline{W}_n$. Since B is the union of the W_n, the section f of λ over B which is equal to f_n on W_n for all n will have the required properties.

Suppose then that f_n has been defined. We shall show that there exists a C^r-section h_{n+1} of λ over A_{n+1} which agrees with f_n on $A_{n+1} \cap \overline{W}_n$ and with g on $A_{n+1} \cap S$. Then the section f_{n+1} which is equal to f_n on $\overline{W}_n$ and to h_{n+1} on $\overline{B}_{n+1}$ will satisfy the conditions (1) and (2). For at a point $x \in \overline{W}_n \cup (U_n \cap S)$ which does not belong to $\overline{B}_{n+1}$, there exists by hypothesis a neighborhood $T \subset U_n$ of x which does not intersect $\overline{B}_{n+1}$, and a C^r-section of λ over T which is equal to f_n (hence also to f_{n+1}) on $T \cap \overline{W}_n = T \cap \overline{W}_{n+1}$, and is equal to g on $T \cap S$; and at a point of $\overline{B}_{n+1}$ or of $A_{n+1} \cap S$, the section h_{n+1} over the neighborhood A_{n+1} of this point agrees with f_{n+1} on $A_{n+1} \cap \overline{W}_{n+1}$ and with g on $A_{n+1} \cap S$.

It remains to define h_{n+1}. By virtue of the choice of the A_n and the hypothesis on the fibers of λ, we may limit ourselves to the case where $\pi^{-1}(A_{n+1}) = A_{n+1} \times \mathbf{R}^N$, so that sections over A_{n+1} may be identified with mappings of A_{n+1} into $\mathbf{R}^N$. Consider now the function u_{n+1}, defined on

$$(A_{n+1} \cap \overline{W}_n) \cup (A_{n+1} \cap S),$$

which is equal to f_n on $A_{n+1} \cap \overline{W}_n$ and to g on $A_{n+1} \cap S$. This function is of class C^r on this closed subset of A_{n+1}. Indeed, this is obvious at a point of $A_{n+1} \cap S$ which does not belong to $\overline{W}_n$, and at a point $x \in A_{n+1} \cap \overline{W}_n$, there exists by hypothesis a neighborhood $T \subset U_n \cap A_{n+1}$ of x and a function of class C^r on T which agrees with f_n on $T \cap \overline{W}_n$ and with g on $T \cap S$, and hence agrees with u_{n+1} on $T \cap (\overline{W}_n \cup S)$. We can now apply (16.4.3) to the N components of u_{n+1} and hence extend u_{n+1} to a C^r-function h_{n+1} on A_{n+1}. The proof is now complete.

In particular:

(16.12.12) *If the fibers of* $\lambda = (X, B, \pi)$ *are diffeomorphic to* $\mathbf{R}^N$, *then there exists a* C^∞-*section of* λ *over* X.

We have only to apply (16.12.11) with $S = \emptyset$.

Remark

(16.12.13) A real-analytic fibration can also be regarded as a differential fibration: the differential fibration so obtained is said to *underlie* the given real-analytic fibration. In the same way we define the real-analytic fibration *underlying* a given complex-analytic fibration. One point that should be emphasized is that the words "trivializable" and "trivialization" signify different things for a real-analytic (resp. complex-analytic) fibration and the underlying differential (resp. real-analytic) fibration.

PROBLEMS

1. Let X and Y be two connected differential manifolds of the same dimension and let $f : X \to Y$ be a local diffeomorphism. Show that the following properties are equivalent:

(a) f is proper (Section 12.7, Problem 2).
(b) f is a closed mapping (that is, the image under f of any closed subset of X is a closed subset of Y).
(c) For each $y \in Y$, the fiber $f^{-1}(y)$ is a finite set, whose number of elements is independent of y.
(d) (X, Y, f) is a covering of Y, all of whose fibers are finite.
(To prove that (a) implies (b) and that (b) implies (c), argue by contradiction.)

Deduce that for a C^∞-mapping $f : \mathbf{R}^n \to \mathbf{R}^n$ to be a diffeomorphism of $\mathbf{R}^n$ onto itself it is necessary and sufficient that f should be a local diffeomorphism and that $\|f(x)\| \to \infty$ as $\|x\| \to \infty$, where $\|x\|$ is any norm on $\mathbf{R}^n$. (Use the fact that $\mathbf{R}^n$ is simply connected (16.28.3).)

2. Let X and Y be two pure complex-analytic manifolds of dimensions m and n, respectively, and let $f : X \times Y \to \mathbf{C}^n$ be a holomorphic mapping. Let

$$Z_0 = \{(x, y) \in X \times Y : f(x, y) = 0\}$$

and let S be the set of points $(x, y) \in Z_0$ at which the tangent linear mapping $T_y f(x, \cdot) : T_y(Y) \to \mathbf{C}^n$ is not bijective (i.e., S is the set of "singular points" of Z_0).

(a) Show that $Z = Z_0 - S$ is a submanifold of $X \times Y$ of dimension m (if not empty), that the restriction p of pr_1 to Z is a local diffeomorphism and that the fiber $p^{-1}(x)$ is discrete for each $x \in X$.
(b) Suppose that $X = \mathbf{C}^m$ and $Y = \mathbf{C}$, so that f is an entire function on $\mathbf{C}^{m+1}$. For each point $(x_1, y_1) \in Z$, let u_1 be the unique holomorphic function on a neighborhood of x_1 such that $f(x, u_1(x)) = 0$ for all x in this neighborhood, and $u_1(x_1) = y_1$. If Z'_1 is the analytic manifold defined by u_1 (Section 16.8 Problem 12) and $p_1 : Z'_1 \to \mathbf{C}^m$ is the canonical mapping, show that there exists a unique isomorphism h of Z'_1 onto the connected component Z_1 of Z containing (x_1, y_1), such that $p \circ h = p_1$. (Lift up to Z_1 a path in X which is the projection of a path in Z'_1.)
(c) Suppose that $X = Y = \mathbf{C}$ and take $f(x, y) = xy - \sin y$. Then $\text{pr}_1(S)$ is not closed, $\text{pr}_1(Z) = X$, and (Z, X, p) is not a covering.
(d) Suppose again that $X = Y = \mathbf{C}$, and take $f(x, y) = x - 2e^y + e^{2y}$. Then $\text{pr}_1(S) = \{1\}$ and $\text{pr}_1(Z) = X - \{1\}$, but $(Z, \text{pr}_1(Z), p)$ is not a covering.
(e) Suppose that $Y = \mathbf{C}^n$ and that the restriction of pr_1 to Z_0 is a *proper* mapping (Section 12.7, Problem 2). Let T be the open set $X - \text{pr}_1(S)$, and let Z_1 be a connected component of $p^{-1}(T)$ in Z. Show that $(Z_1, p(Z_1), p)$ is a covering of $p(Z_1)$ with a finite number of sheets. (Use Problem 1.)
(f) Suppose that X is connected, $Y = \mathbf{C}$, and take $f(x, y)$ to be a polynomial in y,

$$f(x, y) = y^r + g_1(x)y^{r-1} + \cdots + g_r(x),$$

with coefficients which are holomorphic functions on X. Suppose also that the discriminant of this polynomial does not vanish identically on X. Then $\text{pr}_1(S)$ is closed, and the open set $T = X - pr_1(S)$ is dense and connected; hence the results of (e) are applicable, and moreover we have $p(Z_1) = T$.

(g) Suppose that $X = Y = \mathbf{C}$, and take $f(x, y) = x - \int_0^y \exp(t^2)\, dt$. Then S is empty, $pr_1(Z) = X$, but (Z, X, p) is not a covering. (Use Picard's theorem (Section **10.3**, Problem 8(b)), and consider the values of $y \in \mathbf{C}$ of the form $e^{i\pi/4}t$, where $t \in \mathbf{R}$.)

3. Let (X, Y, p) be a covering, in which the differential manifold Y is compact. Show that there exists a finite open covering $(U_i)_{1 \leq i \leq m}$ of Y such that, for each i, each connected component of $p^{-1}(U_i)$ meets at most one connected component of $p^{-1}(U_j)$ for each $j \neq i$. (Use (**3.16.6**).)

4. Let $\lambda = (X, B, \pi)$ be a fibration and let $f: B_1 \to B$ and $g: B_2 \to B_1$ be two mappings of class C^∞. Define a B_2-isomorphism of $(f \circ g)^*(\lambda)$ onto $g^*(f^*(\lambda))$.

5. Let $\lambda = (X, B, \pi)$ and $\lambda' = (X', B, \pi')$ be two fibrations with base B, and

$$\lambda'' = (X \times_B X', B, \pi'')$$

their fiber product over B.

(a) Consider the fibration $\pi^*(\lambda')$, with base X and projection p, and the fibration $\pi'^*(\lambda)$ with base X′ and projection p'. The space of each fibration is $X \times_B X'$. Show that $\pi'' = \pi \circ p = \pi' \circ p'$.

(b) Let $\mu = (Y, B, \varpi)$ be a fibration with base B, and let $f: Y \to X$ and $f: Y \to X'$ be B-morphisms. Show that there exists a unique B-morphism $f'': Y \to X \times_B X'$ such that $p \circ f'' = f$ and $p' \circ f'' = f'$. We have $f''_b = (f_b, f'_b)$ for all $b \in B$.

6. Let X, Y be two pure differential manifolds, of dimensions p, q respectively. Let (U, φ, p) and (V, ψ, q) be charts on X and Y, respectively. For each integer $r \geq 0$ show that the mapping

$$J_x^r(f) \mapsto J_{f(x)}^r(\psi) \circ J_x^r(f) \circ J_{\varphi(x)}^r(\varphi^{-1})$$

is a bijection $j_{\varphi,\psi}^r$ of $J^r(U, V)$ onto $J^r(\varphi(U), \psi(V))$. Show that $J^r(\varphi(U), \psi(V))$ may be canonically identified with $\varphi(U) \times \psi(V) \times L_{p,q}^r$ (Section **16.5**, Problem 9) and is hence canonically endowed with a structure of differential manifold, induced by that of $\mathbf{R}^p \times \mathbf{R}^q \times L_{p,q}^r$. Show that the charts $(J^r(U, V), j_{\varphi,\psi}^r, N)$ (where $N = p + q + \dim(L_{p,q}^r)$) form an atlas on $J^r(X, Y)$ which defines a structure of differential manifold. If π (resp. π') is the mapping which to each jet associates its source (resp. its target), then

$$(J^r(X, Y), X, \pi), \qquad (J^r(X, Y), Y, \pi'), \qquad (J^r(X, Y), X \times Y, (\pi, \pi'))$$

are fibrations in which the fibers are diffeomorphic to

$$J_0^r(\mathbf{R}^p, Y), \qquad J^r(X, \mathbf{R}^q)_0, \qquad L_{p,q}^r,$$

respectively. If $f: X \to Y$ is a C^∞-mapping, the mapping $J^r(f): x \mapsto J_x^r(f)$ is a C^∞-section of $J^r(X, Y)$ considered as a fiber bundle over X.

If $s \leq r$, the canonical mapping $J^r(X, Y) \to J^s(X, Y)$ defined in Section **16.9**, Problem 1 is a morphism for the fibrations over X, Y and $X \times Y$. If $f: X \to Y$ is a C^∞-mapping, the jet of order $r - s$ of the mapping $x \mapsto J_x^s(f)$ depends only on $J^r(f)$. Hence we have a mapping $u \mapsto J^{r-s}(u)$ of $J^r(X, Y)$ into $J^{r-s}(X, J^s(X, Y))$. Show that this mapping is an embedding and a morphism for the fibrations over X. Is it a diffeomorphism?

Let X, Y, X′, Y′ be pure differential manifolds and let $u: X \to X'$, $v: Y' \to Y$ be C^∞-mappings. For $x \in X$ and $y' \in Y'$, put $x' = u(x)$ and $y = v(y')$, and define a mapping of $J^r_{x'}(X', Y')$ into $J^r_x(X, Y)$ by $w \mapsto J^r_{y'}(v) \circ w \circ J^r_x(u)$. This gives rise to a C^∞-mapping

$$j^r(u, v): u^*(J^r(X', Y')) = X \times_{X'} J^r(X', Y') \to J^r(X, Y).$$

If $X' = X$ (resp. $Y' = Y$), the mapping $j^r(1_X, v)$ (resp. $j^r(u, 1_Y)$) is a morphism for the fibrations over X (resp. Y).

13. DEFINITION OF FIBRATIONS BY MEANS OF CHARTS

(16.13.1) Let $\lambda = (X, B, \pi)$ be a fibration. By hypothesis, there exists an open covering (U_α) of B such that the fibrations induced by λ on each U_α (16.12.9) are trivializable. This property is then *a fortiori* true for each open covering which is *finer* than (U_α) (12.6). For each α, let F_α be the fiber at an (arbitrarily chosen) point of U_α. Then by hypothesis there exists a diffeomorphism

$$\varphi_\alpha : U_\alpha \times F_\alpha \to \pi^{-1}(U_\alpha)$$

satisfying (LT) (16.12.1). For each pair of indices (α, β), we denote by $\varphi_{\beta\alpha}$ the restriction $(U_\alpha \cap U_\beta) \times F_\alpha \to \pi^{-1}(U_\alpha \cap U_\beta)$ of φ_α. Then we have a diffeomorphism (called a "transition function")

$$\psi_{\beta\alpha} = \varphi_{\alpha\beta}^{-1} \circ \varphi_{\beta\alpha} : (U_\alpha \cap U_\beta) \times F_\alpha \to (U_\alpha \cap U_\beta) \times F_\beta,$$

which is of the form

$$(b, t) \mapsto (b, \theta_{\beta\alpha}(b, t)),$$

where $\theta_{\beta\alpha}$ is a C^∞-mapping. Moreover, it follows directly from this definition that if α, β, γ are any three indices and if we denote by $\psi^\gamma_{\beta\alpha}$, $\psi^\alpha_{\gamma\beta}$, and $\psi^\beta_{\gamma\alpha}$ the restrictions of $\psi_{\beta\alpha}$, $\psi_{\gamma\beta}$, and $\psi_{\gamma\alpha}$ to

$$(U_\alpha \cap U_\beta \cap U_\gamma) \times F_\alpha, \quad (U_\alpha \cap U_\beta \cap U_\gamma) \times F_\beta, \quad \text{and} \quad (U_\alpha \cap U_\beta \cap U_\gamma) \times F_\alpha,$$

respectively, then we have

$$\psi^\beta_{\gamma\alpha} = \psi^\alpha_{\gamma\beta} \circ \psi^\gamma_{\beta\alpha}. \tag{16.13.1.1}$$

(16.13.2) Now consider two fibrations $\lambda = (X, B, \pi)$ and $\lambda' = (X', B, \pi')$ with the same base, and a B-morphism (resp. a B-isomorphism) g of X into X′ (resp. onto X′). Then there exists an open covering (U_α) of B such that for each α the fibrations induced on U_α by both λ and λ' are trivializable, so that we have diffeomorphisms

$$\varphi_\alpha : U_\alpha \times F_\alpha \to \pi^{-1}(U_\alpha), \qquad \varphi_\alpha : U_\alpha \times F'_\alpha \to \pi'^{-1}(U_\alpha)$$

satisfying (LT). The composite mapping

$$g_\alpha = \varphi_\alpha'^{-1} \circ (g \,|\, \pi^{-1}(U_\alpha)) \circ \varphi_\alpha : U_\alpha \times F_\alpha \to U_\alpha \times F'_\alpha$$

is then of the form

$$(b, t) \mapsto (b, \sigma_\alpha(b, t)),$$

where σ_α is a mapping of class C^∞ (resp. a mapping of class C^∞ such that $\sigma_\alpha(b, \cdot)$ is a diffeomorphism of F_α onto F'_α for each $b \in U_\alpha$). The mapping g_α is called the *local expression* of g corresponding to φ_α and φ'_α.

Also, with the notation of **(16.13.1)** and analogous notation for the fibration λ', if we put $g_{\beta\alpha} = g_\alpha | ((U_\alpha \cap U_\beta) \times F_\alpha)$, the diagram

(16.13.2.1)

$$\begin{array}{ccc} (U_\alpha \cap U_\beta) \times F_\alpha & \xrightarrow{g_{\beta\alpha}} & (U_\alpha \cap U_\beta) \times F'_\alpha \\ \downarrow{\scriptstyle \psi_{\beta\alpha}} & & \downarrow{\scriptstyle \psi'_{\beta\alpha}} \\ (U_\alpha \cap U_\beta) \times F_\beta & \xrightarrow[g_{\alpha\beta}]{} & (U_\alpha \cap U_\beta) \times F'_\beta \end{array}$$

is commutative for each pair of indices (α, β).

(16.13.3) *Conversely*, consider a differential manifold B and an open covering (U_α) of B; suppose that for each index α we are given a differential manifold F_α, and for each pair of indices (α, β) a mapping

$$\psi_{\beta\alpha} : (U_\alpha \cap U_\beta) \times F_\alpha \to (U_\alpha \times U_\beta) \times F_\beta$$

of the form

$$(b, t) \mapsto (b, \theta_{\beta\alpha}(b, t)),$$

where $\theta_{\beta\alpha}$ is of class C^∞. Suppose also that:

(1) for each $b \in U_\alpha \cap U_\beta$, the mapping $\theta_{\beta\alpha}(b, \cdot) : F_\alpha \to F_\beta$ is a diffeomorphism (which implies **(16.12.2.1)** that $\psi_{\beta\alpha}$ is a diffeomorphism);

(2) the "patching condition" **(16.13.1.1)** (with the notation used there) is satisfied for each triple of indices (α, β, γ).

This latter condition, together with the facts that the $\psi_{\beta\alpha}$ are homeomorphisms and $(U_\alpha \cap U_\beta) \times F_\alpha$ is open in $U_\alpha \times F_\alpha$, allows us to define first of all a *topological space* X by *patching together* the topological spaces $U_\alpha \times F_\alpha$

along the open sets $(U_\alpha \cap U_\beta) \times F_\alpha$ by means of the homeomorphisms $\psi_{\beta\alpha}$ (12.2). Hence (*loc. cit.*) we have homeomorphisms $\varphi_\alpha : U_\alpha \times F_\alpha \to X_\alpha$, where the X_α are open subsets of X which cover X, such that if $\varphi_{\beta\alpha}$ is the restriction of φ_α to $(U_\alpha \cap U_\beta) \times F_\alpha$, we have

$$\varphi_{\beta\alpha}((U_\alpha \cap U_\beta) \times F_\alpha) = X_\alpha \cap X_\beta, \qquad \psi_{\beta\alpha} = \varphi_{\alpha\beta}^{-1} \circ \varphi_{\beta\alpha}.$$

Let us first show that X is *metrizable, separable, and locally compact.* There exists (12.6.1) a denumerable open covering (A_n) of B which is finer than the covering (U_α); hence (12.6.2) a denumerable open covering (B_n) of B such that $\bar{B}_n \subset A_n$ for all n. For each n, let $\alpha(n)$ be an index such that $A_n \subset U_{\alpha(n)}$, and put $Y_n = \varphi_{\alpha(n)}(\bar{B}_n \times F_{\alpha(n)}) \subset X_{\alpha(n)}$. Since the interior $\mathring{Y}_n$ of Y_n in X contains $\varphi_{\alpha(n)}(B_n \times F_{\alpha(n)})$, the open sets $\mathring{Y}_n$ cover X. By (12.4.7), it is enough to show that the sets Y_n are *closed* in X, and for this it is enough (12.2.2) to show that $Y_n \cap X_\beta$ is closed in X_β, for each index β. This is evident if $X_{\alpha(n)} \cap X_\beta = \varnothing$, and if $X_{\alpha(n)}$ meets X_β, then $Y_n \cap X_\beta$ is the image under $\varphi_{\alpha(n),\beta}$ of the set

$$(\bar{B}_n \cap U_\beta) \times F_\beta,$$

which is closed in $U_\beta \times F_\beta$.

Next we define a mapping $\pi : X \to B$ as follows. Each $x \in X$ belongs to some X_α, hence is of the form $\varphi_\alpha(b_\alpha, t_\alpha)$ with $(b_\alpha, t_\alpha) \in U_\alpha \times F_\alpha$; we define $\pi(x) = b_\alpha$, and from the hypotheses it is immediate that this definition is independent of the choice of the index α. Finally, we transport to X_α by means of φ_α the structure of (product) differential manifold on $U_\alpha \times F_\alpha$; the fact that the $\psi_{\beta\alpha}$ are diffeomorphisms ensures that the structures induced on $X_\alpha \cap X_\beta$ by those on X_α and X_β are the same. Hence we have defined a *structure of differential manifold* on X (16.2.5). It is now clear that $\lambda = (X, B, \pi)$ is a fibration; it is said to be obtained by *patching together the trivial fibrations* $(U_\alpha \times F_\alpha, U_\alpha, pr_1)$ *by means of the* $\psi_{\beta\alpha}$.

(16.13.4) Keeping the hypotheses and notation of (16.13.3), consider another open covering (U'_γ) of B which is *finer* than the covering (U_α). For each index γ let $\alpha(\gamma)$ be an index such that $U'_\gamma \subset U_{\alpha(\gamma)}$, and put $F'_\gamma = F_{\alpha(\gamma)}$. For each pair of indices (γ, δ), let

$$\psi'_{\delta\gamma} : (U'_\gamma \cap U'_\delta) \times F'_\gamma \to (U'_\gamma \cap U'_\delta) \times F'_\delta$$

denote the restriction of $\psi_{\alpha(\delta),\alpha(\gamma)}$ to $(U'_\gamma \cap U'_\delta) \times F'_\gamma$. It is clear that the $\psi'_{\delta\gamma}$ satisfy the same conditions as the $\psi_{\beta\alpha}$, and therefore define a fibration $\lambda' = (X', B, \pi')$ by patching together the trivial fibrations $(U'_\gamma \times F'_\gamma, U'_\gamma, pr_1)$ by means of the $\psi'_{\delta\gamma}$. This fibration λ' is B-*isomorphic* to λ. For if $x' \in X'$, then (with the obvious notation) we have $x' = \varphi'_\gamma(b, t)$ with $b \in U'_\gamma$ and $t \in F'_\gamma$ for some index γ; to x' corresponds the point $x = \varphi_{\alpha(\gamma)}(b, t)$ of X, and it is immediately seen that this point x does not depend on the choice of γ, and that in this way we have defined a B-morphism $g : X' \to X$. Conversely, for each point $x \in X$ we have $x = \varphi_\alpha(b, t)$ with $b \in U_\alpha$ and $t \in F_\alpha$, for some index α; there exists an index γ such that $b \in U'_\gamma$, and to x corresponds the point $x' = \varphi'_\gamma(b, t)$. Once again, this point $x' \in X'$ does not depend on the choices of α and γ, and thus we have defined a B-morphism $h : X \to X'$. Finally, it is straightforward to check that $g \circ h$ and $h \circ g$ are the identity mappings, and so our assertion is proved.

(16.13.5) Still keeping the hypotheses and notation of **(16.13.3)**, suppose that we are given, for each index α, a differential manifold F'_α, and for each pair of indices (α, β), a mapping

$$\psi'_{\beta\alpha} : (U_\alpha \cap U_\beta) \times F'_\alpha \to (U_\alpha \cap U_\beta) \times F'_\beta$$

such that the conditions of **(16.13.3)** are satisfied by these mappings. Let $\lambda' = (X', B, \pi')$ be the corresponding fibration. Suppose further that we are given, for each α, a mapping of class C^∞:

$$\sigma_\alpha : U_\alpha \times F_\alpha \to F'_\alpha,$$

and that, if $g_\alpha : U_\alpha \times F_\alpha \to U_\alpha \times F'_\alpha$ is the mapping defined by $g_\alpha(b, t) = (b, \sigma_\alpha(b, t))$, the diagrams **(16.13.2.1)** are commutative. Then there exists a unique B-morphism $g : X \to X'$ such that $g_\alpha = \varphi'^{-1}_\alpha \circ g \circ \varphi_\alpha$ for each α. For if x is any point of X, there exists an index α such that $x = \varphi_\alpha(b, t)$ with $b \in U_\alpha$ and $t \in F_\alpha$; we put $g(x) = \varphi'_\alpha(g_\alpha(b, t))$ and the commutativity of the diagrams **(16.13.2.1)** guarantees that this point does not depend on the choice of index α. The fact that g is a B-morphism is clear. In particular, if $\sigma_\alpha(b, \cdot)$ is a diffeomorphism for each α and each $b \in U_\alpha$, then g is a B-*isomorphism*.

Another particular case in which the preceding method may be applied is the definition of a C^∞-*section* of the fibration λ: for such a section may be regarded as a B-morphism of the trivial fibration $(B, B, 1_B)$ into λ.

14. PRINCIPAL FIBER BUNDLES

We recall that a group G is said to act *freely* (or *without fixed points*) on a set E (cf. **(12.10)**) if for each $x \in E$ the stabilizer S_x of x consists only of the identity element of G: in other words, if for each $x \in E$ the canonical mapping $s \mapsto s \cdot x$ of G into the orbit $G \cdot x$ is *bijective*. The group G then acts faithfully on E.

(16.14.1) *Let* X *be a differential manifold and* G *a Lie group acting differentiably and freely on* X; *suppose that the orbit manifold* X/G *exists* **(16.10.3)**, *and let* $\pi : X \to X/G$ *be the canonical submersion. Then*:

(i) *$(X, X/G, \pi)$ is a fibration. More precisely, each point of* X/G *has an open neighborhood* U *for which there exists a* C^∞*-mapping* $\sigma : U \to X$ *such that* $\pi(\sigma(u)) = u$ *for all* $u \in U$ *and such that the mapping* $(u, s) \mapsto s \cdot \sigma(u)$ *is a diffeomorphism of* $U \times G$ *onto* $\pi^{-1}(U)$.

(ii) *Let* $R \subset X \times X$ *be the set of pairs* (x, y) *such that* x *and* y *belong to the same orbit. For each* $(x, y) \in R$, *let* $\tau(x, y)$ *be the unique element of* G *such that* $y = \tau(x, y) \cdot x$. *Then* τ *is a submersion of the submanifold* R (16.10.3) *into* G.

(i) Since π is a submersion, it follows from (16.8.3) that every point of X/G admits an open neighborhood U for which there exists a mapping $\sigma : U \to X$ of class C^∞ such that, for each $u \in U$, we have $\pi(\sigma(u)) = u$ and $T_{\sigma(u)}(\sigma(U))$ is a supplement of $T_{\sigma(u)}(\pi^{-1}(u))$ in $T_{\sigma(u)}(X)$. Since by hypothesis the mapping $\varphi : U \times G \to \pi^{-1}(U)$ defined by $\varphi(u, s) = s \cdot \sigma(u)$ is *bijective*, it is enough to show that φ is a submersion (16.8.8(iv)). This is a consequence of the following more general result:

(16.14.1.1) *Let* X *be a differential manifold and* G *a Lie group which acts differentiably on* X *such that the orbit manifold* X/G *exists. Let* $\pi: X \to X/G$ *be the canonical submersion, and suppose that there exists a* C^∞*-mapping* $\sigma : X/G \to X$ *such that* $\pi \circ \sigma = 1_{X/G}$. *Then* σ *is an immersion, and the mapping* $\varphi : (X/G) \times G \to X$ *defined by* $\varphi(u, s) = s \cdot \sigma(u)$ *is a surjective submersion.*

The fact that σ is an immersion follows from the relation

$$T_{\sigma(u)}(\pi) \circ T_u(\sigma) = 1_{T_u(X/G)}.$$

Next, we shall show that φ is a submersion at a point of the form (u_0, e). Put $x_0 = \sigma(u_0)$, so that $\pi^{-1}(u_0)$ is the orbit $G \cdot x_0$. By virtue of (16.10.7), the canonical mapping $G \to G \cdot x_0$ is a submersion of G onto the submanifold $\pi^{-1}(u_0)$ of X, and we can apply (16.6.6) and (16.8.8). If now (u_0, s_0) is any point of $(X/G) \times G$, we remark that φ is the composition of the three mappings

$$x \mapsto s_0 \cdot x, \qquad (u, t) \mapsto t \cdot \sigma(u), \qquad (u, s) \mapsto (u, s_0^{-1} s),$$

the first of which is a diffeomorphism of X onto itself (16.10), the third a diffeomorphism of $(X/G) \times G$ onto itself, and the second a submersion at the point (u_0, e). From this it follows that φ is a submersion at the point (u_0, s_0).

(ii) Since the question is local with respect to $B = X/G$, we may assume that there exists a C^∞-section $\sigma : B \to X$ and that $\varphi : (b, s) \mapsto s \cdot \sigma(b)$ is a diffeomorphism of $B \times G$ onto X. Then the mapping $\rho : x \mapsto \mathrm{pr}_2(\varphi^{-1}(x))$ is of class C^∞, and hence the mapping τ, which is the restriction to R of

$$(x, y) \mapsto \rho(y)\rho(x)^{-1},$$

is of class C^∞. Moreover, for each $x \in X$, the restriction of τ to

$$\{x\} \times (G \cdot x) \subset R$$

is a diffeomorphism of this submanifold onto G (16.10.7). Hence τ is a submersion of R into G.

Examples

(16.14.2) Let G be a Lie group and H a Lie subgroup of G. It is clear that H acts *freely* on G (on the right) by the action $(s, x) \mapsto xs$; hence it follows from (16.10.6) and (16.14.1) that $(G, G/H, \pi)$, where $\pi : G \to G/H$ is the canonical mapping, is a fibration. As another example, we have seen (16.11.11) that the group G consisting of the identity mapping and the symmetry $x \mapsto -x$ acts freely on $\mathbf{S}_n$ and has the projective space $\mathbf{P}_n(\mathbf{R})$ as orbit manifold; since $\mathbf{S}_n$ is connected and G discrete, the fibration so defined is *not trivializable*. Again, with the notation of (16.11.10), the group $\mathbf{GL}(p, \mathbf{R})$ acts freely on the right on the space $L_{n,p}$, and hence defines a fibration of $L_{n,p}$ whose base is the Grassmannian $\mathbf{G}_{n,p}$.

It should be remarked that it can happen that a Lie group acts freely on a manifold X but that the orbit space X/G is not a manifold, even if G is discrete (cf. (16.10.3.4)).

When the conditions of (16.14.1) are fulfilled, the manifold X endowed with the action of G is said to be a *differential principal fiber bundle* (or simply a *principal bundle*) with *structure group* G; the manifold $B = X/G$ is the *base* of the bundle, and the fibers are the orbits of the points of x, and are diffeomorphic to G. Usually we shall regard the structure group of a principal bundle as acting on the *right*.

The Riemann surface of the logarithm (16.12.4) is a principal bundle, with base $\mathbf{C}^*$ and structure group $\mathbf{Z}$.

(16.14.3) Let X, X′ be two principal bundles, B, B′ their bases, π, π' their projections, and G, G′ their structure groups. A *morphism* of X into X′ is by definition a pair (u, ρ), where $u : X \to X'$ is a C^∞-mapping and $\rho : G \to G'$ is a Lie group homomorphism, such that

(16.14.3.1) $$u(x \cdot s) = u(x) \cdot \rho(s)$$

for all $s \in G$ and $x \in X$. The image under u of an orbit $x \cdot G$ is therefore contained in the orbit $u(x) \cdot G'$; in other words, there exists a mapping $v : B \to B'$ such that $\pi' \circ u = v \circ \pi$, and it follows from (16.10.4) that v is of class C^∞. The mapping v is said to be *associated* with the morphism (u, ρ); it is clear that (v, u) is a morphism of fibrations (16.12). When ρ is an *isomorphism* of

G onto G′, it follows from (16.14.3) that the restriction of u to an orbit $x \cdot G$ is a diffeomorphism of $x \cdot G$ onto $u(x) \cdot G'$. If moreover v is a diffeomorphism of B onto B′, then (v, u) is an isomorphism of fibrations (16.12.2). In these conditions, (u, ρ) is said to be an *isomorphism* of the principal bundle X onto the principal bundle X′. When $G = G'$ and $\rho = 1_G$, we shall say simply that u is a *morphism* of X into X′.

Example

(16.14.4) Given a differential manifold B and a Lie group G, we define a right action of G on B × G by the rule

$$(b, t) \cdot s = (b, ts).$$

Since the orbits of this action are the sets $\mathrm{pr}_2^{-1}(t)$ for $t \in G$, and since pr_2 is a submersion, if follows that the orbit manifold exists and may be identified with B (16.10.3). Moreover, it is clear that G acts freely on B × G; hence, with the above action, B × G is a principal bundle, called a *trivial* principal bundle. A principal bundle X with structure group G is said to be *trivializable* if it is isomorphic to a principal bundle of the form B × G. An isomorphism of X onto B × G is called a *trivialization* of X.

(16.14.5) *A (differentiable) principal bundle is trivializable if and only if it admits a* C^∞*-section. In particular, a principal bundle whose structure group is diffeomorphic to* $\mathbf{R}^N$ *is trivializable* (16.12.11).

The condition is clearly necessary. Conversely, if a principal bundle X with structure group G and base $B = X/G$ admits a C^∞-section $\sigma : B \to X$, then it follows from (16.14.1.1) that the mapping $(b, s) \mapsto \sigma(b) \cdot s$ is a bijective submersion, hence a diffeomorphism (16.8.8(iv)) and consequently an isomorphism of the principal bundle B × G onto X.

(16.14.6) *Let X be a principal bundle with structure group G, base* $B = X/G$ *and projection* $\pi : X \to B$. *Let B′ be a differential manifold and let* $f : B' \to B$ *be a* C^∞*-mapping. The group G acts differentiably and freely on the manifold* $X' = B' \times_B X$ (16.12.8) *by the rule* $(b', x) \cdot s = (b', x \cdot s)$. *With respect to this action, X′ is a principal bundle with structure group G, and the fibration of X′ may be identified with the inverse image under f of the fibration* $\lambda = (X, B, \pi)$.

Furthermore, if Y′ is a principal bundle with structure group G and base B′, and if $u : Y' \to X$ *is a morphism for which f is the associated mapping, then there exists a unique B′-isomorphism* $w : Y' \to B' \times_B X$ *such that* $u = f' \circ w$, *where* $f' : B' \times_B X \to X$ *is the restriction of* pr_2.

The first assertion is obvious; with the notation of (16.12.8) the orbits of G in X′ are the fibers $\pi'^{-1}(b')$ of the fibration $\lambda' = (B' \times_B X, B', \pi') = f^*(\lambda)$. Since π' is a submersion, the orbit manifold X′/G exists and the corresponding fibration of X′ may be identified with λ', by virtue of (16.10.3). The last assertion follows from (16.12.8).

The principal bundle X′ defined in (16.14.6) is called the *inverse image* of X by f. In particular, if B′ is a *submanifold* of B and if $j : B' \to B$ is the canonical injection, then the inverse image of X by j may be identified with the submanifold $\pi^{-1}(B')$ of X, the action of G on this submanifold being the restriction of the action of G on X. This principal bundle is also called the bundle *induced* by X over B′. In this terminology, (16.14.1(i)) states that each point of B admits an open neighborhood U over which the induced principal bundle is *trivializable* (16.14.4).

(16.14.7) *Let* X *be a principal bundle with structure group* G *(acting on the right); let* $B = G\backslash X$ *be the base and* $\pi : X \to B$ *the projection. Also let* F *be a differential manifold on which* G *acts differentiably (on the left). Then* G *acts differentiably and freely on the right on the product* X × F *by the rule*

$$(x, y) \cdot s = (x \cdot s, s^{-1} \cdot y).$$

For this action:

(i) *The orbit-manifold* $G\backslash(X \times F)$ *exists. We denote it by* $X \times^G F$, *and the projection* $X \times F \to X \times^G F$ *by* $(x, y) \mapsto x \cdot y$.

(ii) *For each orbit* $z \in G\backslash(X \times F)$, *let* $\pi_F(z)$ *be the element of* B *which is equal to* $\pi(x)$ *for all* $(x, y) \in z$. *Then* $(X \times^G F, B, \pi_F)$ *is a fibration in which all the fibers are diffeomorphic to* F. *More precisely, if* U *is an open set in* B *such that* $\pi^{-1}(U)$ *is trivializable and if* $\sigma : U \to \pi^{-1}(U)$ *is a* C^∞*-section of* $\pi^{-1}(U)$, *then the mapping* $(b, y) \mapsto \sigma(b) \cdot y$ *is a* U*-isomorphism of* U × F *onto* $\pi_F^{-1}(U)$ *(which is therefore trivializable).*

(i) Let R′ be the set of points $(x, x', y, y') \in X \times X \times F \times F$ (identified with the product manifold $(X \times F) \times (X \times F)$) such that (x, y) and (x', y') belong to the same orbit. With the notation of (16.14.1), R′ is identified with the set of points $(r, y, \tau(r) \cdot y) \in R \times F \times F$, that is to say with the graph of the mapping $(r, y) \mapsto \tau(r) \cdot y$ of R × F into F. By (16.14.1(ii)) and (16.8.13), this is a closed submanifold of R × F × F, hence also of X × X × F × F. By virtue of (16.10.3), this establishes (i).

(ii) It is sufficient to prove the second assertion, for which we may assume that U = B and that X is trivial. Then it follows from (16.10.4) that π_F is a surjective mapping of class C^∞. Next, for each $x \in X$, put

$$s(x) = \tau(x, \sigma(\pi(x))) \in G$$

(with the notation of (16.14.1)), so that $\sigma(\pi(x)) = x \cdot s(x)$. If $f: X \times F \to B \times F$ is the mapping $(x, y) \mapsto (\pi(x), s(x)^{-1} \cdot y)$, we have $f(x \cdot t, t^{-1} \cdot y) = f(x, y)$ for all $t \in G$, because $s(x \cdot t) = t^{-1}s(x)$ by definition. Since f is of class C^∞ (16.14.1(ii)) there exists by virtue of (16.10.4) a mapping $g : X \times^G F \to B \times F$ of class C^∞ such that $f(x, y) = g(x \cdot y)$, and it is immediately verified that g is the inverse of the mapping $(b, y) \mapsto \sigma(b) \cdot y$.

When $X = B \times G$ is trivial, so that we may take $\sigma(b) = b \cdot e$ and identify $X \times^G F$ with $B \times F$ by means of g, we have $(b, s) \cdot y = (b, s \cdot y)$.

(16.14.7.1) With the same notation as above, every section φ of $X \times^G F$ over U may be uniquely expressed in the form $\varphi : b \mapsto \sigma(b) \cdot \psi(b)$, where ψ is a mapping of U into F. The section φ is of class C^r (r an integer or $+\infty$) if and only if ψ is of class C^r. Since σ is a diffeomorphism of U onto a submanifold $\sigma(U)$ of X, the inverse of σ being the restriction of π, we may also write $\psi(b) = \Phi(\sigma(b))$, where $\Phi = \psi \circ (\pi | \sigma(U))$; the mapping Φ is of class C^r if and only if ψ is of class C^r. Moreover, by taking U to be sufficiently small, we may suppose that Φ is defined on a neighborhood of $\sigma(U)$ in X, and is of class C^r in this neighborhood if φ is of class C^r (16.4.3).

(16.14.7.2) For $x \in X$, $y \in F$, and $t \in G$ we have $(x \cdot t) \cdot y = x \cdot (t \cdot y)$. The relation $x \cdot y = x \cdot y'$ signifies that $x = x \cdot t$ and $y' = t^{-1} \cdot y$ for some $t \in G$; hence $y' = y$, so that $y \mapsto x \cdot y$ is a *diffeomorphism* of F onto the fiber $\pi_F^{-1}(\pi(x))$. It should be noted carefully that the group G does not act *canonically* on a fiber $\pi_F^{-1}(b)$ of $X \times^G F$: we can make G act on this fiber by choosing a point x_0 in $\pi^{-1}(b)$ and putting $t \cdot (x_0 \cdot y) = x_0 \cdot (t \cdot y)$; but this action depends in general on the choice of x_0; for if $x_0' = x_0 \cdot t_0$, we have $x_0' \cdot y = x_0 \cdot (t_0 \cdot y)$, so that on replacing x_0 by x_0' the new action of G on $\pi_F^{-1}(b)$ is

$$(t, x_0 \cdot y) \mapsto x_0 \cdot ((t_0 t t_0^{-1}) \cdot y),$$

which is not the same as the previous action unless the commutator subgroup of G acts trivially on F. This condition will be satisfied in particular if G is commutative.

As in (16.10), putting $m(x, y) = x \cdot y$, we denote the tangent linear mappings $T_x(m(\cdot, y))$ and $T_y(m(x, \cdot))$ by

(16.14.7.3)
$$\mathbf{h}_x \mapsto \mathbf{h}_x \cdot y, \qquad \mathbf{k}_y \mapsto x \cdot \mathbf{k}_y,$$

respectively. Then we have

(16.14.7.4)
$$\mathbf{h}_x \cdot (t \cdot y) = (\mathbf{h}_x \cdot t) \cdot y, \qquad x \cdot (t \cdot \mathbf{k}_y) = (x \cdot t) \cdot \mathbf{k}_y$$

for all $t \in G$, and the mapping $\mathbf{k}_y \mapsto x \cdot \mathbf{k}_y$ is bijective. It follows (16.6.6) that

(16.14.7.5) $$T_{(x,y)}(m) \cdot (\mathbf{h}_x, \mathbf{k}_y) = \mathbf{h}_x \cdot y + x \cdot \mathbf{k}_y$$

which implies that m is a submersion.

The space $X \times^G F$ is called the *bundle of fiber-type* F *associated* with X and the action of G on F. This notion will be especially useful in Chapter XX. At this point, we shall make use of it to prove the following proposition:

(16.14.8) *Let* X *be a principal bundle with structure group* G, *and let* H *be a Lie subgroup of* G. *Then* H *acts on* X *(on the right) by restricting the action of* G. *The orbit-manifold* $H\backslash X$ *exists, so that* X *is a principal bundle with base* $H\backslash X$ *and group* H (16.14.1). *Also if* $\pi : H\backslash X \to G\backslash X$ *is the mapping which associates with each* H*-orbit the unique* G*-orbit containing it, then* $(H\backslash X, G\backslash X, \pi)$ *is a fibration whose fibers are diffeomorphic to the homogeneous space* G/H.

If $R \subset X \times X$ (resp. $R' \subset X \times X$) is the set of pairs (x, y) which belong to the same G-orbit (resp. the same H-orbit), then in the notation of (16.14.1) we have $R' = \tau^{-1}(H)$, which shows that R' is a closed submanifold, because τ is a submersion ((16.14.1) and (16.8.12)).

Next we remark that G acts differentiably on the *left* on G/H, so that we can define the associated bundle $X \times^G (G/H)$ over $G\backslash X$. Let

$$\pi_0 : X \times^G (G/H) \to G\backslash X$$

be the projection. We shall define a diffeomorphism

$$u : X \times^G (G/H) \to H\backslash X$$

such that the diagram

(16.14.8.1) $$\begin{array}{ccc} X \times^G (G/H) & \xrightarrow{u} & H\backslash X \\ & {\scriptstyle \pi_0} \searrow \quad \swarrow {\scriptstyle \pi} & \\ & G\backslash X & \end{array}$$

is commutative; this will prove the proposition. Let $\varphi : G \to G/H$ and $\rho : X \to H\backslash X$ be the canonical projections. Let $f : X \times G \to H\backslash X$ be the composite mapping $(x, s) \mapsto \rho(x \cdot s)$. For each $t \in H$, we have $f(x, st) = f(x, s)$, so that we may write $f(x, s) = g(x, \varphi(s))$, where $g : X \times (G/H) \to H\backslash X$ is a mapping of class C^∞ (16.10.4). Further, for $s' \in G$, we have

$$g(x \cdot s', s'^{-1} \cdot \varphi(s)) = f(x \cdot s', s'^{-1}s) = f(x, s) = g(x, \varphi(s))$$

so that we may write $g(x, \varphi(s)) = u(x \cdot \varphi(s))$, where $u : \mathrm{X} \times^{\mathrm{G}} (\mathrm{G}/\mathrm{H}) \to \mathrm{H}\backslash\mathrm{X}$ is a C^{∞}-mapping (16.10.4). Next, for each $x \in \mathrm{X}$, put $f'(x) = x \cdot \varphi(e)$, which defines a C^{∞}-mapping $f' : \mathrm{X} \to \mathrm{X} \times^{\mathrm{G}} (\mathrm{G}/\mathrm{H})$. For each $t \in \mathrm{H}$, we have

$$f'(x \cdot t) = (x \cdot t) \cdot \varphi(e) = x \cdot (t \cdot \varphi(e)) = x \cdot \varphi(e) = f'(x)$$

because $t \cdot \varphi(e) = \varphi(t) = \varphi(e)$ since $t \in \mathrm{H}$. Hence we may write $f'(x) = u'(\rho(x))$ with u' a mapping of class C^{∞} (16.10.4). It remains to verify that u and u' are inverses of each other and that the diagram (16.14.8.1) is commutative, which is straightforward.

In particular, and changing the notation:

(16.14.9) *Let* G *be a Lie group and* H, K *be two Lie subgroups of* G *such that* $\mathrm{K} \subset \mathrm{H}$. *Let* $\pi : \mathrm{G}/\mathrm{K} \to \mathrm{G}/\mathrm{H}$ *be the mapping which associates with each left coset of* K *the left coset of* H *which contains it. Then* $(\mathrm{G}/\mathrm{K}, \mathrm{G}/\mathrm{H}, \pi)$ *is a fibration with fibers diffeomorphic to the homogeneous space* H/K. *If* K *is a normal subgroup of* H, *then* G/K *is a principal bundle over* G/H *with structure group* H/K.

The last assertion follows from the fact that H/K acts freely on G/K on the right, because $x\mathrm{K}t = xt\mathrm{K}$ for all $x \in \mathrm{G}$ and $t \in \mathrm{H}$.

Examples

(16.14.10) It follows in particular from (16.14.9) and from (16.11.4) and (16.11.6) that for $p = 2, \dots, n$ the Stiefel manifold $\mathbf{S}_{n,p}(\mathbf{R})$ (resp. $\mathbf{S}_{n,p}(\mathbf{C})$, resp. $\mathbf{S}_{n,p}(\mathbf{H})$) is *fibered* over $\mathbf{S}_{n,p-1}(\mathbf{R})$ (resp. $\mathbf{S}_{n,p-1}(\mathbf{C})$, resp. $\mathbf{S}_{n,p-1}(\mathbf{H})$) with fibers diffeomorphic to the sphere $\mathbf{S}_{n-p}$ (resp. $\mathbf{S}_{2(n-p)+1}$, resp. $\mathbf{S}_{4(n-p)+3}$). Again, by virtue of (16.11.9), $\mathbf{S}_{n,p}(\mathbf{R})$ (resp. $\mathbf{S}_{n,p}(\mathbf{C})$, resp. $\mathbf{S}_{n,p}(\mathbf{H})$) is a *principal bundle* over the Grassmannian $\mathbf{G}_{n,p}(\mathbf{R})$ (resp. $\mathbf{G}_{n,p}(\mathbf{C})$, resp. $\mathbf{G}_{n,p}(\mathbf{H})$) with structure group $\mathbf{O}(p, \mathbf{R})$ (resp. $\mathbf{U}(p, \mathbf{C})$, resp. $\mathbf{U}(p, \mathbf{H})$). In particular, the sphere $\mathbf{S}_n$ (resp. $\mathbf{S}_{2n+1}$, resp. $\mathbf{S}_{4n+3}$) is a principal bundle over the projective space $\mathbf{P}_n(\mathbf{R})$ (resp. $\mathbf{P}_n(\mathbf{C})$, resp. $\mathbf{P}_n(\mathbf{H})$) with structure group $\{-1, +1\}$ (resp. $\mathbf{U}(1, \mathbf{C})$, which is isomorphic to the multiplicative group $\mathbf{U}$ of complex numbers of absolute value 1, hence also isomorphic to $\mathbf{T}$, resp. $\mathbf{U}(1, \mathbf{H})$, which is isomorphic to the multiplicative group of quaternions of norm 1).

More particularly, if we take $n = 1$ (having regard to (16.11.12)) we obtain a fibration of $\mathbf{S}_1$ over $\mathbf{S}_1$ with fiber-type $\{-1, 1\}$; a fibration of $\mathbf{S}_3$ over $\mathbf{S}_2$ with fiber-type $\mathbf{S}_1$; and a fibration of $\mathbf{S}_7$ over $\mathbf{S}_4$ with fiber-type $\mathbf{S}_3$. Since $\mathbf{S}_1$ is connected, it is clear that the first of these three fibrations is not trivializable, and it can be shown that the same is true of the other two ("Hopf fibrations").

If F is any differential manifold on which the group $G = \{-1, +1\}$ acts, we obtain from the first of these three principal bundles an associated bundle with fiber-type F and base $\mathbf{U} = \mathbf{S}_1$. Taking $F = \mathbf{R}$, the action of -1 on $\mathbf{R}$ being $t \mapsto -t$, we obtain the orbit-manifold $(\mathbf{U} \times \mathbf{R})/G$, where -1 acts by $(z, t) \mapsto (-z, -t)$. This manifold is called the *Möbius strip*. Taking $F = \mathbf{U}$, with -1 acting by $z \mapsto -z$, we obtain the orbit-manifold $(\mathbf{U} \times \mathbf{U})/G$, where -1 acts by $(z, z') \mapsto (-z, -z')$. This manifold is called the *twisted torus*. Finally, taking $F = \mathbf{U}$ again, with -1 now acting by complex conjugation $z \mapsto \bar{z}$, we obtain the orbit-manifold $(\mathbf{U} \times \mathbf{U})/G$, with -1 acting by

$$(z, z') \mapsto (-z, \bar{z}').$$

This manifold is called the *Klein bottle*.

PROBLEMS

1. Let X be a principal bundle with structure group G, base $B = X/G$, and projection π, and let (U_α) be an open covering of B such that for each index α there exists a section σ_α of X over U_α for which $\varphi_\alpha : (b, s) \mapsto \sigma_\alpha(b) \cdot s$ is a diffeomorphism of $U_\alpha \times G$ onto $\pi^{-1}(U_\alpha)$. For each pair of indices (α, β), let $\varphi_{\beta\alpha}$ denote the restriction of φ_α to $U_\alpha \cap U_\beta$, and put $\psi_{\beta\alpha} = \varphi_{\alpha\beta}^{-1} \circ \varphi_{\beta\alpha}$, which is a transition diffeomorphism of the form

$$(b, s) \mapsto (b, \theta_{\beta\alpha}(b)s) \tag{1}$$

where $\theta_{\beta\alpha} : U_\alpha \cap U_\beta \to G$ is a C^∞-mapping. Further, for each triple of indices (α, β, γ) and each point $b \in U_\alpha \cap U_\beta \cap U_\gamma$ we have the "cocycle condition"

$$\theta_{\gamma\alpha}(b) = \theta_{\gamma\beta}(b)\theta_{\beta\alpha}(b). \tag{2}$$

Conversely, let B be a differential manifold and (U_α) an open covering of B; and suppose that we are given a C^∞-mapping $\theta_{\beta\alpha} : U_\alpha \cap U_\beta \to G$ for each pair of indices α, β, these mappings satisfying the cocycle condition (2). Show that there exists a principal bundle X with structure group G and base B, and for each index α a section σ_α of X over U_α such that the transition diffeomorphisms are of the form (1).

If for the same covering (U_α) of B we are given another family of mappings $\theta'_{\beta\alpha} : U_\alpha \cap U_\beta \to G$ satisfying the condition (2), and hence defining a principal bundle X′ over B with structure group G, show that for X and X′ to be isomorphic it is necessary and sufficient that there should exist for each index α a C^∞-mapping $\mu_\alpha : U_\alpha \to G$ such that, for each pair of indices (α, β)

$$\theta'_{\beta\alpha}(b) = \mu_\beta(b)^{-1}\theta_{\beta\alpha}(b)\mu_\alpha(b) \tag{3}$$

for all $b \in U_\alpha \cap U_\beta$.

Describe the relations between two families $(\theta_{\alpha\beta})$ and $(\theta'_{\gamma\delta})$ defining the same principal bundle, corresponding, respectively, to an open covering (U_α) and a finer open covering (U'_γ).

When G is *commutative*, given two principal bundles X, X′ over the same base B and with G as structure group, we can define (up to isomorphism) their *composition* $X \cdot X'$ as follows: for a given open covering (U_α) of B, suppose that X (resp. X′) is defined by the family $(\theta_{\beta\alpha})$ (resp. $(\theta'_{\beta\alpha})$) satisfying the cocycle condition (2); then the family $(\theta_{\beta\alpha}\theta'_{\beta\alpha})$ also satisfies (2) and hence defines a principal bundle with structure group G, denoted by $X \cdot X'$. Verify that up to isomorphism this bundle is independent of the choice of families $(\theta_{\beta\alpha})$ and $(\theta'_{\beta\alpha})$ defining X and X′. In this way the set of isomorphism classes of principal bundles with base B and structure group G is endowed with a commutative group structure.

2. Let X be a connected complex-analytic manifold. Show that the ring O(X) of (complex-valued) holomorphic functions on X is an integral domain (Section 16.3, Problem 3(a)). Let $R_0(X)$ denote the field of fractions of O(X). If $u, v \in O(X)$ and $v \neq 0$, the function $x \mapsto u(x)/v(x)$ is defined and holomorphic on a dense open subset of X. If $u/v = u_1/v_1$ in the field $R_0(X)$ (i.e., if the holomorphic function $uv_1 - u_1 v$ is identically zero), then the functions $x \mapsto u(x)/v(x)$ and $x \mapsto u_1(x)/v_1(x)$ are defined and equal on a dense open subset of X. Hence, for each $f \in R_0(X)$, there is a largest dense open set $\delta(f)$ in X with the property that for each point $x_0 \in \delta(f)$ there exist two elements u, v in O(X) such that $u/v = f$ and $v(x_0) \neq 0$, so that $u(x)/v(x)$ is defined and holomorphic in a neighborhood of x_0. Put $\tilde{f}(x_0) = u(x_0)/v(x_0)$; then the complex number $\tilde{f}(x_0)$ depends only on f (and x_0). Hence we have defined a holomorphic mapping $\tilde{f}: \delta(f) \to \mathbf{C}$, and the mapping $f \mapsto \tilde{f}$ is bijective. Usually therefore we shall identify f and $\tilde{f}$, and say that f is an *elementary meromorphic function* on X (by abuse of language), whose *domain of definition* is $\delta(f)$. If f and g are two elements of $R_0(X)$, then $f+g$, fg, and $1/f$ (if $f \neq 0$) are defined as elements of $R_0(X)$; but all that can be said about their domains of definition is that $\delta(f+g)$ and $\delta(fg)$ contain $\delta(f) \cap \delta(g)$, and in general neither of the sets $\delta(f)$, $\delta(1/f)$ is contained in the other.

A *meromorphic function* on X is by definition a function f which is defined and holomorphic on a dense open subset U of X, such that for each $x \in X$ there exists a connected open neighborhood V_x of x and an elementary meromorphic function $f_x \in R_0(V_x)$ such that $\delta(f_x) = U \cap V_x$ and such that f_x agrees with f on $U \cap V_x$; then there exists no holomorphic function on an open set U′ strictly containing U, which extends f. Show that the set R(X) of meromorphic functions on X can be endowed with a field structure which induces the field structure of each $R_0(V_x)$.

If X is compact and connected, then $O(X) = R_0(X) = \mathbf{C}$ (Section 16.3, Problem 3(b)). If $X = \mathbf{P}_n(\mathbf{C})$ (which is compact and connected), show that, for each pair P, Q of nonzero homogeneous polynomials of the same degree on $\mathbf{C}^{n+1}$, there exists a meromorphic function f on X such that, if $\pi : \mathbf{C}^{n+1} - \{0\} \to \mathbf{P}_n(\mathbf{C})$ is the canonical mapping, we have $P(z)/Q(z) = f(\pi(z))$ at all points $z \neq 0$ in $\mathbf{C}^{n+1}$ such that $\pi(z) \in \delta(f)$. When $n = 1$, we obtain *all* meromorphic functions on $\mathbf{P}_1(\mathbf{C})$ in this way (use Liouville's theorem (9.11.1)).

3. Let X be a connected complex-analytic manifold. A *predivisor* on X is a pair consisting of a covering (U_α) of X by connected open sets and a family (f_α), where f_α is meromorphic on U_α (Problem 2) and not identically zero, such that for each pair of indices α, β, there exists a holomorphic function $g_{\beta\alpha} : U_\alpha \cap U_\beta \to \mathbf{C}^*$ such that

$$f_\beta(x) = g_{\beta\alpha}(x) f_\alpha(x)$$

at all points $x \in U_\alpha \cap U_\beta$ at which f_α and f_β are defined. Two predivisors $((U_\alpha), (f_\alpha))$ and $((U'_\lambda), (f'_\lambda))$ are said to be *equivalent* if, for each $x \in X$, there exists an open neighborhood V_x of x contained in some U_α and in some U'_λ, and a holomorphic function $h_x : V_x \to \mathbf{C}^*$, such that $f_\alpha(y) = h_x(y) f'_\lambda(y)$ at all points $y \in V_x$ at which f_α and f'_λ are defined. A *divisor* on X is an equivalence class of predivisors.

If D, D′ are two divisors, then there exist two predivisors belonging to D and D′, respectively, and corresponding to the same open covering (U_α). If $((U_\alpha), (f_\alpha))$ and $((U_\alpha), (f'_\alpha))$ are two such predivisors, then we denote by $D + D'$ the divisor containing the predivisor $((U_\alpha), (f_\alpha f'_\alpha))$. Show that $D + D'$ does not depend on the choice of predivisors in D and D′. The mapping $(D, D') \mapsto D + D'$ defines a commutative group structure on the set Div(X) of divisors on X. The neutral element of this group (denoted by 0) is the divisor containing the predivisor consisting of X and the constant function 1.

A *principal divisor* is a divisor containing a predivisor of the form (X, f), where f is a meromorphic function on X, not identically zero. The divisor containing this predivisor is called the *divisor of f* and is denoted by Div(f). Two meromorphic functions f, g, neither of which is identically zero, have the same divisor if and only if there exists a holomorphic function u on X *which does not vanish at any point of* X, such that $f(x) = u(x)g(x)$ at all $x \in X$, where f and g are both defined. The principal divisors form a subgroup Princ(X) of Div(X), isomorphic to $R^*(X)/O^*(X)$, where $R^*(X)$ is the multiplicative group of the field R(X) and $O^*(X)$ is the group of invertible elements of the ring O(X).

4. We retain the hypotheses and notation of Problem 3. If $((U_\alpha), (f_\alpha))$ is a predivisor on X, the functions $g_{\beta\alpha}$ define (Problem 1) a principal bundle over X with structure group $\mathbf{C}^*$, and equivalent predivisors give rise in this way to isomorphic principal bundles, so that to each divisor D on X there corresponds, up to isomorphism, a principal bundle P(D). The bundle P(D) is trivializable (as a holomorphic fiber bundle) if and only if D is principal. Two bundles P(D) and P(D′) are X-isomorphic if and only if $D - D' = \mathrm{Div}(f)$, where f is a meromorphic function on X. In this way we obtain an isomorphism of the quotient group Div(X)/Princ(X) onto a subgroup of the multiplicative group of isomorphism classes of principal bundles over X with structure group $\mathbf{C}^*$ (Problem 1).

5. Let π be the canonical mapping $\mathbf{C}^{n+1} - \{0\} \to \mathbf{P}_n(\mathbf{C})$, and for $j = 0, 1, \ldots, n$ let U_j be the image in $\mathbf{P}_n(\mathbf{C})$ of the open set consisting of the points $\mathbf{z} = (z^0, z^1, \ldots, z^n) \in \mathbf{C}^{n+1}$ such that $z^j \neq 0$. The U_j are connected open sets which cover $\mathbf{P}_n(\mathbf{C})$. For each j, let f_j be the holomorphic function on U_j whose value at $\pi(\mathbf{z}) \in U_j$ is z^0/z^j. Then $((U_j), (f_j))$ is a predivisor, and the corresponding divisor D_1 or $D_1(\mathbf{C})$ is called the *fundamental* divisor on $\mathbf{P}_n(\mathbf{C})$. Show that when $n \geqq 1$, this is not a principal divisor (Section **16.3**, Problem 3(b)) and that its class in Div(X)/Princ(X) generates a subgroup isomorphic to **Z**. It follows that there are infinitely many nonisomorphic holomorphic principal bundles over $\mathbf{P}_n(\mathbf{C})$ with structure group $\mathbf{C}^*$.

Show that the principal bundle $P(D_1(\mathbf{C}))$ is that defined by the action of the multiplicative group $\mathbf{C}^* = \mathbf{GL}(1, \mathbf{C})$ on the space $L_{n+1,1}(\mathbf{C}) = \mathbf{C}^{n+1} - \{0\}$, (**16.11.10**).

Consider the analogs for real-analytic manifolds of the definitions and results of Problems 1–5. In analogous notation, show that the fundamental divisor $D_1(\mathbf{R})$ on $\mathbf{P}_n(\mathbf{R})$ is such that $2D_1(\mathbf{R})$ is principal. (Consider the function $\left(\sum_{j=0}^{n} x_j^2\right)^{-1}$ on $\mathbf{R}^{n+1} - \{0\}$.)

6. Let B be a differential manifold, G an at most denumerable discrete group, and U, V two open subsets of B. Describe all the isomorphism classes of principal bundles over B with structure group G, such that the induced principal bundles over U and V are trivializable (cf. Problem 1). In particular, if $U \cap V$ has exactly two connected components, then the isomorphism classes in question are in bijective correspondence with the conjugacy classes in G.

7. Define the product of two principal bundles X, X′ with structure groups G, G′ and bases B, B′, respectively; also define the fiber product of two principal bundles X, X′ over the same base, with structure groups G, G′, respectively. Show that if X is a principal bundle over B whose structure group G is the product of two subgroups G′, G″, then X is canonically isomorphic to the fiber product over B of two principal bundles X′, X″ over B with G′, G″ as respective structure groups; and that X is trivializable if and only if X′ and X″ are trivializable.

8. Let X be a principal bundle with base B, structure group G, and projection π, and let $E = X \times^G F$ be a fiber bundle with fiber-type F and projection π_F, associated with X. For each C^∞-section f of E over B, there exists a unique mapping $\varphi_f : X \to F$ such that $x \cdot \varphi_f(x) = f(\pi(x))$; this mapping is of class C^∞ and satisfies the relation

$$\varphi_f(x \cdot s) = s^{-1} \cdot \varphi_f(x)$$

for all $x \in X$ and all $s \in G$. Show that $f \mapsto \varphi_f$ is a bijection of the set of C^∞-sections of E over B onto the set of C^∞-mappings $\varphi : X \to F$ such that $\varphi(x \cdot s) = s^{-1} \cdot \varphi(x)$ for all $x \in X$ and $s \in G$. In particular, if there exists $y_0 \in F$ such that $s \cdot y_0 = y_0$ for all $s \in G$, then the bundle E admits a section over B; if G acts trivially on F, the bundle E is trivializable.

9. Let X, X′ be two principal bundles, with bases B, B′, structure groups G, G′, and projections π, π' respectively; let $E = X \times^G F$, $E' = X' \times^{G'} F'$ be the fiber bundles with fiber-types F, F′ associated with X, X′, respectively, and let $\pi_F, \pi'_{F'}$ be their projections. Let (u, ρ) be a morphism of X into X′ (**16.14.3**) and let $v : B \to B'$ be the mapping associated with u. Show that for each mapping $f: F \to F'$ such that $f(s \cdot y) = \rho(s) \cdot f(y)$ for all $y \in F$ and $s \in G$, there exists a unique mapping $w_f : E \to E'$ such that the pair (v, w_f) is a morphism of the fibration (E, B, π_F) into $(E', B', \pi'_{F'})$, and that $w_f(x \cdot y) = u(x) \cdot f(y)$ for all $x \in X$ and $y \in F$. Such a morphism of fibrations is called a (u, ρ)-morphism.

If $G = G'$, $X = X'$, and if $u = 1_X$, $\rho = 1_G$, then w_f is said to be an X-morphism.

10. Let X be a principal bundle with base B and structure group G, and let $E = X \times^G F$ be a fiber bundle associated with X, with fiber-type F. Let $v : B' \to B$ be a C^∞-mapping, and let $X' = v^*(X) = B' \times_B X$ be the inverse image of X by v. Show that there exists a unique mapping $f: E' = X' \times^G F \to v^*(E) = B' \times_B E$ such that $f((b', x) \cdot y) = (b', x \cdot y)$ for $b' \in B'$, $x \in X$, and $y \in F$, and that f is a B′-isomorphism of fiber bundles. What factorization property analogous to (**16.12.8(iii)**) can be stated in this context?

11. Let $\lambda = (E, B, \pi)$ be a fibration in which B is a connected differential manifold. In order that λ should be B-isomorphic to a fibration $(X \times^G F, B, \pi_F)$ associated with a principal bundle X, it is necessary and sufficient that there should exist: (1) a Lie group G acting differentiably and faithfully on a differential manifold F diffeomorphic to

the fibers of E; (2) an open covering (U_α) of B and for each α a diffeomorphism $\varphi_\alpha : U_\alpha \times F \to \pi^{-1}(U_\alpha)$ such that $\pi(\varphi_\alpha(x, y)) = x$ for all $(x, y) \in U_\alpha \times F$; (3) for each pair of indices (α, β) a C^∞-mapping $g_{\beta\alpha} : U_\alpha \cap U_\beta \to G$ such that if $\varphi_{\beta\alpha}$ is the restriction of φ_α to $(U_\alpha \cap U_\beta) \times F$, then the transition diffeomorphism $\psi_{\beta\alpha} = \varphi_{\alpha\beta}^{-1} \circ \varphi_{\beta\alpha}$ is of the form $(x, y) \mapsto (x, g_{\beta\alpha}(x) \cdot y)$ (cf. Problem 1).

12. Let X, Y be pure differential manifolds, of dimensions p and q, respectively. Let $R^r(X)$ (resp. $R^r_x(X)$) denote the set of invertible jets of order r from X to $\mathbf{R}^p$ (resp. invertible jets of order r from X to $\mathbf{R}^p$, with source x) (Section **16.9**, Problem 1). We can define on $R^r(X)$ a structure of differential manifold as in Section **16.12**, Problem 6; with respect to this structure, the group $G^r(p)$ (Section **16.9**, Problem 1) acts differentiably on the right on $R^r(X)$ and defines on $R^r(X)$ a structure of principal bundle over X. Show that the fibration $(J^r(X, Y), X \times Y, (\pi, \pi'))$ (Section **16.12**, Problem 6) is isomorphic to a fibration associated with the principal bundle $R^r(X) \times R^r(Y)$, with fiber-type $L^r_{p,q}$; likewise that the fibration $(J^r(X, Y), X, \pi)$ (resp. $(J^r(X, Y), Y, \pi')$) is isomorphic to a fibration associated with the principal bundle $R^r(X)$ (resp. $R^r(Y)$), with fiber-type $J^r_0(\mathbf{R}^p, Y)$ (resp. $J^r(X, \mathbf{R}^q)_0$).

13. Show that the twisted torus (**16.14.10**) considered as a fiber bundle with base $\mathbf{S}_1$ and fiber-type $\mathbf{S}_1$, is trivializable.

14. With the hypotheses of (**16.14.8**), suppose in addition that H is a normal subgroup of G. Show that the quotient group G/H acts differentiably and freely on the right on the manifold H\X, and that the orbit-manifold may be canonically identified with G\X, so that H\X is a principal bundle with base G\X and group G/H.

15. Let (X, B, π) be a principal bundle with structure group G, and let H be a Lie group acting differentiably on the *right* on X. Suppose that $(x \cdot s) \cdot t = (x \cdot t) \cdot s$ for all $x \in X$, $s \in G$, and $t \in H$; this implies in particular that for each $b \in B$ the set $\pi^{-1}(b) \cdot t$ is a fiber $\pi^{-1}(b')$ for some $b' \in B$, so that if we write $b' = b \cdot t$, then H acts differentiably on B, and equivariantly (**16.10.10**) on X and B. Suppose further that H acts *freely* on B and that the orbit-manifold H\B exists, so that B is a principal bundle over H\B with structure group H (**16.14.1**).

(a) Show that H acts freely on X, that the orbit manifold H\X exists, and that if $\pi' : H\backslash X \to H\backslash B$ is the unique mapping which makes the diagram

$$\begin{array}{ccc} X & \xrightarrow{p} & H\backslash X \\ {\scriptstyle\pi}\downarrow & & \downarrow{\scriptstyle\pi'} \\ B & \xrightarrow[q]{} & H\backslash B \end{array}$$

commutative, then $(H\backslash X, H\backslash B, \pi')$ is a principal bundle with structure group G. (Reduce to the case where $(B, H\backslash B, q)$ is trivial.)

(b) Let F be a differential manifold on which G acts differentiably on the left. Show that there exists a unique differentiable right action of H on $X \times^G F$ such that $(x \cdot y) \cdot t = (x \cdot t) \cdot y$ for $t \in H$, $x \in X$, and $y \in F$. Furthermore, show that the orbit manifold $H\backslash(X \times^G F)$ exists and is canonically diffeomorphic to $(H\backslash X) \times^G F$.

(c) For each pair (x, x') of elements of X such that $\pi(x) = \pi(x')$, let $\tau(x, x')$ denote the element of G such that $x = x' \cdot \tau(x, x')$. Suppose that there exists a C^∞-section σ of X over B and a C^∞-mapping $\beta : H \to G$ such that $\sigma(b \cdot t) \cdot t^{-1} = \sigma(b) \cdot \beta(t)$ for *all* $b \in B$ and all $t \in H$.

(α) Show that β is a homomorphism of H into G.

(β) For each $x \in X$, put $f(x) = (\pi(x), \tau(\sigma(\pi(x)), x))$, so that f is a C^∞-mapping of X into $B \times G$. Show that the unique mapping g which makes the diagram

$$\begin{array}{ccc} X & \longrightarrow & H\backslash X \\ {\scriptstyle f}\downarrow & & \downarrow{\scriptstyle g} \\ B \times G & \longrightarrow & B \times^{H} G \end{array}$$

commutative (where H acts on the left on G by the rule $(t, s) \mapsto s\beta(t^{-1})$) is an isomorphism of principal bundles with base $H\backslash B$ and structure group G.

16. Let (X, B, π) and (X', B', π') be two principal bundles with structure groups G, G′, respectively. Suppose that G acts differentiably on the left on X′ and that

$$s \cdot (x' \cdot t') = (s \cdot x') \cdot t'$$

(which we shall denote by $s \cdot x' \cdot t'$) for all $s \in G$, $x' \in X'$, and $t' \in G'$; then G also acts differentiably on B′, and equivariantly on X′ and B′. Show that the unique mapping $p : X \times^{G} X' \to X \times^{G} B'$, which makes the diagram

$$\begin{array}{ccc} X \times X' & \longrightarrow & X \times^{G} X' \\ {\scriptstyle 1_X \times \pi'}\downarrow & & \downarrow{\scriptstyle p} \\ X \times B' & \underset{q}{\longrightarrow} & X \times^{G} B' \end{array}$$

commutative, is such that $(X \times^{G} X', X \times^{G} B', p)$ is a principal bundle with structure group G′, the action of G′ being such that $(x \cdot x') \cdot t' = x \cdot (x' \cdot t')$ for all $x \in X$, $x' \in X'$, $t' \in G'$. (Reduce to the case where X′ is trivial.) Furthermore, the composite mapping

$$p' : \quad X \times X' \xrightarrow{1_X \times \pi'} X \times B' \xrightarrow{q} X \times^{G} B'$$

is such that $(X \times X', X \times^{G} B', p')$ is a principal bundle with structure group $G \times G'$ (the action of $G \times G'$ on $X \times X'$ being defined by $(x, x') \cdot (s, t') = (x \cdot s, s^{-1} \cdot x' \cdot t')$).

Let F be a differential manifold on which G′ acts differentiably on the left; then G acts differentiably on the left on the fiber bundle $X' \times^{G'} F'$ associated with X′ with fiber-type F′, by the rule $s \cdot (x' \cdot z') = (s \cdot x') \cdot z'$. Show that there exists a unique diffeomorphism (called *canonical*)

$$\beta : \quad (X \times^{G} X') \times^{G'} F' \to X \times^{G} (X' \times^{G'} F')$$

for which the diagram

$$\begin{array}{ccc} & X \times X' \times F' & \\ {\scriptstyle f}\swarrow & & \searrow{\scriptstyle g} \\ (X \times^{G} X') \times^{G'} F' & \underset{\beta}{\longrightarrow} & X \times^{G} (X' \times^{G'} F') \end{array}$$

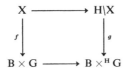

is commutative, where f and g are the canonical mappings. Hence on the space $\mathrm{Y} = (\mathrm{X} \times^{\mathrm{G}} \mathrm{X}') \times^{\mathrm{G}'} \mathrm{F}'$ we have two structures of fiber bundle. one with base $\mathrm{X} \times^{\mathrm{G}} \mathrm{B}'$ and fiber-type F', and the other with base B and fiber-type $\mathrm{X}' \times^{\mathrm{G}'} \mathrm{F}'$.

17. Let $(\mathrm{X}, \mathrm{B}, \pi)$ be a principal bundle with structure group G, and let $\rho : \mathrm{G} \to \mathrm{G}'$ be a homomorphism of the Lie group G into a Lie group G′. Then G acts differentiably on the left on G′ by the rule $(s, t') \mapsto \rho(s)t'$, and we may consider the fiber bundle $\mathrm{X} \times^{\mathrm{G}} \mathrm{G}'$ associated with X by this action. Show that G′ acts on the right on $\mathrm{X} \times^{\mathrm{G}} \mathrm{G}'$ by the rule $(x \cdot s') \cdot t' = x \cdot (s't')$ and that with respect to this action $(\mathrm{X} \times^{\mathrm{G}} \mathrm{G}', \mathrm{B}, \pi')$ (where π' is the canonical mapping) is a principal bundle with base B and group G′; this bundle is called the *ρ-extension* of X. If X′ is a principal bundle with base B and structure group G′, and if $u : \mathrm{X} \to \mathrm{X}'$ is such that (u, ρ) is a morphism (16.14.3) of X into X′, show that there exists a unique isomorphism $v : \mathrm{X} \times^{\mathrm{G}} \mathrm{G}' \to \mathrm{X}'$ such that $u = v \circ \varphi$, where $\varphi : \mathrm{X} \to \mathrm{X} \times^{\mathrm{G}} \mathrm{G}'$ is the canonical mapping $x \mapsto x \cdot e'$ (e' being the neutral element of G′).

18. Let $(\mathbf{e}_k)_{1 \leq k \leq 2n}$ be the canonical basis of $\mathbf{C}^{2n}$, and identify $\mathbf{R}^{2n}$ with the real vector subspace of $\mathbf{C}^{2n}$ spanned by the $\mathbf{e}_k$; then $\mathbf{C}^{2n} = \mathbf{R}^{2n} \oplus i\mathbf{R}^{2n}$, that is to say every vector $\mathbf{z} \in \mathbf{C}^{2n}$ can be written uniquely in the form $\mathbf{z} = \mathbf{x} + i\mathbf{y}$ with $\mathbf{x}, \mathbf{y}$ in $\mathbf{R}^{2n}$. Put $\mathbf{x} = \mathscr{R}\mathbf{z}$.

Let $\mathrm{B}(\mathbf{z}, \mathbf{w})$ be the symmetric bilinear form on $\mathbf{C}^{2n}$ for which $\mathrm{B}(\mathbf{e}_h, \mathbf{e}_k) = \delta_{hk}$ (Kronecker delta). Let V be a totally isotropic subspace of $\mathbf{C}^{2n}$ of maximum (complex) dimension n, relative to the form B. Then the mapping $\mathbf{z} \mapsto \mathscr{R}\mathbf{z}$ is an **R**-linear *bijection* of V onto $\mathbf{R}^{2n}$. Let f_{V} be the inverse bijection, and put $j_{\mathrm{V}}(\mathbf{x}) = \mathscr{R}(if_{\mathrm{V}}(\mathbf{x}))$ for $\mathbf{x} \in \mathbf{R}^{2n}$, so that j_{V} is an **R**-linear bijection of $\mathbf{R}^{2n}$ onto itself.

(a) Show that $j_{\mathrm{V}}^2(\mathbf{x}) = -\mathbf{x}$, $\mathrm{B}(j_{\mathrm{V}}(\mathbf{x}), j_{\mathrm{V}}(\mathbf{y})) = \mathrm{B}(\mathbf{x}, \mathbf{y})$, and that $\mathrm{B}(\mathbf{x}, j_{\mathrm{V}}(\mathbf{x})) = 0$ for all $\mathbf{x}, \mathbf{y} \in \mathbf{R}^{2n}$.

(b) The orthogonal group of the restriction of B to $\mathbf{R}^{2n}$ may be identified with $\mathbf{O}(2n)$. It acts transitively on the set of mappings j_{V} satisfying the relations in (a) by the rule $j_{\mathrm{V}} \mapsto j_{s(\mathrm{V})} = s j_{\mathrm{V}} s^{-1}$ ($\mathbf{O}(2n)$ may be considered as a subgroup of $\mathbf{O}(\mathrm{B})$). The stabilizer of j_{V} (or of the corresponding subspace V) may be identified with the unitary group $\mathbf{U}(\mathrm{S}_{\mathrm{V}})$, where

$$\mathrm{S}_{\mathrm{V}}(\mathbf{x}, \mathbf{y}) = \mathrm{B}(\mathbf{x}, \mathbf{y}) + i\mathrm{B}(\mathbf{x}, j_{\mathrm{V}}(\mathbf{y}))$$

is a positive definite Hermitian form relative to the complex vector space structure on $\mathbf{R}^{2n}$ defined by $(\alpha + \beta i) * \mathbf{x} = \alpha\mathbf{x} + \beta j_{\mathrm{V}}(\mathbf{x})$. The set Γ_n of maximal totally isotropic subspaces of $\mathbf{C}^{2n}$ relative to B is therefore in canonical bijective correspondence with the homogeneous space $\mathbf{O}(2n)/\mathbf{U}(n, \mathbf{C})$, and the structure of real-analytic manifold of this homogeneous space may therefore be transported to Γ_n by means of this correspondence. This manifold has two connected components, one of which corresponds to $\mathbf{SO}(2n)/\mathbf{U}(n, \mathbf{C})$ and is denoted by Γ_n^+.

(c) Show that Γ_n^+ is endowed with a fibration with base S_{2n-2} and fiber diffeomorphic to Γ_{n-1}^+. (Consider all the maximal totally isotropic subspaces V containing a given isotropic vector whose projection on $\mathbf{R}^{2n}$ is $\mathbf{e}_1$.)

(d) Show that the real Stiefel manifold $\mathrm{S}_{2n,\,2n-2}(\mathbf{R})$ is endowed with a fibration whose base is diffeomorphic to Γ_n^+ and whose fibers are diffeomorphic to the complex Stiefel manifold $\mathrm{S}_{n,\,n-1}(\mathbf{C})$. (Observe that a totally isotropic subspace of $\mathbf{C}^{2n}$ of dimension $n - 1$ is contained in a unique subspace belonging to Γ_n^+.)

15. VECTOR BUNDLES

(16.15.1) Let (E, B, π) be a differential fibration such that for each $b \in$ B the fiber $E_b = \pi^{-1}(b)$ is endowed with a structure of a *finite-dimensional real* (resp. *complex*) *vector space*. Then E, endowed with the structure defined by the fibration (E, B, π) and the vector space structures on the fibers E_b, is said to be a *real* (resp. *complex*) *vector bundle* if the following condition is satisfied:

(VB) *For each* $b \in$ B, *there exists an open neighborhood* U *of* b *in* B, *a finite-dimensional real* (resp. *complex*) *vector space* F, *and a diffeomorphism*

$$\varphi : U \times F \to \pi^{-1}(U)$$

such that $\pi(\varphi(y, \mathbf{t})) = y$ *for all* $y \in U$ *and* $\mathbf{t} \in F$, *and such that for each point* $y \in U$ *the partial mapping* $\varphi(y, \cdot)$ *is an* **R**-*linear* (resp. **C**-*linear*) *bijection of the vector space* F *onto the vector space* E_y.

This condition is equivalent to the following:

(VB′) For each $b \in$ B, there exists an open neighborhood U of b, an integer n (depending on b) and n mappings $\mathbf{s}_i : U \to E$ of class C^∞ such that $\pi \circ \mathbf{s}_i = 1_U$ for each i and such that the mapping

$$\varphi : (y, \xi^1, \ldots, \xi^n) \mapsto \xi^1 \mathbf{s}_1(y) + \cdots + \xi^n \mathbf{s}_n(y)$$

is a diffeomorphism of $U \times \mathbf{R}^n$ (resp. $U \times \mathbf{C}^n$) onto $\pi^{-1}(U)$.

For if (VB) is satisfied and if $(\mathbf{a}_i)_{1 \leq i \leq n}$ is a basis of F over **R** (resp. **C**), then the mappings $y \mapsto \mathbf{s}_i(y) = \varphi(y, \mathbf{a}_i)$ $(1 \leq i \leq n)$ satisfy (VB′). Conversely, it is clear that the mapping φ defined in (VB′) satisfies (VB) with $F = \mathbf{R}^n$ (resp. $\mathbf{C}^n$).

If V is an open subset of B, the induced fibration $(\pi^{-1}(V), V, \pi | \pi^{-1}(V))$ (16.12.9) and the vector space structures on the fibers E_b for $b \in V$ clearly define a structure of a real (resp. complex) vector bundle on $\pi^{-1}(V)$. This vector bundle is said to be *induced* by E on V and is sometimes denoted by $E | V$.

A mapping φ satisfying (VB) is called a *framing* over U, and the sections $\mathbf{s}_i$ $(1 \leq i \leq n)$ over U are said to form a *frame* for the vector bundle E over U. Sometimes we shall call a basis of E_b over **R** (resp. **C**) a *frame for* E (or E_b) *at the point* b. For n sections $\mathbf{s}_i$ $(1 \leq i \leq n)$ of class C^∞ of E over U to form a frame, it is necessary and sufficient that, for each point $y \in U$, the vectors $\mathbf{s}_i(y)$ should be *linearly independent* over **R** (resp. **C**) in E_y; this follows from (16.12.2.1).

Let $\varphi : U \times F \to \pi^{-1}(U)$ and $\varphi' : U' \times F \to \pi^{-1}(U')$ be two framings. Then the restrictions ψ and ψ' of φ and φ' to $(U \cap U') \times F$ are framings over $U \cap U'$, and the transition diffeomorphism

$$\psi \circ \psi'^{-1} : (U \cap U') \times F \to (U \cap U') \times F$$

is of the form

(16.15.1.1) $$(x, \mathbf{t}) \mapsto (x, A(x) \cdot \mathbf{t}),$$

where $x \mapsto A(x)$ is a C^∞-mapping of $U \cap U'$ into $\mathbf{GL}(F)$.

We denote by $\mathbf{0}_b$ the zero element of the vector space E_b, for each $b \in B$. The *support* of a section $\mathbf{s}$ of E over an open set U is by definition the *closure in* U of the set of points $y \in U$ at which $\mathbf{s}(y) \neq \mathbf{0}_y$. The mapping $b \mapsto \mathbf{0}_b$ is a C^∞-*section* of E, by virtue of (VB), and it is the only section with empty support. It is called the *zero section* of E and is denoted by $\mathbf{0}_E$ or $\mathbf{0}$.

(16.15.1.2) *Every* C^r-*section of a vector bundle* E *over a closed subset* S *of* B *can be extended to a* C^r-*section over the whole of* B. *For each* $b \in B$ *and each* $\mathbf{u}_b \in E_b$, *there exists a* C^∞-*section* $\mathbf{s}$ *of* E *over* B *such that* $\mathbf{s}(b) = \mathbf{u}_b$.

The first assertion is a particular case of (16.12.11). The second follows from the first, because by virtue of (VB) there exists a C^∞-section of E over U taking an arbitrary value at the point b.

Suppose that U and φ satisfy the conditions of (VB), and let (V, ψ, m) be a chart on B at the point b such that $V \subset U$; also let ρ be an isomorphism of F onto $\mathbf{R}^n$ (resp. $\mathbf{C}^n$). Then it is clear that $(\pi^{-1}(V), (\psi \times \rho) \circ \varphi^{-1}, m + n)$ (resp. $(\pi^{-1}(V), (\psi \times \rho) \circ \varphi^{-1}, m + 2n)$) is a *chart* on E at each point of the fiber $\pi^{-1}(b)$. A chart on E obtained in this fashion is called a *fibered chart*. The local expression (16.3) of a section $\mathbf{s}$ of E over V relative to this chart and the chart (V, ψ, m) on B is a mapping of the form

(16.15.1.3) $$x \mapsto (x, \mathbf{f}(x))$$

of $\psi(V)$ into $\psi(V) \times \mathbf{R}^n$ (resp. $\psi(V) \times \mathbf{C}^n$). The mapping $\mathbf{f} : \psi(V) \to \mathbf{R}^n$ (resp. $\mathbf{f} : \psi(V) \to \mathbf{C}^n$) is sometimes called the *vector part* of $\mathbf{s}$ (relative to the charts under consideration). The support of $\mathbf{s}$ is equal to the support of $\mathbf{f}$, and $\mathbf{s}$ is of class C^r if and only if $\mathbf{f}$ is of class C^r.

The dimension of the vector space E_b over $\mathbf{R}$ (resp. over $\mathbf{C}$) is called the *rank of* E *at* b and is denoted by $\mathrm{rk}_b(E)$. By (16.8.8), for each $x \in E_b$ we have

(16.15.1.4) $\begin{cases} \dim_x(E) = \dim_b(B) + \mathrm{rk}_b(E) & \text{if E is a real vector bundle,} \\ \dim_x(E) = \dim_b(B) + 2\,\mathrm{rk}_b(E) & \text{if E is a complex vector bundle.} \end{cases}$

It follows from (VB) that the rank $\mathrm{rk}_b(E)$ is *locally constant on* B, hence constant on each connected component of B. When $\mathrm{rk}_b(E)$ is constant its value is called the *rank* of E. A vector bundle E of rank 1 is called a *line-bundle*.

When E is a *complex* vector bundle over B, it is clear that the fibration of E, together with the *real* vector space structures of the fibers E_b (underlying their complex vector space structures) define on E a structure of a *real* vector bundle. This real vector bundle E_0 is called the real vector bundle *underlying* E; we have $\mathrm{rk}_b(E_0) = 2\,\mathrm{rk}_b(E)$ for each $b \in B$.

(16.15.2) If E and E′ are two real (resp. complex) vector bundles, (E, B, π) and (E', B', π') the corresponding fibrations, then a *vector bundle morphism* (or simply a *morphism*) of E into E′ is by definition a morphism (f, g) of (E, B, π) into (E', B', π') (16.12) such that for each $b \in B$ the restriction g_b of g to E_b is an **R**-*linear* (resp. **C**-*linear*) mapping of E_b into $E'_{f(b)}$. When $B = B'$ and $f = 1_B$, we say that g is a *(linear)* B-*morphism* of vector bundles. The set of linear B-morphisms of E into E′ is then denoted by Mor(E, E′) (cf. (16.15.8)). An isomorphism (f, g) of E onto E′ is a morphism of vector bundles which is an isomorphism for the corresponding fibrations (16.12). For (f, g) to be an isomorphism it is sufficient that f should be a diffeomorphism and g_b *bijective* for each point $b \in B$ (16.12.2).

The local expression of a morphism of E into E′ relative to the fibered charts corresponding to charts (U, ψ, m) and (U', ψ', m') on B and B′, respectively (16.15.1), is therefore of the form

(16.15.2.1) $$(x, \mathbf{t}) \mapsto (F(x), A(x) \cdot \mathbf{t}),$$

where F is a C^∞-mapping of $\psi(U)$ into $\psi'(U')$ and $x \mapsto A(x)$ is a C^∞-mapping of $\psi(U)$ into the set $\mathrm{Hom}(\mathbf{R}^n, \mathbf{R}^{n'})$ (resp. $\mathrm{Hom}(\mathbf{C}^n, \mathbf{C}^{n'})$), (identified, if we prefer, with the set of matrices of type (n', n) over **R** (resp. **C**)).

Example

(16.15.3) Let B be a differential manifold and F a real (resp. complex) vector space of dimension n, and consider the trivial fibration $(B \times F, B, \mathrm{pr}_1)$ (16.12.3). By transporting the vector space structure of F onto the set

$$\{b\} \times F = \mathrm{pr}_1^{-1}(b)$$

by means of the mapping $\mathbf{t} \mapsto (b, \mathbf{t})$, we obtain on $B \times F$ a structure of a real (resp. complex) vector bundle of rank n corresponding to the trivial fibration, because the condition (VB) is satisfied by taking $U = B$ and φ to be the identity mapping. Such a vector bundle is said to be *trivial.* A vector bundle E over B is said to be *trivializable* if there exists a B-isomorphism of E onto a trivial vector bundle; such an isomorphism is called a *trivialization* of E, and it is precisely the inverse of a *framing* over B (16.15.1).

(16.15.4) *The tangent bundle of a differential manifold.*

Let M be a differential manifold and let T(M) be the union of the (pairwise disjoint) tangent spaces $T_x(M)$ as x runs through M†. Let $o_M : T(M) \to M$ be the mapping which associates with each tangent vector $\mathbf{h}_x \in T_x(M)$ the point $x \in M$ (the "origin" of $\mathbf{h}_x$). We shall show that there exists a *unique structure of differential manifold* on T(M) such that $\tau = (T(M), M, o_M)$ is a *fibration* (the tangent spaces $T_x(M)$ being the fibers) and such that the following condition is satisfied:

(TB) *For each chart* $c = (U, \varphi, n)$ *on* M, *the mapping* (cf. (16.5.3))

$$\psi_c : (x, \mathbf{h}) \mapsto (d_x \varphi)^{-1} \cdot \mathbf{h} \tag{16.15.4.1}$$

of $U \times \mathbf{R}^n$ *onto* $o_M^{-1}(U)$ *is a diffeomorphism.*

An equivalent condition is that, for each $\mathbf{h} \in \mathbf{R}^n$, the mapping

$$x \mapsto (d_x \varphi)^{-1} \cdot \mathbf{h}$$

is a C^∞-*section* of T(M) over U. For if this is so, and if for each vector $\mathbf{e}_i$ of the canonical basis of $\mathbf{R}^n$ we put $X_i(x) = (d_x \varphi)^{-1} \cdot \mathbf{e}_i$, and $\mathbf{h} = \sum_{i=1}^{n} \xi^i \mathbf{e}_i$, then the mapping $(x, \xi^1, \dots, \xi^n) \mapsto \sum_{i=1}^{n} \xi^i X_i(x)$ is a diffeomorphism, by virtue of (16.12.2), hence (TB) is satisfied. The existence of these sections shows at the same time that the vector space structures of the $T_x(M)$ and the fibration τ define on T(M) a structure of a *real vector bundle*; this vector bundle is called the *tangent vector bundle*, or more briefly the *tangent bundle*, of the differential manifold M (cf. Section 16.12, Problem 6).

† By definition (16.5), an element of $T_x(M)$ is an equivalence class of mappings of $\mathbf{R}$ into M, hence a subset of the set $\mathscr{F}(\mathbf{R}, M)$ of all mappings of $\mathbf{R}$ into M; consequently $T_x(M)$ is a subset of $\mathfrak{P}(\mathscr{F}(\mathbf{R}, M))$, and the sets $\{x\} \times T_x(M)$ are pairwise disjoint in the set $M \times \mathfrak{P}(\mathscr{F}(\mathbf{R}, M))$. As a *set*, T(M) is defined to be their union.

To establish the existence of the differential manifold T(M), consider an open covering (U_α) of M such that for each α there exists a chart $(U_\alpha, \varphi_\alpha, n_\alpha)$ on M. Let α, β be two indices such that $U_\alpha \cap U_\beta \neq \varnothing$, which implies that $n_\alpha = n_\beta = n$ say; let $\varphi_{\beta\alpha}$ and $\varphi_{\alpha\beta}$ be the restrictions of φ_α and φ_β, respectively, to $U_\alpha \cap U_\beta$, and let $f_{\beta\alpha} = \varphi_{\beta\alpha} \circ \varphi_{\alpha\beta}^{-1}$ be the transition diffeomorphism. Then (16.5.7) we have

$$Df_{\beta\alpha}(\varphi_\beta(x)) = (d_x \varphi_{\beta\alpha}) \circ (d_x \varphi_{\alpha\beta})^{-1}$$

for $x \in U_\alpha \cap U_\beta$, and the mapping

$$\psi_{\beta\alpha} : (U_\alpha \cap U_\beta) \times \mathbf{R}^n \to (U_\alpha \cap U_\beta) \times \mathbf{R}^n$$

defined by

$$\psi_{\beta\alpha}(x, \mathbf{h}) = (x, Df_{\beta\alpha}(\varphi_\beta(x)) \cdot \mathbf{h})$$

is evidently of class C^∞, hence a diffeomorphism by (16.12.2.1). Further, if for each triple of indices (α, β, γ) we denote by $f_{\beta\alpha}^\gamma$, $f_{\gamma\beta}^\alpha$, and $f_{\gamma\alpha}^\beta$ the restrictions of the mappings $f_{\beta\alpha}$, $f_{\gamma\beta}$, and $f_{\gamma\alpha}$ to

$$\varphi_\beta(U_\alpha \cap U_\beta \cap U_\gamma),\ \varphi_\gamma(U_\alpha \cap U_\beta \cap U_\gamma), \quad \text{and} \quad \varphi_\gamma(U_\alpha \cap U_\beta \cap U_\gamma),$$

respectively, then we have $f_{\gamma\alpha}^\beta = f_{\alpha\beta}^\gamma \circ f_{\gamma\beta}^\alpha$, and by (8.2.1) the patching condition (16.13.1.1) is satisfied. This establishes the existence of the fibration τ (16.13.3); moreover, it follows from the construction in (16.13.3) that this fibration satisfies (TB) and is the unique fibration with this property; for the condition (TB) implies that each mapping (16.15.4.1) is a *framing* of T(M) over U. This framing ψ_c is said to be *associated* with the chart c.

(16.15.4.2) A *section* of T(M) over a subset A of M (16.12.9) is called a *tangent vector field* (or simply a *vector field*) over A. With the preceding notation, the mappings $x \mapsto X_i(x) = (d_x \varphi)^{-1} \cdot \mathbf{e}_i = \theta_{c,x}^{-1}(\mathbf{e}_i)$ (16.5.1) for $1 \leqq i \leqq n$ are C^∞-vector fields on U which form a *frame* of T(M) over U. These vector fields are called the vector fields *associated* with the chart c, and the frame they form is the frame *associated* with c. Every vector field on U is uniquely expressible in the form

(16.15.4.3) $$x \mapsto X(x) = \sum_{i=1}^{n} a^i(x) \cdot X_i(x),$$

where the a^i are n scalar-valued functions on U. For X to be of class C^r, it is necessary and sufficient that the a^i should be of class C^r.

(16.15.4.4) With the notation of (16.15.4.1), we see that

$$(o_M^{-1}(U), (\varphi \times 1_{\mathbf{R}^n}) \circ \psi_c^{-1}, 2n)$$

is a *fibered chart* on T(M) (16.15.1), called the fibered chart *associated* with the chart $c = (\mathrm{U}, \varphi, n)$ on M. If $(\mathrm{U}, \varphi', n)$ is another chart on M with the same domain of definition, and if $u = \varphi \circ \varphi'^{-1} : \varphi'(\mathrm{U}) \to \varphi(\mathrm{U})$ is the transition diffeomorphism, then the transition diffeomorphism for the associated fibered charts on T(M) is the mapping

(16.15.4.5) $$(x, \mathbf{h}) \mapsto (u(x), \mathrm{D}u(x) \cdot \mathbf{h})$$

of $\varphi'(\mathrm{U}) \times \mathbf{R}^n$ onto $\varphi(\mathrm{U}) \times \mathbf{R}^n$.

If a vector field X on U is given by (16.15.4.3), its *local expression* relative to the fibered chart associated with (U, φ, n) (16.15.1.3) is the mapping

(16.15.4.6) $$z \mapsto \left(z, \sum_{i=1}^{n} a^i(\varphi^{-1}(z))\mathbf{e}_i\right)$$

of $\varphi(\mathrm{U})$ into $\varphi(\mathrm{U}) \times \mathbf{R}^n$.

(16.15.5) When M is an open subset of $\mathbf{R}^n$, the inverse of the framing associated with the chart $(\mathrm{M}, 1_\mathrm{M}, n)$ is a *trivialization* of T(M), called the *canonical* trivialization and given by $\mathbf{h}_x \mapsto (x, \tau_x(\mathbf{h}_x))$ (16.5.2). Usually we shall identify T(M) with $\mathrm{M} \times \mathbf{R}^n$ by means of this trivialization. A vector field on M is then of the form $x \mapsto (x, \mathbf{f}(x))$, where $\mathbf{f}$ is a mapping of M into $\mathbf{R}^n$.

(16.15.6) Let M, N be two differential manifolds and let $f : \mathrm{M} \to \mathrm{N}$ be a mapping of class C^r (where r is an integer $\geqq 1$, or $+\infty$). Then the mapping

(16.15.6.1) $$\mathrm{T}(f) : \mathbf{h} \mapsto \mathrm{T}_{\rho_\mathrm{M}(\mathbf{h})}(f) \cdot \mathbf{h}$$

(which we shall write in the more legible form

(16.15.6.2) $$\mathbf{h}_x \mapsto \mathrm{T}_x(f) \cdot \mathbf{h}_x)$$

is a mapping of class C^{r-1} (with the convention that $r - 1 = \infty$ if $r = \infty$) of T(M) into T(N). For if (U, φ, m) and (V, ψ, n) are charts on M and N, respectively, such that $f(\mathrm{U}) \subset \mathrm{V}$, and if F is the local expression of f relative to these charts (16.3), then the local expression of T(f) relative to the associated fibered charts (16.15.4) is the mapping

(16.15.6.3) $$(x, \mathbf{h}) \mapsto (\mathrm{F}(x), \mathrm{F}'(x) \cdot \mathbf{h})$$

of $\varphi(\mathrm{U}) \times \mathbf{R}^m$ into $\psi(\mathrm{V}) \times \mathbf{R}^n$. If f is of class C^∞, then $(f, \mathrm{T}(f))$ is a *morphism of vector bundles* (16.15.2). It is clear that

(16.15.6.4) $$\operatorname{rk}_{\mathbf{h}_x}(\mathrm{T}(f)) = 2\operatorname{rk}_x(f).$$

Hence if f is a subimmersion (resp. an immersion, resp. a submersion), so also is $\mathrm{T}(f)$.

Also it is clear that $\mathrm{T}(1_{\mathrm{M}}) = 1_{\mathrm{T(M)}}$, and that if $g : \mathrm{N} \to \mathrm{P}$ is another C^r-mapping, we have $\mathrm{T}(g \circ f) = \mathrm{T}(g) \circ \mathrm{T}(f)$. If f is a diffeomorphism, then so is $\mathrm{T}(f)$, and $\mathrm{T}(f^{-1}) = \mathrm{T}(f)^{-1}$.

If M and N are two differential manifolds and pr_1, pr_2 are the projections of $\mathrm{M} \times \mathrm{N}$ onto M and N, respectively, then the mapping $(\mathrm{T}(\mathrm{pr}_1), \mathrm{T}(\mathrm{pr}_2))$ is a canonical isomorphism of $\mathrm{T}(\mathrm{M} \times \mathrm{N})$ onto $\mathrm{T}(\mathrm{M}) \times \mathrm{T}(\mathrm{N})$ (16.6.2); usually we shall identify these two vector bundles over $\mathrm{M} \times \mathrm{N}$.

As an application of these results, we remark that if a Lie group G acts on a differential manifold X, then the mapping (16.10.1.2)

$$(\mathbf{h}_s, \mathbf{k}_x) \mapsto s \cdot \mathbf{k}_x + \mathbf{h}_s \cdot x$$

of $\mathrm{T}(\mathrm{G}) \times \mathrm{T}(\mathrm{X})$ into $\mathrm{T}(\mathrm{X})$ is of class C^∞. Likewise, if $(\mathrm{X}, \mathrm{B}, \pi)$ is a principal bundle with structure group G and if $\mathrm{E} = \mathrm{X} \times^{\mathrm{G}} \mathrm{F}$ is the associated fiber bundle with fiber-type F, then the mapping (16.14.7.2)

$$(\mathbf{h}_x, \mathbf{k}_y) \mapsto \mathbf{h}_x \cdot y + x \cdot \mathbf{k}_y$$

of $\mathrm{T}(\mathrm{X}) \times \mathrm{T}(\mathrm{F})$ into $\mathrm{T}(\mathrm{E})$ is of class C^∞.

(16.15.7) *The tangent bundle of a vector bundle.*

If $\lambda = (\mathrm{X}, \mathrm{B}, \pi)$ is a fibration, then $(\mathrm{T}(\mathrm{X}), \mathrm{T}(\mathrm{B}), \mathrm{T}(\pi))$ is also a *fibration*. For if U is an open subset of B such that there exists a diffeomorphism $\varphi : \mathrm{U} \times \mathrm{F} \to \pi^{-1}(\mathrm{U})$ with $\pi \circ \varphi = \mathrm{pr}_1$, then it follows that $\mathrm{T}(\varphi)$ is a diffeomorphism of $\mathrm{T}(\mathrm{U}) \times \mathrm{T}(\mathrm{F})$ onto $\mathrm{T}(\pi^{-1}(\mathrm{U}))$ such that $\mathrm{T}(\pi) \circ \mathrm{T}(\varphi) = \mathrm{pr}_1$. We recall (16.12.1) that the tangent vectors $\mathbf{h}_x \in \mathrm{T}_x(\mathrm{X})$ such that $\mathrm{T}(\pi) \cdot \mathbf{h}_x = 0$ in $\mathrm{T}_{\pi(x)}(\mathrm{B})$ are said to be *vertical*, or *along the fiber*.

Now let E be a *vector bundle* with base B and projection π. Then $\mathrm{T}(\mathrm{E})$ is also a *vector bundle* with base $\mathrm{T}(\mathrm{B})$ and projection $\mathrm{T}(\pi)$. Since the question is local with respect to B, we may assume that E is trivializable, and hence we reduce to the case where $\mathrm{E} = \mathrm{B} \times \mathrm{F}$, with F a vector space of dimension n and B an open set in $\mathbf{R}^m$. Then (16.15.6) $\mathrm{T}(\mathrm{E})$ may be identified with $\mathrm{T}(\mathrm{B}) \times \mathrm{T}(\mathrm{F})$, and the canonical trivializations of $\mathrm{T}(\mathrm{B})$ and $\mathrm{T}(\mathrm{F})$ (16.15.5) therefore finally identify $\mathrm{T}(\mathrm{E})$ with $(\mathrm{B} \times \mathbf{R}^m) \times (\mathrm{F} \times \mathbf{R}^n)$. Since $\mathrm{F} \times \mathbf{R}^n$ is canonically endowed with the structure of a vector space of dimension $2n$, it follows that $\mathrm{T}(\mathrm{E})$ is endowed with a vector bundle structure over $\mathrm{T}(\mathrm{B})$; but it has to be shown that this structure is independent of the trivialization of E from which we started.

Now a transition diffeomorphism from one trivialization to another is of the form

$$\psi : (b, \mathbf{y}) \mapsto (b, A(b) \cdot \mathbf{y}),$$

where $A : \mathrm{B} \to \mathbf{GL}(\mathrm{F}) \subset \mathrm{End}(\mathrm{F})$ is a C^∞-mapping; and then $\mathrm{T}(\psi)$ is the diffeomorphism

$$((b, \mathbf{h}), (\mathbf{y}, \mathbf{k})) \mapsto ((b, \mathbf{h}), (A(b) \cdot \mathbf{y}, A(b) \cdot \mathbf{k} + (A'(b) \cdot \mathbf{h}) \cdot \mathbf{y})) \tag{16.15.7.1}$$

((8.1.3) and (8.9.1)), and our assertion follows from the fact that the mapping

$$(\mathbf{y}, \mathbf{k}) \mapsto A(b) \cdot \mathbf{k} + (A'(b) \cdot \mathbf{h}) \cdot \mathbf{y}$$

is *linear*.

Hence we have *two* vector bundle structures on T(E): one with base E and projection o_{E}, the other with base T(B) and projection $\mathrm{T}(\pi)$; moreover we have $o_{\mathrm{B}} \circ \mathrm{T}(\pi) = \pi \circ o_{\mathrm{E}}$. Furthermore, if $\varpi = o_{\mathrm{B}} \circ \mathrm{T}(\pi) = \pi \circ o_{\mathrm{E}}$, then $(\mathrm{T}(\mathrm{E}), \mathrm{B}, \varpi)$ is a *fibration* over B; but although, for a *given* trivialization of E, each fiber of this fibration can be endowed with the structure of a vector space of dimension $2n + m$, yet it is not possible to define in this way a *vector bundle* structure on the fibration. For the preceding calculation shows that the right-hand side of (16.15.7.1) *is not linear in* $(\mathbf{h}, \mathbf{y}, \mathbf{k})$, although it is linear in $(\mathbf{y}, \mathbf{k})$ and in $(\mathbf{h}, \mathbf{k})$, which correspond to the two vector bundle structures on T(E) previously described.

In the particular case where $\mathrm{E} = \mathrm{T}(\mathrm{M})$, the tangent bundle of a differential manifold M, the two vector bundle structures on T(T(M)) both have the *same base* T(M), but are quite distinct from each other.

(16.15.8) Let E be a real (resp. complex) vector bundle with base B and projection π, and let $\mathscr{E}(\mathrm{B}; \mathbf{R})$ (resp. $\mathscr{E}(\mathrm{B}; \mathbf{C})$) be the set of C^∞-mappings of B into **R** (resp. into **C**), which is an **R**-algebra (resp. a **C**-algebra). Consider also the set Mor(E, E′) of B-morphisms of E into another real (resp. complex) vector bundle E′ over B. If $u', u'' \in \mathrm{Mor}(\mathrm{E}, \mathrm{E}')$, we define $u' + u''$ to be the mapping of E into E′ such that $(u' + u'')_b = u'_b + u''_b$ for all $b \in \mathrm{B}$, and it follows from (VB) that $u' + u''$ is a B-morphism. Again, for each function $f \in \mathscr{E}(\mathrm{B}; \mathbf{R})$ (resp. $f \in \mathscr{E}(\mathrm{B}; \mathbf{C})$) and each $u \in \mathrm{Mor}(\mathrm{E}, \mathrm{E}')$, we define a mapping $f \cdot u : \mathrm{E} \to \mathrm{E}'$ by the rule $(f \cdot u)_b = f(b) u_b$, and $f \cdot u$ is again a B-morphism. Hence we have defined on the set Mor(E, E′) a structure of $\mathscr{E}(\mathrm{B}; \mathbf{R})$-*module* (resp. $\mathscr{E}(\mathrm{B}; \mathbf{C})$-*module*).

In particular, the set $\Gamma(\mathrm{B}, \mathrm{E})$ or $\Gamma(\mathrm{E})$ of all C^∞-*sections* of E may be considered as the set of B-morphisms of the trivial bundle $(\mathrm{B} \times \{0\}, \mathrm{B}, \mathrm{pr}_1)$ into E, and therefore $\Gamma(\mathrm{E})$ is an $\mathscr{E}(\mathrm{B}; \mathbf{R})$-module (resp. an $\mathscr{E}(\mathrm{B}; \mathbf{C})$-module). If there exists a *frame* over B (in other words, if E is trivializable), then this module is free, and every frame over B is a basis of it.

By applying these remarks to the vector bundle induced on $\pi^{-1}(\mathrm{U})$, where U is open in B, we define a structure of $\mathscr{E}(\mathrm{U}; \mathbf{R})$-module (resp. $\mathscr{E}(\mathrm{U}; \mathbf{C})$-module) on the set $\Gamma(\mathrm{U}, \mathrm{E})$ of C^∞-*sections of* E *over* U. For each $u \in \mathrm{Mor}(\mathrm{E}, \mathrm{E}')$, the mapping $s \mapsto u \circ s$ of $\Gamma(\mathrm{U}; \mathrm{E})$ into $\Gamma(\mathrm{U}; \mathrm{E}')$ is $\mathscr{E}(\mathrm{U}; \mathbf{R})$-linear (resp. $\mathscr{E}(\mathrm{U}; \mathbf{C})$-linear).

Remark

(16.15.9) In exactly the same way we can define the notion of a real (resp. complex) vector bundle over a *real-analytic manifold* B; such a bundle is called a *real-* (resp. *complex-*) *analytic vector bundle* over B. Everything goes through as before, with the exception of (16.15.1.2). When the base B is a *complex-analytic manifold*, the corresponding notion of vector bundle is of interest only when the fibers are *complex* vector spaces; such vector bundles are called *holomorphic* vector bundles. When M is a real- (resp. complex-) analytic manifold, the tangent bundle T(M) is a real-analytic (resp. holomorphic) vector bundle over M. All the developments of Sections 16.16 and 16.17 (with the exception of (16.17.3)) extend immediately to real-analytic and holomorphic vector bundles.

PROBLEMS

1. A differential manifold M is said to be *parallelizable* if the tangent bundle T(M) is trivializable. Show that the differential manifold underlying a Lie group is parallelizable. In particular, the spheres $\mathbf{S}_1$ and $\mathbf{S}_3$ are parallelizable.

2. Let $\mathbf{m}$ be a bilinear mapping of $\mathbf{R}^k \times \mathbf{R}^n$ into $\mathbf{R}^n$ such that

$$\|\mathbf{m}(\mathbf{y}, \mathbf{x})\| = \|\mathbf{y}\| \cdot \|\mathbf{x}\|$$

(the norms being Euclidean norms).

(a) For all $\mathbf{y}, \mathbf{y}'$ in $\mathbf{R}^k$, we have $(\mathbf{m}(\mathbf{y}, \mathbf{x}) \mid \mathbf{m}(\mathbf{y}', \mathbf{x})) = (\mathbf{y} \mid \mathbf{y}')\|\mathbf{x}\|^2$, and for all $\mathbf{x}, \mathbf{x}'$ in $\mathbf{R}^n$ we have $(\mathbf{m}(\mathbf{y}, \mathbf{x}) \mid \mathbf{m}(\mathbf{y}, \mathbf{x}')) = \|\mathbf{y}\|^2(\mathbf{x} \mid \mathbf{x}')$. For each $\mathbf{y} \neq 0$, the mapping $\mathbf{x} \mapsto \mathbf{m}(\mathbf{y}, \mathbf{x})$ therefore belongs to $\mathbf{O}(n, \mathbf{R})$. If $(\mathbf{e}_i)_{1 \leq i \leq k}$ is the canonical basis of $\mathbf{R}^k$ and if we put $\mathbf{v}(\mathbf{x}) = \mathbf{m}(\mathbf{e}_k, \mathbf{x})$, then the mapping $(\mathbf{x}, \mathbf{y}) \mapsto \mathbf{m}_0(\mathbf{y}, \mathbf{x}) = \mathbf{m}(\mathbf{y}, \mathbf{v}^{-1}\mathbf{x})$ has the same property as $\mathbf{m}$, and we have $\mathbf{m}_0(\mathbf{e}_k, \mathbf{x}) = \mathbf{x}$. In order that there should exist such a bilinear mapping $\mathbf{m}_0$, it is necessary and sufficient that there should exist $k-1$ elements $u_i \in \mathbf{O}(n, \mathbf{R})$ $(1 \leq i \leq k-1)$ such that $u^2 = -1$, $u_i u_j + u_j u_i = 0$ for $1 \leq i,j \leq n$ and $i \neq j$ (in the algebra $\mathrm{End}(\mathbf{R}^n)$). For each $x \in \mathbf{S}_{n-1}$, the $k-1$ vectors $(x, u_i(x)) \in \mathrm{T}_x(\mathbf{S}_{n-1})$ (where $\mathrm{T}_x(\mathbf{S}_{n-1})$ is identified with a subspace of $\mathrm{T}_x(\mathbf{R}^n)$) are then linearly independent, and so we have $k-1$ *vector fields* on $\mathbf{S}_{n-1}$ which are *linearly independent* at each point of $\mathbf{S}_{n-1}$.

(b) We shall assume the existence of an (associative) algebra $\mathbf{C}_m$ over $\mathbf{R}$, of dimension 2^m, generated by the unit element 1 and m elements c_i ($1 \leqq i \leqq m$) such that $c_i^2 = -1$, $c_i c_j + c_j c_i = 0$ for $i \neq j$, so that the element 1 and the products $c_{i_1} c_{i_2} \cdots c_{i_p}$ for

$$1 \leqq i_1 < i_2 < \cdots < i_p \leqq m$$

form a basis of $\mathbf{C}_m$. The algebra $\mathbf{C}_m$ is called the *Clifford algebra* of index m. In order that there should exist a mapping $\mathbf{m}$ with the property considered in (a) above, it is necessary and sufficient that there should exist a homomorphism of $\mathbf{C}_{k-1}$ into $\operatorname{End}(\mathbf{R}^n)$. (Observe that if Γ is the (finite) group generated by the images u_i of the c_i in $\mathbf{GL}(n, \mathbf{R})$, then the bilinear form $\sum_{s \in \Gamma} (s \cdot \mathbf{x} \mid s \cdot \mathbf{y})$ is positive-definite on $\mathbf{R}^n$ and invariant under Γ.)

(c) It can be shown that for $0 \leqq m \leqq 7$ the Clifford algebras $\mathbf{C}_m$ are given by the following table:

$m = 0$	1	2	3	4	5	6	7
$\mathbf{R}$	$\mathbf{C}$	$\mathbf{H}$	$\mathbf{H} \times \mathbf{H}$	$\mathbf{M}_2(\mathbf{H})$	$\mathbf{M}_4(\mathbf{C})$	$\mathbf{M}_8(\mathbf{R})$	$\mathbf{M}_8(\mathbf{R}) \times \mathbf{M}_8(\mathbf{R})$

and that $\mathbf{C}_{m+8}$ is isomorphic to $\mathbf{C}_m \otimes_{\mathbf{R}} \mathbf{M}_{16}(\mathbf{R})$. Deduce that if each integer $n \geqq 1$ is expressed in the form $n' \cdot 2^{c(n)} 16^{d(n)}$ with $0 \leqq c(n) \leqq 3$, $d(n) \geqq 0$, and n' odd, then there are $\rho(n) - 1 = 2^{c(n)} + 8d(n) - 1$ vector fields on $\mathbf{S}_{n-1}$ which are linearly independent at each point of $\mathbf{S}_{n-1}$. (Use the fact that, if K is a field or a division ring, the simple $\mathbf{M}_n(K)$-modules are of dimension n over K.) In particular, $\mathbf{S}_7$ is a parallelizable manifold (Problem 1). It can be shown that there do not exist $\rho(n)$ vector fields of class C^0 on $\mathbf{S}_{n-1}$ which are linearly independent at each point; in particular, the only parallelizable spheres are $\mathbf{S}_1$, $\mathbf{S}_3$, and $\mathbf{S}_7$.

3. The notion of a *bundle of* $\mathbf{R}$- (resp. $\mathbf{C}$-) *algebras* over B is defined as in (16.15.1), replacing F by a finite-dimensional $\mathbf{R}$- (resp. $\mathbf{C}$-) algebra, and the linear bijections in the condition (VB) by algebra isomorphisms. If X is any differential manifold, define a canonical algebra bundle structure on $\mathrm{P}^r(\mathrm{X}) = \mathrm{J}^r(\mathrm{X}, \mathbf{R})$ (Section 16.5, Problem 9).

4. Let B be a pure differential manifold and $\lambda = (\mathrm{X}, \mathrm{B}, p)$ a differential fibration over B, such that X is a pure manifold. For each $b \in \mathrm{B}$ let $\mathrm{P}_b^r(\mathrm{B}, \lambda)$ denote the subset of $\mathrm{J}_b^r(\mathrm{B}, \mathrm{X})$ consisting of jets of sections of X over a neighborhood of b, and let $\mathrm{P}^r(\mathrm{B}, \mathrm{X})$ denote the union of the $\mathrm{P}_b^r(\mathrm{B}, \lambda)$. Show that $\mathrm{P}^r(\mathrm{B}, \lambda)$ is a closed submanifold of $\mathrm{J}^r(\mathrm{B}, \lambda)$ (Section 16.12, Problem 6), and that the restrictions to $\mathrm{P}^r(\mathrm{B}, \lambda)$ of the projections of the fibrations of $\mathrm{J}^r(\mathrm{B}, \mathrm{X})$ define fibrations on $\mathrm{P}^r(\mathrm{B}, \lambda)$ over the graph of p in $\mathrm{X} \times \mathrm{B}$, B, and X, respectively.

If $s \leqq r$, the canonical mapping of $\mathrm{J}^r(\mathrm{B}, \mathrm{X})$ into $\mathrm{J}^{r-s}(\mathrm{B}, \mathrm{J}^s(\mathrm{B}, \mathrm{X}))$ (Section 16.12, Problem 6), restricted to $\mathrm{P}^r(\mathrm{B}, \lambda)$, is a B-morphism of $\mathrm{P}^r(\mathrm{B}, \lambda)$ into $\mathrm{P}^{r-s}(\mathrm{B}, \mu)$, where μ is the fibration $(\mathrm{B}, \mathrm{P}^s(\mathrm{B}, \lambda), \pi)$.

If $\lambda' = (\mathrm{X}', \mathrm{B}, p')$ is another fibration over B, and if $g : \mathrm{X}' \to \mathrm{X}$ is a B-morphism, then g defines canonically a B-morphism $\mathrm{P}^r(g) : \mathrm{P}^r(\mathrm{B}, \lambda) \to \mathrm{P}^r(\mathrm{B}, \lambda')$.

If E is a vector bundle over B and $\lambda = (\mathrm{E}, \mathrm{B}, p)$ is the corresponding fibration, we write $\mathrm{P}^r(\mathrm{B}, \mathrm{E})$ in place of $\mathrm{P}^r(\mathrm{B}, \lambda)$. Then $\mathrm{P}^r(\mathrm{B}, \mathrm{E})$ is canonically endowed with a vector bundle structure over B, and also with a module bundle structure over the algebra bundle $\mathrm{P}^r(\mathrm{B})$ (Problem 3; show how to define a module bundle over an algebra bundle). If $\mathrm{E} = \mathrm{B} \times \mathrm{F}$ is trivial, $\mathrm{P}^r(\mathrm{B}, \mathrm{E})$ is canonically identified with $\mathrm{J}^r(\mathrm{B}, \mathrm{F})$.

5. Let G be a Lie group acting differentiably on a differential manifold M. Show that if the orbit manifold M/G exists, then so does the orbit manifold $\mathrm{T}(\mathrm{M})/\mathrm{G}$ (cf. Section 16.19, Problem 5).

16. OPERATIONS ON VECTOR BUNDLES

In this section and the following one we shall consider only *real* vector bundles. The extension of the definitions and results to *complex* vector bundles is left to the reader.

(16.16.1) Let E', E'' be two vector bundles over the same base B, with projections π', π''. Let $E' \oplus E''$ (resp. $E' \otimes E''$) be the disjoint union of the sets $E'_b \oplus E''_b$ (resp. $E'_b \otimes E''_b$) as b runs through B, and let σ (resp. μ) denote the mapping $E' \oplus E'' \to B$ (resp. $E' \otimes E'' \to B$) which sends each element of $E'_b \oplus E''_b$ (resp. $E'_b \otimes E''_b$) to b. If U is open in B and if $\mathbf{s}'$, $\mathbf{s}''$ are sections of E', E'', respectively, over U, let $\mathbf{s}' \oplus \mathbf{s}''$ (resp. $\mathbf{s}' \otimes \mathbf{s}''$) denote the mapping $b \mapsto \mathbf{s}'(b) \oplus \mathbf{s}''(b)$ (resp. $b \mapsto \mathbf{s}'(b) \otimes \mathbf{s}''(b)$) of U into $E' \oplus E''$ (resp. $E' \otimes E''$). We shall show that $E' \oplus E''$ (resp. $E' \otimes E''$) carries a unique vector bundle structure with base B and projection σ (resp. μ), satisfying the following condition: for each open set U in B and each pair of C^∞-sections $\mathbf{s}'$, $\mathbf{s}''$, of E' and E'', respectively, over U, $\mathbf{s}' \oplus \mathbf{s}''$ (resp. $\mathbf{s}' \otimes \mathbf{s}''$) is a C^∞-section of $E' \oplus E''$ (resp. $E' \otimes E''$) over U. The bundle $E' \oplus E''$ is called the *sum* (or *Whitney sum*) of E' and E'', and $E' \otimes E''$ is called the *tensor product* of E' and E''.

Let us prove for example the existence of $E' \otimes E''$. Consider an open covering (U_α) of B such that for each α there exists a finite-dimensional vector space F'_α (resp. F''_α) and a diffeomorphism

$$\varphi'_\alpha : U_\alpha \times F'_\alpha \to \pi'^{-1}(U_\alpha) \qquad (\text{resp. } \varphi''_\alpha : U_\alpha \times F''_\alpha \to \pi''^{-1}(U_\alpha))$$

satisfying the condition (VB). For each pair of indices α, β we have transition diffeomorphisms (16.13.1)

(16.16.1.1)
$$\begin{cases} \psi'_{\beta\alpha} : (U_\alpha \cap U_\beta) \times F'_\alpha \to (U_\alpha \cap U_\beta) \times F'_\beta, \\ \psi''_{\beta\alpha} : (U_\alpha \cap U_\beta) \times F''_\alpha \to (U_\alpha \cap U_\beta) \times F''_\beta, \end{cases}$$

which are of the form $(b, \mathbf{t}') \to (b, f'_{\beta\alpha}(b, \mathbf{t}'))$ and $(b, \mathbf{t}'') \mapsto (b, f''_{\beta\alpha}(b, \mathbf{t}''))$, respectively, where for each $b \in U_\alpha \cap U_\beta$ the mappings $f'_{\beta\alpha}(b, \cdot)$ and $f''_{\beta\alpha}(b, \cdot)$ are *linear*. Now consider the mappings

(16.16.1.2)
$$\psi_{\beta\alpha} : (U_\alpha \cap U_\beta) \times (F'_\alpha \otimes F''_\alpha) \to (U_\alpha \cap U_\beta) \times (F'_\beta \otimes F''_\beta)$$

defined by

$$(b, \mathbf{t}) \mapsto (b, f_{\beta\alpha}(b, \mathbf{t})),$$

where, for each $b \in U_\alpha \cap U_\beta$, $f_{\beta\alpha}(b, \cdot) = f'_{\beta\alpha}(b, \cdot) \otimes f''_{\beta\alpha}(b, \cdot)$ (tensor product of linear mappings). It is immediately seen (by taking bases in F'_α and F''_α) that the $\psi_{\beta\alpha}$ are diffeomorphisms and satisfy the patching condition (16.13.1.1).

They therefore define (16.13.3) a fibration $(E' \otimes E'', B, \mu)$, and it is clear that $E' \otimes E''$ is thus endowed with a vector bundle structure satisfying the condition on the sections stated above. Conversely, suppose that $E' \otimes E''$ is endowed with a vector bundle structure satisfying this condition. With the notation used above, let $(\mathbf{e}'_{\alpha i})$ be a basis of F'_α and $(\mathbf{e}''_{\alpha j})$ a basis of F''_α. Then the sections $b \mapsto \varphi'_\alpha(b, \mathbf{e}'_{\alpha i})$ (resp. $b \mapsto \varphi''_\alpha(b, \mathbf{e}''_{\alpha j})$) form a *frame* $(\mathbf{s}'_{\alpha i})$ (resp. $(\mathbf{s}''_{\alpha j})$) of E' (resp. E'') over U_α, and the condition on the sections shows that the $\mathbf{s}'_{\alpha i} \otimes \mathbf{s}''_{\alpha j}$ form a *frame* for $E' \otimes E''$ over U_α. Let $\varphi_\alpha : U_\alpha \times (F'_\alpha \otimes F''_\alpha) \to \mu^{-1}(U_\alpha)$ be the diffeomorphism corresponding to this frame. It is then immediately verified that for any two indices α, β the transition diffeomorphisms

$$(U_\alpha \cap U_\beta) \times (F'_\alpha \otimes F''_\alpha) \to (U_\alpha \cap U_\beta) \times (F'_\beta \otimes F''_\beta)$$

corresponding to the φ_α are precisely the mappings (16.16.1.2). This establishes the uniqueness of the vector bundle structure on $E' \otimes E''$.

It follows moreover that if U is any open set in B such that $\pi'^{-1}(U)$ and $\pi''^{-1}(U)$ are trivializable, then $\Gamma(U, E' \otimes E'')$ is an $\mathscr{E}(U; \mathbf{R})$ module *isomorphic* to $\Gamma(U, E') \otimes_{\mathscr{E}(U; \mathbf{R})} \Gamma(U, E'')$.

If F', F'' are two vector bundles over B and if $u' : E' \to F'$ and $u'' : E'' \to F''$ are two B-morphisms, then one shows in the same way that there exists a unique B-morphism $u' \otimes u'' : E' \otimes E'' \to F' \otimes F''$ such that, if $\mathbf{s}'$ and $\mathbf{s}''$ are sections of E', E'', respectively, over an open set U in B, and if u'_U, u''_U and $(u' \otimes u'')_U$ are the restrictions of u', u'' and $u' \otimes u''$ to $\pi'^{-1}(U)$, $\pi''^{-1}(U)$, and $\mu^{-1}(U)$, respectively, then we have

$$(u' \otimes u'')_U \circ (\mathbf{s}' \otimes \mathbf{s}'') = (u'_U \circ \mathbf{s}') \otimes (u''_U \circ \mathbf{s}'').$$

The restriction of $u' \otimes u''$ to a fiber $(E' \otimes E'')_b = E'_b \otimes E''_b$ is the tensor product $u'_b \otimes u''_b$ of the linear mappings $u'_b : E'_b \to F'_b$ and $u''_b : E_b \to F''_b$. If u' and u'' are B-isomorphisms, then so is $u' \otimes u''$.

The proofs of the corresponding assertions for $E' \oplus E''$ are analogous and simpler. For each open $U \subset B$ such that $\pi'^{-1}(U)$ and $\pi''^{-1}(U)$ are trivializable, $\Gamma(U, E' \oplus E'')$ is an $\mathscr{E}(U; \mathbf{R})$-module isomorphic to

$$\Gamma(U, E') \oplus \Gamma(U, E'').$$

There are canonical injective B-morphisms $j' : E' \to E' \oplus E''$, $j'' : E'' \to E' \oplus E''$ and canonical surjective B-morphisms $p' : E' \oplus E'' \to E'$, $p'' : E' \oplus E'' \to E''$ such that, with the same notation as above,

$$j'_U \circ \mathbf{s}' = \mathbf{s}' \oplus \mathbf{0}_{E''}, \qquad j''_U \circ \mathbf{s}'' = \mathbf{0}_{E'} \oplus \mathbf{s}'',$$
$$p'_U \circ (\mathbf{s}' \oplus \mathbf{s}'') = \mathbf{s}', \qquad p''_U \circ (\mathbf{s}' \oplus \mathbf{s}'') = \mathbf{s}''.$$

Furthermore, if F is a vector bundle over B and if $u' : E' \to F$ and $u'' : E'' \to F$ are B-morphisms, then there exists a unique B-morphism $u : E' \oplus E'' \to F$

such that $u_U \circ (\mathbf{s}' \oplus \mathbf{s}'') = u'_U \circ \mathbf{s}' + u''_U \circ \mathbf{s}''$. We write $u = u' + u''$. If G', G'' are vector bundles over B and if $v' : G' \to E'$ and $v'' : G'' \to E''$ are B-morphisms, we denote by $v' \oplus v''$ the morphism $(j' \circ v') + (j'' \circ v'')$ of $G' \oplus G''$ into $E' \oplus E''$. For each $b \in B$ we have $(u' + u'')_b = u'_b + u''_b$, $(v' \oplus v'')_b = v'_b \oplus v''_b$.

Finally we remark that the *fibration* $(E' \oplus E'', B, \sigma)$ is isomorphic to the *fiber product* $E' \times_B E''$ defined in (16.12.10).

(16.16.2) The definitions of Whitney sum and tensor product generalize immediately to the case of any finite number of vector bundles over B. In particular, for each integer $m > 0$, we define a *multiple* mE (resp. a *tensor power* $E^{\otimes m}$) of a vector bundle E as the sum (resp. tensor product) of m copies of E. We use an analogous notation for B-morphisms.

Likewise, for each integer $m > 0$, the *exterior power* $\bigwedge^m E$ is defined: this space has for its underlying set the disjoint union of the sets $\bigwedge^m E_b$ as b runs through B, and its projection $\lambda : \bigwedge^m E \to B$ sends each element of $\bigwedge^m E_b$ to the point $b \in B$. For each sequence $(\mathbf{s}_j)_{1 \leq j \leq m}$ of m sections of E over an open set U in B, we denote by $\mathbf{s}_1 \wedge \mathbf{s}_2 \wedge \cdots \wedge \mathbf{s}_m$ the mapping

$$b \mapsto \mathbf{s}_1(b) \wedge \mathbf{s}_2(b) \wedge \cdots \wedge \mathbf{s}_m(b);$$

the vector bundle structure on $\bigwedge^m E$ is defined by the condition that, for each sequence $(\mathbf{s}_j)_{1 \leq j \leq m}$ of m sections of class C^∞ of E over an open subset U of B, the mapping $\mathbf{s}_1 \wedge \mathbf{s}_2 \wedge \cdots \wedge \mathbf{s}_m$ is a C^∞-section of $\bigwedge^m E$. If π is the projection $E \to B$, and if we are given an open covering (U_α) of B together with diffeomorphisms $\varphi_\alpha : U_\alpha \times F_\alpha \to \pi^{-1}(U_\alpha)$ satisfying (VB), then the fiber bundle $\bigwedge^m E$ may be constructed as follows: we take the transition homomorphisms $(b, \mathbf{t}) \mapsto (b, f_{\beta\alpha}(b, \mathbf{t}))$ corresponding to the φ_α, and we form the mappings $(b, \mathbf{t}) \mapsto (b, g_{\beta\alpha}(b, \mathbf{t}))$, where for each $b \in U_\alpha \cap U_\beta$, $g_{\beta\alpha}(b, \cdot) = \bigwedge^m f_{\beta\alpha}(b, \cdot)$.

If $(\mathbf{a}_i)_{1 \leq i \leq n}$ is a frame of E over an open set $U \subset B$, then the $\binom{n}{m}$ sections

$$\mathbf{a}_{i_1} \wedge \mathbf{a}_{i_2} \wedge \cdots \wedge \mathbf{a}_{i_m}$$

of $\bigwedge^m E$ such that $i_1 < i_2 < \cdots < i_m$ form a frame of $\bigwedge^m E$ over U.

If $u : E \to F$ is a B-morphism of vector bundles, there exists a unique B-morphism $\bigwedge^m u : \bigwedge^m E \to \bigwedge^m F$ such that, if $\mathbf{s}_1, \ldots, \mathbf{s}_m$ are sections of E over U, we have

$$(\bigwedge^m u) \circ (\mathbf{s}_1 \wedge \mathbf{s}_2 \wedge \cdots \wedge \mathbf{s}_m) = (u \circ \mathbf{s}_1) \wedge (u \circ \mathbf{s}_2) \wedge \cdots \wedge (u \circ \mathbf{s}_m).$$

For each $b \in \mathrm{B}$ we have $(\bigwedge^m u)_b = \bigwedge^m u_b$. If u is an isomorphism, then so is $\bigwedge^m u$.

Conventionally $\mathrm{E}^{\otimes 0}$, or $\bigwedge^0 \mathrm{E}$, or I, denotes the trivial line-bundle $\mathrm{B} \times \mathbf{R}$.

(16.16.3) With the notation of (16.16.1), let $\mathrm{Hom}(\mathrm{E}', \mathrm{E}'')$ denote the disjoint union of the sets $\mathrm{Hom}(\mathrm{E}'_b, \mathrm{E}''_b)$ (the set of all *linear mappings* of E'_b into E''_b) as b runs through B, and let $\eta : \mathrm{Hom}(\mathrm{E}', \mathrm{E}'') \to \mathrm{B}$ be the mapping which sends each element of $\mathrm{Hom}(\mathrm{E}'_b, \mathrm{E}''_b)$ to b. If $\mathbf{s}'$ (resp. $\mathbf{u}$) is a section of E' (resp. $\mathrm{Hom}(\mathrm{E}', \mathrm{E}'')$) over an open subset U of B, let $\mathbf{u}(\mathbf{s}')$ or $\mathbf{u} \cdot \mathbf{s}'$ denote the section $b \mapsto (\mathbf{u}(b))(\mathbf{s}'(b))$ of E'' over U. Then there exists on $\mathrm{Hom}(\mathrm{E}', \mathrm{E}'')$ a unique structure of a vector bundle over B with projection η, satisfying the following condition: for each open subset U of B and each pair of sections $\mathbf{s}'$, $\mathbf{u}$ of E' and $\mathrm{Hom}(\mathrm{E}', \mathrm{E}'')$, respectively, over U, if $\mathbf{s}'$ and $\mathbf{u}(\mathbf{s}')$ are C^∞-sections of E' and E'', respectively, over U, then $\mathbf{u}$ is a C^∞-section of $\mathrm{Hom}(\mathrm{E}', \mathrm{E}'')$ over U.

The proof is analogous to that given in (16.16.1) and we shall merely sketch it. With the same notation as before, consider this time the mappings

$$\omega_{\beta\alpha} : (\mathrm{U}_\alpha \cap \mathrm{U}_\beta) \times \mathrm{Hom}(\mathrm{F}'_\alpha, \mathrm{F}''_\alpha) \to (\mathrm{U}_\alpha \cap \mathrm{U}_\beta) \times \mathrm{Hom}(\mathrm{F}'_\beta, \mathrm{F}''_\beta)$$

of the form $(b, u) \mapsto (b, h_{\beta\alpha}(b, u))$, where for each $b \in \mathrm{U}_\alpha \cap \mathrm{U}_\beta$ we have $h_{\beta\alpha}(b, u) = f''_{\beta\alpha}(b, \cdot) \circ u \circ f'_{\beta\alpha}(b, \cdot)^{-1}$. We leave it to the reader to check the details, which are analogous to those in (16.16.1).

Consider in particular the trivial bundle $\mathrm{B} \times \mathbf{R}$ (16.15.3). If E is any vector bundle over B, the bundle $\mathrm{Hom}(\mathrm{E}, \mathrm{B} \times \mathbf{R})$ is called the *dual* of the bundle E and is denoted by E^*. If $\mathbf{s}$ and $\mathbf{s}^*$ are sections of E and E^*, respectively, over an open $\mathrm{U} \subset \mathrm{B}$, we write $\langle \mathbf{s}, \mathbf{s}^* \rangle$ or $\langle \mathbf{s}^*, \mathbf{s} \rangle$ in place of $\mathbf{s}^*(\mathbf{s})$. If $(\mathbf{a}_i)_{1 \leq i \leq n}$ is a frame of E over an open set $\mathrm{U} \subset \mathrm{B}$ (16.15.1), there exists a unique frame $(\mathbf{a}_i^*)_{1 \leq i \leq n}$ of E^* over U such that $\langle \mathbf{a}_i^*, \mathbf{a}_j \rangle = 0$ if $i \neq j$ and $\langle \mathbf{a}_i^*, \mathbf{a}_i \rangle = 1$ for all i. The frame $(\mathbf{a}_i^*)$ is called the *dual* of $(\mathbf{a}_i)$.

If $u' : \mathrm{F}' \to \mathrm{E}'$ and $u'' : \mathrm{E}'' \to \mathrm{F}''$ are two B-morphisms, there exists a unique B-morphism $w : \mathrm{Hom}(\mathrm{E}', \mathrm{E}'') \to \mathrm{Hom}(\mathrm{F}', \mathrm{F}'')$, denoted by $\mathrm{Hom}(u', u'')$, such that for each section $\mathbf{f}$ of $\mathrm{Hom}(\mathrm{E}', \mathrm{E}'')$ over an open $\mathrm{U} \subset \mathrm{B}$ and each section $\mathbf{s}'$ of F' over U, we have (in the notation of (16.16.1)) $(w_\mathrm{U} \circ \mathbf{f})(\mathbf{s}') = u''_\mathrm{U} \circ \mathbf{f}(u'_\mathrm{U} \circ \mathbf{s}')$. If u' and u'' are isomorphisms, then so is $\mathrm{Hom}(u', u'')$. In particular, for each B-morphism $u : \mathrm{E} \to \mathrm{F}$ of vector bundles, we define its *transpose* ${}^t u : \mathrm{F}^* \to \mathrm{E}^*$ by the condition

$$\langle {}^t u_\mathrm{U} \circ \mathbf{s}^*, \mathbf{r} \rangle = \langle \mathbf{s}^*, u_\mathrm{U} \circ \mathbf{r} \rangle$$

for any two sections $\mathbf{r}$ of E and $\mathbf{s}^*$ of F^* over an open subset U of B.

For each $b \in \mathrm{B}$ we have $\mathrm{Hom}(u', u'')_b = \mathrm{Hom}(u'_b, u''_b) : \mathbf{f}_b \mapsto u''_b \circ \mathbf{f}_b \circ u_b$, and in particular $({}^t u)_b = {}^t(u_b)$.

(16.16.4) *With the notation of* (16.16.3), *for each open* $\mathrm{U} \subset \mathrm{B}$ *and each* C^∞-*section* $\mathbf{u}$ *of* $\mathrm{Hom}(\mathrm{E}', \mathrm{E}'')$ *over* U, *let* $\tilde{\mathbf{u}}$ *denote the mapping*

$$(b, \mathbf{t}') \mapsto (b, (\mathbf{u}(b))(\mathbf{t}'))$$

of $\pi'^{-1}(\mathrm{U})$ *into* $\pi''^{-1}(\mathrm{U})$. *Then* $\tilde{\mathbf{u}}$ *is a* U-*morphism of* $\pi'^{-1}(\mathrm{U})$ *into* $\pi''^{-1}(\mathrm{U})$, *and the mapping* $\mathbf{u} \mapsto \tilde{\mathbf{u}}$ *is an isomorphism of the* $\mathscr{E}(\mathrm{U}; \mathbf{R})$-*module* $\Gamma(\mathrm{U}, \mathrm{Hom}(\mathrm{E}', \mathrm{E}''))$ *onto the* $\mathscr{E}(\mathrm{U}; \mathbf{R})$-*module* $\mathrm{Mor}(\pi'^{-1}(\mathrm{U}), \pi''^{-1}(\mathrm{U}))$.

Since the question is local with respect to B, we reduce immediately to the situation in which U = B and E′, E″ are trivial; in which case Hom(E′, E″) is also trivial, and then the result is obvious.

(16.16.5) We denote by $\mathbf{T}^p_q(\mathrm{E})$ the vector bundle $((\mathrm{E}^*)^{\otimes q}) \otimes (\mathrm{E}^{\otimes p})$. This bundle is called the *tensor bundle of type* (p, q) *over* E (p is the *contravariant index* and q the *covariant index*). Conventionally we put $\mathbf{T}^1_0(\mathrm{E}) = \mathrm{E}$, $\mathbf{T}^0_1(\mathrm{E}) = \mathrm{E}^*$, $\mathbf{T}^0_0(\mathrm{E}) = \mathrm{B} \times \mathbf{R}$. A section of $\mathbf{T}^p_q(\mathrm{E})$ over a subset A of B is sometimes called a *tensor field of type* (p, q) *on* A (relative to the bundle E). If $(\boldsymbol{a}_i)_{1 \leqq i \leqq n}$ is a frame of E over the open set $\mathrm{U} \subset \mathrm{B}$, and $(\boldsymbol{a}_i^*)_{1 \leqq i \leqq n}$ the dual frame (16.16.3), then the n^{p+q} tensor fields

$$\boldsymbol{a}^*_{k_1} \otimes \boldsymbol{a}^*_{k_2} \otimes \cdots \otimes \boldsymbol{a}^*_{k_q} \otimes \boldsymbol{a}_{j_1} \otimes \boldsymbol{a}_{j_2} \otimes \cdots \otimes \boldsymbol{a}_{j_p}$$

form a frame of $\mathbf{T}^p_q(\mathrm{E})$ over U, called the frame *induced* by $(\boldsymbol{a}_i)$.

If $u : \mathrm{E} \to \mathrm{F}$ is a B-*isomorphism*, we define $\mathbf{T}^p_q(u)$ to be the B-isomorphism of $\mathbf{T}^p_q(\mathrm{E})$ onto $\mathbf{T}^p_q(\mathrm{F})$ induced from u by transport of structure, that is to say the tensor product $(({}^t u^{-1})^{\otimes q}) \otimes (u^{\otimes p})$. (If $p > 0$ and $q > 0$, we cannot define $\mathbf{T}^p_q(u)$ for an arbitrary B-*morphism* u, because ${}^t u$ is a B-morphism of F* into E*, *not* of E* into F*.)

Remark

(16.16.6). It is also necessary to consider morphisms such as $u' \otimes u''$, $\bigwedge^p u$, $\mathrm{Hom}(u', u'')$ in the context of isomorphisms of vector bundles over different base manifolds. For example, let E′, E″ be two vector bundles over B_1, and F′, F″ two vector bundles over B_2; let (f, u') be an isomorphism of E′ onto F′, and (f, u'') an isomorphism of E″ onto F″, corresponding to the *same* diffeomorphism f of B_1 onto B_2. Then $(f, \mathrm{Hom}(u', u''))$ will be the isomorphism of Hom(E′, E″) onto Hom(F′, F″) which sends $\mathbf{h}_b \in \mathrm{Hom}(\mathrm{E}'_b, \mathrm{E}'')$ to the element $\mathbf{g}_{f(b)} : \mathbf{y}'_{f(b)} \mapsto u''_b(\mathbf{h}_b(u_b'^{-1}(\mathbf{y}'_{f(b)})))$ of $\mathrm{Hom}(\mathrm{F}'_b, \mathrm{F}''_b)$. We leave it to the reader to write down the definitions in the other cases.

PROBLEMS

1. For each differential manifold B, let I or I(B) denote the trivial real line-bundle $B \times \mathbf{R}$; it is an identity element for the tensor product of vector bundles over B. Likewise $I_{\mathbf{C}}$ or $I_{\mathbf{C}}(B)$ denotes the trivial complex line-bundle $B \times \mathbf{C}$. The sum mI (resp. $mI_{\mathbf{C}}$) of m copies of I (resp. $I_{\mathbf{C}}$) may be identified with the trivial bundle $B \times \mathbf{R}^m$ (resp. $B \times \mathbf{C}^m$).

(a) Consider the Grassmannian $\mathbf{G}_{n,p} = \mathbf{G}_{n,p}(\mathbf{R})$ and the subset $U_{n,p} = U_{n,p}(\mathbf{R})$ of the trivial bundle $nI(\mathbf{G}_{n,p}) = \mathbf{G}_{n,p} \times \mathbf{R}^n$ consisting of pairs (V, x) where V is a vector subspace of dimension p in $\mathbf{R}^n$, and $x \in V$. Show that $U_{n,p}$ is a vector subbundle of $nI(\mathbf{G}_{n,p})$, isomorphic to the bundle associated to the principal bundle $L_{n,p}$ (**16.14.2**) with structure group $\mathbf{GL}(p, \mathbf{R})$, with fiber-type $\mathbf{R}^p$ (for the canonical action of $\mathbf{GL}(p, \mathbf{R})$ on $\mathbf{R}^p$ on the left). Define in the same way the complex vector bundle $U_{n,p}(\mathbf{C})$ over $\mathbf{G}_{n,p}(\mathbf{C})$. The bundle $U_{n,p}(\mathbf{R})$ (resp. $U_{n,p}(\mathbf{C})$) is called the *canonical* (or *tautological*) *vector bundle* over $\mathbf{G}_{n,p}(\mathbf{R})$ (resp. $\mathbf{G}_{n,p}(\mathbf{C})$).

In particular, when $p = 1$ (so that $\mathbf{G}_{n,1}(\mathbf{R}) = \mathbf{P}_{n-1}(\mathbf{R})$, $\mathbf{G}_{n,1}(\mathbf{C}) = \mathbf{P}_{n-1}(\mathbf{C})$), we write $L_{n-1}(1)$ or $L_{n-1,\mathbf{R}}(1)$ in place of $U_{n,1}(\mathbf{R})$, and $L_{n-1,\mathbf{C}}(1)$ in place of $U_{n,1}(\mathbf{C})$. The principal bundle $P(D_1(\mathbf{R}))$ (resp. $P(D_1(\mathbf{C}))$) (Section **16.14**, Problem 5) may be identified with the complement of the zero section in $L_{n-1}(1)$ (resp. $L_{n-1,\mathbf{C}}(1)$). We denote by $L_{n-1}(k)$ (resp. $L_{n-1,\mathbf{C}}(k)$), for each integer $k > 0$, the tensor product of k copies of $L_{n-1}(1)$ (resp. $L_{n-1,\mathbf{C}}(1)$).

(b) If $n \geq 1$, the bundles $L_n(2)$ are trivializable as real-analytic bundles, but $L_n(1)$ is not trivializable as a differential bundle. (Argue by contradiction, by lifting a section which is $\neq 0$ at each point of the space $\mathbf{R}^{n+1} - \{0\}$; show that, for each $x \in \mathbf{P}_n(\mathbf{R})$, there exists a real-analytic section of $L_n(1)$ over the whole of $\mathbf{P}_n(\mathbf{R})$ which is $\neq 0$ at the point x.) On the other hand, the *holomorphic* bundles $L_{n,\mathbf{C}}(k)$ for $k \geq 1$ are pairwise non-isomorphic and admit *no* holomorphic section over $\mathbf{P}_n(\mathbf{C})$ other than the zero section (same method, using Section **9.10**, Problem 5). However, for each $x \in \mathbf{P}_n(\mathbf{C})$ there exists a real-analytic section of $L_{n,\mathbf{C}}(1)$ over $\mathbf{P}_n(\mathbf{C})$ which does not vanish at x.

(c) If we endow $\mathbf{R}^n$ with the Euclidean scalar product, then the mapping which sends each p-dimensional subspace $V \subset \mathbf{R}^n$ to its orthogonal supplement $V^{\perp}$, of dimension $n - p$, is a real-analytic isomorphism ω of $\mathbf{G}_{n,p}$ onto $\mathbf{G}_{n,n-p}$. Show that the direct sum $U_{n,p} \oplus \omega^*(U_{n,n-p})$ is a trivializable bundle over $\mathbf{G}_{n,p}$. Define in the same way a holomorphic isomorphism ω of $\mathbf{G}_{n,p}(\mathbf{C})$ onto $\mathbf{G}_{n,n-p}(\mathbf{C})$, but show that the analogous assertion about the canonical bundles is false (consider the isotropic vector subspaces of $\mathbf{C}^n$). On the other hand, there exists a *real*-analytic isomorphism ω_0 of $\mathbf{G}_{n,p}(\mathbf{C})$ onto $\mathbf{G}_{n,n-p}(\mathbf{C})$ such that $U_{n,p}(\mathbf{C}) \oplus \omega_0^*(U_{n,n-p}(\mathbf{C}))$ is trivializable as a real-analytic bundle.

2. (a) Let L be a real or complex line-bundle over a differential manifold B. Show that the tensor product $L \otimes L^*$ is trivializable (cf. (**16.18.3.5**)). Likewise for real- or complex-analytic line-bundles over a real-analytic manifold, and for holomorphic line-bundles over a complex manifold. By reason of this fact, vector bundles of rank 1 are also called *invertible* vector bundles; we write $L^{\otimes(-1)} = L^*$, $L^{\otimes(-k)} = (L^*)^{\otimes k}$ for all $k > 0$, with the convention that $L^{\otimes 0} = I$.

(b) The tensor product of k copies of $L_n(1)^*$ (resp. $L_{n,\mathbf{C}}(1)^*$) (Problem 1) is denoted by $L_n(-k)$ (resp. $L_{n,\mathbf{C}}(-k)$). Show that for each $k \geq 1$ and each $x \in \mathbf{P}_n(\mathbf{C})$ there exists a holomorphic section of $L_{n,\mathbf{C}}(-k)$ over $\mathbf{P}_n(\mathbf{C})$ which does not vanish at x.

(c) Let L be a (differential) real line-bundle over a differential manifold B. Show that $L \otimes L$ is trivializable. (Observe that if $(f_j)_{1 \leqq j \leqq r}$ is a finite sequence of sections of L over B, then $\sum_j f_j \otimes f_j$ is a section of $L \otimes L$ over B which vanishes only at the points where all the f_j vanish.) What is the analogous result for complex line-bundles?

17. EXACT SEQUENCES, SUBBUNDLES, AND QUOTIENT BUNDLES

All vector bundles considered in this section have the *same base* B, and all morphisms are B-morphisms.

(16.17.1) Let E be a vector bundle, E′ a subset of E. For each $b \in B$ let $E_b' = E' \cap E_b$. Then E′ is said to be a *subbundle* of E if the following two conditions are satisfied:

(1) For each $b \in B$, E_b' is a vector subspace of E_b.

(2) For each $b \in B$, there exists an open neighborhood U of b, a frame $(\mathbf{s}_1, \dots, \mathbf{s}_n)$ of E over U and an integer $m \leqq n$ such that, for each $y \in U$, the vectors $\mathbf{s}_1(y), \dots, \mathbf{s}_m(y)$ form a basis of E_y'.

Under these conditions:

(i) E′ is a closed submanifold of E.

(ii) If π' is the restriction to E′ of the projection $\pi : E \to B$, then the space E′ together with the fibration (E', B, π') and the vector space structures on the fibers E_b' is a *vector bundle* over B (which justifies the terminology introduced above).

It is enough to prove these assertions when E′ is replaced by $E' \cap \pi^{-1}(U)$, where U is as in (2) above; but then assertion (i) follows from (16.8.7(ii)), and (ii) is obvious, having regard to the condition (VB′).

It is clear that the canonical injection $j : E' \to E$ is a morphism of vector bundles.

Example

(16.17.1.1) Let (X, B, π) be a fibration, and consider the tangent bundle $E = T(X)$ of X. For each $x \in X$, let $V_x \subset E_x = T_x(X)$ be the subspace of *vertical* tangent vectors at the point x (16.12.1). Then the union V(X) of the V_x is a *subbundle* of T(X). For, since the question is local with respect to X, we may assume that $X = B \times F$; if $x = (b, z)$, let U be an open neighborhood of b in B, let W be an open neighborhood of z in F, and consider tangent vector fields Y_i $(1 \leqq i \leqq n)$ on U (resp. Z_j $(1 \leqq j \leqq m)$ on W) forming a frame of T(B) (resp. T(F)) over U (resp. W). Identifying T(X) with T(B) × T(F), we have a frame of T(X) over U × W, obtained by taking the vector fields $(b', z') \mapsto (Y_i(b'), 0)$ and $(b', z') \mapsto (0, Z_j(z'))$. For each $x' = (b', z') \in U \times W$, the vectors $(0, Z_j(z'))$ form a basis of $V_{x'}$, which proves our assertion (cf. Section **16.19**, Problem 4).

(16.17.2) With the hypotheses and notation of (**16.17.1**), let $E''_b = E_b/E'_b$ for all $b \in B$, and let E″ be the disjoint union of the sets E''_b as b runs through B. Let $\pi'' : E'' \to B$ be the mapping which sends each element of E''_b to b, and let $p : E \to E''$ be the mapping whose restriction to E_b is the canonical mapping $E_b \to E''_b$, for each $b \in B$. Then there exists on E″ a unique structure of a vector bundle over B with projection π'', such that p is a morphism of vector bundles. For if U is an open set in B satisfying condition (2) of (**16.17.1**), and if F_y is the vector subspace of E_y spanned by $\mathbf{s}_{m+1}(y), \ldots, \mathbf{s}_n(y)$, for each $y \in U$, then it is clear that the union F of the spaces F_y for $y \in U$ is a subbundle of $\pi^{-1}(U)$, and the restriction q of p to F must be an *isomorphism* of F onto $\pi''^{-1}(U)$, in view of (**16.12.2.1**). From this follows the uniqueness of the bundle structure on E″. On the other hand, if we put $\mathbf{s}''_k(y) = p(\mathbf{s}_k(y))$ for $m + 1 \leqq k \leqq n$ and all $y \in U$, it follows from above that $(\mathbf{s}''_{m+1}, \ldots, \mathbf{s}''_n)$ must be a *frame* of E″ over U. The existence of a vector bundle structure on E″ possessing these frames is then verified by the same method as in (**16.16.1**), and we leave the details to the reader.

The vector bundle E″ thus constructed is denoted by E/E′ and is called the *quotient* of E by the vector subbundle E′. It is clear that the canonical morphism $p : E \to E''$ is a submersion (**16.12.2.1**).

(16.17.3) With the notation and hypotheses of (**16.17.1**) and (**16.17.2**), *there exists a morphism* $r : E'' \to E$ *such that* $p \circ r = 1_{E''}$. Consider a locally finite denumerable open covering (U_α) of B such that each U_α satisfies condition (2) of (**16.17.1**). Let $E_\alpha = \pi^{-1}(U_\alpha)$, $E''_\alpha = \pi''^{-1}(U_\alpha)$, and let $p_\alpha : E_\alpha \to E''_\alpha$ be the restriction of p; then it is immediately clear that there exists a morphism $r_\alpha : E''_\alpha \to E_\alpha$ such that $p_\alpha \circ r_\alpha = 1_{E''_\alpha}$. Let (f_α) be a partition of unity subordinate to the covering (U_α) and consisting of C^∞-functions (**16.4.1**). For each index α, let $r'_\alpha : E'' \to E$ be the morphism whose restriction to E''_b is $f_\alpha(b)(r_\alpha | E''_b)$ when $b \in U_\alpha$, and 0 when $b \notin U_\alpha$. Then the morphism $r = \sum_\alpha r'_\alpha$ has the required property.

It follows immediately that the morphism $j + r : E' \oplus E'' \to E$ is an *isomorphism*; $F = r(E'')$ is a subbundle of E, such that E'_b and F_b are supplementary subspaces of E_b for each $b \in B$. Every vector subbundle of E with this property is called a *supplement* of E′ in E.

(16.17.4) Let E, F be two vector bundles over B, and $u : \mathrm{E} \to \mathrm{F}$ a B-morphism. For each $b \in \mathrm{B}$ let $u_b : \mathrm{E}_b \to \mathrm{F}_b$ denote the linear mapping which is the restriction of u to E_b. The rank $\mathrm{rk}(u_b)$ is called the *rank of u at b*. Put $\mathrm{N}_b = \mathrm{Ker}(u_b)$, $\mathrm{I}_b = \mathrm{Im}(u_b)$ and

$$\mathrm{N} = \bigcup_{b \in \mathrm{B}} \mathrm{N}_b \subset \mathrm{E}, \qquad \mathrm{I} = \bigcup_{b \in \mathrm{B}} \mathrm{I}_b \subset \mathrm{F}.$$

With this notation we have:

(16.17.5) (i) *The mapping* $b \mapsto \mathrm{rk}(u_b)$ *of* B *into the discrete space* **N** *is lower semicontinuous.*

(ii) *The following conditions are equivalent*:

(a) *The function* $b \mapsto \mathrm{rk}(u_b)$ *is continuous* (and therefore constant on each connected component of B).

(b) N *is a vector subbundle of* E.

(c) I *is a vector subbundle of* F.

Since the question is local with respect to B, we may assume that E and F are trivial, of the form $\mathrm{B} \times \mathrm{P}$ and $\mathrm{B} \times \mathrm{Q}$, respectively, where P and Q are vector spaces. Then u is of the form $(b, \mathbf{t}) \mapsto (b, \mathbf{f}(b, \mathbf{t}))$ where, for each $b \in \mathrm{B}$, $\mathbf{f}(b, \cdot)$ is a linear mapping of P into Q, and moreover the elements of the matrix of $\mathbf{f}(b, \cdot)$ relative to bases of P and Q are C^∞-functions on B. Since $\mathrm{rk}(u_b)$ is the rank of $\mathbf{f}(b, \cdot)$, it is the largest integer p such that there exists at least one nonzero $p \times p$ minor in the matrix of $\mathbf{f}(b, \cdot)$. Assertion (i) now follows immediately.

Since $\mathrm{rk}(u_b) = \dim(\mathrm{I}_b) = \mathrm{codim}(\mathrm{N}_b)$, it is clear that the conditions (b) and (c) in (ii) each imply (a). Conversely, suppose that (a) is satisfied; since the question is again local with respect to B, we may assume in addition that $\mathrm{rk}(u_b) = p$ is constant on B, and that there exists a basis $(\mathbf{e}_i)_{1 \leqq i \leqq n}$ of P, and a basis $(\mathbf{e}'_j)_{1 \leqq j \leqq m}$ of Q, such that for each $b \in \mathrm{B}$, I_b is a supplement of the subspace of $\{b\} \times \mathrm{Q}$ generated by the $(b, \mathbf{e}'_j)$ for $p + 1 \leqq j \leqq m$, and such that the vectors $(b, \mathbf{f}(b, \mathbf{e}_i))$ $(1 \leqq i \leqq p)$ form a basis of I_b. If

$$\mathbf{f}(b, \mathbf{e}_i) = \sum_{j=1}^{m} \alpha_{ij}(b)\mathbf{e}'_j \qquad \text{for} \quad 1 \leqq i \leqq n,$$

this signifies that the $p \times p$ minor $\Delta(b)$ in the matrix $(\alpha_{ij}(b))$, formed by the elements with both indices $\leqq p$, is nonzero. Then the fact that I is a vector subbundle of F follows because the sections

$$\begin{cases} b \mapsto (b, \mathbf{f}(b, \mathbf{e}_i)) & (1 \leqq i \leqq p), \\ b \mapsto (b, \mathbf{e}'_j) & (p + 1 \leqq j \leqq m) \end{cases}$$

form a frame of F satisfying condition (2) of (16.17.1) relative to I. On the other hand, Cramer's formulas show that

$$\mathbf{f}(b, \mathbf{e}_k) = \sum_{i=1}^{p} \beta_{ki}(b)\mathbf{f}(b, \mathbf{e}_i) \qquad (p+1 \leqq k \leqq n),$$

where the β_{ki} are of class C^∞; and the fact that N is a vector subbundle of E follows because the sections

$$\begin{cases} b \mapsto \left(b, \mathbf{e}_k - \sum_{i=1}^{p} \beta_{ki}(b)\mathbf{e}_i\right) & (p+1 \leqq k \leqq n), \\ b \mapsto (b, \mathbf{e}_i) & (1 \leqq i \leqq p) \end{cases}$$

form a frame of E satisfying condition (2) of (16.17.1) relative to N.

When the conditions of (16.17.5(ii)) are satisfied, the vector bundles N and I are called, respectively, the *kernel* and the *image* of u and are written $\mathrm{Ker}(u)$ and $\mathrm{Im}(u)$. If $p : E \to E/\mathrm{Ker}(u)$ and $j : \mathrm{Im}(u) \to F$ are the canonical morphisms, then the unique mapping $v : E/\mathrm{Ker}(u) \to \mathrm{Im}(u)$ such that $u = j \circ v \circ p$ is an *isomorphism* of vector bundles (16.15.2).

(16.17.6) Let $E \xrightarrow{u} F \xrightarrow{v} G$ be a sequence of two morphisms of vector bundles over B. The sequence is said to be *exact* if, for each $b \in B$, the sequence

$$E_b \xrightarrow{u_b} F_b \xrightarrow{v_b} G_b$$

of linear mappings is exact. A finite sequence

$$E_0 \xrightarrow{u_1} E_1 \to \cdots \to E_{n-1} \xrightarrow{u_n} E_n$$

of morphisms of vector bundles over B is said to be *exact* if each of the sequences $E_k \xrightarrow{u_{k+1}} E_{k+1} \xrightarrow{u_{k+2}} E_{k+2}$ $(0 \leqq k \leqq n-2)$ is exact.

If we denote by 0 the trivial bundle $B \times \{0\}$, then a morphism $u : E \to F$ of vector bundles over B is *injective* (resp. *surjective*) if and only if the sequence $0 \to E \xrightarrow{u} F$ (resp. $E \xrightarrow{u} F \to 0$) is exact. For each vector subbundle E′ of E, the sequence

$$0 \to E' \xrightarrow{j} E \xrightarrow{p} E/E' \to 0$$

(with the notation of (16.17.1) and (16.17.2)) is exact.

(16.17.7) *If* $E \xrightarrow{u} F \xrightarrow{u} G$ *is an exact sequence of morphisms of vector bundles over* B, *then* u *and* v *satisfy the equivalent conditions of* (16.17.5(ii)), *and the bundles* $\mathrm{Im}(u)$ *and* $\mathrm{Ker}(v)$ *are therefore defined and equal.*

It is enough to observe that $\mathrm{rk}(u_b) + \mathrm{rk}(v_b) = \dim F_b$ is a continuous function of b; since $\mathrm{rk}(u_b)$ and $\mathrm{rk}(v_b)$ are lower semicontinuous as functions of b, it follows that they are continuous.

In particular, for each *injective* (resp. *surjective*) morphism $u : E \to F$, $\mathrm{Im}(u)$ (resp. $\mathrm{Ker}(u)$) is defined, and E is isomorphic to $\mathrm{Im}(u)$ (resp. F is isomorphic to $E/\mathrm{Ker}(u)$).

(16.17.8) If E', E'' are two vector bundles over B, then E' (resp. E'') may be identified with a subbundle of $E' \oplus E''$ by means of the canonical injection j' (resp. j'') (16.16.1), and also with a quotient bundle of $E' \oplus E''$ by means of the canonical surjection p'' (resp. p').

(16.17.9) The properties of tensor products and spaces of linear mappings relative to exact sequences in the category of vector spaces give rise to analogous properties of vector bundles: if $E' \to E \to E''$ is an exact sequence of vector bundles over B and if F is a vector bundle over B, then the sequences

$$E' \otimes F \to E \otimes F \to E'' \otimes F,$$
$$\mathrm{Hom}(E'', F) \to \mathrm{Hom}(E, F) \to \mathrm{Hom}(E', F),$$
$$\mathrm{Hom}(F, E') \to \mathrm{Hom}(F, E) \to \mathrm{Hom}(F, E'')$$

are exact.

Examples

(16.17.10) Let E be a vector bundle over B and let m be an integer >0. For each $b \in B$, let $\mathbf{S}_m(E_b)$ (resp. $\mathbf{A}_m(E_b)$) be the subspace of $E_b^{\otimes m}$ consisting of *symmetric* (resp. *antisymmetric*) tensors. By considering a frame of E over an open subset U of B, and the corresponding frame of $E^{\otimes m}$, it is immediately verified that the union $\mathbf{S}_m(E)$ (resp. $\mathbf{A}_m(E)$) of the $\mathbf{S}_m(E_b)$ (resp. the $\mathbf{A}_m(E_b)$) is a vector subbundle of $E^{\otimes m}$, called the *bundle of symmetric* (resp. *antisymmetric*) (*contravariant*) *tensors of index m*.

18. CANONICAL MORPHISMS OF VECTOR BUNDLES

Except in (16.18.5), we shall again restrict our discussion to real vector bundles, and leave it to the reader to develop the corresponding results for complex vector bundles.

(16.18.1) Let E′, E″, F be three vector bundles over the same base manifold B, and let π', π'', π be their projections onto B. Consider the vector bundle $E' \oplus E''$ and its projection σ. A B-morphism u of the *fibration* $(E' \oplus E'', B, \sigma)$ into the *fibration* (F, B, π) is said to be *bilinear* if, for each $b \in B$, the restriction $u_b : E'_b \oplus E''_b \to F_b$ is a bilinear mapping.

For each open subset U of B, the mapping $(\mathbf{s}', \mathbf{s}'') \mapsto u_U \circ (\mathbf{s}' \oplus \mathbf{s}'')$ is an $\mathscr{E}(U; \mathbf{R})$-*bilinear* mapping of $\Gamma(U, E') \times \Gamma(U, E'')$ into $\Gamma(U, F)$.

Multilinear B-morphisms are defined in the same way.

In particular, there exists a unique bilinear B-morphism

$$m : E' \oplus E'' \to E' \otimes E''$$

such that, for each pair of sections $\mathbf{s}'$, $\mathbf{s}''$ of E′ and E″ over an open set $U \subset B$, we have $m(\mathbf{s}' \oplus \mathbf{s}'') = \mathbf{s}' \otimes \mathbf{s}''$. This follows immediately from the definitions of the vector bundles $E' \oplus E''$ and $E' \otimes E''$ (16.16.1) and the local definition of a morphism of fibrations (16.13.5). The morphism m is called *canonical.*

Furthermore, given any *bilinear* B-morphism $u : E' \oplus E'' \to F$, there exists a unique *linear* B-morphism $v : E' \otimes E'' \to F$ such that $u = v \circ m$. Here again, by virtue of (16.13.5), the proof reduces to the case of trivial bundles and rests in the last analysis on the corresponding algebraic proposition and the fact that polynomials are functions of class C^∞ in $\mathbf{R}^n$.

(16.18.2) The preceding argument applies in the same way to all the *canonical* linear or multilinear mappings defined in algebra, and provides *canonical* linear or multilinear B-morphisms correspondingly. Moreover, whenever a canonical linear mapping defined in algebra is *bijective* for finite-dimensional vector spaces, the corresponding B-morphism is an *isomorphism* (16.15.2).

We shall restrict the following discussion to defining the most important of these canonical B-morphisms by characterizing their effect on sections. First of all, we have the *isomorphisms* of *associativity* and *distributivity*:

(16.18.2.1) $$E_1 \otimes (E_2 \otimes E_3) \to (E_1 \otimes E_2) \otimes E_3,$$

which maps a section $\mathbf{s}_1 \otimes (\mathbf{s}_2 \otimes \mathbf{s}_3)$ over an open set U to the section

$$(\mathbf{s}_1 \otimes \mathbf{s}_2) \otimes \mathbf{s}_3;$$

(16.18.2.2) $$(E_1 \oplus E_2) \otimes E_3 \to (E_1 \otimes E_3) \oplus (E_2 \otimes E_3),$$

which maps $(\mathbf{s}_1 \oplus \mathbf{s}_2) \otimes \mathbf{s}_3$ to $(\mathbf{s}_1 \otimes \mathbf{s}_3) \oplus (\mathbf{s}_2 \otimes \mathbf{s}_3)$;

(16.8.2.3) $$\mathrm{Hom}(E \otimes F, G) \to \mathrm{Hom}(E, \mathrm{Hom}(F, G))$$

such that if $\mathbf{s}'$, $\mathbf{s}''$, $\mathbf{u}$ are sections of E, F, $\mathrm{Hom}(E \otimes F, G)$, respectively, over an open set U, then the image of $\mathbf{u}$ is the section $\mathbf{v}$ of $\mathrm{Hom}(E, \mathrm{Hom}(F, G))$ such that

$$(\mathbf{v}(\mathbf{s}''))(\mathbf{s}') = \mathbf{u}(\mathbf{s}' \otimes \mathbf{s}'');$$

(16.18.2.4) $$\mathrm{Hom}(E' \oplus E'', F) \to \mathrm{Hom}(E', F) \oplus \mathrm{Hom}(E'', F),$$

such that if $\mathbf{s}'$, $\mathbf{s}''$, $\mathbf{u}$ are sections of E′, E″, $\mathrm{Hom}(E' \oplus E'', F)$, respectively, over an open set U, then the image of $\mathbf{u}$ is the section $\mathbf{v}' \oplus \mathbf{v}''$, where $\mathbf{v}'(\mathbf{s}') = \mathbf{u}(\mathbf{s}')$, $\mathbf{v}''(\mathbf{s}'') = \mathbf{u}(\mathbf{s}'')$;

(16.18.2.5) $$\mathrm{Hom}(E, F') \oplus \mathrm{Hom}(E, F'') \to \mathrm{Hom}(E, F' \oplus F'')$$

such that if $\mathbf{s}$, $\mathbf{v}'$, $\mathbf{v}''$ are sections of E, $\mathrm{Hom}(E, F')$, and $\mathrm{Hom}(E, F'')$, respectively, over an open set U, then the image of $\mathbf{v}' \oplus \mathbf{v}''$ is the section $\mathbf{v}$ such that $\mathbf{v}(\mathbf{s}) = \mathbf{v}'(\mathbf{s}) \oplus \mathbf{v}''(\mathbf{s})$;

(16.18.2.6) $$\mathrm{Hom}(E', F') \otimes \mathrm{Hom}(E'', F'') \to \mathrm{Hom}(E' \otimes E'', F' \otimes F'')$$

such that if $\mathbf{s}'$, $\mathbf{s}''$, $\mathbf{u}'$, $\mathbf{u}''$ are sections of E′, E″, $\mathrm{Hom}(E', F')$, and $\mathrm{Hom}(E'', F'')$, respectively, over an open set U, then to $\mathbf{u}' \otimes \mathbf{u}''$ there corresponds the section $\mathbf{u}$ such that $\mathbf{u}(\mathbf{s}' \otimes \mathbf{s}'') = \mathbf{u}'(\mathbf{s}') \otimes \mathbf{u}''(\mathbf{s}'')$.

(16.18.3) Second, we have the isomorphisms related to *duality*:

(16.18.3.1) $$E \to E^{**}$$

such that, if $\mathbf{s}$, $\mathbf{s}^*$ are sections of E, E*, respectively, over an open set U, then to $\mathbf{s}$ there corresponds the section $\tilde{\mathbf{s}}$ of E** such that $\langle \mathbf{s}^*, \mathbf{s} \rangle = \langle \tilde{\mathbf{s}}, \mathbf{s}^* \rangle$;

(16.18.3.2) $$\mathrm{Hom}(E, F) \to \mathrm{Hom}(F^*, E^*)$$

such that, if $\mathbf{r}$, $\mathbf{s}^*$, $\mathbf{u}$ are sections of E, F*, and $\mathrm{Hom}(E, F)$, respectively, then to $\mathbf{u}$ there corresponds the section ${}^t\mathbf{u}$ such that $\langle \mathbf{u}(\mathbf{r}), \mathbf{s}^* \rangle = \langle \mathbf{r}, {}^t\mathbf{u}(\mathbf{s}^*) \rangle$;

(16.18.3.3) $$(E \otimes F)^* \to E^* \otimes F^*,$$

which is a particular case of (16.18.2.6);

(16.18.3.4) $$E^* \otimes F \to \mathrm{Hom}(E, F)$$

such that, if $\boldsymbol{r}$, $\boldsymbol{r}^*$, and $\boldsymbol{s}$ are sections of E, E*, and F, respectively, then $\boldsymbol{r}^* \otimes \boldsymbol{s}$ is mapped to the section $\boldsymbol{u}$ of Hom(E, F) such that $\boldsymbol{u}(\boldsymbol{r}) = \langle \boldsymbol{r}, \boldsymbol{r}^* \rangle \boldsymbol{s}$.

In particular, we have a canonical isomorphism of End(E) on $E^* \otimes E$, under which the morphism 1_E corresponds to a section of $E^* \otimes E = \mathbf{T}_1^1(E)$, canonically associated with E, and called the *Kronecker tensor field*; if $(\boldsymbol{a}_i)_{1 \leq i \leq n}$ is a frame of E over an open set U, and $(\boldsymbol{a}_i^*)$ the dual frame (16.16.2), then the Kronecker tensor field is

$$\sum_{i,j} \delta_i^j \boldsymbol{a}_i^* \otimes \boldsymbol{a}_j,$$

where $\delta_i^j = 0$ if $i \neq j$ and $\delta_i^i = 1$ for all i.

Finally, there is a canonical B-morphism

$$E^* \otimes E \to B \times \mathbf{R} \tag{16.18.3.5}$$

such that if $\mathbf{s}^*$, $\mathbf{s}$ are sections of E*, E, respectively, then $\mathbf{s}^* \otimes \mathbf{s}$ is mapped to the section $\langle \mathbf{s}^*, \mathbf{s} \rangle$ of the trivial bundle.

These morphisms may be combined with each other and with the preceding ones to give canonical morphisms between tensor bundles over a vector bundle E (16.16.5). First of all we have *tensor multiplication*, which is a *bilinear* morphism

$$\mathbf{T}_q^p(E) \oplus \mathbf{T}_s^r(E) \to \mathbf{T}_{q+s}^{p+r}(E) \tag{16.18.3.6}$$

which maps a pair of sections

$$\mathbf{t}_1^* \otimes \cdots \otimes \mathbf{t}_q^* \otimes \mathbf{t}_1 \otimes \cdots \otimes \mathbf{t}_p, \qquad \mathbf{t}_1'^* \otimes \cdots \otimes \mathbf{t}_s'^* \otimes \mathbf{t}_1' \otimes \cdots \otimes \mathbf{t}_r'$$

to the section

$$\mathbf{t}_1^* \otimes \cdots \otimes \mathbf{t}_q^* \otimes \mathbf{t}_1'^* \otimes \cdots \otimes \mathbf{t}_s'^* \otimes \mathbf{t}_1 \otimes \cdots \otimes \mathbf{t}_p \otimes \mathbf{t}_1' \otimes \cdots \otimes \mathbf{t}_r'.$$

This bilinear morphism corresponds canonically (16.18.1) to the associativity isomorphism (16.18.2.1):

$$\mathbf{T}_q^p(E) \otimes \mathbf{T}_s^r(E) \to \mathbf{T}_{q+s}^{p+r}(E). \tag{16.18.3.7}$$

Next, we have a canonical *isomorphism*

$$\mathbf{T}_p^q(E) \otimes \mathbf{T}_s^r(E) \to \mathrm{Hom}(\mathbf{T}_q^p(E), \mathbf{T}_s^r(E)) \tag{16.18.3.8}$$

which is obtained from (16.18.3.4) by using the canonical isomorphism

(16.18.3.9) $$(\mathbf{T}_q^p(E))^* \to \mathbf{T}_p^q(E)$$

which in turn comes from (16.18.3.3), (16.18.3.1), and the associativity isomorphisms. From all this it follows that a *bilinear* B-*morphism* of

$$\mathbf{T}_q^p(E) \oplus \mathbf{T}_s^r(E)$$

into $\mathbf{T}_n^m(E)$ corresponds canonically to a *global* C^∞-*section of the tensor bundle* $\mathbf{T}_{p+r+n}^{q+s+m}(E)$.

Let $p, q \geqq 1$ and let j, k be such that $1 \leqq j \leqq p$, $1 \leqq k \leqq q$. Then there is a canonical linear morphism which generalizes (16.18.3.5), called *contraction of the indices j and k*:

(16.18.3.10) $$c_k^j : \mathbf{T}_q^p(E) \to \mathbf{T}_{q-1}^{p-1}(E)$$

which maps a section $\mathbf{t}_1^* \otimes \cdots \otimes \mathbf{t}_q^* \otimes \mathbf{t}_1 \otimes \cdots \otimes \mathbf{t}_p$ to the section

$$\langle \mathbf{t}_k^*, \mathbf{t}_j \rangle \mathbf{t}_1^* \otimes \cdots \otimes \mathbf{t}_{k-1}^* \otimes \mathbf{t}_{k+1}^* \otimes \cdots \otimes \mathbf{t}_q^* \otimes \mathbf{t}_1 \otimes \cdots \otimes \mathbf{t}_{j-1} \otimes \mathbf{t}_{j+1} \otimes \cdots \otimes \mathbf{t}_p.$$

In particular, the bundle $\text{End}(E) = \text{Hom}(E, E)$ is identified with $\mathbf{T}_1^1(E) = E^* \otimes E$, and a B-morphism $u : E \to E$ is identified with a global section of $\mathbf{T}_1^1(E)$ (16.16.4). The contraction $c_1^1(u)$, which is a section of the trivial line-bundle $B \times \mathbf{R}$, is called the *trace* of u and is written $\text{Tr}(u)$; its value at $b \in B$ is the number $\text{Tr}(u_b)$. Likewise, the section $u(\mathbf{s})$ of E, where $\mathbf{s}$ is a given section of E, is identified with $c_1^2(u \otimes \mathbf{s})$ ($u \otimes \mathbf{s}$ being a section of $\mathbf{T}_1^2(E)$).

(16.18.4) We shall now consider B-morphisms connected with the exterior algebra. First of all, there is a canonical B-morphism

(16.18.4.1) $$\lambda : E^{\otimes m} \to \bigwedge^m E$$

such that, if $\mathbf{s}_1, \ldots, \mathbf{s}_m$ are sections of E over an open set $U \subset B$, the section $\mathbf{s}_1 \otimes \mathbf{s}_2 \otimes \cdots \otimes \mathbf{s}_m$ is mapped to the section $\mathbf{s}_1 \wedge \mathbf{s}_2 \wedge \cdots \wedge \mathbf{s}_m$, so that λ is *surjective*. There exists a canonical B-morphism $\boldsymbol{a} : E^{\otimes m} \to E^{\otimes m}$, whose image is a supplement of $\text{Ker}(\lambda)$ (16.17.3), which is such that $\boldsymbol{a} \circ \boldsymbol{a} = \boldsymbol{a}$ and which maps a section $\mathbf{s}_1 \otimes \cdots \otimes \mathbf{s}_m$ of $E^{\otimes m}$ to its *antisymmetrization*

$$\frac{1}{m!} \sum_{\sigma \in \mathfrak{S}_m} \varepsilon_\sigma \mathbf{s}_{\sigma(1)} \otimes \mathbf{s}_{\sigma(2)} \otimes \cdots \otimes \mathbf{s}_{\sigma(m)}$$

(here $\mathfrak{S}_m$ denotes the symmetric group of all permutations of the set $\{1, 2, \ldots, m\}$). This B-morphism enables us to identify canonically $\bigwedge^m E$ with the *subbundle* $\mathbf{A}_m(E)$ of $E^{\otimes m}$ (16.17.10).

Next, let $\mathbf{s}$, $\mathbf{t}$ be sections of $\bigwedge^p \mathrm{E}$, $\bigwedge^q \mathrm{E}$ over U (where $p \geqq 1$, $q \geqq 1$). We denote by $\mathbf{s} \wedge \mathbf{t}$ the section of $\bigwedge^{p+q} \mathrm{E}$ over U such that $(\mathbf{s} \wedge \mathbf{t})(b) = \mathbf{s}(b) \wedge \mathbf{t}(b)$ for all $b \in \mathrm{U}$; $\mathbf{s} \wedge \mathbf{t}$ is called the *exterior product* of $\mathbf{s}$ and $\mathbf{t}$. We then have a canonical B-morphism

(16.18.4.2) $$\Big(\bigwedge^p \mathrm{E}\Big) \otimes \Big(\bigwedge^q \mathrm{E}\Big) \to \bigwedge^{p+q} \mathrm{E}$$

which maps $\mathbf{s} \otimes \mathbf{t}$ to $\mathbf{s} \wedge \mathbf{t}$. If we denote by $\bigwedge \mathrm{E}$ the direct sum of the bundles $\bigwedge^p \mathrm{E}$ for $0 \leqq p \leqq \mathrm{rk}(\mathrm{E})$, then $\Gamma(\mathrm{U}, \bigwedge \mathrm{E})$ is the direct sum of the $\Gamma(\mathrm{U}, \bigwedge^p \mathrm{E})$ and is endowed by the B-morphisms (16.18.4.2) with a structure of an *anti-commutative graded* $\mathscr{E}(\mathrm{U}; \mathbf{R})$-*algebra*. If E is trivializable over U, so that $\Gamma(\mathrm{U}, \mathrm{E})$ is a *free* $\mathscr{E}(\mathrm{U}; \mathbf{R})$-module, then $\Gamma(\mathrm{U}, \bigwedge \mathrm{E})$ is canonically identified with the exterior algebra $\bigwedge(\Gamma(\mathrm{U}, \mathrm{E}))$ of this module.

Next, there is a canonical isomorphism

(16.18.4.3) $$\delta : \bigwedge^m(\mathrm{E}^*) \to \Big(\bigwedge^m \mathrm{E}\Big)^*$$

such that, if $\mathbf{s}_1, \ldots, \mathbf{s}_m$ are sections of E and $\mathbf{s}_1^*, \ldots, \mathbf{s}_m^*$ are sections of E^* over an open set $\mathrm{U} \subset \mathrm{B}$, we have

(16.18.4.4) $$\langle \mathbf{s}_1 \wedge \mathbf{s}_2 \wedge \cdots \wedge \mathbf{s}_m, \delta(\mathbf{s}_1^* \wedge \mathbf{s}_2^* \wedge \cdots \wedge \mathbf{s}_m^*)\rangle = \det(\langle \mathbf{s}_i, \mathbf{s}_j^* \rangle).$$

We shall henceforth identify $\bigwedge^m(\mathrm{E}^*)$ with $(\bigwedge^m \mathrm{E})^*$ by means of δ, and we shall write them both as $\bigwedge^m \mathrm{E}^*$.

Finally, if $\mathbf{s}$ is a section of E and $\mathbf{z}^*$ a section of $\bigwedge^p \mathrm{E}^*$ over U (where $p \geqq 1$), we denote by $i(\mathbf{s})\mathbf{z}^*$ or $i_{\mathbf{s}} \cdot \mathbf{z}^*$, the section of $\bigwedge^{p-1} \mathrm{E}^*$ whose value at each $b \in \mathrm{U}$ is the *interior product* $\mathbf{s}(b) \,\lrcorner\, \mathbf{z}^*(b)$ in $\bigwedge^{p-1} \mathrm{E}_b^*$; this section is called the *interior product* of $\mathbf{s}$ and $\mathbf{z}^*$. If $\mathbf{s}_1, \ldots, \mathbf{s}_{p-1}$ are sections of E over U, we have

(16.18.4.5)
$$\langle i(\mathbf{s})\mathbf{z}^*, \mathbf{s}_1 \wedge \mathbf{s}_2 \wedge \cdots \wedge \mathbf{s}_{p-1}\rangle = \langle \mathbf{z}^*, \mathbf{s} \wedge \mathbf{s}_1 \wedge \cdots \wedge \mathbf{s}_{p-1}\rangle$$

and in particular, when $p = 1$,

(16.18.4.6) $$i(\mathbf{s})\mathbf{s}^* = \langle \mathbf{s}^*, \mathbf{s}\rangle$$

for any section $\mathbf{s}^*$ of E^*. For arbitrary p and sections $\mathbf{s}_1^*, \ldots, \mathbf{s}_p^*$ of E^* over U, we have

(16.18.4.7)

$$i(\mathbf{s})(\mathbf{s}_1^* \wedge \mathbf{s}_2^* \wedge \cdots \wedge \mathbf{s}_p^*) = \sum_{j=1}^{p} (-1)^{j+1} \langle \mathbf{s}_j^*, \mathbf{s} \rangle \mathbf{s}_1^* \wedge \cdots \wedge \hat{\mathbf{s}}_j^* \wedge \cdots \wedge \mathbf{s}_p^*$$

(with the usual convention that the symbol under the circumflex is to be omitted). Also we have

(16.18.4.8) $$i(\mathbf{s}) \circ i(\mathbf{s}) = 0,$$

(16.18.4.9) $$i(\mathbf{s})(\mathbf{z}_p^* \wedge \mathbf{z}_q^*) = (i(\mathbf{s})\mathbf{z}_p^*) \wedge \mathbf{z}_q^* + (-1)^p \mathbf{z}_p^* \wedge (i(\mathbf{s})\mathbf{z}_q^*),$$

where $\mathbf{z}_p^*$, $\mathbf{z}_q^*$ are sections of $\bigwedge^p E^*$, $\bigwedge^q E^*$, respectively, over U. This last formula may be expressed by saying that for each section $\mathbf{s}$ of E over U, $i(\mathbf{s})$ is an *antiderivation* of degree -1 of the $\mathscr{E}(U; \mathbf{R})$-algebra $\Gamma(U, \bigwedge E^*)$.

We have then, for each integer $p \geqq 1$, a canonical B-morphism

(16.18.4.10) $$E \otimes \left(\bigwedge^p E^*\right) \to \bigwedge^{p-1} E^*$$

such that, if $\mathbf{s}$ and $\mathbf{z}^*$ are two sections as above, then $\mathbf{s} \otimes \mathbf{z}^*$ is mapped to the section $i(\mathbf{s})\mathbf{z}^*$.

(16.18.5) Finally, we consider B-morphisms connected with extension of scalars (from $\mathbf{R}$ to $\mathbf{C}$). The (real) vector bundle E may be canonically identified with the tensor product $E \otimes (B \times \mathbf{R})$; it follows immediately from the definitions that the tensor product $E \otimes (B \times \mathbf{C})$ (wherein $B \otimes \mathbf{C}$ is considered as a *real* vector bundle) admits a canonical structure of a *complex* vector bundle of the same rank as E at each point of B. This complex vector bundle is called the *complexification* of E and is written $E_{(\mathbf{C})}$. We have a canonical injective B-morphism

(16.18.5.1) $$E \to E_{(\mathbf{C})},$$

under which a section $\mathbf{s}$ of E over U is mapped to the section $\mathbf{s} \otimes 1$ of $E_{(\mathbf{C})}$. We remark that the *real* vector bundle underlying $E_{(\mathbf{C})}$ is the *sum* $E \oplus iE$ of its two real subbundles E and iE.

Let E, E′ be two real vector bundles over B. Then we have canonical isomorphisms

(16.18.5.2) $$E_{(C)} \otimes_C E'_{(C)} \to (E \otimes E')_{(C)}$$

(16.18.5.3) $$(\mathrm{Hom}(E, E'))_{(C)} \to \mathrm{Hom}(E_{(C)}, E'_{(C)}),$$

(16.18.5.4) $$(E^*)_{(C)} \to (E_{(C)})^*,$$

(16.18.5.5) $$\Big(\bigwedge^m E\Big)_{(C)} \to \bigwedge^m (E_{(C)}),$$

which on the sections reduce to the usual linear mappings. For example, if $\boldsymbol{u}$ is a section of Hom(E, E′) over U, then to $\boldsymbol{u} \otimes 1$ there corresponds under (16.18.5.3) the section $\mathbf{v}$ of $\mathrm{Hom}(E_{(C)}, E'_{(C)})$ defined as follows: if $\mathbf{s}$ is a section of E and ξ a section of $B \times \mathbf{C}$ over U, then $\mathbf{v}(\mathbf{s} \otimes \xi) = \mathbf{u}(\mathbf{s}) \otimes \xi$. In particular, if $\mathbf{s}^*$ is a section of E^*, $\mathbf{s}$ a section of E, λ and μ two sections of $B \times \mathbf{C}$, the identification of $\mathbf{s}^* \otimes \lambda$ with a section of $(E_{(C)})^*$ identifies $\langle \mathbf{s}^* \otimes \lambda, \mathbf{s} \otimes \mu \rangle$ with $\langle \mathbf{s}^*, \mathbf{s} \rangle \lambda\mu$.

19. INVERSE IMAGE OF A VECTOR BUNDLE

Let E be a real (resp. complex) vector bundle over B, and let $\lambda = (E, B, \pi)$ be the corresponding fibration. Let B′ be a differential manifold, $f: B' \to B$ a C^∞-mapping, and consider the *inverse image* $f^*(\lambda) = (E', B', \pi')$ of λ under f, and the canonical morphism (f, f') of $f^*(\lambda)$ into λ (16.12.8). We recall that, for each $b' \in B'$, the restriction $f'_{b'} : E'_{b'} \to E_{f(b')}$ of f' is a bijection; we can therefore transport to $E'_{b'}$ the real (resp. complex) vector space structure on $E_{f(b')}$ by means of the inverse bijection $f'^{-1}_{b'}$. We shall show that, with these vector space structures and the fibration $f^*(\lambda)$, E′ is a real (resp. complex) *vector bundle* over B′. We reduce immediately to the situation where $E = B \times F$ is trivial, and then it is immediate that $E' = G \times F$, where $G \subset B' \times B$ is the *graph* of f. If p is the restriction to G of the projection $\mathrm{pr}_1 : B' \times B \to B'$, then the mapping $p^{-1} \times 1_F : B' \times F \to G \times F = E'$ is a diffeomorphism which evidently satisfies the condition (VB) of (16.15); this proves our assertion. The bundle E′ is called the *inverse image of the vector bundle* E *by the mapping* f and is denoted by $f^*(E)$. It has the following "universal property" relative to f:

(16.19.1) *Let* E_1 *be a real* (resp. *complex*) *vector bundle over* B′, *and let* $g : E_1 \to E$ *be a* C^∞-*mapping such that* (f, g) *is a morphism of vector bundles* (16.15.2). *Then there exists a unique* B′-*morphism* $u : E_1 \to f^*(E)$ *such that* $g = f' \circ u$. *If* $g_{b'} : (E_1)_{b'} \to E_{f(b')}$ *is bijective for all* $b' \in B'$, *then* u *is a* B′-*isomorphism.*

This follows immediately from (16.12.8(iii)) and the definition of the vector space structures on the fibers of $f^*(\mathrm{E})$.

If $\mathbf{s}_1, \ldots, \mathbf{s}_n$ are sections of E over U which form a *frame* (16.15.1) over U, then their inverse images $\mathbf{s}'_1, \ldots, \mathbf{s}'_n$ under f (16.12.8) form a frame of $f^*(\mathrm{E})$ over $f^{-1}(\mathrm{U})$.

If E, F are two vector bundles over B and if $v : \mathrm{E} \to \mathrm{F}$ is a B-morphism (resp. a B-isomorphism) of vector bundles, then $f^*(v) : f^*(\mathrm{E}) \to f^*(\mathrm{F})$ (16.12.8) is a B′-morphism (resp. a B′-isomorphism) of vector bundles.

A particular case is that in which B′ is a *submanifold* of B and f is the canonical injection; in that case $f^*(\mathrm{E})$ is said to be the vector bundle *induced* by E on B′.

Examples

(16.19.2) Let X be a differential manifold, Y a submanifold of X, $j : \mathrm{Y} \to \mathrm{X}$ the canonical injection, and consider the morphism $(j, \mathrm{T}(j)) : \mathrm{T}(\mathrm{Y}) \to \mathrm{T}(\mathrm{X})$ of vector bundles (16.15.5), which is an immersion; hence there exists a unique Y-morphism $u : \mathrm{T}(\mathrm{Y}) \to j^*(\mathrm{T}(\mathrm{X}))$ such that $\mathrm{T}(j) = j' \circ u$, where

$$j' : j^*(\mathrm{T}(\mathrm{X})) \to \mathrm{T}(\mathrm{X})$$

is the canonical mapping. It is clear that the mapping $y \mapsto \mathrm{rk}(u_y) = \dim_y(\mathrm{Y})$ is locally constant, so that u is an isomorphism of T(Y) onto a *vector subbundle* of $j^*(\mathrm{T}(\mathrm{X}))$. Usually we shall identify T(Y) with its image in $j^*(\mathrm{T}(\mathrm{X}))$; the *quotient bundle* $j^*(\mathrm{T}(\mathrm{X}))/\mathrm{T}(\mathrm{Y})$ is called the *normal bundle of* Y *in* X (this name will be justified in Chapter XX) (cf. Problem 5).

(16.19.3) Let E_1, E_2 be two vector bundles over B. With the notation introduced at the beginning of this section, consider the canonical injections $j_1 : \mathrm{E}_1 \to \mathrm{E}_1 \oplus \mathrm{E}_2$, $j_2 : \mathrm{E}_2 \to \mathrm{E}_1 \oplus \mathrm{E}_2$ (16.16.1). To them correspond injections $f^*(j_1) : f^*(\mathrm{E}_1) \to f^*(\mathrm{E}_1 \oplus \mathrm{E}_2)$ and $f^*(j_2) : f^*(\mathrm{E}_2) \to f^*(\mathrm{E}_1 \oplus \mathrm{E}_2)$, and hence we have a B′-morphism

$$f^*(j_1) + f^*(j_2) : f^*(\mathrm{E}_1) \oplus f^*(\mathrm{E}_2) \to f^*(\mathrm{E}_1 \oplus \mathrm{E}_2).$$

It is immediately verified (by reducing to the case where E_1 and E_2 are trivial bundles) that this is a B′-*isomorphism*. We shall usually *identify* $f^*(\mathrm{E}_1) \oplus f^*(\mathrm{E}_2)$ with $f^*(\mathrm{E}_1 \oplus \mathrm{E}_2)$ by means of this isomorphism.

If $m : \mathrm{E}_1 \oplus \mathrm{E}_2 \to \mathrm{E}_1 \otimes \mathrm{E}_2$ is the canonical bilinear B-morphism (16.18.1), then $f^*(m) : f^*(\mathrm{E}_1) \oplus f^*(\mathrm{E}_2) \to f^*(\mathrm{E}_1 \otimes \mathrm{E}_2)$ is a bilinear B-morphism, and therefore factorizes into

$$f^*(\mathrm{E}_1) \oplus f^*(\mathrm{E}_2) \xrightarrow{m'} f^*(\mathrm{E}_1) \otimes f^*(\mathrm{E}_2) \xrightarrow{v} f^*(\mathrm{E}_1 \otimes \mathrm{E}_2),$$

where v is a linear B′-morphism. It is immediately checked that v is a B′-isomorphism, which permits us to *identify* $f^*(E_1) \otimes f^*(E_2)$ with $f^*(E_1 \otimes E_2)$. Under this identification, if $\mathbf{s}_1, \mathbf{s}_2$ are sections of E_1, E_2, respectively, over an open subset U of B, $f^*(\mathbf{s}_1) \otimes f^*(\mathbf{s}_2)$ is identified with $f^*(\mathbf{s}_1 \otimes \mathbf{s}_2)$ (sections over $f^{-1}(U)$).

Likewise, we have a canonical B′-isomorphism

$$f^*(\mathrm{Hom}(E_1, E_2)) \to \mathrm{Hom}(f^*(E_1), f^*(E_2));$$

if $\boldsymbol{u}$ is any section of $\mathrm{Hom}(E_1, E_2)$ over U, this isomorphism identifies $f^*(\boldsymbol{u})$ with the section $\mathbf{v}$ of $\mathrm{Hom}(f^*(E_1), f^*(E_2))$ such that $\mathbf{v}(f^*(\mathbf{s}_1)) = f^*(\boldsymbol{u}(\mathbf{s}_1))$ for all sections $\mathbf{s}_1$ of E_1 over U. In particular, $f^*(E^*)$ is thus identified with the dual bundle $(f^*(E))^*$, for any vector bundle E over B.

Finally, with the same notation, for each integer $p \geqq 1$ we have a B′-isomorphism $f^*(\bigwedge^p E) \to \bigwedge^p f^*(E)$, which identifies $f^*(\mathbf{s}_1 \wedge \mathbf{s}_2 \wedge \cdots \wedge \mathbf{s}_p)$ with $f^*(\mathbf{s}_1) \wedge \cdots \wedge f^*(\mathbf{s}_p)$ ($\mathbf{s}_1, \ldots, \mathbf{s}_p$ being sections of E over U).

PROBLEMS

1. (a) Let $U_{n,p}^{\perp}$ be the subset of the trivial bundle nI over the Grassmannian $\mathbf{G}_{n,p}$ consisting of all pairs (V, x), where V is a p-dimensional subspace of $\mathbf{R}^n$ and x is orthogonal to V (with respect to the Euclidean scalar product on $\mathbf{R}^n$). Show that $U_{n,p}^{\perp}$ is a vector subbundle of nI and that $U_{n,p}^{\perp} \oplus U_{n,p} = n\mathrm{I}$, in the notation of Section 16.16, Problem 1.
(b) Show that the inverse image $\omega^*(U_{n,n-p})$ (Section 16.16, Problem 1(c)) is isomorphic to $U_{n,p}^{\perp}$.

2. (a) Define a canonical injection of $\mathbf{G}_{n,p}$ into $\mathbf{G}_{n+m,p}$ and show that the inverse image of $U_{n+m,p}$ under this injection is isomorphic to $U_{n,p}$.
(b) Define a canonical injection of $\mathbf{G}_{n,p}$ into $\mathbf{G}_{n+m,p+m}$ and show that the inverse image of $U_{n+m,p+m}$ under this injection is isomorphic to $U_{n,p} \oplus m\mathrm{I}$.

3. (a) Show that the normal bundle of $\mathbf{S}_n$ (considered as a submanifold of $\mathbf{R}^{n+1}$) is trivializable.
(b) Define a canonical injection of $\mathbf{S}_n$ into $\mathbf{S}_{n+p}$ and show that the normal bundle of $\mathbf{S}_n$ in $\mathbf{S}_{n+p}$ is trivializable.

4. Let (E, B, π), (E', B', π') be two real (resp. complex) vector bundles and let (f, g) be a morphism of the first into the second (16.15.2). Suppose that the kernel of $g_b : E_b \to E'_{f(b)}$ has locally constant dimension. Show that the union of the $\mathrm{Ker}(g_b)$ for $b \in B$ is a vector subbundle of E (factorize g by using (16.19.1)). This bundle is called the *kernel* of the morphism (f, g).
In particular, let M, M′ be differential manifolds, and f: M → M′ a *submersion* of M into M′. Then the morphism $(f, \mathrm{T}(f))$ of vector bundles (16.15.6) has a kernel.

More particularly, let (X, B, π) be a fibration. Then the kernel of the morphism (π, T(π)) is the bundle V(X) of vertical tangent vectors (**16.17.1.1**), and the quotient bundle T(X)/V(X) over X is isomorphic to the inverse image T(π)*(T(B)).

5. (a) Let M be a submanifold of $\mathbf{R}^n$, so that for each $x \in$ M the tangent space $T_x(M)$ is contained in $T_x(\mathbf{R}^n)$. Let $N_x(M)$ be the orthogonal supplement of $T_x(M)$ in $T_x(\mathbf{R}^n)$ (with respect to the scalar product obtained by transporting the Euclidean scalar product on $\mathbf{R}^n$ by means of the canonical isomorphism $\boldsymbol{\tau}_x^{-1}$ (**16.5.2**)). Show that the $N_x(M)$ are the fibers of a vector subbundle (over M) of $j^*(T(\mathbf{R}^n))$ (where $j: M \to \mathbf{R}^n$ is the canonical injection), isomorphic to the normal bundle of M in $\mathbf{R}^n$.
(b) Suppose that M is a fiber bundle over B. For each $b \in$ B and $x \in M_b$, let $P_x(M)$ be the orthogonal supplement in $T_x(M)$ of the subspace $T_x(M_b)$ tangent to the fiber M_b (with respect to the scalar product on $T_x(\mathbf{R}^n)$ defined in (a)). Show that the $P_x(M)$ are the fibers of a subbundle P(M) of T(M), supplementary to the subbundle of vertical tangent vectors V(M) (**16.17.1.1**).

6. Let M be a submanifold of $\mathbf{R}^n$, and suppose that M is endowed with a fibration (M, B, π) making it a principal bundle for a compact group G; suppose also that G is a subgroup of the orthogonal group $\mathbf{O}(n)$ and that, for each $x \in$ M and $s \in$ G, $x \cdot s$ is the image of $x \in \mathbf{R}^n$ under the orthogonal transformation s^{-1}. Then the group G acts freely on T(M), N(M), V(M), and P(M), in the notation of Problem 5, and orbit manifolds exist for all these actions, and are canonically endowed with vector bundle structures over B (cf. Section **20.1** and Section **16.14**, Problem 15(b)).
Show that G\P(M) may be canonically identified with T(B), and that

$$G\backslash(T(M) \oplus N(M))$$

may be canonically identified with $M \times^G \mathbf{R}^n$. Deduce that

$$T(B) \oplus (G\backslash V(M)) \oplus (G\backslash N(M))$$

is isomorphic to $M \times^G \mathbf{R}^n$.

7. Consider the principal bundle ($\mathbf{S}_n$, $\mathbf{P}_n(\mathbf{R})$, π) with structure group $\mathbf{Z}/2\mathbf{Z}$, identified with the subgroup of $\mathbf{O}(n+1)$ consisting of the identity and the symmetry $x \mapsto -x$. Show that the associated vector bundle $\mathbf{S}_n \times^{\mathbf{Z}/2\mathbf{Z}} \mathbf{R}$ is isomorphic to the canonical vector bundle $L_n(1)$ on $\mathbf{P}_n(\mathbf{R})$ (Section **16.16**, Problem 1). (Consider the mapping $(x, t) \mapsto (\pi(x), tx)$ of $\mathbf{S}_n \times \mathbf{R}$ into $\mathbf{P}_n(\mathbf{R}) \times \mathbf{R}^{n+1}$.) Deduce that $T(\mathbf{P}_n(\mathbf{R})) \oplus I$ (notation of Section **16.16**, Problem 1) is isomorphic to $(n+1)L_n(1)$ (use Problem 6). State and prove the analogous result for $\mathbf{P}_n(\mathbf{C})$.

8. Let E be a real vector bundle of rank k, with base B and projection π.

(a) Show that there exists a canonical one-to-one correspondence between C^∞-mappings $u: E \to \mathbf{R}^m$ whose restriction to each fiber E_b is linear and injective (such mappings are called *Gaussian*) and B-isomorphisms of E into $f^*(U_{m,k})$, where $f: B \to \mathbf{G}_{m,k}$ is a C^∞-mapping (the notation is that of Section **16.16**, Problem 1). (Define $f(b)$ to be $u(E_b)$.)
(b) Deduce from (a) that if V is any relatively compact open subset of B, there exists an integer N such that the bundle $\pi^{-1}(V)$ is B-isomorphic to an inverse image $f^*(U_{N,k})$. (Cover V by a finite number of open sets over which E is trivializable, and use a partition of unity.)

9. Let E be a real vector bundle of rank n over a differential manifold B, with projection π. The multiplicative group $\mathbf{R}^*$ acts differentiably on the open subset $E - \mathbf{O}(B)$ of E by the rule $(c, u_b) \mapsto cu_b$. Show that the orbit manifold $(E - \mathbf{O}(B))/\mathbf{R}^*$ exists. We denote it by $\mathbf{P}(E)$. If $p : \mathbf{P}(E) \to B$ is the mapping obtained from π by passing to the quotient, show that $(\mathbf{P}(E), B, p)$ is a fibration whose fibers are diffeomorphic to $\mathbf{P}_{n-1}(\mathbf{R})$; also that the transition diffeomorphisms are diffeomorphisms of $V \times \mathbf{P}_{n-1}(\mathbf{R})$ onto itself (where V is an open set in B) of the form $(x, z) \mapsto (x, u(x, z))$, where $u(x, \cdot)$ belongs to the projective group $\mathbf{PGL}(n, \mathbf{R}) = \mathbf{GL}(n, \mathbf{R})/\mathbf{R}^*$ ($\mathbf{R}^*$ being identified with the center of $\mathbf{GL}(n, \mathbf{R})$). $\mathbf{P}(E)$ is called the *projective bundle* defined by E.

The *canonical line bundle* over $\mathbf{P}(E)$ is the subbundle $L_E(1)$ of $p^*(E)$ consisting of all pairs $(z, \mathbf{u}) \in \mathbf{P}(E) \times E$ such that $p(z) = \pi(\mathbf{u})$ and such that either $\mathbf{u} = 0$ or else $\mathbf{u}$ has z as its orbit under the action of $\mathbf{R}^*$ on $E - \mathbf{O}(B)$. Show that if we identify one fiber $\mathbf{P}(E)_b$ with $\mathbf{P}_{n-1}(\mathbf{R})$, the bundle induced by $L_E(1)$ on $\mathbf{P}(E)_b$ is isomorphic to the canonical line-bundle $L_{n-1}(1)$ (Section **16.16**, Problem 1).

10. Let E be a vector bundle of rank n, L a subbundle of E of rank 1, and F a supplement of L in E. Show that $\bigwedge^m E$ is canonically isomorphic to $\left(\bigwedge^m F\right) \oplus \left(L \oplus \bigwedge^{m-1} F\right)$.

11. Let E be a vector bundle with base B and projection π. For each $b \in B$ and each $\mathbf{u}_b \in E_b$, let $\tau_{\mathbf{u}_b} : T_{\mathbf{u}_b}(E_b) \to E_b$ denote the canonical bijection (**16.5.2**) of the space of vertical tangent vectors at the point $\mathbf{u}_b$ onto the fiber E_b. Show that for each pair of vectors $\mathbf{u}_b, \mathbf{v}_b$ in E_b there exists a unique vertical tangent vector $\lambda(\mathbf{u}_b, \mathbf{v}_b) \in T_{\mathbf{u}_b}(E_b)$ such that $\tau_{\mathbf{u}_b}(\lambda(\mathbf{u}_b, \mathbf{v}_b)) = \mathbf{v}_b$. Considering $\pi^*(E) = E \times_B E$ as a vector bundle over E, show that $\lambda : E \times_B E \to T(E)$ is an injective E-morphism of vector bundles whose image is the bundle V(E) of vertical tangent vectors (**16.17.1.1**).

For each tangent vector $\mathbf{k}_{\mathbf{u}_b} \in T_{\mathbf{u}_b}(E)$, let $\mu(\mathbf{k}_{\mathbf{u}_b})$ denote the point $(\mathbf{u}_b, T(\pi) \cdot \mathbf{k}_{\mathbf{u}_b})$ of $E \times_B T(B)$. Show that if we consider $E \times_B T(B) = \pi^*(T(B))$ as a vector bundle over E, then $\mu : T(E) \to E \times_B T(B)$ is an E-morphism of vector bundles and that the sequence

$$0 \to E \times_B E \xrightarrow{\lambda} T(E) \xrightarrow{\mu} E \times_B T(B) \to 0$$

is exact.

12. (a) Let (X, B, π), (X', B, π') be two fibrations with the same base B, and let $f : X \to X'$ be a B-morphism which is an *embedding* of X into X'. Show that the inverse image $f^*(V(X'))$ of the bundle of vertical tangent vectors to X' is isomorphic to $V(X) \oplus N(X)$, where N(X) is the normal bundle of X in X'.

(b) Suppose moreover that (X', B, π') is a vector bundle over B. Show that

$$V(X) \oplus N(X)$$

is isomorphic to $\pi^*(X')$ (use Problem 11).

13. The group $G = \{1, -1\}$ acts analytically on $\mathbf{C}^n$, the action of the element -1 being $\mathbf{z} \mapsto -\mathbf{z}$.

(a) Consider the holomorphic mapping $f : \mathbf{C}^n \to \mathbf{C}^N$, where $N = \frac{1}{2}n(n+1)$, defined by $f(\zeta^1, \zeta^2, \dots, \zeta^n) = (\zeta^i \zeta^j)_{1 \leq i \leq j \leq n}$. Show that f factorizes into

$$\mathbf{C}^n \xrightarrow{\pi} \mathbf{C}^n/G \xrightarrow{j} \mathbf{C}^N,$$

where π is the canonical mapping onto the orbit space (12.10.6) and j is a homeomorphism onto a closed subset V_n of $\mathbf{C}^N$. The canonical image of $V_n - \{0\}$ in the projective space $\mathbf{P}_{N-1}(\mathbf{C})$ is analytically isomorphic to $\mathbf{P}_{n-1}(\mathbf{C})$.

(b) The complex manifold obtained by blowing up the point 0 in $\mathbf{C}^N$ (Section 16.11, Problem 3) is analytically isomorphic to the canonical bundle $L_{N-1,\mathbf{C}}(1)$. If

$$q : L_{N-1,\mathbf{C}}(1) \to \mathbf{C}^N$$

is the canonical projection of this blowing-up (it is a local isomorphism everywhere outside the fiber $q^{-1}(0)$), show that $q^{-1}(V_n)$ is analytically isomorphic to the bundle $L_{n-1,\mathbf{C}}(2)$ over $\mathbf{P}_{n-1}(\mathbf{C})$ (Section 16.16, Problem 1).

20. DIFFERENTIAL FORMS

(16.20.1) Let M be a differential manifold. The dual $T(M)^*$ of the tangent bundle $T(M)$ is called the *cotangent bundle* of M. If F is the transition diffeomorphism between two charts on M, then, as we have seen, the transition diffeomorphism between the associated fibered charts of $T(M)$ is

$$(x, \mathbf{h}) \mapsto (F(x), DF(x) \cdot \mathbf{h})$$

(16.15.4.5). Hence the transition diffeomorphism between the associated fibered charts of $T(M)^*$ is $(x, \mathbf{h}^*) \mapsto (F(x), {}^tDF(x)^{-1} \cdot \mathbf{h}^*)$.

We shall write $\mathbf{T}_q^p(M)$ in place of $\mathbf{T}_q^p(T(M))$ when there is no risk of confusion; in particular, therefore, $\mathbf{T}_0^1(M) = T(M)$ and $\mathbf{T}_1^0(M) = T(M)^*$. A section of $\mathbf{T}_q^p(M)$ over a subset A of M is called a *tensor field* (or, by abuse of language, a *tensor*) of *type* (p, q) over A. The set $\Gamma(M, \mathbf{T}_q^p(M))$ of tensor fields of class C^∞ over M is denoted by $\mathscr{T}_q^p(M)$ or $\mathscr{T}_{q,\mathbf{R}}^p(M)$; it is a module over the ring $\mathscr{E}(M) = \mathscr{E}(M; \mathbf{R})$ (also denoted by $\mathscr{E}_{\mathbf{R}}(M)$) of real-valued C^∞-functions on M, and is a *free* module when $T(M)$ is trivializable (16.15.8).

For each $p \geqq 1$, a section over A of the bundle $\bigwedge^p T(M)^*$ of tangent p-covectors is called a *differential p-form* on A (or simply a *differential form* when $p = 1$). The set $\Gamma(M, \bigwedge^p T(M)^*)$ of differential p-forms *of class* C^∞ on M is denoted by $\mathscr{E}_p(M)$ or by $\mathscr{E}_{p,\mathbf{R}}(M)$. It is a module over $\mathscr{E}(M)$, and is free when $T(M)$ is trivializable. We have $\mathscr{T}_1^0(M) = \mathscr{E}_1(M)$, and $\mathscr{E}_p(M)$ may be identified with the module of *antisymmetric p-covariant tensor fields* (16.18.4) of class C^∞ over M.

Example

(16.20.2) Let f be a real-valued function of class C^r ($r \geqq 1$) on M, so that for each point $x \in M$ the vector $d_x f \in T_x(M)^*$ is a tangent covector at x (16.5.7). Then the mapping $x \mapsto d_x f$ is a *differential form of class* C^{r-1} on M (with the convention that $r - 1 = \infty$ if $r = \infty$); it is denoted by df and is called the *differential* of f. By means of a chart we reduce immediately to the case where M is an open subset of $\mathbf{R}^n$, and hence T(M) is identified with $M \times \mathbf{R}^n$ (16.15.5). If we denote by $(\mathbf{e}_i^*)$ the basis dual to the canonical basis $(\mathbf{e}_i)$ of $\mathbf{R}^n$, then it follows from (16.5.7.1) that the differential form df is the mapping

(16.20.2.1) $$x \mapsto \left(x, \sum_{i=1}^{n} \mathrm{D}_i f(x) \cdot \mathbf{e}_i^*\right)$$

which proves that it is of class C^{r-1}. Hence, by (16.5.7)

(16.20.2.2) $$\langle df(x), \mathbf{h}_x \rangle = \boldsymbol{\tau}_{f(x)}(\mathrm{T}_x(f) \cdot \mathbf{h}_x) \in \mathbf{R}.$$

If f and g are two real-valued C^∞-functions on M, then clearly we have

(16.20.2.3) $$d(fg) = g \cdot df + f \cdot dg.$$

(16.20.3) Let $c = (U, \varphi, n)$ be a chart on M, where $\varphi = (\varphi^i)_{1 \leqq i \leqq n}$. Then it follows from (16.5.7) and (16.5.4) that the n differential forms $d\varphi^i$ $(1 \leqq i \leqq n)$ form a *frame* of $T(M)^*$ over U, called the frame *associated* with the chart c. This frame is the *dual* of the frame $(X_i)_{1 \leqq i \leqq n}$ of T(M) associated with c ((16.15.4.2) and (16.16.3)); in other words,

(16.20.3.1) $$\langle d\varphi^i, X_j \rangle = \delta_i^j \qquad \text{(Kronecker delta).}$$

It follows that every differential form ω on U is uniquely expressible in the form

(16.20.3.2) $$x \mapsto \omega(x) = \sum_{i=1}^{n} a_i(x) \cdot d\varphi^i(x),$$

where the a_i are n scalar functions on U. The form ω is of class C^r if and only if the functions a_i are of class C^r.

The framing defined by the sections $d\varphi^i$ is

(16.20.3.3) $$\psi_c^* : \quad (x, \xi_1, \ldots, \xi_n) \mapsto \sum_{i=1}^{n} \xi_i \, d\varphi^i(x)$$

and $(o_M^{-1}(U), (\varphi \times 1_{\mathbf{R}^n}) \circ (\psi^*)_c^{-1}, 2n)$ is a *fibered chart* of $T(M)^*$ (16.15.1), associated with the chart c.

If a differential form ω on U is given by (16.20.3.2), its *local expression* relative to the fibered chart associated with (U, φ, n) is the mapping

(16.20.3.4) $$z \mapsto \left(z, \sum_{i=1}^{n} a_i(\varphi^{-1}(z))\mathbf{e}_i^*\right)$$

of $\varphi(\mathrm{U})$ into $\varphi(\mathrm{U}) \times (\mathbf{R}^n)^*$.

(16.20.4) With the same notation, every tensor field $\boldsymbol{Z}$ of type (p, q) on U is uniquely expressible in the form

(16.20.4.1)
$$\boldsymbol{Z} = \sum_{(i_k),(j_h)} a_{i_1 i_2 \cdots i_q}^{j_1 j_2 \cdots j_p}\, d\varphi^{i_1} \otimes d\varphi^{i_2} \otimes \cdots \otimes d\varphi^{i_q} \otimes X_{j_1} \otimes X_{j_2} \otimes \cdots \otimes X_{j_p},$$

the summation running over all sequences $(i_k)_{1 \leqq k \leqq q}$ and $(j_h)_{1 \leqq h \leqq p}$ of indices from 1 to n. Likewise, every differential p-form α on U is uniquely expressible in the form

(16.20.4.2) $$\alpha = \sum_{\mathrm{H}} a_{\mathrm{H}}\, d\varphi^{\mathrm{H}},$$

where $\mathrm{H} = \{i_1, i_2, \ldots, i_p\}$ runs through the set of all subsets of p elements of the set $\{1, 2, \ldots, n\}$ (the sequence (i_k) being *strictly increasing*), and where

(16.20.4.3) $$d\varphi^{\mathrm{H}} = d\varphi^{i_1} \wedge d\varphi^{i_2} \wedge \cdots \wedge d\varphi^{i_p}.$$

The coefficients in (16.20.4.1) (resp. (16.20.4.2)) are scalar functions on U which are of class C^r if and only if $\boldsymbol{Z}$ (resp. α) is of class C^r. Identifying $\bigwedge^p \mathrm{T(M)}^*$ with the dual of the bundle $\bigwedge^p \mathrm{T(M)}$ (16.18.4), we see that if $Y_1, \ldots, Y_p$ are p vector fields on U, then we have

(16.20.4.4) $$\langle \alpha, Y_1 \wedge Y_2 \wedge \cdots \wedge Y_p \rangle = \sum_{\mathrm{H}} a_{\mathrm{H}} \cdot \det(\langle d\varphi^{i_h}, Y_k \rangle),$$

where h and k run from 1 to p in each determinant.

(16.20.5) When M is an open set in $\mathbf{R}^n$, the preceding remarks may be applied to the canonical chart $(\mathrm{M}, 1_{\mathrm{M}}, n)$, so that $\varphi^i = \mathrm{pr}_i$ $(1 \leqq i \leqq n)$. By abuse of notation, we denote the coordinates of a point x by $\xi^1, \xi^2, \ldots, \xi^n$, and the differential forms $d(\mathrm{pr}_i)$ by $d\xi^i$ $(1 \leqq i \leqq n)$, so that a differential p-form on M is written

(16.20.5.1) $$(\xi^1, \xi^2, \ldots, \xi^n) \mapsto \sum_{\mathrm{H}} a_{\mathrm{H}}(\xi^1, \ldots, \xi^n)\, d\xi^{i_1} \wedge d\xi^{i_2} \wedge \cdots \wedge d\xi^{i_p}.$$

For each integer $r \geqq 0$ (or $r = \infty$), the differential p-forms on an open subset M of $\mathbf{R}^n$

(16.20.5.2)
$$d\xi^{\mathrm{H}} = d\xi^{i_1} \wedge d\xi^{i_2} \wedge \cdots \wedge d\xi^{i_p}$$

form a basis (called the *canonical* basis) of the module of differential p-forms of class C^r on M, over the ring $\mathscr{E}^{(r)}(\mathrm{M})$ of real-valued functions of class C^r on M.

Example

(16.20.6) Let M be any differential manifold. We shall define a differential 1-form κ_{M} (or κ) of class C^∞ *on the manifold* $\mathrm{T(M)}^*$ (the cotangent bundle of M), called the *fundamental form*, as follows. Let $o_{\mathrm{M}}^* : \mathrm{T(M)}^* \to \mathrm{M}$ be the projection map for the bundle $\mathrm{T(M)}^*$, and consider the mapping

$$\mathrm{T}(o_{\mathrm{M}}^*) : \mathrm{T}(\mathrm{T(M)}^*) \to \mathrm{T(M)}$$

(16.15.6): for each covector $\mathbf{h}_x^* \in \mathrm{T}_x(\mathrm{M})^*$ and each tangent vector

$$\mathbf{k}_{\mathbf{h}_x^*} \in \mathrm{T}_{\mathbf{h}_x^*}(\mathrm{T(M)}^*),$$

the vector $\mathrm{T}_{\mathbf{h}_x^*}(o_{\mathrm{M}}^*) \cdot \mathbf{k}_{\mathbf{h}_x^*}$ is a tangent vector belonging to $\mathrm{T}_x(\mathrm{M})$. We may there fore consider the linear form $\mathbf{k}_{\mathbf{h}_x^*} \mapsto \langle \mathbf{h}_x^*, \mathrm{T}_{\mathbf{h}_x^*}(o_{\mathrm{M}}^*) \cdot \mathbf{k}_{\mathbf{h}_x^*} \rangle$, which is a *covector* $\kappa_{\mathrm{M}}(\mathbf{h}_x^*)$ at the point $\mathbf{h}_x^* \in \mathrm{T}_x(\mathrm{M})^*$. To show that this does in fact define a differential form of class C^∞ on $\mathrm{T(M)}^*$, we may assume that M is an open set in $\mathbf{R}^n$, so that $\mathrm{T(M)}^*$ is identified by the canonical trivialization with $\mathrm{M} \times (\mathbf{R}^n)^*$, and $\mathrm{T}(\mathrm{T(M)}^*)$ with $\mathrm{T(M)} \times \mathrm{T}((\mathbf{R}^n)^*)$ and therefore with

$$(\mathrm{M} \times (\mathbf{R}^n)^*) \times (\mathbf{R}^n \times \mathbf{R}^n)$$

after permuting the factors. If $\mathbf{h}_x^* = (x, \mathbf{h}^*)$ and $\mathbf{k}_{\mathbf{h}_x^*} = ((x, \mathbf{h}^*), (\mathbf{y}, \mathbf{k}))$, then the mapping $\mathrm{T}_{\mathbf{h}_x^*}(o_{\mathrm{M}})$ is

$$((x, \mathbf{h}^*), (\mathbf{y}, \mathbf{k})) \mapsto (x, \mathbf{y})$$

and hence $\kappa_{\mathrm{M}}(\mathbf{h}_x^*)$ is the covector

$$((x, \mathbf{h}^*), (\mathbf{y}, \mathbf{k})) \mapsto \langle \mathbf{h}^*, \mathbf{y} \rangle$$

on $\mathrm{T}_{\mathbf{h}_x^*}(\mathrm{T(M)}^*)$. With the notation of (16.20.5), the fundamental form is

(16.20.6.1)
$$((\xi^i), (\eta_i)) \mapsto \sum_{i=1}^{n} \eta_i \, d\xi^i$$

on $\mathrm{M} \times (\mathbf{R}^n)^*$.

(16.20.7) Consider now a mapping $f : M' \to M$ of class C^∞, and the bundle $f^*(T(M^*))$ on M', the inverse image under f of the bundle $T(M)^*$ (16.19.1). We shall show that there exists a unique M'-morphism of vector bundles

$$w : f^*(T(M)^*) \to T(M')^* \tag{16.20.7.1}$$

such that, for each $x' \in M'$, $w_{x'}$ is the composite linear mapping

$$(f^*(T(M)^*))_{x'} \xrightarrow{v_{x'}} T_{f(x')}(M)^* \xrightarrow{{}^t T_{x'}(f)} T_{x'}(M')^*,$$

$v_{x'}$ being the canonical isomorphism of fibers defined in (16.12.8(ii)). This property defines the mapping w uniquely, and it remains to show that w is of class C^∞. Now, if (U, φ, n) and (U', φ', m) are charts on M and M′, respectively, and if $F : \varphi(U) \to \varphi'(U')$ is the local expression of f, then it is easily checked, by virtue of the definition of $v_{x'}$ (16.19.1) and the local expression of $T_{x'}(f)$ (16.15.6.3) that the local expression of w is

$$(x', \mathbf{h}^*) \mapsto (x', {}^t DF(x') \cdot \mathbf{h}^*), \tag{16.20.7.2}$$

whence the assertion follows.

(16.20.8) For each differential 1-form ω on M, the differential form $w \circ f^*(\omega)$ is called the *inverse image* of ω by f and is denoted by ${}^t f(\omega)$. In view of (16.20.7.2), the form ${}^t f(\omega)$ is equivalently defined by the condition that, for each $x' \in M'$ and each tangent vector $\mathbf{h}_{x'} \in T_{x'}(M)$, we have

$$\langle {}^t f(\omega)(x'), \mathbf{h}_{x'} \rangle = \langle \omega(f(x')), T_{x'}(f) \cdot \mathbf{h}_{x'} \rangle$$

or again

$${}^t f(\omega)(x') = {}^t T_{x'}(f) \cdot \omega(f(x')). \tag{16.20.8.1}$$

In particular, if g is a real-valued function of class C^1 on M, the formula (16.20.8.1) and the definition of dg (16.20.2.2) give

$$\begin{aligned}\langle {}^t f(dg)(x'), \mathbf{h}_{x'} \rangle &= \langle dg(f(x')), T_{x'}(f) \cdot \mathbf{h}_{x'} \rangle \\ &= T_{f(x')}(g) \cdot (T_{x'}(f) \cdot \mathbf{h}_{x'}) = T_{x'}(g \circ f) \cdot \mathbf{h}_{x'}\end{aligned}$$

by (16.5.4). Hence we have the formula

$${}^t f(dg) = d(g \circ f). \tag{16.20.8.2}$$

Let $c = (U, \varphi, n)$ be a chart on M, where $\varphi = (\varphi^i)_{1 \leqq i \leqq n}$. If a differential form on U is given by

$$\omega = \sum_{i=1}^{n} a_i \, d\varphi^i$$

(16.20.3.1), then ${}^t f(\omega)$ on $f^{-1}(\mathrm{U})$ will be given by

$$(16.20.8.3) \qquad {}^t f(\omega) = \sum_{i=1}^{n} (a_i \circ f) \, d(\varphi^i \circ f).$$

Consider also a chart (V, ψ, m) on M' such that $f(\mathrm{U}) \subset \mathrm{V}$, and let F be the local expression of f relative to these two charts, and $x \mapsto (x, \mathbf{w}^*(x))$ the local expression of ω (where $\mathbf{w}^*(x) \in (\mathbf{R}^n)^*$); then the local expression of ${}^t f(\omega)$ is

$$(16.20.8.4) \qquad x' \mapsto (x', {}^t\mathrm{DF}(x') \cdot \mathbf{w}^*(\mathrm{F}(x'))).$$

In the particular case where M (resp. M′) is an open set in $\mathbf{R}^n$ (resp. $\mathbf{R}^m$), then with the notation introduced in (16.20.5), the inverse image of the form

$$\omega : (\xi^1, \ldots, \xi^n) \mapsto \sum_{i=1}^{n} a_i(\xi^1, \ldots, \xi^n) \, d\xi^i$$

by the mapping $\mathrm{F} = (\mathrm{F}^1, \ldots, \mathrm{F}^n)$ of M′ into M will be given by

$${}^t\mathrm{F}(\omega) \cdot (\xi'^1, \ldots, \xi'^m) \mapsto$$

$$\sum_{i=1}^{m} \left(\sum_{j=1}^{n} (a_j(\mathrm{F}^1(\xi'^1, \ldots, \xi'^m), \ldots, \mathrm{F}^n(\xi'^1, \ldots, \xi'^m)) \mathrm{D}_i \mathrm{F}^j(\xi'^1, \ldots, \xi'^m)) \right) d\xi'^i.$$

(16.20.9) Under the hypotheses of (16.20.7), by replacing each of the linear mappings $w_{x'}$ by a tensor or exterior power, we may define in the same way the *inverse image* under f of a *covariant tensor field* $\mathbf{Z}$ or of a *differential p-form* α on M. These inverse images are denoted by ${}^t f(\mathbf{Z})$ and ${}^t f(\alpha)$, respectively. If (U, φ, n) is a chart on M, and if $\mathbf{Z}$ and α are given on U by

$$(16.20.9.1) \qquad \begin{cases} \mathbf{Z} = \displaystyle\sum_{(i_1, \ldots, i_p)} a_{i_1 i_2 \cdots i_p} \, d\varphi^{i_1} \otimes d\varphi^{i_2} \otimes \cdots \otimes d\varphi^{i_p}, \\ \alpha = \displaystyle\sum_{\mathrm{H}} a_{\mathrm{H}} \, d\varphi^{i_1} \wedge d\varphi^{i_2} \wedge \cdots \wedge d\varphi^{i_p} \qquad (\mathrm{H} = \{i_1, i_2, \ldots, i_p\}), \end{cases}$$

then we have

$$(16.20.9.2) \qquad \begin{cases} {}^t f(\mathbf{Z}) = \displaystyle\sum_{(i_1, \ldots, i_p)} (a_{i_1 i_2 \cdots i_p} \circ f) \, d(\varphi^{i_1} \circ f) \otimes \cdots \otimes d(\varphi^{i_p} \circ f), \\ {}^t f(\alpha) = \displaystyle\sum_{\mathrm{H}} (a_{\mathrm{H}} \circ f) \, d(\varphi^{i_1} \circ f) \wedge \cdots \wedge d(\varphi^{i_p} \circ f). \end{cases}$$

We may also write

(16.20.9.3) $$^t\!f(\alpha)(x') = \Big(\bigwedge^p {}^t\mathrm{T}_{x'}(f)\Big) \cdot \alpha(f(x')).$$

If M and M′ are pure, this shows in particular that if $\dim(\mathrm{M}) = n$ and $\dim(\mathrm{M}') = n'$, then for a p-form α on M we have ${}^t\!f(\alpha) = 0$ if $n' < p \leqq n$. Suppose that $n' = n$, and consider two charts on M and M′, respectively. If $x \mapsto (x, \mathrm{V}(x))$ (where $\mathrm{V}(x) \in \mathbf{R}$) is the local expression of a differential n-form υ on M, and if F (a C^∞-mapping of an open set in $\mathbf{R}^n$ into an open set in $\mathbf{R}^m$) is the local expression of f, relative to the two charts under consideration, then the local expression of ${}^t\!f(\upsilon)$ is

(16.20.9.4) $$x' \mapsto (x', \mathrm{J}(x')\mathrm{V}(\mathrm{F}(x'))),$$

where $\mathrm{J}(x') = \det(\mathrm{DF}(x'))$ is the Jacobian of F at the point x' (8.10.1).

It is clear that if α is a differential p-form and β a differential q-form, then

(16.20.9.5) $$^t\!f(\alpha \wedge \beta) = {}^t\!f(\alpha) \wedge {}^t\!f(\beta).$$

Finally, if $g : \mathrm{M}'' \to \mathrm{M}'$ is another C^∞-mapping, we have

(16.20.9.6) $$^t(f \circ g)(\alpha) = {}^t g({}^t\!f(\alpha)).$$

There are analogous formulas for covariant tensor fields.

(16.20.10) If M′ is a *submanifold* of M and $j : \mathrm{M}' \to \mathrm{M}$ is the canonical injection, then ${}^t\!j(\mathbf{Z})$ (resp. ${}^t\!j(\alpha)$) is said to be the covariant tensor field *induced* by $\mathbf{Z}$ on M′ (resp. the differential p-form *induced* by α on M′); we have ${}^t\!j(\alpha) = 0$ if $p > \dim(\mathrm{M}')$. If M is an open set in $\mathbf{R}^n$, and M′ an open set in $\mathbf{R}^m$ (a situation to which we may reduce by means of charts), and if we identify the vectors of the canonical basis of $\mathbf{R}^m$ with the first m vectors of the canonical basis of $\mathbf{R}^n$, then we pass from the p-form

$$\alpha = \sum_{\mathrm{H}} a_{\mathrm{H}}(\xi^1, \ldots, \xi^n)\, d\xi^{i_1} \wedge d\xi^{i_2} \wedge \cdots \wedge d\xi^{i_p}$$

to the induced form ${}^t\!j(\alpha)$ *by replacing the ξ^k and the $d\xi^k$ with $k > m$ by zero.*

(16.20.11) It should be noted carefully that there is no equivalent of the preceding developments for *contravariant tensor fields* or *mixed tensor fields.* For such fields *neither* inverse images *nor* direct images relative to a C^∞-mapping $f : \mathrm{M}' \to \mathrm{M}$ can be defined. The reasons behind this are that the mappings $\mathrm{T}_{x'}(f)$ are not necessarily bijective (which prevents us defining the inverse image of a tangent vector) and that f itself need not be bijective (which prevents us defining the direct image of a section (16.12.6)). Of course, if f is a *diffeomorphism* of M′ onto M, we can define the image $f(\mathbf{Z})$ of *any* tensor field $\mathbf{Z}$ on M′, by transport of structure (16.16.6); if α' is a differential p-form on M′, then $f(\alpha')$ is the inverse image ${}^t f^{-1}(\alpha')$ of α' under f^{-1}, as defined above.

Examples

(16.20.12) Let M, M′ be two differential manifolds. A *homogeneous contact transformation* from M to M′ is by definition a diffeomorphism f of an open set in T(M)* onto an open set in T(M′)*, such that the image under f of the form induced by the fundamental form κ_{M} (16.20.6) is the form induced by the fundamental form $\kappa_{\mathrm{M}'}$. If $g : \mathrm{M} \to \mathrm{M}'$ is a diffeomorphism, we obtain canonically from g by transport of structure a contact transformation

$$\mathrm{T}(g)^* : \mathrm{T}(\mathrm{M})^* \to \mathrm{T}(\mathrm{M}')^*:$$

for each $x \in \mathrm{M}$ and each covector $\mathbf{h}_x^* \in \mathrm{T}_x(\mathrm{M})$, we have

(16.20.12.1) $$\mathrm{T}(g)^* \cdot \mathbf{h}_x^* = {}^t\mathrm{T}_x(g)^{-1} \cdot \mathbf{h}_x^*,$$

and it is immediately verified that the image of κ_{M} under $\mathrm{T}(g)^*$ is $\kappa_{\mathrm{M}'}$.

However, it is easy to define contact transformations which are not of this type. Take for example $\mathrm{M} = \mathrm{M}' = \mathbf{R}^n$, so that T(M)* may be identified with $\mathbf{R}^n \times (\mathbf{R}^n)^*$, and let U be the open set in T(M)* consisting of the $((\xi^i), (\eta_i))$ such that $\eta_n \neq 0$. Then the mapping

$$((\xi^i), (\eta_i)) \mapsto \left(\left(\frac{\eta_1}{\eta_n}, \ldots, \frac{\eta_{n-1}}{\eta_n}, \frac{1}{\eta_n}\sum_{i=1}^{n} \xi^i \eta_i\right), (-\eta_n \xi^1, \ldots, -\eta_n \xi^{n-1}, \eta_n)\right)$$

may be verified to be a contact transformation of U onto U; it is a particular case of a *Legendre transformation*.

(16.20.13) Let G be a Lie group acting differentiably (on the left) on a differential manifold M. We have already seen that there is a canonical differentiable action of G on the tangent bundle T(M) (16.15.6). We shall show that there is also a canonical action of G on the cotangent bundle T(M)*. Since $\gamma(s) : x \mapsto s \cdot x$ is a diffeomorphism of M, we obtain by transport of structure a diffeomorphism of T(M)*, which transforms a covector $\mathbf{h}_x^*$ at the point $x \in \mathrm{M}$ into the covector ${}^t\mathrm{T}_x(\gamma(s))^{-1} \cdot \mathbf{h}_x^*$ at the point $s \cdot x$. If we denote this covector by $s \cdot \mathbf{h}_x^*$ we shall have, for every tangent vector $\mathbf{k}_{s \cdot x}$ at the point $s \cdot x$,

(16.20.13.1) $$\langle s \cdot \mathbf{h}_x^*, \mathbf{k}_{s \cdot x} \rangle = \langle \mathbf{h}_x^*, s^{-1} \cdot \mathbf{k}_{s \cdot x} \rangle$$

and since the right-hand side is a C^∞-function on the submanifold of

$$G \times T(M) \times T(M)^*$$

where it is defined, it follows that the action of G on $T(M)^*$ so defined is indeed differentiable. If ω is a differential form on M, its image $s \cdot \omega$ or $\gamma(s)\omega$ under the diffeomorphism $\gamma(s)$ is therefore the form defined by

(16.20.13.2) $$\langle (s \cdot \omega)(x), \mathbf{h}_x \rangle = \langle \omega(s^{-1} \cdot x), s^{-1} \cdot \mathbf{h}_x \rangle.$$

Remark

(16.20.14) The definitions (16.20.8.1) and (16.20.9.3) retain their validity under the weaker hypothesis that the mapping f: $M' \to M$ is of class C^r for some $r \geqq 1$; the formula (16.20.8.4) then shows that the inverse image under f of a form of class C^s is of class $\inf(r-1, s)$.

(16.20.15) *Vector-valued differential forms.*

Instead of considering in (16.20.1) the vector bundle $T(M^*) = \mathrm{Hom}(T(M), M \times \mathbf{R})$ or

$$\bigwedge^p T(M)^* = \mathrm{Hom}\Big(\bigwedge^p T(M), M \times \mathbf{R}\Big),$$

we may more generally consider the vector bundle $\mathrm{Hom}(T(M), M \times F)$ or $\mathrm{Hom}\Big(\bigwedge^p T(M), M \times F\Big)$, where F is a real vector space of finite dimension. A section of $\mathrm{Hom}\Big(\bigwedge^p T(M), M \times F\Big)$ over M is called a *vector-valued differential p-form on* M, *with values in* F. If $(\mathbf{a}_i)_{1 \leq i \leq r}$ is a basis of F, such a p-form is uniquely expressible as $\boldsymbol{\alpha} = \sum_{i=1}^{r} \alpha_i \mathbf{a}_i$, where the α_i are differential p-forms in the sense defined earlier, or (as we shall sometimes call them) *scalar-valued* differential p-forms. For each $x \in M$, $\boldsymbol{\alpha}(x)$ takes its values in the space of linear mappings of $\bigwedge^p T_x(M)$ into F (identified with $(M \times F)_x$); if $\mathbf{h}_1, \ldots, \mathbf{h}_p$ are p tangent vectors belonging to $T_x(M)$, we denote by $\boldsymbol{\alpha}(x) \cdot (\mathbf{h}_1 \wedge \cdots \wedge \mathbf{h}_p)$ the value of $\boldsymbol{\alpha}(x)$ at the p-vector $\mathbf{h}_1 \wedge \mathbf{h}_2 \wedge \cdots \wedge \mathbf{h}_p$, so that we have

(16.20.15.1) $$\boldsymbol{\alpha}(x) \cdot (\mathbf{h}_1 \wedge \cdots \wedge \mathbf{h}_p) = \sum_{i=1}^{r} \langle \alpha_i(x), \mathbf{h}_1 \wedge \cdots \wedge \mathbf{h}_p \rangle \mathbf{a}_i.$$

If $X_1, \ldots, X_p$ are p vector fields on M, we denote by

(16.20.15.2) $$\boldsymbol{\alpha} \cdot (X_1 \wedge X_2 \wedge \cdots \wedge X_p)$$

the function on M, *with values in* F,

$$x \mapsto \boldsymbol{\alpha}(x) \cdot (X_1(x) \wedge X_2(x) \wedge \cdots \wedge X_p(x)).$$

In particular, if $\mathbf{f} = \sum_i f_i \mathbf{a}_i$ is a C^1-mapping of M into F, then $d\mathbf{f} = \sum_i (df_i)\mathbf{a}_i$ is a differential 1-form with values in F, called the *differential* of $\mathbf{f}$; it is precisely the mapping $x \mapsto d_x \mathbf{f}$ (16.5.7).

For example, if E is a finite-dimensional real vector space, the mapping $x \mapsto \boldsymbol{\tau}_x$ (16.5.2) is the differential $d(1_E)$ of the identity mapping.

If $f : M' \to M$ is a C^∞-mapping, we define the *inverse image* ${}^t f(\boldsymbol{\alpha})$ of $\boldsymbol{\alpha}$ under f by the same condition as in (16.20.8): for each $x' \in M'$ and each p-vector $\mathbf{h}_1' \wedge \mathbf{h}_2' \wedge \cdots \wedge \mathbf{h}_p'$, where the $\mathbf{h}_j'$ belong to $T_{x'}(M')$, we put

(16.20.15.3) $$\begin{aligned} {}^t f(\boldsymbol{\alpha})(x') \cdot (\mathbf{h}_1' \wedge \cdots \wedge \mathbf{h}_p') \\ = \boldsymbol{\alpha}(f(x')) \cdot (T_{x'}(f) \cdot \mathbf{h}_1' \wedge \cdots \wedge T_{x'}(f) \cdot \mathbf{h}_p'). \end{aligned}$$

Equivalently, we have

(16.20.15.4) $${}^t f(\boldsymbol{\alpha}) = \sum_{i=1}^{n} {}^t f(\alpha_i)\mathbf{a}_i$$

which brings us back to scalar-valued p-forms.

There is no "exterior calculus" for vector-valued differential p-forms analogous to the exterior algebra of scalar-valued differential forms, but we shall be led to consider in Chapter XX operations analogous to the exterior product in certain particular cases. Consider three finite-dimensional real vector spaces F', F'', F, and let $B : F' \times F'' \to F$ be a bilinear mapping. If we are given vector-valued differential 1-forms $\boldsymbol{\omega}'$, $\boldsymbol{\omega}''$ on M, with values in F' and F'', respectively, we can associate with them a differential 2-form with values in F, in the following manner: the mapping

(16.20.15.5) $$(\mathbf{h}_x, \mathbf{k}_x) \mapsto B(\boldsymbol{\omega}'(x) \cdot \mathbf{h}_x, \boldsymbol{\omega}''(x) \cdot \mathbf{k}_x) - B(\boldsymbol{\omega}'(x) \cdot \mathbf{k}_x, \boldsymbol{\omega}''(x) \cdot \mathbf{h}_x)$$

of $T_x(M) \times T_x(M)$ into F is bilinear and alternating, and hence there exists a unique linear mapping of $\bigwedge^2 T_x(M)$ into F, for which the right-hand side of (16.20.15.5) is the value at the bivector $\mathbf{h}_x \wedge \mathbf{k}_x$. We denote this mapping by $(\boldsymbol{\omega}' \wedge_B \boldsymbol{\omega}'')(x)$, and thus we have defined a 2-form $\boldsymbol{\omega}' \wedge_B \boldsymbol{\omega}''$ with values in F. (When there is no risk of confusion we shall write $\boldsymbol{\omega}' \wedge \boldsymbol{\omega}''$, but it should be remarked that $\boldsymbol{\omega}'' \wedge_B \boldsymbol{\omega}'$ in general has no meaning.) If $(\mathbf{a}_i')_{1 \leq i \leq r}$ is a basis of F' and $(\mathbf{a}_j'')_{1 \leq j \leq s}$ a basis of F'', and if we put

(16.20.15.6) $$\boldsymbol{\omega}' = \sum_i \omega_i' \mathbf{a}_i', \qquad \boldsymbol{\omega}'' = \sum_j \omega_j'' \mathbf{a}_j'',$$

where the ω_i' and ω_j'' are scalar-valued 1-forms, then we have

(16.20.15.7) $$\boldsymbol{\omega}' \wedge_{\mathrm{B}} \boldsymbol{\omega}'' = \sum_{i,j} (\omega_i' \wedge \omega_j'')\mathrm{B}(\mathbf{a}_i', \mathbf{a}_j'').$$

In the particular case in which $\mathrm{F}' = \mathrm{F}'' = \mathrm{E}$, $\boldsymbol{\omega}' = \boldsymbol{\omega}'' = \boldsymbol{\omega}$, and the bilinear mapping B is *alternating*, we do not use the notation $\boldsymbol{\omega} \wedge_{\mathrm{B}} \boldsymbol{\omega}$. The mapping

(16.20.15.8) $$(\mathbf{h}_x, \mathbf{k}_x) \mapsto \mathrm{B}(\boldsymbol{\omega}(x) \cdot \mathbf{h}_x, \boldsymbol{\omega}(x) \cdot \mathbf{k}_x)$$

is *already* bilinear and alternating, and we denote by $\mathrm{B}(\boldsymbol{\omega}, \boldsymbol{\omega})_x$ the corresponding linear mapping of $\bigwedge^2 \mathrm{T}_x(\mathrm{M})$ into F. If $\boldsymbol{\omega}'$, $\boldsymbol{\omega}''$ are 1-forms with values in E, we have

(16.20.15.9) $$\mathrm{B}(\boldsymbol{\omega}' + \boldsymbol{\omega}'', \boldsymbol{\omega}' + \boldsymbol{\omega}'') = \mathrm{B}(\boldsymbol{\omega}', \boldsymbol{\omega}') + \boldsymbol{\omega}' \wedge_{\mathrm{B}} \boldsymbol{\omega}'' + \mathrm{B}(\boldsymbol{\omega}'', \boldsymbol{\omega}''),$$

and if we put $\boldsymbol{\omega} = \sum_i \omega_i \mathbf{a}_i$, where $(\mathbf{a}_i)$ is a basis of E, then

(16.20.15.10) $$\mathrm{B}(\boldsymbol{\omega}, \boldsymbol{\omega}) = \frac{1}{2} \sum_{i,j} (\omega_i \wedge \omega_j)\mathrm{B}(\mathbf{a}_i, \mathbf{a}_j)$$

(summed over *all* pairs of indices (i, j)), which can also be written in the form

(16.20.15.11) $$\mathrm{B}(\boldsymbol{\omega}, \boldsymbol{\omega}) = \sum_{i<j} (\omega_i \wedge \omega_j)\mathrm{B}(\mathbf{a}_i, \mathbf{a}_j).$$

(16.20.16) *Differential forms on complex manifolds.*

All the definitions in this section can be transposed to the context of a complex manifold M, by replacing C^∞-mappings throughout by holomorphic mappings, and real vector bundles by complex vector bundles. The cotangent bundle $\mathrm{T}(\mathrm{M})^*$ is a holomorphic bundle (16.15.9) of (complex) rank n, if M is a pure manifold of (complex) dimension n.

If we denote by $\mathrm{M}_{|\mathbf{R}}$ the differential manifold of dimension $2n$ underlying M, then the cotangent bundle $\mathrm{T}(\mathrm{M}_{|\mathbf{R}})^*$ is a real vector bundle of rank $2n$ over $\mathrm{M}_{|\mathbf{R}}$, and its complexification $(\mathrm{T}(\mathrm{M}_{|\mathbf{R}})^*)_{(\mathbf{C})}$ is a complex vector bundle of (complex) rank $2n$ over $\mathrm{M}_{|\mathbf{R}}$. The results of (16.5.13) show that there exists a **C**-automorphism ${}^tJ$ of this fiber bundle such that ${}^tJ^2 = -I$; the images of the two morphisms $p' = \frac{1}{2}(I - i\,{}^tJ)$, $p'' = \frac{1}{2}(I + i\,{}^tJ)$ are, respectively, the cotangent bundle $\mathrm{T}(\mathrm{M})^*$ and its complex conjugate $\overline{\mathrm{T}(\mathrm{M})^*}$, so that

$$(\mathrm{T}(\mathrm{M}_{|\mathbf{R}})^*)_{(\mathbf{C})} = \mathrm{T}(\mathrm{M})^* \oplus \overline{\mathrm{T}(\mathrm{M})^*}.$$

To any *complex* function f of class C^1 on $M_{|\mathbf{R}}$ we can associate two sections $d'f = p' \circ df$, $d''f = p'' \circ df$ of $T(M)^*$ and $\overline{T(M)^*}$, respectively, so that the complex differential form df is equal to $d'f + d''f$. The function f is *holomorphic* if and only if $d''f = 0$. If (U, φ, n) is a chart on the complex manifold M, where $\varphi = (\varphi^j)_{1 \leq j \leq n}$, then the $d\varphi^j$ form a frame of $T(M)^*$ over U, and the $\overline{d\varphi^j}$ a frame of $\overline{T(M)^*}$ over U.

PROBLEMS

1. Let A be one of the rings $\mathscr{E}(M)$, $\mathscr{E}^{(r)}(M)$ (where $r \geq 1$) on a differential manifold M. For each $x \in M$, let $\mathfrak{m}_x$ denote the ideal of functions belonging to A which vanish at x. Consider the mapping $f \mapsto d_x f$ of $\mathfrak{m}_x$ into $T_x(M)^*$. Show that, if $A = \mathscr{E}(M)$, the kernel $\mathfrak{n}_x$ of this mapping is equal to $\mathfrak{m}_x^2$, but that if $A = \mathscr{E}^{(r)}(M)$ with r finite, then $\mathfrak{m}_x^2$ is of infinite codimension in $\mathfrak{n}_x$. (Observe that in the latter case the product of two elements of $\mathfrak{m}_x$ has a local expression which admits derivatives of order $r + 1$ at the point of the chart corresponding to x.)

2. Let M be a differential manifold.

(a) For each real-valued C^∞-function f on M, let $\tilde{d}f$ denote the C^∞-function on T(M) defined by $\mathbf{h}_x \mapsto \langle df(x), \mathbf{h}_x \rangle$; then $d(\tilde{d}f)$ is a differential form on T(M). If F is the local expression of f by means of a chart of M, show that the local expression of $d(\tilde{d}f)$, relative to the trivialization of T(T(M)) corresponding to this chart, is

$$(x, \mathbf{h}, \mathbf{u}, \mathbf{k}) \mapsto D^2F(x) \cdot (\mathbf{h}, \mathbf{u}) + DF(x) \cdot \mathbf{k}.$$

(b) Show that there exists one and only one involutory diffeomorphism j of T(T(M)) satisfying the following conditions for all $\mathbf{k} \in T_{\mathbf{u}_x}(T(M))$:

(i) $o_{T(M)}(j(\mathbf{k})) = T(o_M) \cdot \mathbf{k}$;
(ii) $T(o_M) \cdot j(\mathbf{k}) = o_{T(M)}(\mathbf{k})$;
(iii) $\langle d\mathbf{h}_x(\tilde{d}f), \mathbf{k} \rangle = \langle d\mathbf{h}_x(\tilde{d}f), j(\mathbf{k}) \rangle$ for all real-valued C^∞-functions f on M.

(Here $\mathbf{h}_x = T(o_M) \cdot \mathbf{k}$.) (For the proof use (a) and (8.12.2).) The involution j is called the *canonical involution* on T(T(M)); it is an isomorphism of the vector bundle $(T(T(M)), T(M), o_{T(M)})$ onto the vector bundle $(T(T(M)), T(M), T(o_M))$.

3. Let M, M′ be two open subsets of $\mathbf{R}^n$. A point of $\mathbf{R} \times T(M)^*$ will be written $(z, (x^i), (y_i))$ with $1 \leq i \leq n$, and similarly for $\mathbf{R} \times T(M')^*$. A diffeomorphism f of an open subset V of $\mathbf{R} \times T(M)^*$ onto an open subset V′ of $\mathbf{R} \times T(M')^*$ is called a *nonhomogeneous contact transformation* if the image under f of the form $dz - \sum_{i=1}^{n} y_i \, dx^i$ is

$$\rho\left(dz' - \sum_{i=1}^{n} y'_i \, dx'^i\right),$$

where ρ is a real-valued C^∞-function which does not vanish in V.

Let N, N′ be two open sets in $\mathbf{R}^{n+1}$. A point of $\mathrm{T(N)}^*$ will be written $((\xi^i), (\eta_i))$ with $0 \leqq i \leqq n$, and likewise for $\mathrm{T(N')}^*$. Let U (resp. U′) be the set of points of $\mathrm{T(N)}^*$ (resp. $\mathrm{T(N')}^*$) such that $\eta_0 \neq 0$ (resp. $\eta_0' \neq 0$). Let M (resp. M′) be the projection of U (resp. U′) on $\mathbf{R}^n$ (the subspace spanned by the last n vectors of the canonical basis). Define a C^∞-mapping g of U into $\mathbf{R} \times \mathrm{T(M)}^*$ as follows: $g((\xi^i), (\eta_i)) = (z, (x^i), (y_i))$ where $z = \xi^0$, $x^i = \xi^i$ for $1 \leqq i \leqq n$, and $y_i = \eta_i/\eta_0$. Define $g' : \mathrm{U}' \to \mathbf{R} \times \mathrm{T(M')}^*$ likewise. Show that, for each nonhomogeneous contact transformation f of $\mathrm{V} \subset \mathbf{R} \times \mathrm{T(M)}^*$ onto $\mathrm{V}' \subset \mathbf{R} \times \mathrm{T(M')}^*$, there exists a unique homogeneous contact transformation F of $g^{-1}(\mathrm{V})$ onto $g'^{-1}(\mathrm{V}')$, for which the diagram

$$\begin{array}{ccc} g^{-1}(\mathrm{V}) & \xrightarrow{\;g\;} & \mathrm{V} \\ {\scriptstyle \mathrm{F}}\downarrow & & \downarrow{\scriptstyle f} \\ g'^{-1}(\mathrm{V}') & \xrightarrow[\;g'\;]{} & \mathrm{V}' \end{array}$$

is commutative.

4. With the notation of Problem 3, consider the graph Γ_f of f in $\mathrm{V} \times \mathrm{V}'$. Show that the projection S of this graph on $(\mathbf{R} \times \mathrm{M}) \times (\mathbf{R} \times \mathrm{M}')$ contains no nonempty open set.

(a) Suppose that S is a submanifold of dimension $2n + 1$ and that the restrictions p, p' to S of the two projections of $(\mathbf{R} \times \mathrm{M}) \times (\mathbf{R} \times \mathrm{M}')$ are submersions. Suppose that S is defined by an equation

$$\mathrm{H}(z, x^1, \ldots, x^n, z', x'^1, \ldots, x'^n) = 0, \tag{1}$$

where H is of class C^∞. Show that in S we have

$$\rho \frac{\partial \mathrm{H}}{\partial z'} + \frac{\partial \mathrm{H}}{\partial z} = 0, \tag{2}$$

$$\frac{\partial \mathrm{H}}{\partial z'} y_i' + \frac{\partial \mathrm{H}}{\partial x'^i} = 0 \qquad (1 \leqq i \leqq n), \tag{3}$$

$$\frac{\partial \mathrm{H}}{\partial z} y_i + \frac{\partial \mathrm{H}}{\partial x^i} = 0 \qquad (1 \leqq i \leqq n), \tag{4}$$

and deduce that H determines f completely, provided that the functional determinant of the left-hand sides of (1) and (4) with respect to $z', x'^1, \ldots, x'^n$ is not zero.

(b) Determine f when H has one of the following forms:

$$zz' + \sum_{i=1}^{n} x^i x'^i - 1 \qquad \textit{(transformation by reciprocal polars)},$$

$$(z' - z)^2 + \sum_{i=1}^{n} (x'^i - x^i)^2 - r^2 \qquad \textit{(dilatation)}.$$

(c) Generalize to the case where S has dimension $n + k$ $(2 \leqq k \leqq n + 1)$. Consider the case where $n = 2$ and S is one of the submanifolds of dimension 4 defined by the following equations:

$$(x^1)^2 + (x^2)^2 + z^2 = (x'^1)^2 + (x'^2)^2 + z'^2, \qquad zz' + x^1x'^1 + x^2x'^2 = 0$$

(*apsidal transformation*),

$$x^1 z' + z + x'^1 = 0, \qquad x^1 x'^2 + x^2 - z' = 0 \qquad (\textit{Lie's transformation}).$$

(d) Let X be a submanifold of $\mathbf{R} \times M$. For each $x \in X$, let $T_x(X)^\perp$ be the subspace of $T_x(\mathbf{R} \times M)^*$ consisting of the covectors which are orthogonal to $T_x(X)$ (relative to the canonical duality between $T_x(\mathbf{R} \times M)$ and $T_x(\mathbf{R} \times M)^*$). If the projection on $\mathbf{R} \times M'$ of the image under F (Problem 3) of the union of the $T_x(X)^\perp$ is a submanifold X′ of $\mathbf{R} \times M'$, then the union of the $T_{x'}(X')^\perp$ is the image under F of the union of the $T_x(X)^\perp$. Then X′ is said to be the *transform* of X by *f*.

5. Let X be a submanifold of $\mathbf{R}^n$, of dimension $n - p$, defined by p equations $f_j(x) = 0$ $(1 \leqq j \leqq p)$, where the f_j are of class C^∞ in a neighborhood of X and are such that the differential forms df_j are linearly independent at each point of X. Let F be a C^∞-function defined in a neighborhood of X, and let G be its restriction to X. Then in order that $dG(x) = 0$ at a point $x \in X$, it is necessary and sufficient that $dF(x)$ should be a linear combination of the p covectors $df_j(x)$, or again that $dF(x) \wedge df_1(x) \wedge \cdots \wedge df_p(x) = 0$ (*Lagrange's method of undetermined multipliers*).

21. ORIENTABLE MANIFOLDS AND ORIENTATIONS

(16.21.1) *Let X be a pure differential manifold of dimension* $n \geqq 0$ (16.1.5). *Then the following properties are equivalent:*

(a) *There exists a continuous differential n-form* υ *on X such that* $\upsilon(x) \neq 0$ *for all* $x \in X$.

(b) *There exists an atlas* $\mathfrak{A}$ *of X such that, if* (U, φ, n) *and* (U', φ', n) *are any two charts of* $\mathfrak{A}$ *for which* $U \cap U' \neq \varnothing$, *and if we put* $\psi = \varphi | (U \cap U')$, $\psi' = \varphi' | (U \cap U')$ *and* $\theta = \psi' \circ \psi^{-1} : \varphi(U \cap U') \to \varphi'(U \cap U')$, *then the Jacobian* $J(\theta)$ *of* θ (8.10) *is strictly positive at each point of* $\varphi(U \cap U')$.

To show that (a) implies (b), we remark first that there exists an atlas $\mathfrak{A}_0$ of X all of whose charts have a connected domain of definition. For each such chart (U, φ_0, n) we may write, for each $x \in U$,

$$\upsilon(x) = w(x)\, d_x \varphi_0^1 \wedge d_x \varphi_0^2 \wedge \cdots \wedge d_x \varphi_0^n$$

(16.20.4.2), where w is a continuous mapping of U into $\mathbf{R}$. By hypothesis we have $w(x) \neq 0$ for all $x \in U$, hence the *sign* of $w(x)$ is the same for all $x \in U$. Put $\varphi = \varphi_0$ if $w(x) > 0$ for $x \in U$, and put $\varphi = (-\varphi_0^1, \varphi_0^2, \ldots, \varphi_0^n)$ if $w(x) < 0$ for $x \in U$; then it is clear that (U, φ, n) is a chart of X and that the set $\mathfrak{A}$ of these charts is an atlas. We have to show that $\mathfrak{A}$ satisfies condition (b). For each $x \in U \cap U'$ we may write

$$\upsilon(x) = w(x)\, d_x \varphi^1 \wedge d_x \varphi^2 \wedge \cdots \wedge d_x \varphi^n = w'(x)\, d_x \varphi'^1 \wedge d_x \varphi'^2 \wedge \cdots \wedge d_x \varphi'^n,$$

where by hypothesis $w(x) > 0$ and $w'(x) > 0$ at all $x \in U \cap U'$. Since the matrix of transition from the basis $(d_x \varphi^i)$ of $T_x(X)^*$ to the basis $(d_x \varphi'^i)$ is the transpose of the Jacobian matrix of θ, by virtue of (16.5.8.4), it follows that

$$d_x \varphi'^1 \wedge \cdots \wedge d_x \varphi'^n = J(\theta)(\varphi(x)) \cdot d_x \varphi^1 \wedge \cdots \wedge d_x \varphi^n$$

and consequently $w(x) = J(\theta)(\varphi(x)) \cdot w'(x)$, which shows that $J(\theta)$ is positive at $\varphi(x)$.

Conversely, let us show that (b) implies (a). By considering the restrictions of charts of $\mathfrak{A}$ to open sets belonging to a denumerable locally finite covering of X which refines the covering formed by the domains of definition of the charts of $\mathfrak{A}$ (12.6.1), we may suppose that $\mathfrak{A}$ is a finite or denumerable set of charts (U_k, φ_k, n), where the U_k form a *locally finite* open covering of X. Let (f_k) be a C^∞-partition of unity on X, subordinate to the covering (U_k) (16.4.1). Put $v_k(x) = 0$ if $x \notin U_k$, and $v_k(x) = f_k(x)\, d_x \varphi_k^1 \wedge \cdots \wedge d_x \varphi_k^n$ if $x \in U_k$; then v_k is a C^∞-differential n-form on X. Moreover, each $x \in X$ has a neighborhood which meets only finitely many of the U_k; hence the sum $v = \sum_k v_k$ is defined and is a differential n-form of class C^∞ on X. We have to show that $v(x) \neq 0$ for all $x \in X$. Let $h_0, h_1, \ldots, h_m$ be the indices k such that $x \in U_k$, and put $z = \varphi_{h_0}(x)$, $\psi_j = \varphi_{h_j} | (U_{h_j} \cap U_{h_0})$, $\theta_j = \psi_j \circ \psi_0^{-1}$ for $1 \leqq j \leqq m$; then, by definition, we have

$$v(x) = \left(f_{h_0}(x) + \sum_{j=1}^{m} f_{h_j}(x) \cdot J(\theta_j)(z) \right) d_x \varphi_{h_0}^1 \wedge \cdots \wedge d_x \varphi_{h_0}^n .$$

But the $f_k(x)$ are $\geqq 0$, and $f_{h_0}(x) + \sum_{j=1}^{m} f_{h_j}(x) = 1$; hence at least one of the $f_{h_j}(x)$ for $0 \leqq j \leqq m$ must be >0; and since *all* the $J(\theta_j)(z)$ are >0 by hypothesis, we have $v(x) \neq 0$.

If X is a pure differential manifold satisfying the two equivalent conditions of (16.21.1), X is said to be *orientable*. The proof shows that there is then a differential n-form v on X *of class* C^∞ such that $v(x) \neq 0$ for all $x \in X$. Any manifold of dimension 0 is orientable, since the conditions of (16.21.1) are then trivially satisfied.

It is clear that every open submanifold of an orientable manifold is orientable, and that a manifold X is orientable if and only if its connected components are orientable.

(16.21.2) Let X be an orientable pure differential manifold of dimension n, and let v_0 be a C^∞ differential n-form on X such that $v_0(x) \neq 0$ for all $x \in X$. Since the vector space $\bigwedge^n (T_x(X))^*$ is of dimension 1, *every* differential n-form

v on X may be written uniquely as $v = f \cdot v_0$, where f is a real-valued function on X, which is of class C^r if and only if v is of class C^r.

Suppose moreover that X is *connected* and v is *continuous*. In order that $v(x) \neq 0$ for all $x \in X$, it is necessary and sufficient that $f(x) \neq 0$ for all $x \in X$, and then we have either $f(x) > 0$ for all $x \in X$, or else $f(x) < 0$ for all $x \in X$ (**3.19.8**). Let O_1 (resp. O_2) be the set of differential n-forms of class C^∞ on X which satisfy the first (resp. the second) of these conditions. Then O_1 and O_2 are called the *orientations* of the connected orientable manifold X, and the pairs (X, O_1) and (X, O_2) are called the *oriented manifolds* (with orientations respectively O_1 and O_2), having the orientable manifold X as underlying manifold. The definitions of O_1 and O_2 remain unchanged if we replace v_0 by any other n-form of class C^∞ belonging to O_1. The orientations O_1 and O_2 (and the corresponding oriented manifolds) are said to be *opposites* of each other.

If X is an oriented manifold and v_0 is a differential n-form belonging to its orientation, then for any differential n-form $v = f \cdot v_0$ on X, we write $v(x) > 0$ (resp. $v(x) < 0$) if $f(x) > 0$ (resp. $f(x) < 0$). This relation is independent of the form v_0 chosen in the orientation of the manifold. Likewise, we say that an n-covector v_x at a point $x \in X$ is >0 (resp. <0) if it is of the form $cv_0(x)$ with $c > 0$ (resp. $c < 0$).

A sequence $(Z_1, \ldots, Z_n)$ of n vector fields over X is said to be *positive* or *direct* (resp. *negative* or *retrograde*) if we have

$$\langle v_0(x), Z_1(x) \wedge \cdots \wedge Z_n(x) \rangle > 0$$

(resp. <0) for all $x \in X$.

If X is an oriented manifold and U is open in X, then the restriction to U of a differential n-form on X belonging to the orientation of X is $\neq 0$ at all points of U, and therefore defines an orientation on U, called the *induced* orientation.

(**16.21.3**) Let X, X′ be two *oriented* connected differential manifolds of the same dimension n, and let $f: X' \to X$ be a local diffeomorphism (**16.5.6**). If v is a differential n-form on X belonging to the orientation of X, it is clear that ${}^t f(v)(x') \neq 0$ for all $x' \in X'$. We say that f *preserves* (resp. *reverses*) *the orientation* if ${}^t f(v)$ belongs to the orientation of X′ (resp. to the opposite orientation). Let $x' \in X'$ and let (U, φ, n) be a chart of X at the point $x = f(x')$, and (U', ψ, n) a chart of X′ at the point x', and suppose that the differential n-forms $d\varphi^1 \wedge \cdots \wedge d\varphi^n$ and $d\psi^1 \wedge \cdots \wedge d\psi^n$ belong, respectively, to the orientations induced on U and U′ by the orientations of X and X′. We may also suppose that $f = \psi^{-1} \circ F \circ \varphi$, where F is a diffeomorphism of $\varphi(U)$ onto an open subset of $\psi(U')$. Then f preserves (resp. reverses) the orientation according as the Jacobian $J(F)(x)$ is >0 (resp. <0).

Examples

(16.21.4) The spaces $\mathbf{R}^n$ are orientable, for the canonical n-form

$$d\xi^1 \wedge d\xi^2 \wedge \cdots \wedge d\xi^n$$

is $\neq 0$ at every point of $\mathbf{R}^n$. The orientation containing this n-form is called the *canonical* orientation. Whenever we shall consider $\mathbf{R}^n$ as an oriented manifold it is the canonical orientation that is to be understood, unless the contrary is expressly stated.

(16.21.5) Let X_1, X_2 be two orientable pure manifolds; then so is their product $X = X_1 \times X_2$. Let $p = \dim X_1$, $q = \dim X_2$, so that $p + q = \dim X$; let v_1 (resp. v_2) be a p-form on X_1 (resp. a q-form on X_2) of class C^∞, such that $v_1(x_1) \neq 0$ for all $x_1 \in X_1$ (resp. $v_2(x_2) \neq 0$ for all $x_2 \in X_2$). Then it is immediately seen that the $(p+q)$-form $v = {}^t\mathrm{pr}_1(v_1) \wedge {}^t\mathrm{pr}_2(v_2)$ on X is such that $v(x_1, x_2) \neq 0$ for all $(x_1, x_2) \in X$. The orientation to which this form belongs is called the *product* of the orientations of X_1 and X_2 defined by v_1 and v_2, respectively. We remark that the canonical diffeomorphism

$$X_1 \times X_2 \to X_2 \times X_1$$

which interchanges the factors does not preserve the product orientation unless either p or q is *even*.

(16.21.6) Let X be an orientable connected differential manifold, $f: Y \to X$ an *étale* morphism (16.5.6). Then Y is orientable. For if v is a differential n-form belonging to an orientation of X, it is immediate that ${}^tf(v)(y) \neq 0$ for all $y \in Y$, since f is a local diffeomorphism. The orientation defined by ${}^tf(v)$ is said to be *induced* by f from the orientation of X defined by v.

(16.21.7) *Let X, Y be two differential manifolds, $f: X \to Y$ a submersion, x a point of X, and $y = f(x)$; let $n = \dim_x(X)$, $m = \dim_y(Y)$, so that $\dim_x(f^{-1}(y)) = n - m$ (16.8.8). Let $j: f^{-1}(y) \to X$ be the canonical injection, $u = T_x(f)$, $w = T_x(j)$, so that ${}^tu: T_y(Y)^* \to T_x(X)^*$ is injective and*

$${}^tw: T_x(X)^* \to T_x(f^{-1}(y))^*$$

is surjective; consequently $\bigwedge^m({}^tu)$ is injective and $\bigwedge^{n-m}({}^tw)$ is surjective. Then, for each n-covector $v_x \in \bigwedge^n(T_x(X)^*)$ and each m-covector $\zeta_y \neq 0$ in $\bigwedge^m(T_y(Y)^*)$, there exists a unique $(n-m)$-covector $\sigma_x \in \bigwedge^{n-m}(T_x(f^{-1}(y))^*)$ such that

(16.21.7.1) $$v_x = \Big(\bigwedge^m({}^tu)(\zeta_y)\Big) \wedge \sigma'_x$$

for every $(n-m)$*-covector* $\sigma'_x \in \bigwedge^{n-m}(\mathrm{T}_x(\mathrm{X})^*)$ *such that*

(16.21.7.2) $$\sigma_x = \bigwedge^{n-m}({}^t w)(\sigma'_x).$$

By virtue of (16.7.4) we reduce immediately to the case where $\mathrm{X} = \mathbf{R}^n$, $\mathrm{Y} = \mathbf{R}^m$, f is the projection $(\xi^1, \ldots, \xi^n) \mapsto (\xi^1, \ldots, \xi^m)$, and x and y are the origins, so that $f^{-1}(y) = \mathbf{R}^{n-m}$. We can then identify $\mathrm{T}_x(\mathrm{X})^*$ (resp. $\mathrm{T}_y(\mathrm{Y})^*$) with $\mathbf{R}^n$ (resp. $\mathbf{R}^m$), and $\mathrm{T}_x(f^{-1}(y))^*$ with $\mathbf{R}^{n-m}$; ${}^t u$ is the canonical injection $(\xi_1, \ldots, \xi_m) \mapsto (\xi_1, \ldots, \xi_m, 0, \ldots, 0)$, and ${}^t w$ is the canonical projection $(\xi_1, \ldots, \xi_n) \mapsto (\xi_{m+1}, \ldots, \xi_n)$. Moreover, if $(\mathbf{e}_i^*)$ is the basis dual to the canonical basis of $\mathbf{R}^n$, we may suppose that $\zeta_y = \mathbf{e}_1^* \wedge \cdots \wedge \mathbf{e}_m^*$ and

$$\upsilon_x = c \cdot \mathbf{e}_1^* \wedge \cdots \wedge \mathbf{e}_n^*.$$

The $(n-m)$-covectors σ'_x such that $\upsilon_x = \left(\bigwedge^{n-m}({}^t u)(\zeta_y)\right) \wedge \sigma'_x$ are then of the form

$$c \cdot \mathbf{e}_{m+1}^* \wedge \cdots \wedge \mathbf{e}_n^* + \mathbf{z}^*,$$

where $\mathbf{z}^*$ is a linear combination of $(n-m)$-covectors, each of which is an exterior product of certain of the $\mathbf{e}_j^*$ in which *at least one of the factors* has index $j \leqq m$. For all these covectors it is clear that the image $\bigwedge^{n-m}({}^t w)(\sigma'_x)$ is the *same* $(n-m)$-covector $c \cdot \mathbf{e}_{m+1}^* \wedge \cdots \wedge \mathbf{e}_n^* = \sigma_x$.

We denote by υ_x/ζ_y the $(n-m)$-covector σ_x whose existence has just been established. The above proof shows that, for *fixed* y and $\zeta_y \neq 0$, we have:

(16.21.8) *If* $x \mapsto \upsilon(x)$ *is a differential n-form of class* C^r *on* X, *then* $x \mapsto \upsilon(x)/\zeta_y$ *is a differential* $(n-m)$*-form of class* C^r *on* $f^{-1}(y)$.

For, by the use of a chart we may suppose that f coincides in a neighborhood of 0 with the canonical projection $\mathbf{R}^n \to \mathbf{R}^m$ considered in the above proof, and that $\upsilon(x) = h(x)\, d\xi^1 \wedge \cdots \wedge d\xi^n$, where h is of class C^r; it follows that $\upsilon(x)/\zeta_y = h(x)\, d\xi^{m+1} \wedge \cdots \wedge d\xi^n$, h being restricted to $\mathbf{R}^{n-m}$.

Furthermore, it is clear that the relation $\upsilon(x) \neq 0$ implies that $\upsilon(x)/\zeta_y \neq 0$, whence:

(16.21.9) *If* X *is an orientable manifold and* $f: \mathrm{X} \to \mathrm{Y}$ *a submersion, then for each* $y \in f(\mathrm{X})$, *the fiber* $f^{-1}(y)$ *is an orientable submanifold of* X.

Remarks

(16.21.9.1) Under the hypotheses of (16.21.9), suppose that Y also is an orientable manifold, and that $\zeta_y = \zeta(y)$ is the value at y of a differential m-form ζ of class C^r on Y which is $\neq 0$ at all points of Y. Then the most practical method of calculating the form $\sigma = \upsilon/\zeta_y$ on $f^{-1}(y)$ is often the following: determine an $(n-m)$-form σ_0 of class C^r *on* X such that

(16.21.9.2) $$ {}^t f(\zeta) \wedge \sigma_0 = \upsilon $$

(in general there will be infinitely many). Then it follows from (16.21.7) that $\sigma = {}^t j(\sigma_0)$. We shall also write $\upsilon/{}^t f(\zeta_y)$ when there may be ambiguity about f.

For example, if Y is open in $\mathbf{R}^m$ and if $\mathbf{f} = (f^1, \dots, f^m)$, where the f^i are real-valued functions of class C^∞, to say that $\mathbf{f}$ is a submersion signifies that at each point $x \in X$ the m covectors $d_x f^j$ are linearly independent (16.7.1). The form ${}^t\mathbf{f}(\zeta)$ is then (16.20.9.2)

$$ (a \circ \mathbf{f})\, df^1 \wedge df^2 \wedge \cdots \wedge df^m, $$

where a is a real-valued function on Y. To determine σ locally we may, by restricting ourselves to a neighborhood U of a point of the fiber $\mathbf{f}^{-1}(y)$, complete the f^j to a system of local coordinates $f^1, \dots, f^n$. Then we have

$$ \upsilon = b \cdot df^1 \wedge \cdots \wedge df^n, $$

where b is a real-valued function on X, and in U we may take

$$ \sigma_0 = b(a \circ \mathbf{f})^{-1}\, df^{m+1} \wedge \cdots \wedge df^n. $$

(16.21.9.3) If X and Y are two *oriented* manifolds, υ an n-form belonging to the orientation of X, and ζ an m-form belonging to the orientation of Y, then the orientation of $f^{-1}(y)$ determined by υ/ζ_y is said to be *induced by f from the orientations of* X *and* Y.

In view of (16.21.5) and the fact that the projections of $X_1 \times X_2$ are submersions, it follows from (16.21.9) that *a product* $X_1 \times X_2$ *of pure manifolds is orientable if and only if each of the factors* X_1, X_2 *is orientable.*

(16.21.10) *The spheres* S_n *are orientable.*

We may assume that $n \geq 1$, and then the assertion is an immediate consequence of the last remark, because the open set $\mathbf{R}^{n+1} - \{0\}$ is diffeomorphic to the product $\mathbf{R}_+^* \times S_n$ (16.8.10).

For later use we shall construct on the sphere $\mathbf{S}_n$ (where $n \geqq 1$) an n-form which is everywhere nonzero, by the method of (16.21.9.1). Consider $\mathbf{S}_n$ as the submanifold $r^{-1}(1)$, where $r : \mathbf{x} \mapsto \|\mathbf{x}\| = ((\xi^0)^2 + (\xi^1)^2 + \cdots + (\xi^n)^2)^{1/2}$ is a submersion of $\mathbf{R}^{n+1} - \{0\}$ onto $\mathbf{R}_+^*$ (16.8.9). On $\mathbf{R}_+^*$ we take the form $\zeta = \xi^{-1}\, d\xi$, and on $\mathbf{R}^{n+1} - \{0\}$ the canonical $(n+1)$-form

$$\upsilon = d\xi^0 \wedge d\xi^1 \wedge \cdots \wedge d\xi^n,$$

and we shall first of all construct an n-form σ_0 on $\mathbf{R}^{n+1} - \{0\}$ such that $\upsilon = {}^t r(\zeta) \wedge \sigma_0$; then we take $\sigma = {}^t j(\sigma_0)$, where $j : \mathbf{S}_n \to \mathbf{R}^{n+1} - \{0\}$ is the canonical injection. Now, we have ${}^t r(\zeta) = r^{-1}\, dr = r^{-2} \sum_{i=0}^{n} \xi^i\, d\xi^i$, and it is immediately checked that the n-form

(16.21.10.1) $$\sigma_0 = \sum_{i=0}^{n} (-1)^i \xi^i\, d\xi^0 \wedge d\xi^1 \wedge \cdots \wedge \widehat{d\xi^i} \wedge \cdots \wedge d\xi^n$$

(where as usual the caret means that the symbol underneath it is to be omitted) has the required property, and the induced n-form σ on $\mathbf{S}_n$ is just $\upsilon/\zeta(1)$. The function r and the form υ are *invariant* under the rotation group

$$\mathbf{SO}(n+1, \mathbf{R})$$

acting on $\mathbf{R}^{n+1} - \{0\}$. By virtue of the uniqueness of the n-form σ (16.21.7), this n-form is also invariant under the action of $\mathbf{SO}(n+1, \mathbf{R})$ on $\mathbf{S}_n$, and changes sign under an orthogonal transformation of determinant -1. In particular, if $s : \mathbf{x} \mapsto -\mathbf{x}$ is the symmetry transformation on $\mathbf{S}_n$, we have

(16.21.10.2) $${}^t s(\sigma)(\mathbf{x}) = (-1)^{n+1}\sigma(\mathbf{x}).$$

We shall write $\sigma^{(n)}$ in place of σ. Conventionally, when $n = 0$, $\sigma^{(0)}$ is the 0-form (= function) on $\mathbf{S}_0 = \{-1, 1\}$ which takes the value 1 at the point 1 and the value -1 at the point -1, so that the formula (16.21.10.2) remains valid.

When we take on $\mathbf{S}_n$ the orientation induced by the function r from the canonical orientations of $\mathbf{R}^{n+1}$ and $\mathbf{R}$ (16.21.9.2), the sphere $\mathbf{S}_n$ is said to be *oriented toward the outside*. With the opposite orientation, $\mathbf{S}_n$ is said to be *oriented toward the inside*.

(16.21.11) *The projective spaces* $\mathbf{P}_{2n-1}(\mathbf{R})$ *are orientable* ($n \geqq 1$).

We have seen in (16.14.10) that, for each $m \geqq 1$, the sphere $\mathbf{S}_m$ is a *two-sheeted covering* of $\mathbf{P}_m(\mathbf{R})$. If $\pi : \mathbf{S}_m \to \mathbf{P}_m(\mathbf{R})$ is the canonical projection, then for each $z \in \mathbf{P}_m(\mathbf{R})$ the two points of $\pi^{-1}(z)$ are antipodal on $\mathbf{S}_m$. If $m = 2n - 1$ is odd, we shall show that there exists on $\mathbf{P}_{2n-1}(\mathbf{R})$ a $(2n-1)$-form σ' such that ${}^t\pi(\sigma') = \sigma$, in the notation of (16.21.10). Each point $z \in \mathbf{P}_{2n-1}(\mathbf{R})$ has a connected open neighborhood U on which are defined two C^∞-sections, $u_1 : U \to \pi^{-1}(U)$ and $u_2 : U \to \pi^{-1}(U)$, which are diffeomorphisms of U onto two disjoint open subsets U_1, U_2 of $\pi^{-1}(U)$, whose union is $\pi^{-1}(U)$; also we have $u_2(z) = s(u_1(z))$ for all $z \in U$. It follows immediately from (16.21.10.2) that ${}^tu_1(\sigma | U_1) = {}^tu_2(\sigma | U_2)$; if σ'_U denotes this $(2n-1)$-form on U, then it is clear that for each open subset V of $\mathbf{P}_{2n-1}(\mathbf{R})$ over which the covering $\pi^{-1}(V)$ is trivial, the restrictions of σ'_U and σ'_V to $U \cap V$ are the same. Hence the existence of the $(2n-1)$-form σ', which is clearly $\neq 0$ at each point.

(16.21.12) *The projective spaces* $\mathbf{P}_{2n}(\mathbf{R})$ *are not orientable* ($n \geqq 1$).

With the same notation as in (16.21.11), suppose that there does exist a continuous differential $2n$-form ρ on $\mathbf{P}_{2n}(\mathbf{R})$ which is nonzero at every point. Then the same is true of ${}^t\pi(\rho)$ on $\mathbf{S}_{2n}$, and therefore ${}^t\pi(\rho) = f \cdot \sigma$, where f is a continuous real-valued function on $\mathbf{S}_{2n}$ which is never zero. However, by definition we must have ${}^ts({}^t\pi(\rho)) = {}^t\pi(\rho)$, because $\pi = \pi \circ s$; and since by (16.21.10.2) we have ${}^ts(\sigma)(x) = -\sigma(x)$, it follows that $f(-x) = -f(x)$ for all $x \in \mathbf{S}_{2n}$. Since $\mathbf{S}_{2n}$ is connected, this contradicts the fact that $f(x) \neq 0$ for all $x \in \mathbf{S}_{2n}$.

(16.21.13) *Let* X *be a pure complex-analytic manifold. Then the differential manifold* X_0 *underlying* X *is orientable.*

Let $\mathfrak{A}$ be an atlas of X, and consider two charts (U, φ, n) and (U', φ', n) belonging to $\mathfrak{A}$, such that $U \cap U' \neq \varnothing$. Let

$$\psi = \varphi | (U \cap U'), \qquad \psi' = \varphi' | (U \cap U'), \; \theta = \psi' \circ \psi^{-1};$$

θ is a holomorphic mapping of an open set in $\mathbf{C}^n$ onto an open set in $\mathbf{C}^n$. For each $z \in \varphi(U \cap U')$, it follows that $D\theta(z)$ is a **C**-*linear* bijective mapping of $\mathbf{C}^n$ onto $\mathbf{C}^n$. The proposition will therefore result from the following lemma:

(16.21.13.1) *If* $u : \mathbf{C}^n \to \mathbf{C}^n$ *is a* **C**-*linear mapping and if* $u_0 : \mathbf{R}^{2n} \to \mathbf{R}^{2n}$ *is the same mapping u considered as an* **R**-*linear mapping, then*

(16.21.13.2) $$\det(u_0) = |\det(u)|^2.$$

To see this, take a basis $(b_j)_{1 \leqq j \leqq n}$ of $\mathbf{C}^n$ with respect to which the matrix of u is upper triangular (A. 6.10):

$$\begin{pmatrix} r_1 & p_{12} & \cdots & p_{1n} \\ 0 & r_2 & \cdots & p_{2n} \\ \cdots & \cdots & \cdots & \cdots \\ 0 & 0 & \cdots & r_n \end{pmatrix}$$

If $r_j = s_j + it_j$, where s_j and t_j are real, then the matrix of u_0 relative to the basis of $\mathbf{R}^{2n}$ formed by the b_j and the ib_j ($1 \leq j \leq n$) is of the form

$$\begin{pmatrix} R_1 & P_{12} & \cdots & P_{1n} \\ 0 & R_2 & \cdots & P_{2n} \\ \cdots & \cdots & \cdots & \cdots \\ 0 & 0 & \cdots & R_n \end{pmatrix},$$

a triangular array of blocks of order 2, in which

$$R_j = \begin{pmatrix} s_j & -t_j \\ t_j & s_j \end{pmatrix}.$$

The formula (16.21.13.2) follows directly by calculating the determinant of this matrix (A. 7.4).

We remark that there is a *canonical* orientation on X_0, with the property that for each chart (U, φ, n) of the complex manifold X the corresponding chart $(U, \varphi, 2n)$ of X_0 preserves the orientation, where $\mathbf{R}^{2n}$ is endowed with the canonical orientation (16.21.4) and $\mathbf{C}^n$ is identified with $\mathbf{R}^{2n}$ via the mapping $(\zeta^1, \zeta^2, \ldots, \zeta^n) \mapsto (\mathscr{R}\zeta^1, \mathscr{I}\zeta^1, \ldots, \mathscr{R}\zeta^n, \mathscr{I}\zeta^n)$. It is easily verified that the forms which belong to the canonical orientation of X_0 are those which, for each chart (U, φ, n) of X, have a restriction to U which can be written as

$$f \cdot d\varphi^1 \wedge d\overline{\varphi}^1 \wedge d\varphi^2 \wedge d\overline{\varphi^2} \wedge \cdots \wedge d\varphi^n \wedge d\overline{\varphi^n},$$

where $f(x) > 0$ for all $x \in U$.

(16.21.14) *The manifold underlying a Lie group* G *is orientable.*

Suppose that $\dim G = n$, and let $\mathbf{z}_e^*$ be a nonzero n-covector at the identity element e of G. Then $x \mapsto \gamma(x)\mathbf{z}_e^*$ is a C^∞ differential n-form on G (16.20.13) which clearly is everywhere $\neq 0$.

We remark that a homogeneous space of a Lie group is not necessarily orientable; for example, we have just seen that $\mathbf{P}_{2n}(\mathbf{R})$ is not orientable ((16.11.8) and (16.21.12)).

(16.21.15) We have seen in (16.21.12) that the nonorientable manifold $\mathbf{P}_{2n}(\mathbf{R})$ *admits an orientable two-sheeted covering*. This is a general fact:

(16.21.16) *Every pure manifold* X *of dimension n admits a canonical orientable two-sheeted covering.*

In the bundle $\bigwedge^n T(X)^*$ consider the open set Z which is the complement of the zero section. The multiplicative group $\mathbf{R}^*$ acts differentiably and *freely* on Z, because $\bigwedge^n T(X)^*$ is a line bundle, and by taking a fibered chart of $\bigwedge^n T(X)^*$ it is immediately seen that Z is a *principal bundle* over X with structure group $\mathbf{R}^*$. Now apply (16.14.8), taking H to be the subgroup $\mathbf{R}_+^*$ of $\mathbf{R}$ consisting of the positive real numbers; since $\mathbf{R}^*/\mathbf{R}_+^*$ is the group of two elements, it follows that $X' = Z/\mathbf{R}_+^*$ is a two-sheeted covering of X. To show that X′ is orientable, we shall construct an atlas of X′ satisfying condition (b) of (16.21.1).

To do this, we start with an atlas $\mathfrak{A}$ of X such that, for each chart (U, φ, n) belonging to $\mathfrak{A}$, the open set U is connected and the inverse image of U in X′ is the disjoint union of two open sets U′, U″ such that the canonical projections $p' : U' \to U$ and $p'' : U'' \to U$ are diffeomorphisms. If $\pi : Z \to X$, $\pi' : Z \to X'$ are the canonical projections, then by hypothesis there is a canonical morphism of fibrations (16.15.4) $\psi : \varphi(U) \times \mathbf{R}^* \to \pi^{-1}(U)$, and $\pi'^{-1}(U')$ and $\pi'^{-1}(U'')$ are each equal to one of the images under ψ of $\varphi(U) \times \mathbf{R}_+^*$ and $\varphi(U) \times (-\mathbf{R}_+^*)$. Let s be the reflection of $\mathbf{R}^n$ with respect to the hyperplane $\xi^1 = 0$. If $\pi'^{-1}(U') = \psi(\varphi(U) \times \mathbf{R}_+^*)$, we take as chart of U′ the triplet $(U', \varphi \circ p', n)$; otherwise we take $(U', s \circ \varphi \circ p', n)$; and similarly for U″. We have now to show that the condition (b) of (16.21.1) is satisfied by the atlas of X′ so defined. For this we may limit ourselves to considering two charts corresponding to charts (U, φ, n), (U, φ', n) of X having the same domain of definition. Let $\psi' : \varphi'(U) \times \mathbf{R}^* \to \pi^{-1}(U)$ be the canonical morphism corresponding to the second chart; if $F : \varphi'(U) \to \varphi(U)$ is the transition diffeomorphism, then the composite morphism $\psi' \circ \psi^{-1}$ is given (16.20.9.4) by

$$(x, \mathbf{t}) \mapsto (F(x), J(x)^{-1}\mathbf{t}),$$

where $J(x)$ is the Jacobian of F at the point x. Suppose for example that $\pi'^{-1}(U') = \psi(\varphi(U) \times \mathbf{R}_+^*)$. If $J(x) > 0$ in $\varphi'(U)$, then we have also $\pi'^{-1}(U') = \psi(\varphi'(U) \times \mathbf{R}_+^*)$, and the transition diffeomorphism for the charts

$$(U', \varphi \circ p', n), \ (U', \varphi' \circ p', n)$$

is then F. If on the other hand $J(x) < 0$ in $\varphi'(U)$, then we have $\pi'^{-1}(U') = \psi'(\varphi'(U) \times (-\mathbf{R}_+^*))$, and the transition diffeomorphism for the charts $(U', \varphi \circ p', n)$ and $(U', s \circ \varphi' \circ p', n)$ is then $F \circ s$. In both cases, the transition diffeomorphism has Jacobian >0. The argument is similar when

$$\pi'^{-1}(U) = \psi(\varphi(U) \times (-\mathbf{R}_+^*)),$$

and the proof is complete.

We remark that if X is orientable, then the covering X′ is trivializable, because Z admits a section over X and is therefore trivializable.

PROBLEMS

1. Let G be a connected Lie group, H a closed subgroup of G. Suppose that, at the point $x_0 \in G/H$ which is the image of e, the endomorphisms $\mathbf{h}_{x_0} \mapsto t \cdot \mathbf{h}_{x_0}$ of $T_{x_0}(G/H)$, where $t \in H$, have determinant 1. Show that G/H is orientable. (Using the hypothesis, show that there is a differential form on G/H of highest degree which is invariant under the action of G.) Hence give another proof of the orientability of spheres (16.11.5). Generalize to Stiefel manifolds.

   Show in the same way that the homogeneous spaces

$$\mathbf{SO}(n, \mathbf{R})/(\mathbf{SO}(p, \mathbf{R}) \times \mathbf{SO}(n-p, \mathbf{R})) = \mathbf{G}'_{n,p}(\mathbf{R})$$

   are orientable. $\mathbf{G}'_{n,p}(\mathbf{R})$ is in one–one correspondence with the set of *oriented* p-dimensional subspaces of $\mathbf{R}^n$. Show that $\mathbf{G}'_{n,p}(\mathbf{R})$ is a two-sheeted covering of the Grassmannian $\mathbf{G}_{n,p}(\mathbf{R})$.

2. Show that the Möbius strip and the Klein bottle (16.14.10) are not orientable (same method as for projective spaces). Generalize to the situation of a principal bundle over $\mathbf{S}_1$, obtained by making an arbitrary finite subgroup of $\mathbf{S}_1 = \mathbf{U}$ act by translations.

3. Let X be a pure differential manifold of dimension n. Define a canonical differential $(n+1)$-form on the manifold $\bigwedge^n T(X)^*$ which does not vanish at any point.

4. If M is any pure differential manifold, show that the tangent bundle T(M) is orientable (use the chart construction of T(M) (16.15.4)).

22. CHANGE OF VARIABLES IN MULTIPLE INTEGRALS. LEBESGUE MEASURES

(16.22.1) *Let* U, U′ *be two open subsets of* $\mathbf{R}^n$, *and let* u *be a homeomorphism of* U *onto* U′ *such that both* u *and* u^{-1} *are of class* C^1. *For each* $x \in U$, *let* $J(x)$ *be the Jacobian of* u *at* x (8.10). *Let* λ_U *and* $\lambda_{U'}$ *be the measures induced on* U *and* U′ *by Lebesgue measure* λ *on* $\mathbf{R}^n$. *Then the image under* u (13.1.6) *of the measure* $|J| \cdot \lambda_U$ *is equal to* $\lambda_{U'}$.

This means that if f is any function in $\mathscr{K}(\mathbf{R}^n)$ with support contained in U′, then

$$\int_{U'} f(x)\, d\lambda(x) = \int_U f(u(x))\,|J(x)|\, d\lambda(x) \tag{16.22.1.1}$$

("formula for change of variables in a multiple integral").

The proof is in several steps.

(1) Let (U_α) be an open covering of U, so that the $U'_\alpha = u(U_\alpha)$ form an open covering of U′, and let $u_\alpha : U_\alpha \to U'_\alpha$ be the restriction of u to U_α. Suppose that the theorem is true for each u_α. Then it is true for u. For if λ_{U_α} is the restriction of λ to U_α and if $\mu_{U'}$ is the image under u of $|J| \cdot \lambda_U$, then its restriction to U'_α is the image under u_α of $(|J| \mid U_\alpha) \cdot \lambda_{U_\alpha}$, hence is the restriction to U'_α of $\lambda_{U'}$, by hypothesis; hence we conclude from (13.1.9) that $\mu_{U'} = \lambda_{U'}$.

(2) Let u' be a homeomorphism of U′ onto an open subset U″ of $\mathbf{R}^n$, such that u' and u'^{-1} are of class C^1, and put $u'' = u' \circ u$, which is a homeomorphism of U onto U″. If the theorem is true for u and u', then it is true for u''. For if $J'(x)$ is the Jacobian of u' at the point x, then for any function $f \in \mathscr{K}(\mathbf{R}^n)$ with support contained in U″ we have

$$\int f(x)\, d\lambda(x) = \int f(u'(x))\,|J'(x)|\, d\lambda(x)$$

$$= \int f(u'(u(x)))\,|J'(u(x))| \cdot |J(x)|\, d\lambda(x)$$

and the Jacobian of u'' at x is $J'(u(x))J(x)$ (8.10.1).

(3) The theorem is true when u is the restriction to U of an affine mapping $x \mapsto a + w(x)$. For then we have $Du = w$ (8.1.3); whence $J(x) = \det w$ for all $x \in \mathbf{R}^n$, and the formula (16.22.1.1) follows from (14.3.9).

(4) The theorem is true for $n = 1$. We have then $J(x) = Du(x)$, and every point of U is the center of a bounded open interval in which Du is bounded and keeps the same sign. By (1) above we may therefore assume that $U =]a, b[$ is a bounded interval in $\mathbf{R}$, u being the restriction to U of a continuous function, differentiable and monotonic in $[a, b]$. Then we have $U' =]u(a), u(b)[$ if $D(x) > 0$ for $x \in U$, and $U' =]u(b), u(a)[$ otherwise, and the formula (16.22.1.1) reduces to (8.7.4).

(5) The theorem is true for n arbitrary and u of the form

(16.22.1.2) $$(\xi_1, \xi_2, \ldots, \xi_n) \mapsto (\theta(\xi_1, \ldots, \xi_n), \xi_2, \ldots, \xi_n),$$

where θ is of class C^1 and $J(x) = D_1\theta(x) \neq 0$ for all $x \in U$. For each point $x' = (\xi_2, \ldots, \xi_n) \in \mathbf{R}^{n-1}$ such that the section $U(x') \neq \varnothing$, the mapping $\xi_1 \mapsto \theta(\xi_1, \xi_2, \ldots, \xi_n)$ is a homeomorphism of the open set $U(x') \subset \mathbf{R}$ onto an open set in $\mathbf{R}$, and both it and its inverse are of class C^1. Hence, by (4),

$$\int f(\xi_1, \xi_2, \ldots, \xi_n)\, d\xi_1 = \int f(\theta(\xi_1, \ldots, \xi_n), \xi_2, \ldots, \xi_n)\,|D_1\theta(\xi_1, \ldots, \xi_n)|\, d\xi_1$$

and therefore, by virtue of the definition of the product measure on $\mathbf{R}^n$ (13.21.2),

$$\begin{aligned}\int f(x)\,d\lambda(x) &= \int \cdots \int d\xi_2 \cdots d\xi_n \int f(\xi_1, \ldots, \xi_n)\,d\xi_1 \\ &= \int \cdots \int d\xi_2 \cdots d\xi_n \int f(u(x))\,|J(x)|\,d\xi_1 \\ &= \int f(u(x))\,|J(x)|\,d\lambda(x).\end{aligned}$$

(6) The results established so far show immediately that in the general case the theorem will result from the following lemma:

(16.22.1.3) *Under the hypotheses of* (16.22.1), *for each* $x \in \mathrm{U}$ *there exists an open neighborhood* V *of* x *such that the homeomorphism of* V *onto* $u(\mathrm{V})$, *obtained by restricting* u, *is of the form* $u_p \circ u_{p-1} \circ \cdots \circ u_1$, *where each* u_j *is a homeomorphism of an open subset of* $\mathbf{R}^n$ *onto an open subset of* $\mathbf{R}^n$, *of one of the types considered in* (3) *and* (5) *above.*

By replacing u by $t \circ u \circ t'^{-1}$, where t and t' are translations, we may assume that $x = u(x) = 0$. Replacing u by $(\mathrm{D}u(0))^{-1} \circ u$, we may assume moreover that $\mathrm{D}u(0) = 1_{\mathbf{R}^n}$. Hence we may write $u(x) = (u_1(x), \ldots, u_n(x))$, where, for $1 \leq j \leq n$, u_j is a C^1-mapping of U into $\mathbf{R}$ such that $\mathrm{D}_i u_j(0) = \delta_{ij}$ (Kronecker delta). Put

$$v_j(x) = (u_1(x), \ldots, u_j(x), \xi_{j+1}, \ldots, \xi_n)$$

(where $x = (\xi_1, \ldots, \xi_n)$); it follows from the implicit function theorem (10.2.5) that there exists an open neighborhood V of 0 in U such that, for each j, $w_j = v_j | \mathrm{V}$ is a homeomorphism of V onto an open neighborhood of 0. We may therefore write

$$u | \mathrm{V} = w_n = (w_n \circ w_{n-1}^{-1}) \circ (w_{n-1} \circ w_{n-2}^{-1}) \circ \cdots \circ (w_2 \circ w_1^{-1}) \circ w_1$$

and each $w_j \circ w_{j-1}^{-1}$ is of the form

$$(\xi_1, \ldots, \xi_n) \mapsto (\xi_1, \ldots, \xi_{j-1}, \theta_j(x), \xi_{j+1}, \ldots, \xi_n).$$

Hence we obtain a factorization of u of the desired type by taking each u_i to be either a linear transformation of the type

$$s_\sigma : (\xi_1, \ldots, \xi_n) \mapsto (\xi_{\sigma(1)}, \ldots, \xi_{\sigma(n)}),$$

where σ is a permutation of $\{1, 2, \ldots, n\}$, or a mapping of the form

$$s_\sigma \circ (w_j \circ w_{j-1}^{-1}) \circ s_\tau,$$

where the permutations σ and τ are chosen so that this mapping is of the type (16.22.1.2). Q.E.D.

For an important example of the application of (16.22.1) ("change to polar coordinates" in $\mathbf{R}^n$), see (16.24.9).

(16.22.2) *Let* X *be a pure differential manifold of dimension n. Then there exists a positive measure μ on* X *with the following property: for each chart* (U, φ, n) *of* X, *the image under φ of the induced measure μ_U is of the form $f \circ (\lambda_{\varphi(\mathrm{U})})$, where λ is Lebesgue measure on $\mathbf{R}^n$ and f is a function of class* C^∞ *which is $\neq 0$ at every point of $\varphi(\mathrm{U})$. Moreover, any two measures μ, μ' on* X *with this property are equivalent and each has a density of class* C^∞ *with respect to the other.*

There exists a sequence of charts $(\mathrm{U}_k, \varphi_k, n)$ of X such that the U_k are relatively compact and form a locally finite open covering of X. Let ν_k be the image under the homeomorphism φ_k^{-1} of the measure induced by λ on $\varphi_k(\mathrm{U}_k)$. Also let (f_k) be a C^∞-partition of unity subordinate to the covering (U_k) (16.4.1). Then there is a measure μ_k on X which coincides with $f_k \cdot \nu_k$ on U_k and with the zero measure on the complement of $\mathrm{Supp}(f_k)$ (13.1.9), and this measure μ_k is clearly bounded. Moreover, since each compact subset of X meets only finitely many of the U_k, the sum $\mu = \sum_k \mu_k$ is defined and is a positive measure on X. We have to show that it has the property stated above. Let g be a function in $\mathscr{K}(\mathrm{X})$ with support contained in U. By definition, we have

$$\begin{aligned}\int g\,d\mu = \sum_k \int_\mathrm{U} g\,d\mu_k &= \sum_k \int_\mathrm{U} (gf_k)\,d\nu_k \\ &= \sum_k \int_{\varphi_k(\mathrm{U}\cap\mathrm{U}_k)} (g \circ \varphi_k^{-1})(f_k \circ \varphi_k^{-1})\,d\lambda,\end{aligned}$$

the summation being over the finite set of indices k such that U_k intersects U. If $\theta_k : \varphi(\mathrm{U} \cap \mathrm{U}_k) \to \varphi_k(\mathrm{U} \cap \mathrm{U}_k)$ is the transition diffeomorphism and $\mathrm{J}(\theta_k)$ its Jacobian, then by (16.22.1.1) we obtain

$$\int g\,d\mu = \int_{\varphi(\mathrm{U})} (g \circ \varphi^{-1})h\,d\lambda,$$

where $h = \sum_k (f_k \circ \varphi^{-1})|\mathrm{J}(\theta_k)|$ is a function which does not depend on g but only on the charts. Since $\mathrm{J}(\theta_k)$ is nonzero at each point of $\varphi(\mathrm{U} \cap \mathrm{U}_k)$ and

$f_k \circ \varphi^{-1}$ vanishes in a neighborhood of each point of $\varphi(U) - \varphi(U \cap U_k)$, it follows that h is of class C^∞ and $\neq 0$ at each point $z \in \varphi(U)$, because at least one of the functions $f_k \circ \varphi^{-1}$ is $\neq 0$ at this point. The last assertion of **(16.22.2)** is evident, because the inverse of a nonvanishing C^∞-function is of class C^∞.

Measures μ satisfying the condition of **(16.22.2)** are called *Lebesgue measures* on X. The set of negligible functions (resp. measurable functions) is the same for all these measures. So also is the set of locally integrable functions, because the density of one Lebesgue measure relative to another is continuous and locally bounded. Whenever we speak of negligible, measurable, or locally integrable functions on X without specifying the measure, it is always a Lebesgue measure that is meant.

We remark that *submanifolds of* X *of dimension* $<n$ *are negligible*, because vector subspaces of dimension $<n$ in $\mathbf{R}^n$ are negligible for Lebesgue measure (**(14.3.6)** and **(13.21.12)**).

If E is a fiber bundle over X, the notion of a *measurable section* of E over an open subset of X is well defined, independently of the choice of Lebesgue measure on X. So also is the notion of a *locally integrable section* if E is a vector bundle; this is clear when the bundle is trivial, and since the notion is local with respect to X, it is enough to verify that on an open set over which E is trivializable, the notion is independent of the trivialization chosen, and this follows immediately from the definitions **(16.15.3)**.

PROBLEMS

1. Let u be a C^1-mapping of an open subset U of $\mathbf{R}^n$ into $\mathbf{R}^n$, and let $J(x)$ be its Jacobian at $x \in U$.

(a) Let K be a compact subset of U such that $J(x) = 0$ for all $x \in K$. Show that there exists a real number $c > 0$ and, for each sufficiently small $\varepsilon > 0$, a real number $\delta_0(\varepsilon) > 0$ with the following property: for each $x \in K$ and each cube C with center x and side length $2\delta < \delta_0(\varepsilon)$, contained in U, we have $\lambda(u(C)) \leq c\varepsilon\lambda(C)$. (By using the uniform continuity of J on K, show that if the image of $\mathbf{R}^n$ under $Du(x)$ has dimension $p < n$, and if $(\mathbf{b}_j)_{1 \leq j \leq p}$ is an orthonormal basis of this space, and $(\mathbf{c}_k)_{1 \leq k \leq n-p}$ an orthonormal basis of its orthogonal supplement, then $u(C)$ is contained in the parallelotope with center $u(x)$, constructed from the p vectors $2\delta n M \mathbf{b}_j$, where M is the least upper bound of $\|Du(y)\|$ in K, and the $n-p$ vectors $\varepsilon\delta n^{1/2}\mathbf{c}_k$.)

(b) Deduce from (a) that, for each λ-measurable subset A of U, we have

$$\lambda^*(u(A)) \leq \int_A^* |J(x)|\, d\lambda(x).$$

(Reduce to the case where A is relatively compact. Let K be the set of points $x \in \bar{A}$ at which $J(x) = 0$, and let V be a relatively compact open neighborhood of K such that

$\lambda(V) < \lambda(K) + \varepsilon$. Cover K by small closed cubes with pairwise disjoint interiors, and use (a) to show that if W is the union of the concentric open cubes of twice the side length, then $\lambda^*(u(W))$ can be made arbitrarily small. Next, show that there exists a partition of $\bar{A} \cap \complement W$ into a finite number of integrable sets G_j, each of which is contained in an open set U_j, such that the restriction u_j of u to U_j is a homeomorphism of U_j onto an open set $u(U_j)$ such that both u_j and u_j^{-1} are of class C^1. Finally, use (16.22.1) in each U_j.)

(c) Deduce from (b) that the image under u of each set $N \subset U$ of measure zero is of measure zero. If also N is closed in U, then $u(N)$ is a meager set, a denumerable union of nowhere dense compact sets of measure zero.

(d) If E is the closed subset of points $x \in U$ such that $J(x) = 0$, show that $u(E)$ is a meager set, a denumerable union of nowhere dense compact sets of measure zero. Deduce that if M is any meager subset of U, then $u(M)$ is meager in $\mathbf{R}^n$. (Show that if B is any compact and nowhere dense subset of U, then $u(B)$ is nowhere dense. For this purpose, consider a decreasing sequence (V_n) of open neighborhoods of E, whose intersection is E, and consider as in (b) above a suitable partition of $B \cap \complement V_n$ into integrable sets.)

2. With the notation of (16.22.1) show that if u is a homeomorphism of U onto U′, of class C^1 (but whose inverse is *not* necessarily of class C^1), the formula (16.22.1.1) remains valid. (Use Problem 1.) Furthermore, the set $E = \{x \in U : J(x) = 0\}$ is nowhere dense, but not necessarily of measure zero. (To prove this last point, use Problem 4 of Section 13.8.)

3. (a) Let F be a closed subset of $\mathbf{R}^n$. Show that there exists a real-valued function g of class C^∞ on $\mathbf{R}^n$, such that $g(x) = 0$ for all $x \in F$ and $g(x) > 0$ for all $x \notin F$. (Use Problem 4 of Section 16.4.)

(b) Let $f_1, f_2, \ldots, f_n$ be n real-valued functions of class C^1 on an open set $A \subset \mathbf{R}^n$. Show that if the Jacobian of the f_j vanishes on A, then for each compact $B \subset A$ there exists a C^∞-function g on $\mathbf{R}^n$ such that the set $g^{-1}(0)$ is nowhere dense in $\mathbf{R}^n$ and such that $g(f_1(x), \ldots, f_n(x)) = 0$ for all $x \in B$. (Use (a) and Problem 1(d).)

4. (a) Let f be a holomorphic function in an annulus $S : r < |z| < R$ in $\mathbf{C}$, and let $f(z) = \sum_{n=-\infty}^{+\infty} a_n z^n$ be its Laurent expansion (9.14.2). Show that $f(S)$ is open in $\mathbf{C}$ and that

$$\lambda^*(f(S)) \leqq \pi \sum_{-\infty}^{+\infty} n|a_n|^2(R^{2n} - r^{2n})$$

(the right-hand side being interpreted as $+\infty$ if the series does not converge). If f is injective on S, the two sides are equal.

(b) Let f be a holomorphic function for $|z| > 1$, and suppose that its Laurent series is of the form

$$f(z) = z + \sum_{n=0}^{\infty} b_n z^{-n}.$$

If f is injective, show that

$$\sum_{n=1}^{\infty} n|b_n|^2 \leqq 1.$$

(Use (a).) Under what conditions is $|b_1| = 1$?

(c) With the same assumptions on f as in (b), suppose moreover that $f(z) \neq 0$ for $|z| > 1$. Show that $|b_0| \leqq 2$. (Remark that there exists a function g, holomorphic for $|z| > 1$, such that $f(z^2) = (g(z))^2$, by using Section 10.2, Problem 8.)
(d) With the same assumptions on f as in (b), show that

$$|f'(z)| \leqq \frac{1}{1 - |z|^{-2}}$$

for $|z| > 1$.

5. (a) Let f be a function holomorphic in the disk $\mathrm{D} : |z| < 1$, with Taylor series of the form $f(z) = z + a_2 z^2 + \cdots + a_n z^n + \cdots$. Show that if f is injective in D, then $|a_2| \leqq 2$. For which functions is $|a_2| = 2$? (*Bieberbach's theorem*: consider the function $g(z) = f(z^{-1})^{-1}$ for $|z| > 1$ and use Problem 4.)
(b) With the same hypotheses, show that the open set $f(\mathrm{D})$ contains the open disk with center 0 and radius $\frac{1}{4}$. (If $c \notin f(\mathrm{D})$, consider the function $f(z)/(1 - c^{-1}f(z))$.)

6. Let B be a positive definite symmetric bilinear form on $\mathbf{R}^n$, and let $\Delta > 0$ be its discriminant relative to the canonical basis of $\mathbf{R}^n$. Show that

$$\int_{\mathbf{R}^n} \exp(-\mathrm{B}(x, x))\, d\lambda(x) = \pi^{n/2}\Delta^{-1/2}.$$

(Consider an automorphism u of the vector space $\mathbf{R}^n$ such that the matrix of the transformed form $\mathrm{B}(u(x), u(y))$ is diagonal.)

7. Let P_n denote the open subset of $\mathbf{R}^{n^2} = \mathbf{M}_n(\mathbf{R})$ consisting of the positive definite symmetric matrices.

(a) Let

$$X = \begin{pmatrix} x_{11} & {}^t\mathbf{z} \\ \mathbf{z} & X_1 \end{pmatrix},$$

where ${}^t\mathbf{z} = (x_{12} x_{13} \cdots x_{1n})$ is a row matrix and X_1 is a positive definite matrix of order $n - 1$. Show that

$$(\det X_1)^{-1}(\det X) = x_{11} - {}^t\mathbf{z} \cdot X_1^{-1} \cdot \mathbf{z}.$$

(Reduce to the case where X_1 is a diagonal matrix by means of an orthogonal transformation in $\mathbf{R}^{n-1}$.)
(b) If $Y \in \mathrm{P}_n$ and $s > \frac{1}{2}(n + 1)$, show that

$$\int_{\mathrm{P}_n} e^{-\operatorname{tr}(XY)}(\det X)^{s-(n+1)/2}\, d\lambda(X) = \pi^{n(n-1)/4}(\det Y)^{-s}\prod_{i=0}^{n-1}\Gamma\left(s - \frac{i}{2}\right).$$

(Reduce to the case $Y = I$ by means of an orthogonal transformation. Integrate first with respect to x_{11}, by taking as new variable of integration

$$u = (\det X_1)(x_{11} - {}^t\mathbf{z} \cdot X_1^{-1} \cdot \mathbf{z})$$

and then with respect to $x_{12}, \ldots, x_{1n}$ by using Problem 6, and finally with respect to X_1; hence obtain a reduction formula for the integral.)

23. SARD'S THEOREM

Let X, Y be two differential manifolds, $f: X \to Y$ a C^∞-mapping. Generalizing the definition of (16.5.11), a point $x \in X$ is said to be *critical* for f if f is not a submersion at x (16.7.1), in other words if

$$\mathrm{rk}_x(f) < \dim_{f(x)}(Y).$$

If E is the set of critical points of f, then $Y - f(E)$ is called the set of *regular values* of f. For each $y \in Y - f(E)$, the fiber $f^{-1}(y)$ is therefore either empty or a *closed submanifold* of X (16.8.8).

(16.23.1) (Sard's theorem) *Let $f: X \to Y$ be a C^∞-mapping, E the set of critical points of f. Then $f(E)$ is negligible in Y, and $Y - f(E)$ is dense in Y.*

The latter assertion follows from the former and the fact that the support of any Lebesgue measure on Y is the whole of Y. To prove the first assertion of the theorem, observe that if (U_k, φ_k, n_k) is a sequence of charts of X such that the U_k cover X and such that each $f(U_k)$ is contained in a chart of Y, it is enough to show that $f(E \cap U_k)$ is negligible for each k (13.6.2). We may therefore assume that $Y = \mathbf{R}^p$ and that X is an open subset of $\mathbf{R}^n$. The proof will be by induction on n; the case $n = 0$ is trivial.

Put $\mathbf{f} = (f_1, \ldots, f_p)$ where each f_j is a real-valued function of class C^∞ on X. Put $E_0 = E$, and for $m \geqq 1$, let $E_m \subset E$ denote the set of points $\mathbf{x} \in X$ such that *all the derivatives of order $\leqq m$ of all the f_j vanish at* $\mathbf{x}$. The theorem will be established if we prove the following two statements:

(i) *For each m,* $\mathbf{f}(E_m - E_{m+1})$ *is negligible.*
(ii) *For $m \geqq n/p$,* $\mathbf{f}(E_m)$ *is negligible.*

(16.23.1.1) *Proof of* (i). Since the topology of $\mathbf{R}^n$ has a denumerable basis, it is enough to show that, for each $\mathbf{x}_0 \in E_m - E_{m+1}$, there exists an open neighborhood V of $\mathbf{x}_0$ in X such that $\mathbf{f}(E_m \cap V)$ is negligible. By hypothesis, there exists an index j and a derivative $D^\alpha f_j$ of order $|\alpha| = m + 1$ which is $\neq 0$ at $\mathbf{x}_0$. By permuting the coordinates, we may assume that $j = 1$ and that $D^\alpha f_1 = D_1 w$, where w is a C^∞-function. Consider the mapping $\mathbf{h}: X \to \mathbf{R}^n$ defined by

$$\mathbf{h}(\mathbf{x}) = (w(\mathbf{x}), \xi_2, \ldots, \xi_n),$$

where $\mathbf{x} = (\xi_1, \xi_2, \ldots, \xi_n) \in X$. Since $D_1 w(\mathbf{x}_0) \neq 0$, it follows immediately from (16.5.6) that there exists an open neighborhood of $\mathbf{x}_0$ in X such that $\mathbf{h} | V$ is a diffeomorphism of V onto an open subset W of $\mathbf{R}^n$. Let $\mathbf{g} = (g_1, \ldots, g_p)$ denote the restriction of $\mathbf{f} \circ \mathbf{h}^{-1}$ to W, which is therefore a C^∞-mapping of W into $\mathbf{R}^p$.

We distinguish two cases: (a) $m = 0$ and (b) $m \geqq 1$.

(a) $m = 0$. By definition, we may assume that $w = f_1$, so that $g_1(\mathbf{x}) = \xi_1$ for all $\mathbf{x} \in \mathrm{W}$. Next, the set E' of critical points of $\mathbf{g}$ is equal to $\mathbf{h}(\mathrm{E} \cap \mathrm{V})$, hence $\mathbf{f}(\mathrm{E} \cap \mathrm{V}) = \mathbf{g}(\mathrm{E}')$ and it is enough to prove that $\mathbf{g}(\mathrm{E}')$ is negligible. Identify $\mathbf{R}^n$ with $\mathbf{R} \times \mathbf{R}^{n-1}$, and for each $\mathbf{x} = (\zeta, \mathbf{z}) \in \mathrm{W}$, put $\mathbf{g}(\mathbf{x}) = (\zeta, \mathbf{g}_\zeta(\mathbf{z}))$. Then the Jacobian matrix of $\mathbf{g}$ at the point $\mathbf{x}$ is of the form

$$D\mathbf{g}(\mathbf{x}) = \begin{pmatrix} 1 & 0 \\ * & D\mathbf{g}_\zeta(\mathbf{z}) \end{pmatrix};$$

hence in order that $\mathbf{x} \in \mathrm{E}'$ it is necessary and sufficient that $\mathbf{z} \in \mathrm{E}'_\zeta$, where E'_ζ is the set of critical points of $\mathbf{g}_\zeta$. Consequently, for each $\zeta \in \mathbf{R}$,

$$(\{\zeta\} \times \mathbf{R}^{p-1}) \cap \mathbf{g}(\mathrm{E}') = \{\zeta\} \times \mathbf{g}_\zeta(\mathrm{E}'_\zeta).$$

Now the inductive hypothesis implies that $\mathbf{g}_\zeta(\mathrm{E}'_\zeta)$ is negligible in $\mathbf{R}^{p-1}$; on the other hand, E' is closed in W, hence is a denumerable union of compact sets, and consequently $\mathbf{g}(\mathrm{E}')$ is a denumerable union of compact sets, hence is Lebesgue-measurable in $\mathbf{R}^p$ (13.9.3). It follows now from (13.21.10) that $\mathbf{g}(\mathrm{E}')$ is negligible.

(b) $m \geqq 1$. By definition, we may assume that $w(\mathbf{x}) = 0$ for $\mathbf{x} \in \mathrm{E}_m$, hence $\mathbf{h}(\mathrm{E}_m \cap \mathrm{V}) \subset \{0\} \times \mathbf{R}^{-n1}$. For each point $(0, \mathbf{z}) \in \mathrm{W} \cap (\{0\} \times \mathbf{R}^{n-1})$, put $\mathbf{g}(0, \mathbf{z}) = \mathbf{g}_0(\mathbf{z})$. Since all the first derivatives of $\mathbf{g}$ vanish at each point of $\mathbf{h}(\mathrm{E}_m \cap \mathrm{V})$, all these points are critical points of $\mathbf{g}_0$. The inductive hypothesis therefore implies that $\mathbf{g}_0(\mathbf{h}(\mathrm{E}_m \cap \mathrm{V}))$ is negligible in $\mathbf{R}^p$, and this set is precisely $\mathbf{f}(\mathrm{E}_m \cap \mathrm{V})$.

(16.23.1.2) *Proof of* (ii). Let $\|\mathbf{x}\|$ denote the norm $\sup_i |\xi_i|$ on $\mathbf{R}^n$ and on $\mathbf{R}^p$. For each real number $a > 0$ and each $\mathbf{k} = (k_1, \ldots, k_n) \in \mathbf{R}^n$, let $\mathrm{I}(\mathbf{k}, a)$ denote the cube in $\mathbf{R}^n$ defined by the inequalities $k_i \leqq \xi_i \leqq k_i + a$ $(1 \leqq i \leqq n)$. Then it is evidently enough to show that $\mathbf{f}(\mathrm{E}_m \cap \mathrm{I}(\mathbf{k}, a))$ is negligible for some $a > 0$ such that $\mathrm{I}(\mathbf{k}, a) \subset \mathrm{X}$. Let M be the least upper bound of $\|\mathbf{f}^{(m+1)}(\mathbf{x})\|$ in $\mathrm{I}(\mathbf{k}, a)$; it follows from Taylor's formula (8.14.3) that if $\mathbf{x} \in \mathrm{E}_m \cap \mathrm{I}(\mathbf{k}, a)$ and $\mathbf{x} + \mathbf{t} \in \mathrm{I}(\mathbf{k}, a)$, then

(16.23.1.3) $$\|\mathbf{f}(\mathbf{x} + \mathbf{t}) - \mathbf{f}(\mathbf{x})\| \leqq \mathrm{M}\|\mathbf{t}\|^{m+1}.$$

Now observe that, for each integer $\mathrm{N} > 1$, the cube $\mathrm{I}(\mathbf{k}, a)$ is the union of the N^n cubes $\mathrm{I}(\mathbf{s}, a/\mathrm{N})$, where

$$\mathbf{s} = (k_1 + (as_1/\mathrm{N}), \ldots, k_n + (as_n/\mathrm{N}))$$

and the integers s_i range independently from 0 to $\mathrm{N} - 1$. The set $\mathbf{f}(\mathrm{E}_m \cap \mathrm{I}(\mathbf{k}, a))$ is therefore contained in the union of the N^n sets $\mathbf{f}(\mathrm{E}_m \cap \mathrm{I}(\mathbf{s}, a/\mathrm{N}))$; but for each $\mathbf{s}$ such that $E_m \cap I(\mathbf{s}, a/N)$ is nonempty, if $\mathbf{x}_0$ is a point of this set, it follows from (16.23.1.3) that for every other point $\mathbf{x} \in E_m \cap I(\mathbf{s}, a/N)$ we have $\|\mathbf{f}(\mathbf{x}) - \mathbf{f}(\mathbf{x}_0)\| \leqq M(a/N)^{m+1}$, and therefore

$$\lambda(\mathbf{f}(E_m \cap I(\mathbf{s}, a/N))) \leqq M^p(a/N)^{p(m+1)},$$

where λ is Lebesgue measure on $\mathbf{R}^p$. Hence

$$\lambda(\mathbf{f}(E_m \cap I(\mathbf{k}, a))) \leqq M^p a^{p(m+1)} N^{n-p(m+1)}.$$

Since by hypothesis $m \geqq n/p$, the right-hand side of this inequality tends to zero with $1/N$, and the proof is complete.

Sard's theorem implies, in particular:

(16.23.2) *Let* X, Y *be pure differential manifolds of dimensions* n, p, *respectively, such that* $n < p$, *and let* $f: X \to Y$ *be a* C^∞-*mapping. Then* $Y - f(X)$ *is dense in* Y.

In other words, for C^∞-mappings there do not exist phenomena of the type of the "Peano curve" (Section 4.2, Problem 5, or Section 9.12, Problem 5).

PROBLEMS

1. (a) Let m, n, p, r be integers >0. Let $\mathbf{f}$ be a C^∞-mapping of an open set $U \subset \mathbf{R}^m$ into $\mathbf{R}^n$, and let $\mathbf{g}$ be a C^∞-mapping of an open set $V \supset f(U)$ in $\mathbf{R}^n$ into $\mathbf{R}^p$. Put $\mathbf{h} = \mathbf{g} \circ \mathbf{f}$. Show that for each $\mathbf{x} \in U$ we have

$$D^r\mathbf{h}(\mathbf{x}) = \sum_{q=0}^{r} \sum_{(i_1, \ldots, i_q)} \sigma_r(i_1, \ldots, i_q) D^q\mathbf{g}(\mathbf{f}(\mathbf{x})) \circ (D^{i_1}\mathbf{f}(\mathbf{x}), \ldots, D^{i_q}\mathbf{f}(\mathbf{x})),$$

where in the inner sum, $(i_1, \ldots, i_q)$ runs through all sequences of q integers $\geqq 1$ such that $i_1 + \cdots + i_q = r$. Furthermore, in this formula, the constants $\sigma_r(i_1, \ldots, i_q)$ are rational numbers which depend only on the i_j, q, m, n, p, and r, and *not* on the functions $\mathbf{f}$ and $\mathbf{g}$.

(b) Let $\mathbf{x}_0 \in U$. Assume now only that, for some $s < r$, $\mathbf{f}$ is a mapping of class C^{r-s} of U into $\mathbf{R}^n$, and that $\mathbf{g}$ is a mapping of class C^r of $V \supset \mathbf{f}(U)$ into $\mathbf{R}^p$. Suppose also that $D^k\mathbf{g}(\mathbf{f}(x_0)) = 0$ for $k \leqq s$. For each $\mathbf{x} \in U$ and each integer $k \in [0, r]$, let

$$\mathbf{h}_k(x) = \sum_{q=s+1}^{k} \sum_{(i_1, \ldots, i_q)} \sigma_k(i_1, \ldots, i_q) D^q\mathbf{g}(\mathbf{f}(\mathbf{x})) \circ (D^{i_1}\mathbf{f}(\mathbf{x}), \ldots, D^{i_q}\mathbf{f}(\mathbf{x})) \in \mathscr{L}_k(\mathbf{R}^m; \mathbf{R}^p),$$

the second sum being over the same sequences $(i_1, \ldots, i_q)$ as in (a), so that $i_j \leqq k - s$ for all j, and hence the function $\mathbf{h}_k$ is well-defined on U. By applying Taylor's formula, show that $\mathbf{h}_k(\mathbf{x})$ can be written in the form

$$\mathbf{h}_k(\mathbf{x}) = \sum_{j=0}^{r-k} \mathbf{a}_{jk} \cdot (\mathbf{x} - \mathbf{x}_0)^{(j)} + \mathbf{R}_k(\mathbf{x}),$$

where the $\mathbf{a}_{jk}$ are constant elements of $\mathscr{L}_j(\mathbf{R}^m; \mathscr{L}_k(\mathbf{R}^m; \mathbf{R}^p)) = \mathscr{L}_{k+j}(\mathbf{R}^m; \mathbf{R}^p)$, and $\|\mathbf{R}_k(\mathbf{x})\|/\|\mathbf{x} - \mathbf{x}_0\|^{r-k}$ tends to 0 as $\mathbf{x} \to \mathbf{x}_0$. Further, the $\mathbf{a}_{jk}$ depend only on the values of the derivatives of $\mathbf{f}$ at $\mathbf{x}_0$ and of $\mathbf{g}$ at $\mathbf{f}(\mathbf{x}_0)$. Deduce that

$$j!\,\mathbf{a}_{jk} = \mathbf{h}_{k+j}(\mathbf{x}_0)$$

for $j \leqq r - k$. (Use (a).)

(c) More generally, let A be a closed subset of U, B a closed subset of V containing $\mathbf{f}(A)$, and suppose that $D^k\mathbf{g}(\mathbf{y}) = 0$ for all $\mathbf{y} \in B$ and $0 \leqq k \leqq s$. Defining the $\mathbf{h}_k$ as in (b) above, show that for $\mathbf{x}_0 \in A$,

$$\mathbf{h}_k(\mathbf{x}') = \sum_{j=0}^{r-k} \frac{1}{j!}\,\mathbf{h}_{k+j}(\mathbf{x}) \cdot (\mathbf{x}' - \mathbf{x})^{(j)} + \mathbf{R}_k(\mathbf{x}', \mathbf{x}),$$

where $\|\mathbf{R}_k(\mathbf{x}', \mathbf{x})\|/\|\mathbf{x}' - \mathbf{x}\|^{r-k}$ tends to 0 as $\mathbf{x}, \mathbf{x}'$ tend to $\mathbf{x}_0$ whilst remaining in A.

(d) Under the hypotheses of (c), deduce that there exists a mapping $\mathbf{H} : U \to \mathbf{R}^p$ of class C^r, such that $\mathbf{H}(\mathbf{x}) = \mathbf{g}(\mathbf{f}(\mathbf{x}))$ for all $\mathbf{x} \in A$ and $D^k\mathbf{H}(\mathbf{x}) = 0$ for all $\mathbf{x} \in A$ and $k \leqq s$ (*Kneser–Glaeser theorem*). (Use Whitney's extension theorem (Section **16.4**, Problem 6).)

2. Let U be an open subset of $\mathbf{R}^n$, $\mathbf{f}$ a mapping of U into $\mathbf{R}^p$, of class C^r. Show that if $r \geqq \max(1, n - p + 1)$, the image $\mathbf{f}(E)$ of the set of critical points of $\mathbf{f}$ is of measure zero in $\mathbf{R}^p$. (Proceed as in the proof of (**16.23.1**), using the result of Problem 1 to deal with the case $m \geqq 1$ in (i).)

3. The notation is that of Section **13.21**, Problem 2. The simple arc $K = g([0, 7])$ is therefore a subset of $\mathbf{R}^2$ of measure > 0.

Let K_0 be the union of $\{g_0(0)\}$, $\{g_0(7)\}$ and the three sets $K_{01} = g_0([1, 2])$, $K_{02} = g_0([3, 4])$, and $K_{03} = g_0([5, 6])$. Let F_0 be the function on K_0 which takes the value 0 at $g_0(0)$, $\frac{1}{4}$ on K_{01}, $\frac{1}{2}$ on K_{02}, $\frac{3}{4}$ on K_{03}, and 1 at $g_0(7)$. Define sets K_n inductively as follows: K_n is the union of K_{n-1} and the sets $K_\mathbf{s}$, where $\mathbf{s}$ is any sequence of n terms $(i_1, \ldots, i_n)$, each term of which is equal to 0, 2, 4, or 6; for any such $\mathbf{s}$, $K_\mathbf{s}$ is the union of the three sets $K_{\mathbf{s},1} = g_n(v_\mathbf{s}([1, 2]))$, $K_{\mathbf{s},2} = g_n(v_\mathbf{s}([3, 4]))$, and $K_{\mathbf{s},3} = g_n(v_\mathbf{s}([5, 6]))$. Define functions F_n on K_n inductively as follows: $F_n | K_{n-1} = F_{n-1}$, and F_n is constant on each $K_{\mathbf{s},j}$ $(1 \leqq j \leqq 3)$: putting $\alpha = g_n(v_\mathbf{s}(0))$, $\beta = g_n(v_\mathbf{s}(7))$, then

$$\begin{aligned} F_n &= \tfrac{3}{4}F_{n-1}(\alpha) + \tfrac{1}{4}F_{n-1}(\beta) &\quad \text{on } K_{\mathbf{s},1},\\ F_n &= \tfrac{1}{2}F_{n-1}(\alpha) + \tfrac{1}{2}F_{n-1}(\beta) &\quad \text{on } K_{\mathbf{s},2},\\ F_n &= \tfrac{1}{4}F_{n-1}(\alpha) + \tfrac{3}{4}F_{n-1}(\beta) &\quad \text{on } K_{\mathbf{s},3}. \end{aligned}$$

(a) Let G be the function on the union of the K_n which is equal to F_n on K_n for each n. Show that G extends by continuity to a real-valued function F on K.

(b) Show that there exists a constant $c > 0$ such that, for any two points $\mathbf{x}, \mathbf{y} \in K$, we have $|F(\mathbf{x}) - F(\mathbf{y})| \leqq c \cdot \|\mathbf{x} - \mathbf{y}\|^{3/2}$ for a suitable choice of the sequence (α_n). (Consider the least integer n such that the two points $g_n^{-1}(\mathbf{x})$, $g_n^{-1}(\mathbf{y})$ do not belong to the same interval $v_\mathbf{s}([0, 7])$, for some sequence $\mathbf{s}$ of n terms.)

(c) Deduce from (b) and Whitney's extension theorem (Section **16.4**, Problem 6) that there exists a function $f: \mathbf{R}^2 \to \mathbf{R}$ of class C^1 for which all the points of K are critical points, and yet such that $f(K)$ is the interval [0, 1].

4. For $k \leq \inf(p, n)$, let M_k be the subset of the space $\mathbf{R}^{pn}$ of $p \times n$ matrices consisting of the matrices of rank k. Show that M_k is a differential submanifold of $\mathbf{R}^{pn}$, of dimension $k(p+n-k)$. (Observe that an $p \times n$ matrix of the form $\begin{pmatrix} A & B \\ C & D \end{pmatrix}$, where A is a $k \times k$ invertible matrix, is of rank k if and only if $D = CA^{-1}B$; for this purpose, multiply the matrix under consideration on the left by a suitably chosen invertible matrix so that the product is of the form $\begin{pmatrix} A & B \\ 0 & D' \end{pmatrix}$.

5. Let U be an open set in $\mathbf{R}^n$ and let $\mathbf{f} : \mathrm{U} \to \mathbf{R}^p$ be a C^∞ mapping. Suppose that $p \geq 2n$. Show that for each $\varepsilon > 0$ there exists a $p \times n$ matrix $A = (a_{ij})$ with $|a_{ij}| \leq \varepsilon$ for all i, j, such that the mapping $\mathbf{x} \mapsto \mathbf{g}(\mathbf{x}) = \mathbf{f}(\mathbf{x}) + A \cdot \mathbf{x}$ of U into $\mathbf{R}^p$ is an *immersion*. (For each $k < n$, consider the C^∞ mapping $\mathrm{F}_k : \mathrm{M}_k \times \mathrm{U} \to \mathbf{R}^{pn}$ (notation of Problem 4) defined by $\mathrm{F}_k(Z, \mathbf{x}) = Z - \mathbf{Df}(\mathbf{x})$, and remark (16.23.1) that the image of F_k is of measure zero in $\mathbf{R}^{pn}$, by using Problem 4; the complement of the union of the images of the F_k for $0 \leq k \leq n-1$ is therefore dense in $\mathbf{R}^{pn}$.)

24. INTEGRAL OF A DIFFERENTIAL n-FORM OVER AN ORIENTED PURE MANIFOLD OF DIMENSION n

(16.24.1) Let X be an *oriented* pure manifold of dimension $n \geq 0$. Then there exists a C^∞ differential n-form v_0 on X, belonging to the orientation of X. We shall define on X a positive Lebesgue measure (16.22) μ_{v_0} depending on v_0. To do this it is sufficient to define a Lebesgue measure on U, for each chart (U, φ, n) of X with U connected, and then to show that if $(\mathrm{U}', \varphi', n)$ is another chart with U′ connected, and if μ, μ' are the measures defined on U and U′, then the restrictions of μ and μ' to $\mathrm{U} \cap \mathrm{U}'$ are equal (13.1.9). By hypothesis, if $\mathbf{x} = (\xi^1, \ldots, \xi^n) \in \varphi(\mathrm{U})$, we may write

$$v_0(\varphi^{-1}(\mathbf{x})) = f(\xi^1, \ldots, \xi^n)\, d\xi^1 \wedge d\xi^2 \wedge \cdots \wedge d\xi^n,$$

and since U is connected, f is of the *same sign* throughout $\varphi(\mathrm{U})$. More precisely, if $\mathbf{R}^n$ is endowed with the canonical orientation (16.21.4), then $f(\mathbf{x}) > 0$ (resp. $f(\mathbf{x}) < 0$) in $\varphi(\mathrm{U})$ if φ preserves (resp. reverses) the orientation. By definition, the measure μ on U is the image under φ^{-1} of the measure $\varepsilon f \cdot \lambda$, where λ is the measure on $\varphi(\mathrm{U})$ induced by Lebesgue measure on $\mathbf{R}^n$, and ε is $+1$ or -1 according as φ preserves or reverses the orientation. Likewise, if $\mathbf{x} \in \varphi'(\mathrm{U}')$, we have $v_0(\varphi'^{-1}(\mathbf{x})) = f'(\mathbf{x})\, d\xi^1 \wedge d\xi^2 \wedge \cdots \wedge d\xi^n$, and we define the measure μ' on U′ as the image under φ'^{-1} of $\varepsilon' f' \cdot \lambda'$, where λ' is the measure on $\varphi'(\mathrm{U}')$ induced by Lebesgue measure, and ε' is $+1$ or -1 according as φ' preserves or reverses the orientation. Now let

$$\theta : \varphi(\mathrm{U} \cap \mathrm{U}') \to \varphi'(\mathrm{U} \cap \mathrm{U}')$$

be the transition diffeomorphism, and let $J(\theta)$ be its Jacobian, which is >0 in $\varphi(U \cap U')$ if $\varepsilon' = \varepsilon$, and <0 if $\varepsilon' = -\varepsilon$. For each $\mathbf{x} \in \varphi(U \cap U')$ we have

$$f(\mathbf{x}) = f'(\theta(\mathbf{x}))J(\theta)(\mathbf{x}).$$

To prove that the measures induced on $U \cap U'$ by μ and μ' are equal, we may assume that $U = U'$. Then, by (16.22.1), the image under θ of the measure $\varepsilon f \cdot \lambda = (\varepsilon(f' \circ \theta)) \cdot (J(\theta) \cdot \lambda)$ is equal to $\varepsilon f' \cdot \lambda'$ if $\varepsilon' = \varepsilon$, and to $-\varepsilon f' \cdot \lambda'$ if $\varepsilon' = -\varepsilon$. Hence in both cases it is equal to $\varepsilon' f' \cdot \lambda$, which proves our assertion.

The fact that the measure μ_{υ_0} is a Lebesgue measure follows from the fact that $\upsilon_0(x) \neq 0$ for all $x \in X$; moreover it is clear that this measure is positive. If υ_1 is another C^∞ differential n-form on X, belonging to the orientation of X, then $\upsilon_1 = h\upsilon_0$, where h is a C^∞-function which is >0 at all points of X, and it follows immediately that $\mu_{\upsilon_1} = h \cdot \mu_{\upsilon_0}$ (13.14.5).

(16.24.2) With the same hypotheses and notation as above, consider now an arbitrary differential n-form υ on X. We have $\upsilon = g\upsilon_0$, where g is a real-valued function on X. The n-form υ is said to be *integrable* (or *integrable over* X) if g is μ_{υ_0}-integrable, and the number $\int g \, d\mu_{\upsilon_0}$ is called the integral of υ (or the integral of υ over X) and is denoted by $\int \upsilon$, or $\int_X \upsilon$, or $\int_X \upsilon(x)$. We must show that this definition is independent of the C^∞ n-form υ_0 chosen in the orientation of X. Now, if υ_1 is another n-form of class C^∞ in the orientation of X, then $\upsilon_1 = h\upsilon_0$, where h is a C^∞-function on X, everywhere >0. Hence $\upsilon = gh^{-1}\upsilon_1$; but also $\mu_{\upsilon_1} = h \cdot \mu_{\upsilon_0}$, and g is μ_{υ_0}-integrable if and only if gh^{-1} is $h \cdot \mu_{\upsilon_0}$-integrable (13.14.3). Moreover we have

$$\int g \, d\mu_{\upsilon_0} = \int (gh^{-1}) \, d(h \cdot \mu_{\upsilon_0})$$

(13.14.3), which proves the assertion.

The integrable differential n-forms on X clearly form a real vector space, and $\upsilon \mapsto \int \upsilon$ is a linear form on this space, taking positive values on forms $\upsilon \geq 0$ (in the sense defined in (16.21.2)).

If we fix a C^∞ n-form υ in the orientation of X, it is clear that the linear form $f \mapsto \int f\upsilon$ on $\mathscr{K}(X)$ is a positive Lebesgue measure, and conversely all positive Lebesgue measures on X are of this type for a suitable choice of υ. These measures are also called *volumes* on the oriented manifold X, and the corresponding forms υ are called *volume-forms*.

(16.24.3) A differential n-form $\upsilon = g\upsilon_0$ on X is said to be *measurable* (resp. *locally integrable*, resp. *negligible*) if the function g is measurable (resp. locally integrable, resp. negligible) with respect to the measure μ_{υ_0}. It is immediately verified that these notions are independent of the choice of the n-form υ_0 in the orientation of X (13.15.6). If υ is locally integrable and of class C^∞, then $f\upsilon$ is integrable for each measurable function f which is *bounded* and of *compact support*.

Let υ be a locally integrable differential n-form on X, let (U_k) be a locally finite covering of X by relatively compact open sets, and let (u_k) be a continuous partition of unity subordinate to (U_k). Then υ is integrable if and only if the series $\sum_k \int |gu_k| \, d\mu_{\upsilon_0}$ is convergent; in which case the series $\sum_k \int u_k \upsilon$ converges and has $\int \upsilon$ as its sum. If each U_k is the domain of definition of a chart (U_k, φ_k, n), where φ_k is orientation-preserving, and if

$$u_k(\varphi_k^{-1}(\mathbf{x}))\upsilon(\varphi_k^{-1}(\mathbf{x})) = f_k(\mathbf{x}) \, d\xi^1 \wedge \cdots \wedge d\xi^n$$

for $\mathbf{x} = (\xi^1, \ldots, \xi^n) \in \varphi_k(U_k)$, then by definition,

$$\int u_k \upsilon = \int_{\varphi_k(U_k)} f_k(\mathbf{x}) \, d\xi^1 \cdots d\xi^n,$$

and so $\int \upsilon$ may be calculated.

(16.24.4) Let X_0 be an orientable pure manifold of dimension n; let X be the oriented manifold obtained by endowing X_0 with an orientation, and let $-X$ be the oppositely oriented manifold. If υ is an integrable differential n-form on X, then υ is also integrable over $-X$, and

$$\int_{-X} \upsilon = -\int_X \upsilon.$$

(16.24.5) Let $f: X' \to X$ be a diffeomorphism of a connected oriented manifold X′ onto an oriented manifold X. Then, for each integrable n-form υ on X, the form ${}^t f(\upsilon)$ is integrable over X′, and we have

(16.24.5.1) $$\int_X \upsilon = \pm \int_{X'} {}^t f(\upsilon),$$

where the sign is + or − according as f preserves or reverses the orientation. For by the method of calculation indicated above, we reduce immediately to the situation where X and X′ are open subsets of $\mathbf{R}^n$, where the result follows immediately from the definitions and the formula (16.22.1.1) for change of variables.

(16.24.6) Let Y be an oriented manifold of dimension m, X a manifold of dimension $n \geqq m$, and $f: \mathrm{Y} \to \mathrm{X}$ a mapping of class C^r ($r \geqq 1$). We have seen (16.20.9) how to define the *inverse image* ${}^t f(\sigma)$ of an m-form σ on X. If this inverse image is integrable, we may consider its integral $\int_{\mathrm{Y}} {}^t f(\sigma)$. When Y is an oriented submanifold of X, and f is the canonical injection, we shall often write $\int_{\mathrm{Y}} \sigma$ in place of $\int_{\mathrm{Y}} {}^t f(\sigma)$.

Example

(16.24.7) *Integration on a sphere.* Let $n > 0$. Consider the closed parallelotope P in $\mathbf{R}^n$ defined by the inequalities

$$-\tfrac{1}{2}\pi \leqq \theta^j \leqq \tfrac{1}{2}\pi \quad (1 \leqq j \leqq n-1), \qquad -\pi \leqq \theta^n \leqq \pi$$

(the θ^j being the coordinates of a point of P). The interior $\mathring{\mathrm{P}}$ of P is defined by the same inequalities with all the signs $\leqq$ replaced by $<$. It is immediately checked (by induction on n) that the mapping $\psi : \mathrm{P} \to \mathbf{S}_n$ which maps the point $(\theta^1, \ldots, \theta^n)$ of P to the point of $\mathbf{S}_n$ with coordinates

$$\text{(16.24.7.1)} \qquad \begin{cases} \xi^1 = \sin \theta^1, \\ \xi^2 = \cos \theta^1 \sin \theta^2, \\ \quad \vdots \\ \xi^n = \cos \theta^1 \cos \theta^2 \cdots \cos \theta^{n-1} \sin \theta^n, \\ \xi^0 = \cos \theta^1 \cos \theta^2 \cdots \cos \theta^{-n1} \cos \theta^n, \end{cases}$$

is *surjective*; its restriction to $\mathring{\mathrm{P}}$ is a *diffeomorphism* of this open set onto an open subset U_0 of $\mathbf{S}_n$, whose complement is contained in the hyperplane $\xi^n = 0$ and is therefore negligible (16.22.2). Let U denote the open subset of U_0 which is the complement of the intersection of U^0 with the hyperplane $\xi^0 = 0$, so that the complement of U in $\mathbf{S}_n$ is also negligible; U is the image under ψ of the open set $\mathrm{Q} \subset \mathring{\mathrm{P}}$ consisting of the points such that $\theta^n \neq \pm \frac{1}{2}\pi$, and the θ^j form a system of coordinates in U. If V is the complement of the hyperplane $\xi^0 = 0$ in $\mathbf{R}^{n+1}$, then the n-form σ on U defined in (16.21.10) is induced by the n-form on V

$$\text{(16.24.7.2)} \qquad \sigma_0 = \frac{d\xi^1 \wedge d\xi^2 \wedge \cdots \wedge d\xi^n}{\partial r / \partial \xi^0}$$

because in V we have

$$r^{-1}\, dr \wedge d\xi^1 \wedge \cdots \wedge d\xi^n = r^{-1} \frac{\partial r}{\partial \xi^0}\, d\xi^0 \wedge d\xi^1 \wedge \cdots \wedge d\xi^n$$

and $r = 1$ on S_n. The "triangular" form of the equations (16.24.7.1) shows that the form (16.24.7.2) is equal to

$$\frac{1}{\xi^0}\frac{\partial \xi^1}{\partial \theta^1}\frac{\partial \xi^2}{\partial \theta^2}\cdots\frac{\partial \xi^n}{\partial \theta^n}\, d\theta^1 \wedge d\theta^2 \wedge \cdots \wedge d\theta^n$$

and consequently, on U,

(16.24.7.3) $\sigma = \cos^{n-1}\theta^1 \cdot \cos^{n-2}\theta^2 \cdots \cos\theta^{n-1}\, d\theta^1 \wedge \cdots \wedge d\theta^n.$

It follows that for a function f on S_n to be such that the form $f \cdot \sigma$ is integrable, it is necessary and sufficient that the function

$$f(\psi(\theta^1, \ldots, \theta^n)) \cos^{n-1}\theta^1 \cos^{n-2}\theta^2 \cdots \cos\theta^{n-1}$$

should be *Lebesgue-integrable over* P, and then

$$\int_{S_n} f \cdot \sigma$$

$$= \int_{-\pi/2}^{\pi/2} \cos^{n-1}\theta^1\, d\theta^1 \int_{-\pi/2}^{\pi/2} \cdots \int_{-\pi/2}^{\pi/2} \cos\theta^{n-1}\, d\theta^{n-1} \int_{-\pi}^{\pi} f(\psi(\theta^1, \ldots, \theta^n))\, d\theta^n.$$

The form σ (also written $\sigma^{(n)}$) is called the *solid angle* form on the sphere S_n *oriented toward the outside.*

(16.24.8) *Let* X, Y *by two oriented pure differential manifolds, of dimensions n and m, respectively, $f: X \to Y$ a surjective submersion, υ an integrable differential n-form $\geqq 0$ on* X, *and ζ a locally integrable differential m-form on* Y *such that $\zeta(y) > 0$* (16.21.2) *almost everywhere on* Y. *Then for almost all $y \in Y$ the $(n - m)$-form $\upsilon/\zeta(y)$* (16.21.7) *is defined and integrable over $f^{-1}(y)$ (endowed with the orientation induced by f from the orientations of* X *and* Y); *the form $y \mapsto \zeta(y) \int_{f^{-1}(y)} \upsilon/\zeta(y)$ is integrable over* Y, *and*

(16.24.8.1) $$\int_X \upsilon = \int_Y \zeta(y) \int_{f^{-1}(y)} \upsilon/\zeta(y).$$

There exists a denumerable open covering (U_k) of X and charts (U_k, φ_k, n), $(f(U_k), \psi_k, m)$ of X and Y, respectively, such that

$$\varphi_k(U_k) = \psi_k(f(U_k)) \times I^{n-m},$$

where $I =]-1, 1[$, and $f | U_k = \psi_k^{-1} \circ F_k \circ \varphi_k$, where F_k is the restriction to $\varphi_k(U_k)$ of the canonical projection of $\mathbf{R}^n$ onto $\mathbf{R}^m$. Using the integrability criterion of (16.24.3), we reduce immediately to the situation in which Y is an open set in $\mathbf{R}^m$, $X = Y \times I^{n-m}$ and $f = \mathrm{pr}_1$. Then we have

$$v(x) = u(x)\, d\xi^1 \wedge d\xi^2 \wedge \cdots \wedge d\xi^n, \qquad \zeta(y) = w(y)\, d\xi^1 \wedge \cdots \wedge d\xi^m,$$

where u is integrable with respect to Lebesgue measure λ on X, and w is locally integrable and $\neq 0$ almost everywhere with respect to Lebesgue measure λ' on Y. The form $v/\zeta(y)$, defined for almost all $y = (\xi^1, \dots, \xi^m)$, may be written as

$$(\xi^{m+1}, \dots, \xi^n) \mapsto \frac{u(\xi^1, \dots, \xi^n)}{w(\xi^1, \dots, \xi^m)}\, d\xi^{m+1} \wedge \cdots \wedge d\xi^n.$$

Now the function $(\xi^1, \dots, \xi^n) \mapsto u(\xi^1, \dots, \xi^n)/w(\xi^1, \dots, \xi^m)$ is measurable with respect to the measure $w \cdot \lambda$ on $\mathbf{R}^n$, and this measure may be thought of as the product measure $(w \cdot \lambda') \otimes \lambda''$, where λ'' is Lebesgue measure on $\mathbf{R}^{n-m}$ (13.21.16). The proposition is therefore a consequence of the Lebesgue–Fubini theorem (13.21.7) and the definition of the orientation on the fibers $f^{-1}(y)$.

(16.24.9) *Application: calculation of integrals in polar coordinates.* Let $n \geqq 2$. We shall apply (16.24.8) by taking $X = \mathbf{R}^n - \{0\}$, $Y = \mathbf{R}_+^* =]0, +\infty[$, $f(x) = \|x\| = ((\xi^1)^2 + \cdots + (\xi^n)^2)^{1/2}$, so that $f^{-1}(u)$, for $u > 0$, is the sphere $u \cdot \mathbf{S}_{n-1}$ homothetic to $\mathbf{S}_{n-1}$. Also take $\zeta(\xi) = \xi^{-1}\, d\xi$, and take v to be an n-form $g \cdot v_0$, where v_0 is the canonical n-form $d\xi^1 \wedge \cdots \wedge d\xi^n$, and g is Lebesgue-integrable over X. The $(n-1)$-form $v_0/\zeta(u)$ on $u \cdot \mathbf{S}_{n-1}$ may be calculated as follows: Let σ' be an $(n-1)$-form on a neighborhood of $u \cdot \mathbf{S}_{n-1}$ in X, such that $v_0 = {}^t f(\zeta) \wedge \sigma'$. Let $h_u : x \mapsto u \cdot x$ be the homothety of ratio u on X. Then we have

$${}^t h_u(v_0) = {}^t h_u({}^t f(\zeta)) \wedge {}^t h_u(\sigma')$$

in a neighborhood of $\mathbf{S}_{n-1}$; but it is immediately seen that

$${}^t h_u(v_0) = u^n \cdot v_0, \qquad {}^t h_u({}^t f(\zeta)) = {}^t f(\zeta).$$

In view of the uniqueness of $v_0/\zeta(u)$, it follows that ${}^t h_u(v_0/\zeta(u)) = u^n \cdot \sigma^{(n-1)}$ on $\mathbf{S}_{n-1}$, whence by (16.24.5.1)

$$\int_{u \cdot \mathbf{S}_{n-1}} g(v_0/\zeta(u)) = u^n \int_{\mathbf{S}_{n-1}} g(u \cdot z)\sigma^{(n-1)}(z).$$

Hence the formula (16.24.8.1) gives

(16.24.9.1)
$$\int \cdots \int g(\xi^1, \dots, \xi^n)\, d\xi^1\, d\xi^2 \cdots d\xi^n = \int_0^{+\infty} u^{n-1}\, du \int_{\mathbf{S}_{n-1}} g(u \cdot z)\sigma^{(n-1)}(z).$$

Using the formula (16.24.7.4), we obtain finally the *formula for calculating an integral over* $\mathbf{R}^n - \{0\}$ *in polar coordinates*:

(16.24.9.2)

$$\int \cdots \int g(\xi^1, \ldots, \xi^n)\, d\xi^1\, d\xi^2 \cdots d\xi^n$$

$$= \int_0^{+\infty} u^{n-1}\, du \int_{-\pi/2}^{\pi/2} \cos^{n-2}\theta^1\, d\theta^1 \int_{-\pi/2}^{\pi/2} \cdots \int_{-\pi/2}^{\pi/2} \cos\theta^{n-2}\, d\theta^{n-2}$$

$$\cdot \int_{-\pi}^{\pi} g(u \sin\theta^1, \ldots, u\cos\theta^1 \cdots \cos\theta^{n-2}\sin\theta^{n-1},$$

$$u\cos\theta^1 \cdots \cos\theta^{n-2}\cos\theta^{n-1})\, d\theta^{n-1}.$$

Of course, we should arrive at the same formula by calculating the Jacobian of the diffeomorphism of $\mathring{\mathbf{P}} \times \mathbf{R}_+^*$ onto an open set in $\mathbf{R}^n - \{0\}$ with complement of measure zero. In the notation of (16.24.7) (with n replaced by $n - 1$) this diffeomorphism is $u \cdot \psi$.

Applying (16.24.9.1) to the particular case where g is the characteristic function of the unit ball $\mathbf{B}_n : \|x\| \leqq 1$ in $\mathbf{R}^n$, we obtain

$$\mathrm{V}_n = \frac{1}{n}\Omega_n,$$

where $\Omega_n = \int_{\mathbf{S}_{n-1}} \sigma^{(n-1)}$ is called the *solid angle* in $\mathbf{R}^n$ or the *superficial measure* of $\mathbf{S}_{n-1}$. From the preceding calculations and the formula (14.3.11.3), its value is

(16.24.9.3)
$$\Omega_n = \frac{n\pi^{n/2}}{\Gamma(\frac{1}{2}n + 1)},$$

or equivalently,

(16.24.9.4)
$$\begin{cases} \Omega_{2n} = \dfrac{2\pi^n}{(n-1)!}, \\[2ex] \Omega_{2n-1} = \dfrac{2^n\pi^{n-1}}{1 \cdot 3 \cdot 5 \cdots (2n-3)}. \end{cases}$$

We remark that, with the definition of $\sigma^{(0)}$ given in (16.21.10), the formula (16.24.9.1) remains valid for $n = 1$, for it then takes the form

$$\int_{-\infty}^{+\infty} g(\xi)\, d\xi = \int_0^{+\infty} (g(u) - g(-u))\, du.$$

(16.24.10) The interesting thing about the preceding method is that it applies without modification when f is *any positive and positively homogeneous function of degree* 1 *on* $\mathbf{R}^n - \{0\}$ (i.e., such that $f(u \cdot x) = uf(x)$ for all $u > 0$) *of class* C^∞ *and whose differential* df *is nonzero at all points of* $\mathbf{R}^n - \{0\}$. If E_1 is the submanifold given by the equation $f(x) = 1$ in $\mathbf{R}^n$, and σ_f is the form $\upsilon_0/{}^t f(\zeta(1))$ on E_1, then by the same reasoning as before we shall obtain the following generalization of (16.24.9.1):

$$\text{(16.24.10.1)} \qquad \int \cdots \int g(\xi^1, \ldots, {}^n\xi)\, d\xi^1 \cdots d\xi^n = \int_0^{+\infty} u^{n-1}\, du \int_{E_1} g(u \cdot z)\sigma_f(z).$$

From a practical point of view, if the set U in E_1 of points where $\partial f/\partial \xi^i \neq 0$ for some index i has a complement of measure zero, we may determine σ_f by the method of (16.21.9.1): We consider, in a neighborhood of U in $\mathbf{R}^n - \{0\}$ the following $(n-1)$-form:

$$\text{(16.24.10.2)} \qquad \sigma_f = (-1)^{i-1} \frac{d\xi^1 \wedge \cdots \wedge \widehat{d\xi^i} \wedge \cdots \wedge d\xi^n}{\partial f/\partial \xi^i}$$

and we take for σ_f the form induced by σ_f' on U.

(16.24.11) (Stokes' formula, elementary version) *Let* V *be an open subset of* $\mathbf{R}^{n-1}$, U *an open subset of* $\mathbf{R}^n$, F *a function of class* C^∞ *on* U *such that* $D_1F = \partial F/\partial \xi^1 \neq 0$ *in* U *and such that the mapping*

$$\psi : (\xi^1, \ldots, \xi^n) \mapsto (F(\xi^1, \ldots, \xi^n), \xi^2, \ldots, \xi^n)$$

is a diffeomorphism of U *onto* $I \times V$, *where* I *is an open interval in* $\mathbf{R}$. *For each* $u \in I$, *let* E_u *be the closed submanifold of* U *defined by the equation*

$$F(\xi^1, \ldots, \xi^n) = u,$$

whose image under ψ *is* $\{u\} \times V$, *and let* σ_u *be the* $(n-1)$-*form on* E_u *equal to* $\upsilon_0/{}^tF(\mathbf{e}_u^*)$, *where* $\mathbf{e}_u^*$ *is the unit covector in* $T_u(\mathbf{R})^*$. *Let* $[a, b]$ *be a closed interval contained in* I. *Then, for every* C^1-*function* f *on* U, *we have*

$$\text{(16.24.11.1)} \qquad \int_{U_{a,b}} D_1 f(\xi^1, \ldots, \xi^n)\, d\xi^1\, d\xi^2 \cdots d\xi^n$$
$$= \int_{E_b} f(z) D_1 F(z) \sigma_b(z) - \int_{E_a} f(z) D_1 F(z) \sigma_a(z),$$

where $U_{a,b}$ *is the set of* $x \in U$ *such that* $a \leq F(x) \leq b$, *and the orientations on* E_a *and* E_b *are induced by* F *from the canonical orientations of* $\mathbf{R}^{n-1}$ *and* $\mathbf{R}$ (16.21.9.2).

Put $U'_{a,b} =]a, b[\times V$, and let $g = f \circ \psi^{-1}$ on $I \times V$. By the Lebesgue–Fubini theorem, we have

$$\begin{aligned}\int_{U'_{a,b}} D_1 g(u, \xi^2, \ldots, \xi^n)\, du \wedge d\xi^2 \wedge \cdots \wedge d\xi^n \\ = \int_V d\xi^2 \wedge \cdots \wedge d\xi^n \int_a^b D_1 g(u, \xi^2, \ldots, \xi^n)\, du \\ = \int_V (g(b, \xi^2, \ldots, \xi^n) - g(a, \xi^2, \ldots, \xi^n))\, d\xi^2 \wedge \cdots \wedge d\xi^n.\end{aligned}$$

Observe that if $z = \psi^{-1}(b, \xi^2, \ldots, \xi^n)$, the projection of z on $\mathbf{R}^{n-1}$ is

$$(\xi^2, \ldots, \xi^n).$$

The calculation of σ_b indicated in (16.21.9.1) shows that this $(n-1)$-form is induced on E_b by the $(n-1)$-form $x \mapsto d\xi^2 \wedge \cdots \wedge d\xi^n / D_1 F(x)$ on U. There is an analogous calculation for σ_a, and hence we see that the value of the right-hand side of (16.24.11.1) is

$$\int_{U'_{a,b}} D_1 g(u, \xi^2, \ldots, \xi^n)\, du \wedge d\xi^2 \wedge \cdots \wedge d\xi^n.$$

However, we have $D_1 g(\psi(x)) = D_1 f(x) \cdot (D_1 F(x))^{-1}$ (8.2.1) for each $x \in U$, and since ${}^t\psi(v_0)(x) = D_1 F(x) \cdot v_0(x)$, we obtain finally

$$\begin{aligned}\int_{U'_{a,b}} D_1 g(u, \xi^2, \ldots, \xi^n)\, du \wedge d\xi^2 \wedge \cdots \wedge d\xi^n \\ = \int_{U_{a,b}} D_1 f(\xi^1, \ldots, \xi^n)\, d\xi^1 \wedge \cdots \wedge d\xi^n,\end{aligned}$$

which proves (16.24.11.1).

Remark

(16.24.12) In the preceding calculation, replace f by the function

$$x \mapsto \int_a^x f(t, \xi^2, \ldots, \xi^n)\, dt,$$

a and b by t and $t + h$; divide by h and let $h \to 0$; then we obtain

(16.24.12.1) $$\frac{d}{dt} \int_{U_{a,t}} f(\xi^1, \ldots, \xi^n)\, d\xi^1 \cdots d\xi^n = \int_{E_t} f(z)\sigma_t(z).$$

In particular, taking $f = 1$, this gives an interpretation of $\int_{E_t} \sigma_t$ as the *derivative of the volume of* $U_{a,t}$.

(16.24.13) The notion of the integral of a differential n-form on an oriented manifold X is easily extended to the case of a differential n-form with values in a vector space F (16.20.15). For example, let I be an open interval in $\mathbf{R}$, γ_0 a C^1-mapping of I into an open subset A of $\mathbf{C}$, and $f: A \to \mathbf{C}$ a continuous function. If γ is the restriction of γ_0 to a compact subinterval J of I, then the integral denoted by $\int_\gamma f(z)\,dz$ in (9.6) is none other than

$$\int_J {}^t\gamma_0(f\cdot(d\xi^1 + i\,d\xi^2)),$$

in which we have taken the inverse image under γ_0 of the complex-valued differential 1-form $f\cdot(d\xi^1 + i\,d\xi^2)$ defined on $A \subset \mathbf{R}^2$.

PROBLEMS

1. Let H be the subset of $\mathbf{R}^{n+1}$ defined by the inequality $\xi^0 > \left(\sum_{j=1}^{n}(\xi^j)^2\right)^{1/2}$, where $(\xi^0, \ldots, \xi^n)$ are the coordinates in $\mathbf{R}^{n+1}$. If $\mathbf{a} = (\alpha^0, \alpha^1, \ldots, \alpha^n) \in H$, show that there exists a constant $c_n > 0$, independent of $\mathbf{a}$, such that

$$\int_H \exp\left(-\sum_{j=0}^{n}\alpha^j\xi^j\right) d\xi^0\,d\xi^1\cdots d\xi^n = c_n\left((\alpha^0)^2 - \sum_{j=1}^{n}(\alpha^j)^2\right)^{-(n+1)/2}.$$

(Use a suitable Lorentz transformation.)

2. Let H be a hyperplane in $\mathbf{R}^n$. There exists a Euclidean displacement transforming $\mathbf{R}^{n-1}$ (identified with the subspace of $\mathbf{R}^n$ generated by the first $n-1$ vectors of the canonical basis) into H. The image under this mapping of Lebesgue measure λ_{n-1} on $\mathbf{R}^{n-1}$ is a measure on H which does not depend on the displacement chosen, and we denote it again by λ_{n-1}. Show that if $\mathbf{u}$ is a unit vector orthogonal to H, and A an integrable subset of H, then the orthogonal projection $p(A)$ of A on $\mathbf{R}^{n-1}$ has measure

$$\lambda_{n-1}(p(A)) = |(\mathbf{e}_n|\mathbf{u})|\lambda_{n-1}(A).$$

3. Let P be a compact convex polyhedron in $\mathbf{R}^n$ with nonempty interior. If F_k $(1 \leq k \leq r)$ are the *faces* of P (Section 16.5, Problem 6), the *area* (or the $(n-1)$-*dimensional area*) of P (or of the frontier of P) is defined to be the number $\mathscr{A}_{n-1}(P) = \sum_k \lambda_{n-1}(F_k)$. For each vector $\mathbf{u} \in \mathbf{S}_{n-1}$, we denote by $\lambda_{n-1}(P, \mathbf{u})$ the measure of the orthogonal projection of P on the hyperplane $(\mathbf{x}|\mathbf{u}) = 0$. Show that

$$\text{(1)} \qquad V_{n-1}\mathscr{A}_{n-1}(P) = \int_{\mathbf{S}_{n-1}} \lambda_{n-1}(P, \mathbf{u})\cdot\sigma^{(n-1)}(\mathbf{u})$$

(*Cauchy's formula* for convex polyhedra). (Use Problem 2.)

Deduce from this formula that if P′ is another compact convex polyhedron containing P, then $\mathscr{A}_{n-1}(P) \subset \mathscr{A}_{n-1}(P')$.

4. Let C be a compact convex body in $\mathbf{R}^n$ (Section **16.5**, Problem 6). There exists a sequence (P_m) of compact convex polyhedra tending to C with respect to the Hausdorff distance h defined in Section **3.16**, Problem 3 (Section **16.5**, Problem 8(a)). Show that the sequence $(\mathscr{A}_{n-1}(\mathrm{P}_m))_{m \geq 1}$ tends to a limit which does not depend on the particular sequence P_m converging to C. This limit is called the *area* (or the *$(n-1)$-dimensional area*) of C (or of the frontier of C), and is denoted by $\mathscr{A}_{n-1}(\mathrm{C})$. For $\mathbf{u} \in \mathbf{S}_{n-1}$, we denote again by $\lambda_{n-1}(\mathrm{C}, \mathbf{u})$ the measure of the orthogonal projection of C on the hyperplane $(\mathbf{x} \mid \mathbf{u}) = 0$. Show that $\mathbf{u} \mapsto \lambda_{n-1}(\mathrm{C}, \mathbf{u})$ is continuous on $\mathbf{S}_{n-1}$ (use Problem 3) and that

$$\text{(2)} \qquad \mathrm{V}_{n-1} \mathscr{A}_{n-1}(\mathrm{C}) = \int_{\mathbf{S}_{n-1}} \lambda_{n-1}(\mathrm{C}, \mathbf{u}) \cdot \sigma^{(n-1)}(\mathbf{u})$$

(*Cauchy's formula* for convex bodies). If C′ is another compact convex body containing C, then $\mathscr{A}_{n-1}(\mathrm{C}) \subset \mathscr{A}_{n-1}(\mathrm{C}')$. If (C_m) is a sequence of compact convex bodies tending to C with respect to Hausdorff distance, then $\lim_{m \to \infty} \mathscr{A}_{n-1}(\mathrm{C}_m) = \mathscr{A}_{n-1}(\mathrm{C})$. Consider the case $\mathrm{C} = \mathbf{S}_{n-1}$.

5. Let $\mathfrak{K}_n$ be the set of all compact convex bodies in $\mathbf{R}^n$, endowed with the Hausdorff distance (Section **3.16**, Problem 3). Define by induction on n a sequence of $n+1$ real-valued functions W_{in} $(0 \leq i \leq n)$ on $\mathfrak{K}_n$, as follows: $\mathrm{W}_{01}(\mathrm{C}) = \lambda_1(\mathrm{C})$, $\mathrm{W}_{11}(\mathrm{C}) = 2$. For $n > 1$, $\mathrm{W}_{0n}(\mathrm{C}) = \lambda_n(\mathrm{C})$, and for $1 \leq i \leq n$,

$$\mathrm{W}_{in}(\mathrm{C}) = \frac{1}{n\mathrm{V}_{n-1}} \int_{\mathbf{S}_{n-1}} \mathrm{W}_{i-1,\, n-1}(p_{\mathbf{u}}(\mathrm{C})) \cdot \sigma^{(n-1)}(\mathbf{u}),$$

where $p_{\mathbf{u}}$ is the orthogonal projection on the hyperplane $(\mathbf{x} \mid \mathbf{u}) = 0$. Show that each of the functions W_{in} is increasing and continuous on $\mathfrak{K}_n$, and that $\mathrm{W}_{in}(\alpha\mathrm{C}) = \alpha^{n-i}\mathrm{W}_{in}(\mathrm{C})$ for all $\alpha > 0$. If A, B are convex bodies belonging to $\mathfrak{K}_n$, such that $\mathrm{A} \cup \mathrm{B}$ is *convex* and $\mathrm{A} \cap \mathrm{B}$ has a nonempty interior, then

$$\mathrm{W}_{in}(\mathrm{A} \cup \mathrm{B}) + \mathrm{W}_{in}(\mathrm{A} \cap \mathrm{B}) = \mathrm{W}_{in}(\mathrm{A}) + \mathrm{W}_{in}(\mathrm{B}).$$

(Observe that if $a \in \mathrm{A}$ and $b \in \mathrm{B}$, there exists a point of $\mathrm{A} \cap \mathrm{B}$ in the segment with a, b as endpoints. Consequently, if H is a supporting hyperplane of $\mathrm{A} \cap \mathrm{B}$, then H is a supporting hyperplane of either A or B. Deduce that

$$p_{\mathbf{u}}(\mathrm{A} \cap \mathrm{B}) = p_{\mathbf{u}}(\mathrm{A}) \cap p_{\mathbf{u}}(\mathrm{B}).)$$

In particular, $n\mathrm{W}_{1,n}(\mathrm{C}) = \mathscr{A}_{n-1}(\mathrm{C})$ (Problem 4).

6. If C is a compact convex body in $\mathbf{R}^n$ containing the origin, recall that the *function of support* of C is the function $\mathrm{H}(\mathbf{z}) = \sup_{\mathbf{x} \in \mathrm{C}} (\mathbf{x} \mid \mathbf{z})$ (Section **16.5**, Problem 7). For each $\mathbf{u} \in \mathbf{S}_{n-1}$, let $b(\mathrm{C}, \mathbf{u}) = \mathrm{H}(\mathbf{u}) + \mathrm{H}(-\mathbf{u})$ (the "width" of C in the direction $\mathbf{u}$, cf. Section **14.3**, Problem 9(a)). Show that, in the notation of Problem 5,

$$\mathrm{W}_{n-1,\, n}(\mathrm{C}) = \frac{1}{2n} \int_{\mathbf{S}_{n-1}} b(\mathrm{C}, \mathbf{u}) \cdot \sigma^{(n-1)}(\mathbf{u}).$$

(Proof by induction on n. For each $\mathbf{u} \in \mathbf{S}_{n-1}$, let $\mathrm{E}(\mathbf{u})$ denote the hyperplane $(\mathbf{x} \mid \mathbf{u}) = 0$ in $\mathbf{R}^n$, and let $\sigma^{(n-2)}$ be the differential $(n-2)$-form on $\mathbf{S}_{n-1} \cap \mathrm{E}(\mathbf{u})$ which is the image of $\sigma^{(n-2)}$ by a rotation of $\mathbf{R}^n$ transforming $\mathbf{S}_{n-2}$ into $\mathbf{S}_{n-1} \cap \mathrm{E}(\mathbf{u})$. Show that the integral

$$\int_{\mathbf{S}_{n-1}} \sigma^{(n-1)}(\mathbf{u}) \int_{\mathbf{S}_{n-1} \cap \mathrm{E}(\mathbf{u})} b(p_{\mathbf{u}}(\mathrm{C}), \mathbf{v}) \cdot \sigma_{\mathbf{u}}^{(n-2)}(\mathbf{v})$$

can be written in the form

$$\int_{\mathrm{P}} b(\mathrm{C}, \mathbf{v}) \cdot \omega(\mathbf{u}, \mathbf{v}),$$

where P is the submanifold of $\mathbf{S}_{n-1} \times \mathbf{S}_{n-1}$ consisting of pairs $(\mathbf{u}, \mathbf{v})$ such that $(\mathbf{u} \mid \mathbf{v}) = 0$, and ω is a $(2n-1)$-form on P obtained by the procedure of (**16.21.7**); use (**16.24.8**).)

7. (a) Show that if P is a compact convex polyhedron of dimension n in $\mathbf{R}^n$, then the area $\mathscr{A}_{n-1}(\mathrm{P})$ is equal to the Minkowski area of P (Section **14.3**, Problem 10(c)). Deduce that, for each $\rho > 0$, in the notation of Section **3.6**,

$$\text{(1)} \qquad \lambda_n(\mathrm{V}_\rho(\mathrm{P})) = \lambda_n(\mathrm{P}) + \int_0^\rho \mathscr{A}_{n-1}(\mathrm{V}_r(\mathrm{P}))\, dr.$$

(b) Show that, for each compact convex body C in $\mathbf{R}^n$, we have

$$\lambda_n(\mathrm{V}_r(\mathrm{C})) = \sum_{j=0}^{n} \binom{n}{j} \mathrm{W}_{jn}(\mathrm{C}) r^j$$

(*Steiner–Minkowski formula*). (Prove the formula first in the case that $\mathrm{C} = \mathrm{P}$ is a compact convex polyhedron of dimension n, by using (a) and induction on n. Then pass to the limit in $\mathfrak{K}_n$.)

(c) Deduce the formula

$$\mathrm{W}_{in}(\mathrm{V}_r(\mathrm{C})) = \sum_{j=0}^{n-i} \binom{n-i}{j} \mathrm{W}_{j+i,\,n}(\mathrm{C}) r^j$$

for $0 \leq i \leq n$ and $r \geq 0$. (Observe that $\mathrm{V}_{r+s}(\mathrm{C}) = \mathrm{V}_r(\mathrm{V}_s(\mathrm{C}))$ and use the Steiner–Minkowski formula.)

(d) Deduce from (b) and (c) that formula (**1**) is valid for any compact convex body C. In particular, $\mathscr{A}_{n-1}(\mathrm{C})$ is equal to the Minkowski area of C (Section **14.3**, Problem 10(c)).

8. Let $\mathfrak{K}'_n$ be the set of all nonempty compact convex sets in $\mathbf{R}^n$, endowed with the Hausdorff distance (Section **3.16**, Problem 3). $\mathfrak{K}'_n$ is a compact space in which $\mathfrak{K}_n$ is dense. Show that the functions W_{in} defined on $\mathfrak{K}_n$ extend by continuity to $\mathfrak{K}'_n$ (induction on n). If A is a compact convex set of dimension $< n$ in $\mathbf{R}^n$, then

$$\mathrm{W}_{in}(\mathrm{A}) = \frac{i\mathrm{V}_i}{n\mathrm{V}_{i-1}} \mathrm{W}_{i-1,\,n-1}(\mathrm{A}) \qquad (1 \leq i \leq n).$$

9. Let $A \subset \mathbf{R}^p$, $B \subset \mathbf{R}^q$ be compact convex sets. Show that

$$W_{i,\,p+q}(A \times B) = \binom{n}{i}^{-1} V_i \sum_{j=0}^{i} \frac{1}{V_j V_{i-j}} \binom{p}{j} \binom{q}{i-j} W_{j,\,p}(A) W_{i-j,\,q}(B).$$

(Apply the Steiner–Minkowski formula to $A \times B$, and use the Lebesgue–Fubini theorem.) Hence calculate the values of $W_{in}(C)$ when C is a cube in $\mathbf{R}^n$.

10. Let $C \subset \mathbf{R}^n$ be a compact convex body of dimension n. Show that the set $\complement V_r(\complement C)$ is convex (express it as an intersection of translates of C). For each $r > 0$, we have $V_r(\complement V_r(\complement C)) \subset C$. By using the continuity of the functions $W_{in}(C)$ with respect to C, deduce that the function $r \mapsto \lambda_n(\complement V_r(\complement C))$ has a derivative on the right at the origin, equal to $-\mathscr{A}_{n-1}(C)$.

11. Let X, Y be two oriented pure differential manifolds, of respective dimensions n and m; let $f: X \to Y$ be a submersion, x a point of X, $y = f(x)$. We shall use the notation of (16.21.7). Let ζ be a C^∞ m-form on Y belonging to the orientation of Y. For each $k \leq m$, ζ defines a canonical isomorphism $\mathbf{z}_y \mapsto \Phi_{\zeta(y)}(\mathbf{z}_y)$ of $\bigwedge^k T(Y)$ onto $\bigwedge^{m-k} T(Y)^*$, such that $\Phi_{\zeta(y)}(\mathbf{z}_y) = \mathbf{z}_y \,\lrcorner\, \zeta(y)$. Let α be a C^∞ $(n - m + k)$-form on X with compact support. To each k-vector $\mathbf{z}_y \in \bigwedge^k T_y(Y)$ there corresponds an $(n - m)$-form $\beta_{\mathbf{z}_y}$ on $f^{-1}(y)$ such that, for $x \in f^{-1}(y)$, we have

$$\beta_{\mathbf{z}_y}(x) = \Big(\alpha(x) \wedge \Big(\Big(\bigwedge^{m-k} ({}^t u)\Big)(\Phi_{\zeta(y)}(\mathbf{z}_y))\Big)\Big) / \zeta(y)$$

in the notation of (16.21.7). This form is *independent* of the choice of ζ in the orientation of Y. We give each fiber $f^{-1}(y)$ the orientation induced by f from the orientations of X and Y (16.21.9.1). Show that there exists a unique C^∞ k-form γ on Y such that, for all $y \in Y$ and all $\mathbf{z}_y \in \bigwedge^k T_y(Y)$,

$$\langle \gamma(y), \mathbf{z}_y \rangle = \int_{f^{-1}(y)} \beta_{\mathbf{z}_y}(x).$$

This form is denoted by $\alpha^\flat$ and is called the *integral of* α *along the fibers of* f. (Reduce to the case (16.7.4).)

If β' is a C^∞ k'-form on Y, then $\alpha^\flat \wedge \beta' = (\alpha \wedge {}^t f(\beta'))^\flat$.

25. EMBEDDING AND APPROXIMATION THEOREMS. TUBULAR NEIGHBORHOODS

(16.25.1) *Let* X *be a differential manifold,* U *a relatively compact open subset of* X. *Then there exists an integer* N *and an embedding* (16.8.4) *of* U *in* $\mathbf{R}^N$.

There exist a finite number of charts (U_k, φ_k, n_k) of X $(1 \leq k \leq m)$ such that the U_k cover the compact set $\bar{U}$. Next, there exists a family $(V_k)_{1 \leq k \leq m}$ of open subsets of X which cover $\bar{U}$ and are such that $\bar{V}_k \subset U_k$ for each k

(12.6.2). Finally, there exists a family of C^∞-functions $(f_k)_{1 \leq k \leq m}$ on X, with values in [0, 1], such that $\mathrm{Supp}(f_k) \subset U_k$ and $f_k(x) = 1$ for all $x \in \bar{V}_k$. We shall show that the mapping

$$g : x \mapsto ((f_k(x))_{1 \leq k \leq m}, (f_k(x)\varphi_k(x))_{1 \leq k \leq m})$$

of $\bar{U}$ into $\mathbf{R}^N = \mathbf{R}^m \times \prod_{k=1}^{m} \mathbf{R}^{n_k}$ gives the desired embedding by restriction to U. It is clear that g is of class C^∞ (16.6.4). Next, g is injective. For if x, x' are two distinct points of X such that $f_k(x) = f_k(x')$ for $1 \leq k \leq m$, then since $x \in V_k$ for some k, it follows that $f_k(x) = 1$ and therefore $f_k(x') = 1$, so that $x' \in U_k$; but then, since $x \neq x'$, we have

$$f_k(x)\varphi_k(x) = \varphi_k(x) \neq \varphi_k(x') = f_k(x')\varphi_k(x')$$

so that $g(x) \neq g(x')$. Since $\bar{U}$ is compact, it follows that g is a homeomorphism of $\bar{U}$ onto $g(\bar{U})$ (3.17.12), hence of U onto $g(U)$. Hence it remains to show that g is an immersion (16.8.4) at each point x of U. Let k be such that $x \in V_k$, and let p be the projection of $\mathbf{R}^N$ onto the factor $\mathbf{R}^{n_k}$, so that $p \circ g = \varphi_k$ on V_k. Since $T_x(\varphi_k) = T_{g(x)}(p) \circ T_x(g)$ is of rank n_k, it follows that $T_x(g)$ is of rank n_k. This shows that g is an immersion at x (16.7.1) and finishes the proof.

One can in fact show that there exists an embedding of the whole of X into an $\mathbf{R}^N$, and that if X is pure of dimension n, one can take $N = 2n + 1$ (Problems 2 and 13(c)).

The above embedding theorem will enable us to extend to manifolds the Weierstrass approximation theorem (7.4.1), polynomials being replaced by C^∞-functions. We shall begin by establishing two extremely useful auxiliary results on "tubular neighborhoods" of a submanifold of $\mathbf{R}^N$.

(16.25.2) *Let X be a pure submanifold of $\mathbf{R}^N$, of dimension n. Let $j : X \to \mathbf{R}^N$ be the canonical injection and U a relatively compact open subset of X. For each $x \in X$ let M_x be the n-dimensional subspace of $\mathbf{R}^N$ which is the image of $T_x(X)$ under the mapping $\boldsymbol{\tau}_x \circ T_x(j)$* (16.5.2). *Let d be the distance function on $\mathbf{R}^N$ derived from the scalar product* $(\mathbf{x} | \mathbf{y}) = \sum_j \xi^j \eta^j$, *and let N_x be the orthogonal supplement of M_x in $\mathbf{R}^N$* (6.3.1). *Suppose that we have defined, on an open neighborhood V of $\bar{U}$ in X, $N - n$ mappings $\mathbf{u}_j$ $(1 \leq j \leq N - n)$ of V into $\mathbf{R}^N$, of class C^∞, such that for each $x \in V$ the $\mathbf{u}_j(x)$ form a basis of N_x. Then there exists an open neighborhood T of U in $\mathbf{R}^N$ and a diffeomorphism*

$$y \mapsto (\pi(y), \theta(y))$$

of T onto $U \times \mathbf{R}^{N-n}$ such that, for each $y \in T$, $\pi(y)$ is the unique point of X whose distance from y is equal to $d(y, X)$.

Consider the mapping $g : V \times \mathbf{R}^{N-n} \to \mathbf{R}^N$ defined by

$$g(x; t_1, \ldots, t_{N-n}) = x + \sum_{j=1}^{N-n} t_j \mathbf{u}_j(x).$$

Clearly g is of class C^∞. Moreover, for each point $a \in V$, the tangent mapping $T_{(a,0)}(g)$ is bijective, hence (16.5.6) there exists an open neighborhood $W_a \subset V$ of a in X and an open ball B_a with center at the origin in $\mathbf{R}^{N-n}$ such that the restriction of g to $W_a \times B_a$ is a diffeomorphism of this open set onto an open neighborhood T_a of a in $\mathbf{R}^N$. Let $y \mapsto (\pi_a(y), \theta_a(y))$ be the inverse diffeomorphism.

We shall show that there exists a ball $S_a \subset T_a$ with center a such that, for each $y \in S_a$, $\pi_a(y)$ is the only point $x \in X$ such that $d(y, x) = d(y, X)$.

We shall use the following lemma:

(16.25.2.1) *For each $a \in X$ there exists a ball S'_a in $\mathbf{R}^N$ with center a such that, for each $y \in S'_a$, there exists at least one point $x \in X$ for which $d(x, y) = d(X, y)$. For such a point x, the vector $y - x$ is orthogonal to M_x.*

Since X is locally closed in $\mathbf{R}^N$, there exists a closed ball S''_a with center a and radius r_a in $\mathbf{R}^N$ such that $X \cap S''_a$ is closed in S''_a, and therefore compact. Let S'_a be the closed ball with center a and radius $\frac{1}{3}r_a$. For each $y \in S'_a$, we have $d(y, a) \leqq \frac{1}{3}r_a$, and for each $z \in X$ such that $z \notin S'_a$, we have $d(y, z) \geqq \frac{2}{3}r_a$, so that $d(y, X) = d(y, X \cap S''_a)$. Hence $X \cap S''_a$ contains a point x such that $d(y, x) = d(y, X)$ (3.17.10). Moreover, the function $z \mapsto h(z) = (d(y, z))^2 = \sum_{j=1}^{n} (\eta^j - \zeta^j)^2$ is of class C^∞ on X and admits a minimum at the point x; hence $d_x h = 0$ (16.5.10). However, relation $d_x h = 0$ may be written as

$$\sum_{j=1}^{N} (\eta^j - \zeta^j)\, d\zeta^j = 0;$$

also, for each vector $\mathbf{t} \in M_z$, the numbers $\langle \mathbf{t}, d\zeta^j \rangle$ $(1 \leqq j \leqq N)$ are the components of $\mathbf{t}$ with respect to the canonical basis of $\mathbf{R}^N$. Hence the vector $y - x$ is orthogonal to M_x.

Having established the lemma, we shall now argue by contradiction. Suppose that there exists a sequence (y_ν) of points of T_a tending to a, and a sequence (x_ν) of points of X such that $x_\nu \neq \pi_a(y_\nu)$ and $d(y_\nu, x_\nu) = d(y_\nu, X)$. Since $d(y_\nu, x_\nu) \leqq d(y_\nu, \pi_a(y_\nu))$, it follows that $d(y_\nu, x_\nu) \to 0$ and hence that $x_\nu \to a$; so we may suppose that $x_\nu \in W_a$. By virtue of the lemma (16.25.2.1), we can write $y_\nu - x_\nu = \sum_{j=1}^{N-n} t_{j\nu} \mathbf{u}_j(x_\nu)$; and since $x_\nu \neq \pi_\nu(y_\nu)$, the point

$$(t_{1\nu}, t_{2\nu}, \ldots, t_{N-n,\nu})$$

does not belong to the ball B_a. If we put $r_\nu^2 = \sum_{j=1}^{N-n} t_{j\nu}^2$, the sequence (r_ν^{-1}) is therefore bounded. Since $\sum_{j=1}^{N-n} (r_\nu^{-1} t_{j\nu})^2 = 1$, we may, by passing to a subsequence of the sequence (y_ν), assume that each of the sequences $(r_\nu^{-1} t_{j\nu})$ has a limit t'_j $(1 \leqq j \leqq N - n)$; we have then $\sum_j t_j'^2 = 1$; but on the other hand, $r_\nu^{-1}(y_\nu - x_\nu) \to 0$ as $\nu \to \infty$, and hence $\sum_j t'_j \mathbf{u}_j(a) = 0$. This is impossible, since the t'_j are not all zero and the $\mathbf{u}_j(a)$ are linearly independent.

If b is another point of V such that $S_a \cap S_b \neq \varnothing$, then it follows from above that π_a and π_b agree on $S_a \cap S_b$. Hence there is a unique function π, defined on the union S of the open sets S_a $(a \in V)$, which extends each of the functions π_a. For each $a \in V$, let $W'_a \subset W_a$ be an open neighborhood of a in X, and $B'_a \subset B_a$ an open ball with center 0 such that $g(W'_a \times B'_a) \subset S_a$. Cover $\bar{U}$ by a finite number of open neighborhoods W'_{a_i} $(1 \leqq i \leqq r)$, and let B'' be an open ball with center 0 in $\mathbf{R}^{N-n}$ (hence diffeomorphic to $\mathbf{R}^{N-n}$) contained in the intersection of the B'_{a_i}. Then $g(U \times B'')$ is an open subset of $\mathbf{R}^N$ contained in S and containing U, and for each $(x, \mathbf{t}) \in U \times B''$ we have

$$\pi(g(x, \mathbf{t})) = x \qquad \text{and} \qquad \mathbf{t} = g(x, \mathbf{t}) - x.$$

This proves that $g|(U \times B'')$ is a bijection of $U \times B''$ onto $g(U \times B'')$, and hence a diffeomorphism of $U \times B''$ onto an open neighborhood of U in $\mathbf{R}^N$ (**16.8.8**(iv)).

(**16.25.3**) There does not always exist a system $(\mathbf{u}_j)$ of mappings of V into $\mathbf{R}^N$ having the properties of the statement of (**16.25.2**). For example, if X is a nonorientable compact manifold embedded in $\mathbf{R}^N$ (**16.25.1**), the existence of an open neighborhood of X in $\mathbf{R}^N$ diffeomorphic to $X \times \mathbf{R}^{N-n}$ would contradict (**16.21.9.1**), since every open subset of $\mathbf{R}^N$ is an orientable manifold.

However, there is the following weaker result:

(**16.25.4**) *Let* X *be a pure submanifold of* $\mathbf{R}^N$ *of dimension n. Then there exists an open neighborhood* T *of* X *in* $\mathbf{R}^N$ *and a* C^∞ *submersion* π *of* T *onto* X *with the following property: for each* $y \in T$, $\pi(y)$ *is the only point of* X *whose distance from y is equal to* $d(y, X)$, *and for each* $x \in X$, *the fiber* $\pi^{-1}(x)$ *is the intersection of* T *and the linear manifold* $x + N_x$ (the space N_x being defined as in (**16.25.2**)).

It is enough to prove that, for each $a \in X$, there exists an open neighborhood T_a with center a in $\mathbf{R}^N$ such that $X \cap T_a$ is closed in T_a, and a surjective submersion $\pi_a : T_a \to X \cap T_a$ such that, for each $y \in T_a$, $\pi_a(y)$ is the unique point of X whose distance from y is equal to $d(y, X)$, and such that, for each

$x \in X \cap T_a$, the fiber $\pi_a^{-1}(x)$ is the intersection of T_a with $x + N_x$; the union T of the T_a will then have the required properties. By virtue of (16.25.2), it is enough to show that there exists a relatively compact open neighborhood V of a in X on which $N - n$ functions $\mathbf{u}_j$ can be defined so as to satisfy the conditions of (16.25.2). By means of a displacement we may assume that $a = 0$ and that M_a is the space $\mathbf{R}^n$ spanned by the first n vectors of the canonical basis $(\mathbf{e}_i)_{1 \leq i \leq N}$ of $\mathbf{R}^N$. Consequently (16.8.3.2) there exists a relatively compact open neighborhood U of 0 in $\mathbf{R}^n$ such that a neighborhood V of 0 in X is formed by the points of $U \times \mathbf{R}^{N-n}$ which satisfy the equations $\xi^{n+j} = f_j(\xi^1, \ldots, \xi^n)$ $(1 \leqq j \leqq N - n)$, where the f_j are C^∞-functions defined on U which vanish, together with their first derivatives, at the origin. For each point

$$x = (\xi^1, \ldots, \xi^n, f_1(\xi^1, \ldots, \xi^n), \ldots, f_{N-n}(\xi^1, \ldots, \xi^n))$$

of V, the space M_x is defined by the $N - n$ linear equations

$$\zeta^{n+j} - \sum_{i=1}^{n} D_i f_j(\xi^1, \ldots, \xi^n) \cdot \zeta^i = 0 \qquad (1 \leqq j \leqq N - n)$$

and consequently the functions

$$\mathbf{u}_j(x) = \mathbf{e}_{n+j} - \sum_{i=1}^{n} D_i f_j(\xi^1, \ldots, \xi^n)\mathbf{e}_i \qquad (1 \leqq j \leqq N - n)$$

satisfy the required conditions.

We can now state the *approximation theorem*:

(16.25.5) *Let* X, Y *be two differential manifolds*, K *a compact subset of* X, *and* $f : K \to Y$ *a continuous map. Let* d *be any distance which defines the topology of* Y, *and let* $\varepsilon > 0$. *Then there exists an open neighborhood* U *of* K *in* X *and a* C^∞*-mapping* $g : U \to Y$ *such that*

$$d(f(x), g(x)) \leqq \varepsilon$$

for all $x \in K$.

Let U_0 be a relatively compact open neighborhood of K in X, and V_0 a relatively compact open neighborhood of $f(K)$ in Y (3.18.2). We may assume that U_0 is embedded in $\mathbf{R}^m$ and V_0 in $\mathbf{R}^n$ (16.25.1). Let T be an open neighborhood of V_0 in $\mathbf{R}^n$ having the properties of (16.25.4), and let $\delta < \frac{1}{2}\varepsilon$ be a positive real number such that every point of $\mathbf{R}^n$ whose distance from $f(K)$ is $\leqq \delta$ lies in T (3.17.11). By the Weierstrass approximation theorem, there exist n polynomials h_j $(1 \leqq j \leqq n)$ in m variables such that if $h = (h_1, \ldots, h_n)$, we have

$d(f(x), h(x)) \leqq \delta$ for all $x \in K$ (7.4.1). Since $h(K) \subset T$, there exists an open neighborhood $U \subset U_0$ of K in X such that $h(U) \subset T$. For each $x \in U$, put $g(x) = \pi(h(x)) \in V_0$. Since $f(x) \in V_0$, we have by definition (16.25.4)

$$d(h(x), \pi(h(x))) \leqq d(h(x), f(x)) \leqq \tfrac{1}{2}\varepsilon$$

and therefore $d(f(x), g(x)) \leqq \varepsilon$ for all $x \in K$.

PROBLEMS

1. Let M be a pure submanifold of $\mathbf{R}^m$, of dimension n, and let A be a compact subset of M. Let Ω_1 be the set of linear mappings $u : \mathbf{R}^m \to \mathbf{R}^{2n+1}$ such that $u|M$ is of rank n at every point of A, and let Ω_2 be the set of linear mappings $u : \mathbf{R}^m \to \mathbf{R}^{2n+1}$ such that $u|A$ is injective. Show that, if $m \geqq 2n+1$, Ω_1 and Ω_2 are dense open sets in the vector space $\mathscr{L}(\mathbf{R}^m; \mathbf{R}^{2n+1})$. (Show first that the complements Φ_1, Φ_2 of Ω_1, Ω_2 are closed, by using the compactness of A; then show that Φ_1, Φ_2 are meager subsets of $\mathscr{L}(\mathbf{R}^m; \mathbf{R}^{2n+1})$, by covering A with a finite number of charts, and using Sard's theorem.)

2. Let M be a pure manifold of dimension n, and let $(U_k)_{k \geqq 1}$ be a locally finite denumerable open covering of M such that each U_k is relatively compact and is the domain of a chart (U_k, φ_k, n) of M, where $\varphi_k = (\varphi_k^1, \ldots, \varphi_k^n)$ is such that $\varphi_k(U_k)$ is the cube in $\mathbf{R}^n$ defined by $|\xi^j - 1| < \frac{1}{2}$ for $1 \leqq j \leqq n$. For each k, let V_k be an open set such that $\bar{V}_k \subset U_k$ and the V_k cover M. Let g_k be a C^∞-mapping of M into [0,1], with support contained in U_k, and equal to 1 on $\bar{V}_k$.

(a) Show that the function $u_0 = \sum_{k=1}^{\infty} k g_k$ is of class C^∞ and is a *proper* mapping (Section 12.7, Problem 2) of M into $\mathbf{R}_+$.
(b) Let $(u_h)_{h \geqq 1}$ denote the sequence of functions g_k and $g_k \varphi_k^i$ ($k \geqq 1$, $1 \leqq i \leqq n$) arranged in some order. For each $x \in M$, let $u(x)$ denote the point of $E = \mathbf{R}^{(\mathbf{N})}$ whose coordinates (all but a finite number of which are zero) are the $u_h(x)$ for $h \geqq 0$. Show that u is injective.
(c) Let A_k (resp. B_k) denote the union of the U_h (resp. $\bar{V}_h$) for $h \leqq k$. Then $u(A_k) \supset u(B_k)$ is contained in a vector subspace E_k of E, of finite dimension which we may assume to be $\geqq 2n+1$. Show that the restriction of u to A_k is an embedding of A_k in E_k.
(d) Let E'_+ denote the topological product $\mathbf{R}_+^{\mathbf{N}}$, and let $F_+ = E_+'^{2n+1}$. An element $v \in F_+$ is therefore of the form $v = (v_1, \ldots, v_{2n+1})$, where $v_j = (\xi_{jh})_{h \geqq 0}$, the sequence $(\xi_{jh})_{h \geqq 0}$ of numbers $\geqq 0$ being arbitrary. We may identify v with the linear mapping of E into $\mathbf{R}^{2n+1}$ which maps the point $(\eta_h)_{h \geqq 0}$ of E (in which all but a finite number of the η_h are zero) to the point $\left(\sum_{h=0}^{\infty} \xi_{jh} \eta_h\right)_{1 \leqq j \leqq 2n+1}$ of $\mathbf{R}^{2n+1}$. Let Ω_k be the subset of F_+ consisting of the $v \in F_+$ such that the restriction of v to $u(A_k)$ is an immersion at each point of $u(B_k)$, and such that the restriction of v to $u(B_k)$ is injective. Use Problem 1 to show that Ω_k is a dense open subset of F_+. Deduce that the intersection of the Ω_k is dense in F_+. In particular, there exists $w = (w_1, \ldots, w_{2n+1})$ in this intersection such that the coordinate of index 0 in w_1 is $\neq 0$. Show that the mapping $f = w \circ u$ is a *proper embedding* of M in $\mathbf{R}^{2n+1}$ (*Whitney's embedding theorem*).

3. Let M be a submanifold of dimension n in $\mathbf{R}^m$. Show that for each point $\mathbf{a} \in M$ there exists a number $\varepsilon > 0$ with the following property: for each $\mathbf{b} \in B \cap M$, where B is the open ball in $\mathbf{R}^m$ with center $\mathbf{a}$ and radius ε, the orthogonal projection of $B \cap M$ on the linear manifold $L_{\mathbf{b}}$ tangent to M at $\mathbf{b}$ (16.8.6) is a bijection of $B \cap M$ onto a convex open subset of $L_{\mathbf{b}}$. (Reduce to the case where $\mathbf{a} = 0$ and $T_{\mathbf{a}}(M)$ is the subspace $\mathbf{R}^n$ of $\mathbf{R}^m$ generated by the first n vectors of the canonical basis $(\mathbf{e}_i)_{1 \leq i \leq m}$, so that in a neighborhood of the origin M is defined by $n - m$ equations $\xi^{n+j} = f_j(\xi^1, \ldots, \xi^n)$ $(1 \leq j \leq n - m)$. Let $\mathbf{u}_i$ $(1 \leq i \leq n)$ be the vectors in $\tau_{\mathbf{b}}(T_{\mathbf{b}}(M))$ which project orthogonally onto the $\mathbf{e}_i$ $(1 \leq i \leq n)$. Orthonormalize the sequence $(\mathbf{u}_1, \ldots, \mathbf{u}_n, \mathbf{e}_{n+1}, \ldots, \mathbf{e}_m)$, thus obtaining an orthonormal basis $(\mathbf{v}_i)_{1 \leq i \leq m}$ of $\mathbf{R}^m$ whose elements are C^∞-functions of $\mathbf{b}$, and the first n vectors of which form a basis of $\tau_{\mathbf{b}}(T_{\mathbf{b}}(M))$. Using the implicit function theorem, show that if ε is sufficiently small, $B \cap M$ is identical with the set of points $\sum_{i=1}^{m} \eta^i \mathbf{v}_i$, where the point $\mathbf{y} = (\eta^i)_{1 \leq i \leq m}$ runs through an open set $H_{\mathbf{b}}$ in $\mathbf{R}^n$, and the $\eta^{n+i} = F_i(\mathbf{y}, \mathbf{b})$ for $1 \leq i \leq m - n$ are C^∞-functions of $(\mathbf{y}, \mathbf{b})$ in a neighborhood of the origin in $\mathbf{R}^n \times M$. Finally show that if ε is sufficiently small, the function $G(\mathbf{y}) = \sum_{i=1}^{n} (\eta^i - \alpha^i)^2 + \sum_{j=1}^{m-n} (F_j(\mathbf{y}, \mathbf{b}) - F_j(\mathbf{a}, \mathbf{b}))^2$ is convex in a neighborhood of 0 in $\mathbf{R}^n$, where $\mathbf{a} = (\alpha^i) \in \mathbf{R}^n$ is sufficiently small.)

4. Let M be a pure manifold of dimension n. Show that there exists a locally finite denumerable open covering (A_k) of M such that every nonempty intersection of a finite number of the sets A_k is diffeomorphic to $\mathbf{R}^n$. (Use Problems 2 and 3.)

5. Let X be a pure submanifold of $\mathbf{R}^m$, and let $\pi : X \to B$ be a surjective submersion. Let Y be the graph of π, which is a submanifold of $X \times B$ diffeomorphic to X, and consider Y as a submanifold of $\mathbf{R}^m \times B$. Show that there exists an open neighborhood T of Y in $\mathbf{R}^m \times B$ and a submersion p of T onto Y with the following property: for each $b \in B$ and $y \in T \cap \text{pr}_2^{-1}(b)$, $p(y)$ is the only point of $Y \cap \text{pr}_2^{-1}(b)$ whose distance from y in $\mathbf{R}^m \times \{b\}$ is equal to $d(y, Y \cap \text{pr}_2^{-1}(b))$ (d being the Euclidean distance on $\mathbf{R}^m \times \{b\}$). (Use (16.7.4).)

6. Let M be a pure differential manifold of dimension n and f a continuous mapping of M into $\mathbf{R}^m$. Let F be a closed subset of M such that the restriction of f to F is *of class* C^r (where r is an integer > 0, or $+\infty$) in the following sense: at each point $x \in F$ there is a chart (V, φ, n) such that $f \circ \varphi^{-1}$ is of class C^r at $\varphi(x)$, in the sense of Section 16.4, Problem 6. Let S be a neighborhood of the graph of f in $M \times \mathbf{R}^m$. Show that there exists a mapping $g : M \to \mathbf{R}^m$ of class C^r, which coincides with f on F and is such that $(x, g(x)) \in S$ for all $x \in M$. (We may assume that $M \subset \mathbf{R}^{2n+1}$ and extend f to $\mathbf{R}^{2n+1}$ by the Tietze–Urysohn theorem. Let S′ be a neighborhood of the graph of f in $\mathbf{R}^{2n+1} \times \mathbf{R}^m$ such that $S' \cap (M \times \mathbf{R}^m) \subset S$, and for each $x \in \mathbf{R}^{2n+1}$ let $r(x)$ be the distance from $(x, f(x))$ to $\complement S'$. Show first, using Weierstrass' theorem and a partition of unity, that there exists a mapping $h : \mathbf{R}^{2n+1} \to \mathbf{R}^m$ of class C^∞ such that $\|f(x) - h(x)\| < \frac{1}{4}r(x)$. Using Whitney's extension theorem (Section 16.4, Problem 6) show that there exists a mapping $u : \mathbf{R}^{2n+1} \to \mathbf{R}^m$ of class C^r which is equal to $f - h$ on F and is such that $\|u(x)\| < \frac{1}{2}r(x)$ for all x.)

7. Let (X, B, π) be a fibration, F a closed subset of B, s a *continuous* section of X over B. Suppose that s is of class C^r (r an integer > 0, or $+\infty$) *in* F, in the following sense: for

each $b \in F$ there exist charts (U, φ, n) of B and (V, ψ, m) of X such that $s(U) \subset V$ and such that $\psi \circ s \circ \varphi^{-1}$ is of class C^r at $\varphi(b)$ in the sense of Section 16.4, Problem 6. Show that for each neighborhood S of the graph of s in $B \times X$, there exists a section s_1 of X over B, of class C^r *in* B, such that $(s(b), s_1(b)) \in S$ for all $b \in F$. (Embed X in $\mathbf{R}^m$ (for some m) and apply the results of Problems 5 and 6.)

8. A C^0-*fibration* $\lambda = (X, B, \pi)$ is defined by replacing, in the definition (16.12.1), differential manifolds by topological spaces, and diffeomorphisms by homeomorphisms. X is also said to be a C^0-*fiber bundle*. If (X, B, π) is a fibration in the sense of (16.12.1), it defines a C^0-fibration (called the *underlying* C^0-fibration) by regarding X and B as topological spaces. Generalize to C^0-fiber bundles the definitions and results of Section 16.12. Define likewise the notion of a C^0-*principal bundle* and a C^0-*vector bundle*, and generalize the definitions and results of Sections 16.14–16.19.

Show that if a vector bundle E (in the sense of (16.15)) is such that the underlying C^0-vector bundle is C^0-trivializable, then E is trivializable. (Use Problem 7.)

Show likewise that if two vector bundles (in the sense of (16.15)) E, F over B are such that the underlying C^0-bundles are isomorphic, then E, F are isomorphic. (Consider the bundle Hom(E, F) and use (16.16.4).)

9. Let (X, B, π) be a C^0-fibration. It is said to have the *section extension property* if every continuous section of X over a closed set A which is the restriction of a continuous section of X over an open neighborhood of A is also the restriction of a continuous global section of X (which need not agree with the preceding one on the open neighborhood of A).

Suppose that B is separable, metrizable, and locally compact. Let (U_α) be an open covering of B such that for each α the fibration induced on $\pi^{-1}(U_\alpha)$ has the section extension property. Then (X, B, π) has the section extension property. (Let A be a closed subset of B, and let V_0 be an open neighborhood of A which is equal to the set of points at which a continuous mapping $g_0 : B \to [0, 1]$ is strictly positive, the function g_0 being equal to 1 at all points of A. Let $(V_n)_{n \geqq 1}$ be a locally finite open covering of B, and let $(g_n)_{n \geqq 1}$ be a continuous partition of unity such that $V_n = \{x \in B : g_n(x) > 0\}$ for all $n \geqq 1$. Then the functions $h_0 = g_0$ and $h_n = (1 - g_0)g_n$ $(n \geqq 1)$ form a continuous partition of unity. Let s be a continuous section of X over A which extends to a continuous section of X over V_0. Proceed by induction on n by introducing the functions $f_n = h_0 + h_1 + \cdots + h_n$, and reduce the problem to the following one: Let u, v be two continuous functions on B with values in [0, 1], and let U, V be, respectively, the open sets on which $u(x) > 0$, $v(x) > 0$; suppose that $\pi^{-1}(V)$ has the section extension property. At the points $x \in U \cap A$ (resp. $V \cap A$) we have $u(x) = 1$ (resp. $v(x) = 1$) and there is a continuous section s over A such that $s|(U \cap A)$ extends to a continuous section s' over U. Show that $s|((U \cup V) \cap A)$ extends to a continuous section s'' over $U \cup V$. For this purpose, consider the function w on $U \cup V$ which is equal to 1 at points x such that $u(x) \geqq v(x)$, and equal to $u(x)/v(x)$ otherwise; observe that the set $V \cap w^{-1}(1)$ is closed in V, and extend the section which is equal to s' in $V \cap w^{-1}(1)$ and to s in $V \cap A$, to a continuous section over V.

10. Let M be a pure differential manifold, and (U_α) an open covering of M. Show that there exists an integer N, depending only on the dimension of M, and a denumerable locally finite refinement (V_n) of the covering (U_α), consisting of relatively compact open sets V_n, such that no point of M belongs to more than N of the sets V_n. (Embed M in some

$\mathbf{R}^m$ (Problem 2) and consider the U_α as the intersections of M with open sets U'_α in $\mathbf{R}^m$; this reduces us to considering only the case $M = \mathbf{R}^m$. Let K_n be the closed cube in $\mathbf{R}^m$ defined by $|z^j| \leqq n$ for $1 \leqq j \leqq m$; decompose each set $K_n - \mathring{K}_{n-1}$ into equal closed cubes, sufficiently small that each one is contained in some U'_α (**3.16.6**); then enlarge each of the cubes slightly to a concentric open cube.)

11. Let $(g_n)_{n \geqq 0}$ be a continuous partition of unity in a separable, metrizable, locally compact space B. Let V_n be the set of $b \in B$ at which $g_n(b) > 0$. For each finite subset J of N let W(J) be the set of all $b \in B$ such that $g_i(b) > g_j(b)$ for all $i \in J$ and $j \notin J$. If J and J′ are two distinct subsets of **N** having the same number of elements, then $W(J) \cap W(J') = \varnothing$. For each $b \in B$, let J(b) be the set of integers n such that $g_n(b) > 0$, so that J(b) is finite. Then $W(J(b)) \subset V_n$ for all $n \in J(b)$. For each integer $m > 0$, if W_m is the union of all the sets W(J(b)) such that J(b) has m elements, the W_m form an open covering of B.

Using these results and Problem 10, show that if B is a pure differential manifold and X a fiber bundle over B, then there exists a *finite* open covering $(W_i)_{1 \leqq i \leqq m}$ of B such that X is trivializable over each W_i.

Generalize Problem 5 of Section **16.19** by replacing the relatively compact open subset U of B by B itself.

12. (a) Let M be a differential manifold of dimension n, let $\mathbf{f}: M \to \mathbf{R}^p$ be a C^∞-mapping, and let N be a compact subset of M such that **f** is of rank n at each point of N. Also let φ be a chart of M with domain W, let V be a relatively compact open set such that $\bar{V} \subset W$, let U be a relatively compact open set such that $\bar{U} \subset V$, and let $u: \mathbf{R}^n \to [0, 1]$ be a C^∞-mapping such that $u(\mathbf{z}) = 1$ for $\mathbf{z} \in \varphi(\bar{U})$ and $u(\mathbf{z}) = 0$ for $\mathbf{z} \notin \varphi(\bar{V})$. Suppose that $p \geqq 2n$. Show that for each $\varepsilon > 0$, there exists a $p \times n$ matrix A such that: (1) $\|A \cdot \mathbf{z}\| \leqq \varepsilon$ for all $\mathbf{z} \in \varphi(V)$; (2) the function $\mathbf{z} \mapsto \mathbf{f}(\varphi^{-1}(\mathbf{z})) + A \cdot \mathbf{z}$ is of rank n in $\varphi(V)$; (3) the function $\mathbf{z} \mapsto \mathbf{f}(\varphi^{-1}(\mathbf{z})) + u(\mathbf{z})(A \cdot \mathbf{z})$ is of rank n in $\varphi(N \cap \bar{V})$. (Use Problem 5 of Section **16.23**.) Deduce that the function $\mathbf{g}: M \to \mathbf{R}^p$ defined by $\mathbf{g}(x) = \mathbf{f}(x)$ for $x \notin \bar{V}$, $\mathbf{g}(x) = \mathbf{f}(x) + u(\varphi(x))(A \cdot \varphi(x))$ for $x \in W$, is of class C^∞, is of rank n at each point of $N \cup \bar{U}$, and is such that $\|\mathbf{g}(x) - \mathbf{f}(x)\| \leqq \varepsilon$ for all $x \in M$.

(b) Let M be a differential manifold of dimension n, let $\mathbf{f}: M \to \mathbf{R}^p$ be a C^∞-mapping, and let N be a compact subset of M such that **f** is of rank n at each point of N. Let δ be a continuous real-valued function on M, everywhere >0. If $p \geqq 2n$, show that there exists an *immersion* $\mathbf{g}: M \to \mathbf{R}^p$ which agrees with **f** on N and is such that $\|\mathbf{g}(x) - \mathbf{f}(x)\| \leqq \delta(x)$ for all $x \in M$. (There exists an open set $T \supset N$ in which **f** is of rank n. Consider a denumerable locally finite open covering (W_k) of M such that the W_k are domains of definition of charts and are contained in either T or M − N. Construct **g** by induction, using (a).)

13. (a) With the same notation as in Problem 12(a), suppose that **f** is an immersion, and that the restriction of **f** to an open set S is injective. Put $v(x) = u(\varphi(x))$, and suppose that $p \geqq 2n + 1$. Show that, for each $\varepsilon > 0$, there exists a vector $\mathbf{a} \in \mathbf{R}^p$ such that: (1) $\|\mathbf{a}\| \leqq \varepsilon$; (2) the function $x \mapsto \mathbf{f}(x) + v(x)\mathbf{a}$ is an immersion of M in $\mathbf{R}^p$; (3) therela tion $\mathbf{g}(x) = \mathbf{g}(y)$ implies $v(x) = v(y)$ and $\mathbf{f}(x) = \mathbf{f}(y)$. (To show that the third condition can be satisfied, consider the open set D in $M \times M$ consisting of pairs (x, y) such that $v(x) \neq v(y)$, and the image of the mapping $(x, y) \mapsto -(\mathbf{f}(x) - \mathbf{f}(y))/(v(x) - v(y))$ of D into $\mathbf{R}^p$.)

(b) Let M be a pure differential manifold of dimension n, let $\mathbf{f}: M \to \mathbf{R}^p$ be an immersion, let S be an open subset of M such that $\mathbf{f}|S$ is injective, let N be a closed subset of M contained in S, and let δ be a continuous real-valued function on M, everywhere >0. If $p \geqq 2n+1$, show that there exists an *embedding* $\mathbf{g}: M \to \mathbf{R}^p$ which agrees with $\mathbf{f}$ on N and is such that $\|\mathbf{g}(x) - \mathbf{f}(x)\| \leqq \delta(x)$ for all $x \in M$. (Consider a denumerable locally finite open covering (W_k) of M, where the W_k are domains of definition of charts, the restrictions $\mathbf{f} | W_k$ are injective, and each W_k is contained either in S or in $M - N$. Construct $\mathbf{g}$ by induction on k, using (a).)

(c) Deduce from Problems 12(b) and 13(b) a new proof of Whitney's embedding theorem. (Problem 2; first define a proper mapping of class C^∞ of M into $\mathbf{R}^{2n+1}$.)

14. (a) Let M be a manifold of dimension n. Show that there exist $2n+1$ real-valued functions f_j of class C^∞ on M $(1 \leqq j \leqq 2n+1)$ such that every C^∞-function on M is of the form $F(f_1, \dots, f_{2n+1})$, where F is a C^∞-function on $\mathbf{R}^{2n+1}$. (Use Whitney's embedding theorem.) Show that the $\mathscr{E}(M)$-module $\mathscr{E}_p(M)$ of C^∞ differential p-forms on M is generated by the p-forms $df_{j_1} \wedge \cdots \wedge df_{j_p}$ $(1 \leqq j_1 < j_2 < \cdots < j_p \leqq 2n+1)$. (Same method.)

(*b*) Let M, N be two differential manifolds. Show that, for each p, the $\mathscr{E}(M \times N)$-module $\mathscr{E}_p(M \times N)$ of C^∞ p-forms on $M \times N$ is the direct sum of the $(p+1)$ $\mathscr{E}(M \times N)$-modules $\mathscr{E}_{r,p-r}$, where $\mathscr{E}_{r,\,p-r}$ is the $\mathscr{E}(M \times N)$-module generated by the p-forms of the type ${}^t\mathrm{pr}_1(\alpha) \wedge {}^t\mathrm{pr}_2(\beta)$, where $\alpha \in \mathscr{E}_r(M)$ and $\beta \in \mathscr{E}_{p-r}(N)$ $(0 \leqq r \leqq p)$.

15. Show that there exists an immersion of the Klein bottle (**16.14.10**) into $\mathbf{R}^3$, the image of which is the set of points (ξ^1, ξ^2, ξ^3), where

$$\begin{aligned} \xi^1 &= (a + \cos \tfrac{1}{2}u \sin t - \sin \tfrac{1}{2}u \sin 2t) \cos u, \\ \xi^2 &= (a + \cos \tfrac{1}{2}u \sin t - \sin \tfrac{1}{2}u \sin 2t) \sin u, \\ \xi^3 &= \sin \tfrac{1}{2}u \sin t + \cos \tfrac{1}{2}u \sin 2t \end{aligned}$$

$(0 \leqq t \leqq 2\pi, 0 \leqq u \leqq 2\pi)$.

16. Define a real-analytic mapping $\mathbf{f}: \mathbf{R}^{n+1} \to \mathbf{R}^{2n+1}$ as follows: If $\mathbf{x} = (\xi^0, \dots, \xi^n)$, then $\mathbf{z} = \mathbf{f}(\mathbf{x}) = (\zeta^0, \dots, \zeta^{2n})$ is given by the formulas

$$\zeta^r = \sum_{s=0}^{r} \xi^s \xi^{r-s}.$$

The restriction $\mathbf{f}_0$ of $\mathbf{f}$ to $\mathbf{S}_n$ factorizes into

$$\mathbf{S}_n \xrightarrow{\pi} \mathbf{P}_n(\mathbf{R}) \xrightarrow{g} \mathbf{R}^{2n+1},$$

where π is the canonical mapping. Show that $h: \pi(\mathbf{x}) \mapsto g(\pi(\mathbf{x}))/\|g(\pi(\mathbf{x}))\|$ is an *embedding* of $\mathbf{P}_n(\mathbf{R})$ into $\mathbf{S}_{2n}$. Define similarly an embedding of $\mathbf{P}_n(\mathbf{C})$ in $\mathbf{S}_{4n-1}$ and of $\mathbf{P}_n(\mathbf{H})$ in $\mathbf{S}_{8n-3}$ (replace $\xi^s \xi^{r-s}$ by $\bar{\xi}^s \xi^{r-s}$). (*James embedding*.)

17. (a) Let M, N be pure differential manifolds of dimensions n, p, respectively; let $f: M \to N$ be a C^∞-mapping, and Z a submanifold of N of dimension $p - q$. Let A be a closed subset of M such that f is transversal over Z at all points of $A \cap f^{-1}(Z)$ (Section **16.8**, Problem 9). Let ψ be a chart of N, with domain T, such that $\psi(T) = H \times I$, where H is an open set in $\mathbf{R}^{p-q}$, I an open set in $\mathbf{R}^q$, and $\psi(T \cap Z) = H$ (**16.8.3**). Also let φ be a chart of M with domain $W \subset f^{-1}(T)$, let U and V be relatively compact open sets such that $\bar{U} \subset V$ and $\bar{V} \subset W$, and let u be a C^∞-mapping of $\mathbf{R}^n$ into $[0, 1]$ such that $u(z) = 1$ for all $z \in \varphi(\bar{U})$ and $u(z) = 0$ for all $z \notin \varphi(\bar{V})$. Show that for each $\varepsilon > 0$ there exists a vector $\mathbf{b} \in \mathbf{R}^q$ such that: (1) $\|\mathbf{b}\| \leq \varepsilon$; (2) the function

$$x \mapsto \psi(f(x)) + u(\varphi(x))\mathbf{b}$$

takes its values in $\psi(T)$ for $x \in \bar{V}$; (3) the mapping $g = g_{\mathbf{b}} : M \to N$ defined by $g(x) = \psi^{-1}(\psi(f(x)) + u(\varphi(x))\mathbf{b})$ for $x \in W$, $g(x) = f(x)$ for $x \in M - V$ is transversal over Z at the points of $g^{-1}(Z) \cap \bar{U}$; (4) the mapping g is transversal over Z at the points of $g^{-1}(Z) \cap A$. (To satisfy (3), use Sard's theorem (16.23.1) for the mapping

$$z \mapsto \text{pr}_2(\psi(f(\varphi^{-1}(z))))$$

of $\varphi(V)$ into $\mathbf{R}^q$. To satisfy (4), it is sufficient that it should be satisfied at the points of $g^{-1}(z) \cap K$, where $K = A \cap (\bar{V} - U)$. For this, consider the mapping

$$(x, \mathbf{b}) \mapsto (g_{\mathbf{b}}(x), D(\text{pr}_2 \circ \psi \circ g_{\mathbf{b}} \circ \varphi^{-1})(\varphi(x)))$$

of $K \times \mathbf{R}^q$ into $N \times \mathbf{R}^{qn}$; observe that it is continuous and transforms $K \times \{0\}$ into a subset of the open set $((N - A) \times \mathbf{R}^{qn}) \cup (A \times M_q)$ of $N \times \mathbf{R}^{qn}$, in the notation of Section 16.23, Problem 4.)

(b) Let M, N be differential manifolds of dimensions m, n respectively, let $f: M \to N$ be a C^∞-mapping, let Z be a submanifold of N of dimension $p - q$, and let A be a closed subset of M such that f is transversal over Z at all points of $A \cap f^{-1}(Z)$. Finally let d be a distance defining the topology of N, and let δ be a continuous function on M, everywhere >0. Show that there exists a C^∞-mapping $g: M \to N$ which is transversal over Z, coincides with f on A, and is such that $d(f(x), g(x)) \leq \delta(x)$ for all $x \in M$ (*Thom's transversality theorem*). (There exists an open neighborhood S of A such that f is transversal over Z at the points of $f^{-1}(Z) \cap S$. Consider a denumerable locally finite open covering $(T_k)_{k \geq 0}$ of N, where $T_0 = N - Z$ and the T_k ($k \geq 1$) are domains of definition of charts ψ_k of N, such that $\psi_k(T_k) = H_k \times I_k$, where H_k is open in $\mathbf{R}^{p-q}$ and I_k is open in $\mathbf{R}^q$, and $\psi_k(T_k \cap Z) = H_k$. Then take a locally finite open covering (W_k) of M, for which the W_k are domains of definition of charts of M, and which refines the covering formed by the intersections of the open sets $f^{-1}(T_k)$ with $M - A$ and S. Construct g by induction on k, using (a).)

26. DIFFERENTIABLE HOMOTOPIES AND ISOTOPIES

(16.26.1) The notion of *homotopy*, defined in (9.6) for paths, extends to arbitrary continuous mappings. Given two continuous mappings f, g of a topological space X into a topological space Y, a *homotopy* of f into g is a continuous mapping φ of $X \times [\alpha, \beta]$ (where $\alpha < \beta$ in $\mathbf{R}$) into Y such that $\varphi(x, \alpha) = f(x)$ and $\varphi(x, \beta) = g(x)$ for all $x \in X$. The mapping g is said to be *homotopic* to f if there exists a homotopy of f into g. Just as in (9.6) it is immediately shown that if g is homotopic to f, then f is homotopic to g; and that if h is homotopic to g, and g is homotopic to f, then h is homotopic to f; in other words, the relation "f is homotopic to g" is an *equivalence relation* on the set of continuous mappings of X into Y.

(16.26.2) Now suppose that X and Y are *differential manifolds* and that f and g are mappings of class C^p (where p is an integer $\geqq 1$, or $+\infty$) of X into Y. Then a *C^p-homotopy* (or a *homotopy of class* C^p) of f into g is a mapping φ of class C^p of $X \times J$ into Y, where J is a nonempty *open* interval in **R**, such that for two points $\alpha < \beta$ in J we have $\varphi(x, \alpha) = f(x)$ and $\varphi(x, \beta) = g(x)$ for all $x \in X$. The mapping g is said to be *C^p-homotopic* to f if there exists a C^p-homotopy of f into g.

(16.26.3) *Let* X, Y *be differential manifolds. Then the relation "f is C^p-homotopic to g" is an equivalence relation on the set of C^p-mappings of* X *into* Y.

It is immediate that the relation is reflexive and symmetric. To show that it is transitive, consider three mappings f, g, h of class C^p from X to Y, a C^p-homotopy φ of f into g, and a C^p-homotopy ψ of g into h. By a linear change of variable we may assume that φ and ψ are both defined on the same space $X \times J$, where J is an open interval in **R**, and that there exist $\alpha < \beta$ in J such that $\varphi(x, \alpha) = f(x)$, $\varphi(x, \beta) = g(x)$, $\psi(x, \alpha) = g(x)$, $\psi(x, \beta) = h(x)$, for all $x \in X$. There exists a C^∞-mapping $\lambda : \mathbf{R} \to [\alpha, \beta]$ such that $\lambda(t) = \alpha$ for $t < \frac{1}{3}$, $\lambda(t) = \beta$ for $t > \frac{2}{3}$ (16.4.2). The functions of class C^p,

$$(x, t) \mapsto \varphi(x, \lambda(t)) \qquad \text{and} \qquad (x, t) \mapsto \psi(x, \lambda(t-1))$$

defined on $X \times \mathbf{R}$ agree on $X \times]\frac{2}{3}, \frac{4}{3}[$. We may therefore define a C^p-homotopy θ of f into h by putting $\theta(x, t) = \varphi(x, \lambda(t))$ for $t \leqq 1$ and $\theta(x, t) = \psi(x, \lambda(t-1))$ for $t \geqq 1$.

(16.26.4) *Let* X, Y *be two differential manifolds,* K *a compact subset of* X, *and f*: K → Y *a continuous mapping. If d is a distance defining the topology of* Y, *then there exists $\varepsilon > 0$ such that every continuous mapping $g : K \to Y$ satisfying $d(f(x), g(x)) \leqq \varepsilon$ for all $x \in K$ is homotopic to f. Moreover, if the restrictions of f and g to $\mathring{K}$ are of class C^p, then f and g are C^p-homotopic.*

Let V be a relatively compact open neighborhood of $f(K)$ in Y. By (16.25.1) we may assume that V is embedded in some $\mathbf{R}^n$. Let T be an open neighborhood of V in $\mathbf{R}^n$ with the properties of (16.25.4). If d' is the Euclidean distance on $\mathbf{R}^n$, there exists a number $\eta > 0$ such that all the points $z \in \mathbf{R}^n$ for which $d'(z, f(K)) \leqq \eta$ belong to T (3.17.11). Next, there exists $\varepsilon > 0$ such that $d(y, f(x)) \leqq \varepsilon$ for $y \in Y$ and $x \in K$ implies that $d'(y, f(x)) \leqq \eta$ (this is easily proved by contradiction, using the compactness of K) and hence $y \in T$. Now let $g : K \to Y$ be a continuous mapping such that $d(f(x), g(x)) \leqq \varepsilon$ for all $x \in K$. For $x \in K$ and $t \in [0, 1]$ we have

$$d'(f(x), tf(x) + (1-t)g(x)) \leqq d'(f(x), g(x)) \leqq \eta$$

and therefore the point $tf(x) + (1 - t)g(x)$ belongs to T. If we put

$$\varphi(x, t) = \pi(tf(x) + (1 - t)g(x)) \in \mathrm{Y},$$

then it is clear that φ is a homotopy of g into f. For by virtue of (3.17.11) applied to $\varphi^{-1}(\mathrm{T})$, there exists $\alpha > 0$ such that $tf(x) + (1 - t)g(x)$ belongs to T for $-\alpha < t < 1 + \alpha$ and all $x \in \mathring{\mathrm{K}}$. Moreover, if $f | \mathring{\mathrm{K}}$ and $g | \mathring{\mathrm{K}}$ are of class C^p, then the restriction of φ to $\mathring{\mathrm{K}} \times]-\alpha, 1 + \alpha[$ is of class C^p, from which the second assertion follows.

Remark

(16.26.4.1) The above proof shows that if $f(x_0) = g(x_0)$ for some $x_0 \in \mathrm{K}$, then the homotopy φ constructed above is such that $\varphi(x_0, t) = f(x_0)$ for all $t \in [0, 1]$.

(16.26.5) *Let* X, Y *be differential manifolds,* K *a compact subset of* X, *and* f *a continuous mapping of* K *into* Y. *Then there exists a relatively compact open neighborhood* U *of* K *in* X *and a* C^∞*-mapping* g *of* U *into* Y *such that* $g | \mathrm{K}$ *is homotopic to* f.

Let d be a distance defining the topology of Y, and let $\varepsilon > 0$ be a real number for which (16.26.4) holds. By virtue of (16.25.5) there exists a relatively compact open neighborhood U of K in X and a C^∞-mapping g of U into Y such that $d(f(x), g(x)) \leqq \varepsilon$ for all $x \in \mathrm{K}$. The result now follows from (16.26.4).

(16.26.6) *Let* X, Y *be differential manifolds,* K *a compact subset of* X, *and* f, g *homotopic continuous mappings of* K *into* Y. *If* $f | \mathring{\mathrm{K}}$ *and* $g | \mathring{\mathrm{K}}$ *are of class* C^p, *then they are* C^p*-homotopic.*

By hypothesis, there exists a compact interval $\mathrm{I} = [\alpha, \beta]$ in **R** and a continuous mapping $\varphi : \mathrm{K} \times \mathrm{I} \to \mathrm{Y}$ such that $\varphi(x, \alpha) = f(x)$ and $\varphi(x, \beta) = g(x)$ for all $x \in \mathrm{K}$. Choose $\varepsilon > 0$ for which (16.26.4) holds for X, Y, f, and K; for X, Y, g, and K; and for $\mathrm{X} \times \mathbf{R}$, Y, φ, and $\mathrm{K} \times \mathrm{I}$. Then there exists an open neighborhood U (resp. J) of K in X (resp. of I in **R**) and a C^∞-mapping $\psi : \mathrm{U} \times \mathrm{J} \to \mathrm{Y}$ such that $d(\varphi(x, t), \psi(x, t)) \leqq \varepsilon$ for all $(x, t) \in \mathrm{K} \times \mathrm{I}$. If we put $f_1(x) = \psi(x, \alpha)$ and $g_1(x) = \psi(x, \beta)$ for $x \in \mathrm{U}$, then ψ is a C^∞-homotopy of f_1 into g_1. However, we have $d(f_1(x), f(x)) \leqq \varepsilon$ and $d(g(x), g_1(x)) \leqq \varepsilon$; hence by (16.26.4) it follows that $f | \mathring{\mathrm{K}}$ and $f_1 | \mathring{\mathrm{K}}$ are C^p-homotopic, and that $g_1 | \mathring{\mathrm{K}}$ and $g | \mathring{\mathrm{K}}$ are C^p-homotopic. Hence, by (16.26.3), $f | \mathring{\mathrm{K}}$ and $g | \mathring{\mathrm{K}}$ are C^p-homotopic.

(16.26.7) If f, g are two *diffeomorphisms* of a differential manifold X onto a differential manifold Y, a C^∞-homotopy φ of f into g, defined on $X \times J$, where J is an open interval in **R**, is said to be a *C^∞-isotopy* (or an *isotopy of class* C^∞) if for each $t \in J$ the mapping $x \mapsto \varphi(x, t)$ is a *diffeomorphism* of X onto Y. The same argument as in (16.26.3) shows that the relation "there exists a C^∞-isotopy of f into g" is an equivalence relation on the set of diffeomorphisms of X onto Y.

(16.26.8) *Let X be a differential manifold, U a connected open subset of X, and a, b two points of U. Then there exists a C^∞-isotopy φ of the identity map 1_X into a diffeomorphism h of X onto X, such that $h(a) = b$ and such that $\varphi(x, t) = x$ for all t and all $x \notin U$.*

(I) We consider first the following particular case: $X = \mathbf{R}^n$, U is the open ball $\|\mathbf{x}\| < \sqrt{n}$ (relative to the Euclidean norm $\|\mathbf{x}\| = \left(\sum_{i=1}^{n} \xi_i^2\right)^{1/2}$), $\mathbf{a} = 0$ and $\|\mathbf{b}\| < 1$. Let $(\mathbf{e}_i)_{1 \leq i \leq n}$ be the canonical basis of $\mathbf{R}^n$, and suppose first that $\mathbf{b} = \beta \mathbf{e}_1$, with $0 < \beta < 1$. We shall use the following lemma:

(16.26.8.1) *Let E be a real Banach space, $f: E \to E$ a bounded and p times continuously differentiable mapping. Then for each $z \in E$ there exists a unique solution $t \mapsto F(t, z)$ of the differential equation $dx/dt = f(x)$, defined on the whole of* **R** *and such that $F(0, z) = z$. Furthermore, for all $s, t \in \mathbf{R}$ we have $F(s, F(t, z)) = F(s + t, z)$, and for each $t \in \mathbf{R}$, the mapping $z \mapsto F(t, z)$ is a homeomorphism of E onto E which together with its inverse is p times continuously differentiable.*

The first assertion is proved as in (10.6.1), by remarking that because of the mean-value theorem every solution v of the differential equation in a relatively compact open interval J of **R** is *bounded* in J, which allows us to apply (10.5.5) because $f(v(t))$ is bounded in J. Every integral of the differential equation which takes the same value as $t \mapsto F(t, z)$ at a point of **R** must be identical with this function, by (10.5.2). Now, for each $s \in \mathbf{R}$, the function $t \mapsto F(t + s, z)$ is a solution of the equation which takes the value $F(s, z)$ when $t = 0$; hence $F(t, F(s, z)) = F(t + s, z)$. In particular, $F(t, F(-t, z)) = z$ for all $t \in \mathbf{R}$ and $z \in E$; also by (10.7.4) we know that $(t, z) \mapsto F(t, z)$ is p times continuously differentiable, hence the proof of the lemma is complete.

To apply this lemma, consider a C^∞-mapping $g : \mathbf{R} \to [0, 1]$ such that $g(\xi) = 0$ for $|\xi| \geq 1$, $g(\xi) > 0$ for $|\xi| < 1$ and $g(0) = 1$ (16.4.1.4). We apply the lemma to the system of differential equations

(16.26.8.2)
$$\begin{cases} \dfrac{d\xi^1}{dt} = g(\xi^1)g(\xi^2)\cdots g(\xi^n), \\ \dfrac{d\xi^i}{dt} = 0 \qquad (2 \leqq i \leqq n). \end{cases}$$

Hence we obtain a mapping $(\mathbf{x}, t) \mapsto \mathbf{F}(\mathbf{x}, t)$ of class C^∞, from $\mathbf{R}^n \times \mathbf{R}$ to $\mathbf{R}^n$, such that $\mathbf{F}(\mathbf{x}, 0) = \mathbf{x}$, and such that $\mathbf{x} \mapsto \mathbf{F}(\mathbf{x}, t)$ is a diffeomorphism of $\mathbf{R}^n$ for all $t \in \mathbf{R}$. If $|\xi^j| \geqq 1$ for some index j, then $\mathbf{F}(\mathbf{x}, t) = \mathbf{x}$ for all t, since the right-hand side of the first equation (16.26.8.2) is then zero. It remains to show that $\mathbf{F}(0, t) = \mathbf{b}$ for a suitable choice of t; but if we put $\mathbf{F}(0, t) = (u(t), 0, \dots, 0)$, then u is strictly increasing on $\mathbf{R}$, and the inverse function is $\xi \mapsto \int_0^\xi g(\zeta)^{-1} d\zeta$, defined for $-1 < \xi < 1$, and which tends to $+\infty$ as $\xi \to 1$ by reason of the fact that all the derivatives of g vanish at the point 1.

It remains to consider the case where $\mathbf{b}$ is arbitrary (subject to $\|\mathbf{b}\| < 1$). There exists a rotation r of $\mathbf{R}^n$ transforming $\mathbf{b}$ into a point of the form $\beta\mathbf{e}_1$, and in place of $\mathbf{F}$ we consider the function $(\mathbf{x}, t) \mapsto r^{-1}(\mathbf{F}(r(\mathbf{x}), t))$.

(II) Now consider the general case. For any two points p, q of U, let $R(p, q)$ denote the relation obtained by replacing a, b by p, q in the statement of (16.26.8). It follows immediately from (16.26.3) that R is an equivalence relation on U. Moreover, each equivalence class of R is *open* in U: for each $p \in U$, there exists a chart (V, ψ, n) of U such that $\psi(p) = 0$ and such that $\psi(V)$ contains the closed ball with center 0 and radius $\sqrt{n}$ in $\mathbf{R}^n$; let $W \subset V$ be the inverse image under ψ of the open ball with center 0 and radius 1. If φ_0 is an isotopy defined on $V \times J$ and such that $\varphi_0(x, t) = x$ for $x \in V - W$, we extend it to an isotopy φ defined on $X \times J$ by putting $\varphi(x, t) = x$ for all $x \in X - V$. Then part (I) of the proof shows that the relation $R(p, q)$ is true for all $q \in W$, which proves our assertion. The equivalence classes of R are therefore both open and closed in U; since U is connected, it follows that there is only one equivalence class. Q.E.D.

(16.26.9) *Let X be a connected differential manifold, $x_0, x_1, \dots, x_p$ distinct points of X, and U a neighborhood of x_0. Then there exists a diffeomorphism h of X onto X such that $h(x_0) = x_0$ and such that $h(x_i) \in U$ for $1 \leqq i \leqq p$.*

Consider a chart (V, ψ, n) of X at x_0 such that $V \subset U$, and such that $\psi(V)$ contains the open ball with center $\psi(x_0)$ and radius 1. Let V_k $(1 \leqq k \leqq p)$ be the inverse image under ψ of the open ball with center $\psi(x_0)$ and radius k/p, so that $\bar{V}_k \subset V_{k+1}$. By induction on k it is enough to show that there exists a diffeomorphism h_{k+1} of X onto X which fixes the points of $\bar{V}_k$ and maps x_{k+1} into V_{k+1}. The only case to be considered is that in which

$$x_{k+1} \notin V_{k+1}.$$

Let W be the connected component of x_{k+1} in $X - \bar{V}_k$, which is open and closed in $X - \bar{V}_k$ (3.19.5). We have $W \cap V_{k+1} \neq \varnothing$, otherwise the closure of W in X would be equal to its closure in $X - \bar{V}_k$, and therefore W would be both open and closed in X, which is absurd. The result now follows from (16.26.8) applied to the connected open set W in X.

(16.26.10) *Let X be a connected differential manifold of dimension* $\geqq 1$, *and let a, b be two points of X. Then there exists an injective* C^∞*-mapping* $f : \mathbf{R} \to X$ *such that f(R) contains both a and b.* (In geometrical terms, a and b can be joined in X by a simple C^∞-arc.)

By means of a diffeomorphism of X onto itself, we may assume that b lies in an open neighborhood U of a such that there exists a chart (U, φ, n) of X for which $\varphi(U) = \mathbf{R}^n$ ((16.26.9) and (16.3.4)). We have then only to take a parametric representation $t \mapsto g(t)$ of the line joining $\varphi(a)$ to $\varphi(b)$, and then take $f = \varphi^{-1} \circ g$.

PROBLEMS

1. Let X, Y be two pure differential manifolds. Show that there exists a neighborhood V of the diagonal in $Y \times Y$ with the following property: If $f, g : X \to Y$ are mappings of class C^p (p an integer >0, or $+\infty$) such that $(f(x), g(x)) \in V$ for all $x \in X$, then f, g are C^p-homotopic. (Embed Y in $\mathbf{R}^m$ (Section 16.25, Problem 2) and use (16.25.4).) Deduce that, for each continuous $f : X \to Y$, there exists a C^∞-mapping $g : X \to Y$ homotopic to f, and that if two C^p-mappings $f, g : X \to Y$ are homotopic, then they are C^p-homotopic.

2. Two topological spaces X, Y are said to have the same *homotopy type* if there exist continuous mappings $f : X \to Y$ and $g : Y \to X$ such that $g \circ f$ is homotopic to 1_X and $f \circ g$ homotopic to 1_Y. In that case f (resp. g) is said to be a *homotopy equivalence*. If X and Y have the same homotopy type, then for two continuous mappings $u : Z \to X$ (resp. $v : X \to Z$), where Z is a topological space, to be homotopic it is necessary and sufficient that $f \circ u$ and $f \circ v$ (resp. $u \circ g$ and $v \circ g$) should be homotopic. If Y is a differential manifold and X is a closed submanifold of Y, show that there exists an open neighborhood U of X in Y such that X and U have the same homotopy type. (Embed Y in some $\mathbf{R}^m$ and use (16.25.4).)

3. (a) Let X be a metrizable topological space, A a closed subset of X, U a neighborhood of A, f a continuous mapping of $U \times [0, 1]$ into a topological space Y. Suppose that there exists a continuous mapping $g : X \to Y$ extending the mapping $x \mapsto f(x, 0)$. Show that there exists a continuous mapping $h : X \times [0, 1] \to Y$ such that $h(x, t) = g(x)$ for $x \in X - U$ and all $t \in [0, 1]$, and such that $h(x, t) = f(x, t)$ for $x \in A$. (Use the Tietze–Urysohn theorem.)
(b) Deduce from (a) that if Y is a differential manifold and if f is a continuous mapping of $A \times [0, 1]$ into Y, such that there exists a continuous extension to X of the mapping $x \mapsto f(x, 0)$ of A into Y, then f can be extended to a continuous mapping of $X \times [0, 1]$ into Y. (Embed Y in some $\mathbf{R}^m$ and use the Tietze–Urysohn theorem and (**16.25.4**).) In particular, the mapping $x \mapsto f(x, 1)$ extends to a continuous mapping of X into Y. Consider the case $A = \varnothing$.
(c) Suppose that X is a differential manifold, A a closed submanifold of X, and Y a topological space. Let $f: A \times [0, 1] \to Y$ be a continuous mapping such that the mapping $x \mapsto f(x, 0)$ admits a continuous extension to X. Then f can be extended to a continuous mapping of $X \times [0, 1]$ into Y. (Embed X in some $\mathbf{R}^m$, then use (a) and (**16.25.4**).) Consider the case $A = \varnothing$.

4. Let $\lambda = (X, B, \pi)$ be a C^0-fibration (Section **16.25**, Problem 9), where B is of the form $A \times [a, b]$, with $[a, b] \subset \mathbf{R}$. Suppose that there exists $c \in]a, b[$ such that the fibrations induced by λ on $A \times [a, c]$ and $A \times [c, b]$ are trivializable. Show that λ is trivializable. (Observe that an A-isomorphism of a trivial bundle $A \times F$ over A can be extended to an $(A \times I)$-isomorphism of $(A \times I) \times F$, for any interval $I \subset \mathbf{R}$.) Consider the analogous theorems for principal bundles and vector bundles.

5. Let $\lambda = (X, B, \pi)$ be a C^0-fibration, where B is of the form $A \times [0, 1]$. Show that there exists an open covering (U_α) of A such that λ is trivializable over each open set $U_\alpha \times [0, 1]$. (Use Problem 4.) Consider the analogous theorems for principal bundles and vector bundles.

6. Let B be a separable, metrizable, locally compact space, A a closed subset of B, and $\lambda = (X, B \times [0, 1], \pi)$ a C^0-fibration. Suppose either that the fibers of X are differential manifolds or that B is a differential manifold and A a closed submanifold of B. Show that every continuous section of X over the closed set $(A \times [0, 1]) \cup (B \times \{0\})$ extends to a continuous section of X over $B \times [0, 1]$. (Use Problems 3 and 5 above, and Section **16.25**, Problem 9.) Consider the case $A = \varnothing$.

7. (a) Let B be a separable, metrizable, locally compact space and let E_1, E_2 be two C^0-vector bundles over $B \times [0, 1]$. Show that if the bundles induced by E_1 and E_2 over $B \times \{0\}$ are isomorphic, then E_1 and E_2 are isomorphic. (Apply Problem 6, with $A = \varnothing$, to the fiber bundle $\mathrm{Isom}(E_1, E_2)$, which is an open subset of the vector bundle $\mathrm{Hom}(E_1, E_2)$ over $B \times [0, 1]$, and the fibers of which consist of the vector space isomorphisms between the corresponding fibers of E_1 and E_2.)
(b) Deduce from (a) that if E is a C^0-vector bundle over $B \times [0, 1]$ and if E_0 is the vector bundle induced by E over $B \times \{0\}$, and $p : (b, t) \mapsto (b, 0)$ the projection onto $B \times \{0\}$, then E is isomorphic to $p^*(E_0)$. In particular, if E_1 is the vector bundle induced by E over $B \times \{1\}$, and if $B \times \{0\}$ and $B \times \{1\}$ are identified canonically with B, then E_0 and E_1 are isomorphic vector bundles over B.

(c) Let f, g be homotopic continuous mappings of B into a separable, metrizable, locally compact space B′. If E′ is a C^0-vector bundle over B′, show that $f^*(E')$ and $g^*(E')$ are B-isomorphic. In particular, if B is *contractible* (16.27.7), then every C^0-vector bundle over B is *trivializable*.

8. Let E be a real C^0-vector bundle of rank k over a separable, metrizable, locally compact space B. Let $\pi : E \to B$ be the projection. Let u, v be two Gaussian mappings of E into $\mathbf{R}^m$ (Section 16.19, Problem 8), and suppose that there exists a Gaussian mapping w of $E \times [0, 1]$ (regarded as a vector bundle over $B \times [0,1]$) into $\mathbf{R}^m$ such that $w(x, 0) = u(x)$ and $w(x, 1) = v(x)$. If f, g are the continuous mappings of B into $\mathbf{G}_{m,k}$ corresponding to u and v, respectively (*loc. cit.*), show that f and g are homotopic.

9. Let q^+ and q^- be the two injective linear mappings of $\mathbf{R}^n$ into $\mathbf{R}^{2n}$ defined by $q^+(\mathbf{e}_i) = \mathbf{e}_{2i}$, $q^-(\mathbf{e}_i) = \mathbf{e}_{2i-1}$ $(1 \leq i \leq n)$.

(a) Show that each of q^+, q^- is homotopic to the injective linear mapping $q : \mathbf{R}^n \to \mathbf{R}^{2n}$ defined by $q(\mathbf{e}_i) = \mathbf{e}_i$ $(1 \leq i \leq n)$.
(b) Consider the Gaussian mappings $q^+ \circ \mathrm{pr}_2$ and $q^- \circ \mathrm{pr}_2$ of $\mathrm{U}_{n,k}$ into $\mathbf{R}^{2n}$; to them correspond canonically (Section 16.19, Problem 8) continuous mappings f^+, f^- of $\mathbf{G}_{n,k}$ into $\mathbf{G}_{2n,k}$. Deduce from (a) and from Problem 8 that f^+, f^- are each homotopic to the canonical injection $j : \mathbf{G}_{n,k} \to \mathbf{G}_{2n,k}$ (Section 16.19, Problem 2(a)).
(c) Let E be a real C^0-vector bundle of rank k, with base B and projection π, and let u, v be two Gaussian mappings of E into $\mathbf{R}^n$ (where $n \geq k$). Then $q^+ \circ u$ and $q^- \circ u$ are Gaussian mappings of E into $\mathbf{R}^{2n}$. Show that the mapping

$$(x, t) \mapsto w(x, t) = (1 - t)q^+(u(x)) + tq^-(v(x))$$

is a Gaussian mapping of $E \times [0, 1]$ into $\mathbf{R}^{2n}$.
(d) Let f, g be two continuous mappings of B into $\mathbf{G}_{n,k}$. Show that if the C^0-vector bundles $f^*(\mathrm{U}_{n,k})$ and $g^*(\mathrm{U}_{n,k})$ are B-isomorphic, then the mappings $j \circ f$ and $j \circ g$ of B into $\mathbf{G}_{2n,k}$ are homotopic. (Consider a vector bundle E of rank k over B and the Gaussian mappings $u, v : E \to \mathbf{R}^n$ corresponding to f, g (Section 16.19, Problem 8); use successively (c), Problem 8, then (b).)

10. Show that there exists $\varepsilon > 0$ such that, for each $\mathbf{a} \in \mathbf{R}^n$ with $\|\mathbf{a}\| \leq \varepsilon$ ($\|\mathbf{a}\|$ being the Euclidean norm) and each endomorphism S of $\mathbf{R}^n$ with $\|I - S\| \leq \varepsilon$ (the norm here being the norm on $\mathrm{End}(\mathbf{R}^n)$ induced by the Euclidean norm (5.7.1)), there exists a C^∞-mapping $\mathbf{F} : \mathbf{R}^n \times \mathbf{R} \to \mathbf{R}^n$, satisfying the following conditions: (1) for each $t \in [0, 1]$, the mapping $\mathbf{x} \mapsto \mathbf{F}(\mathbf{x}, t)$ is a diffeomorphism of $\mathbf{R}^n$ onto $\mathbf{R}^n$ such that $\mathbf{F}(\mathbf{x}, t) = \mathbf{x}$ whenever $\|\mathbf{x}\| > 2$; (2) $\mathbf{F}(\mathbf{x}, 0) = \mathbf{x}$ for all $\mathbf{x} \in \mathbf{R}^n$; (3) $\mathbf{F}(\mathbf{x}, 1) = \mathbf{a} + S \cdot \mathbf{x}$ whenever $\|\mathbf{x}\| < 1$. (Take $\mathbf{F}(\mathbf{x}, t) = \mathbf{x} + th(\|\mathbf{x}\|^2)(\mathbf{a} + S \cdot \mathbf{x} - \mathbf{x})$, where h is a C^∞-mapping of $\mathbf{R}$ into $[0, 1]$ which is equal to 1 for $|\xi| \leq 1$ and zero for $|\xi| \geq 4$. To show that condition (1) is satisfied for all small ε, use Section 16.12, Problem 1.)

11. With the hypotheses of (16.26.8), let $(\mathbf{u}_j)_{1 \leq j \leq n}$ be a basis of $T_a(X)$ and $(\mathbf{v}_j)_{1 \leq j \leq n}$ a basis of $T_b(X)$. If X is orientable, suppose that these two bases are direct. Show that the isotopy φ may be constructed so that $h(a) = b$ and also $T_a(h) \cdot \mathbf{u}_j = \mathbf{v}_j$ for $1 \leq j \leq n$. (Follow the second part of the proof of (16.26.8), but work in the frame bundle R(X) (20.1.1), and use Problem 10.)

12. Let $f_1, \ldots, f_n$ be real-valued C^∞-functions defined on a neighborhood of 0 in $\mathbf{R}^n$, such that the Jacobian matrix of $\mathbf{F}_0 = (f_j)_{1 \leqq j \leqq n}$ at 0 is the identity matrix. Show that there exists a real number $\varepsilon > 0$ and a C^∞-mapping $\mathbf{F} : \mathbf{R}^n \times \mathbf{R} \to \mathbf{R}^n$ satisfying the following conditions: (1) for each $t \in [0, 1]$, the mapping $\mathbf{x} \mapsto \mathbf{F}(\mathbf{x}, t)$ is a diffeomorphism of $\mathbf{R}^n$ onto $\mathbf{R}^n$ such that $\mathbf{F}(\mathbf{x}, t) = \mathbf{x}$ for $\|\mathbf{x}\| \geqq 2\varepsilon$; (2) $\mathbf{F}(\mathbf{x}, 0) = \mathbf{x}$ for all $\mathbf{x} \in \mathbf{R}^n$; (3) $\mathbf{F}(\mathbf{x}, 1) = (f_1(\mathbf{x}), \ldots, f_n(\mathbf{x}))$ for $\|\mathbf{x}\| \leqq \varepsilon$. (Take $\mathbf{F}(\mathbf{x}, t) = \mathbf{x} + th(\varepsilon^{-2}\|\mathbf{x}\|^2)(\mathbf{F}_0(\mathbf{x}) - \mathbf{x})$, where h is chosen as in Problem 10; then use again Problem 1 of Section **16.12**.)

13. Let X be a connected differential manifold, U an open neighborhood of a point $a \in X$, and f a C^∞-mapping of U into X such that $T_a(f)$ is a bijection of $T_a(X)$ onto $T_{f(a)}(X)$. If X is orientable, assume also that $T_a(f)$ preserves the orientation. Show that there exists a C^∞-mapping $F : X \times \mathbf{R} \to X$ satisfying the following conditions: (1) $x \mapsto F(x, t)$ is a diffeomorphism of X onto itself, for all $t \in [0, 1]$; (2) $F(x, 0) = x$ for all $x \in X$; (3) $F(x, 1) = f(x)$ for all x in some neighborhood $V \subset U$ of a. (Using Problem 11, reduce to the case where $f(a) = a$ and $T_a(f)$ is the identity mapping. Then use Problem 12.)

14. Let f, g be two diffeomorphisms of the open ball $B = \{\mathbf{x} \in \mathbf{R}^n : \|\mathbf{x}\| < 1 + \alpha\}$ (where $\|\mathbf{x}\|$ is the Euclidean norm) onto open subsets of a connected differential manifold X. If X is orientable, assume also that f and g are orientation-preserving. Show that there exists a diffeomorphism F of X onto itself such that $F(f(\mathbf{x})) = g(\mathbf{x})$ for $\|\mathbf{x}\| < 1$. (Using Problem 13, show that there exists $\varepsilon > 0$ and a diffeomorphism H of X onto itself such that $H(f(\mathbf{x})) = g(\mathbf{x})$ for $\|\mathbf{x}\| < \varepsilon$. Then remark that there exists a C^∞ real-valued function h on B and diffeomorphisms G_1, G_2 of X onto itself such that $G_1(f(\mathbf{x})) = f(h(\mathbf{x})\mathbf{x})$ and $G_2(g(\mathbf{x})) = g(h(\mathbf{x})\mathbf{x})$ for all $\mathbf{x} \in B$, and such that $|h(\mathbf{x})| \leqq \varepsilon$ for $\|\mathbf{x}\| < 1$, and $G_1(y) = y$ (resp. $G_2(y) = y$) for y not in the image of f (resp. g).)

15. Let X_1, X_2 be two connected differential manifolds of the same dimension n and let u_1 (resp. u_2) be a diffeomorphism of the open ball $B : \|\mathbf{x}\| < 2$ in $\mathbf{R}^n$ onto an open subset of X_1 (resp. X_2). If B' is the closed ball $\|\mathbf{x}\| \leqq \frac{1}{2}$ in $\mathbf{R}^n$, let U_1 (resp. U_2) be the open set in X_1 (resp. X_2) which is the complement of $u_1(B')$ (resp. $u_2(B')$). Let $U_{12} = u_1(B - B')$, $U_{21} = u_2(B - B')$. Let X be the topological space obtained by patching together U_1 and U_2 along U_{12} and U_{21} by means of the homeomorphism $h_{21} : U_{12} \to U_{21}$ defined by

$$h_{21}(x) = u_2 \frac{u_1^{-1}(x)}{\|u_1^{-1}(x)\|^2}.$$

Identify U_1 and U_2 (endowed with their structures of differential manifolds) canonically with their images in X. Show that the structures of differential manifolds induced by U_1 and U_2 on $U_1 \cap U_2$ are the same, and hence are induced by a structure of a connected differential manifold on X. The space X, endowed with this structure of differential manifold, is called the *connected sum* of X_1 and X_2 relative to u_1 and u_2.

Using Problem 14, show that up to diffeomorphism X is independent of the choice of u_1 and u_2, provided that each of the manifolds X_1, X_2 is either nonorientable, or orientable but admitting an orientation-reversing diffeomorphism onto itself. By abuse of notation we write $X = X_1 \# X_2$. If X_3 is another connected differential manifold of dimension n with the same properties, show that (with the same abuse of notation) $(X_1 \# X_2) \# X_3 = X_1 \# (X_2 \# X_3)$. Show also that $X_1 \# S_n = X_1$.

27. THE FUNDAMENTAL GROUP OF A CONNECTED MANIFOLD

The definitions in (9.6) of the notions of *path*, *loop*, *opposite paths*, and *juxtaposition of paths* apply without any alteration when the open subset of $\mathbf{C}$ is replaced by an arbitrary topological space X. An *unending path* in X is by definition a continuous mapping of an *open* interval $\mathrm{J} \subset \mathbf{R}$ into X. If X is a differential manifold, an unending path in X is said to be of *class* C^r ($r \geqq 1$ or $r = \infty$) if it is a C^r-mapping of an open interval $\mathrm{J} \subset \mathbf{R}$ into X. A *path* in X is said to be of *class* C^r if it is the restriction to a compact interval of an unending path of class C^r.

Let X be a connected differential manifold. Given two points a, b in X, we denote by $\Omega_{a,b}$ the set of paths defined on $\mathrm{I} = [0, 1]$ with values in X, with origin a and end-point b. This set $\Omega_{a,b}$ is not empty (16.26.10). We define an equivalence relation $\mathrm{R}_{a,b}$ on $\Omega_{a,b}$ by considering two paths $\gamma_1, \gamma_2 \in \Omega_{a,b}$ as equivalent if there exists a homotopy $\varphi : \mathrm{I} \times [\alpha, \beta] \to \mathrm{X}$ of γ_1 into γ_2 such that $\varphi(0, \xi) = a$ and $\varphi(1, \xi) = b$ for all $\xi \in [\alpha, \beta]$ (for brevity, we say that φ *leaves a and b fixed*).

We have the following lemma:

(16.27.1) *Let ρ be any continuous mapping of* I *into* I *such that* $\rho(0) = 0$ *and* $\rho(1) = 1$. *Then for each path* $\gamma \in \Omega_{a,b}$, *the paths* γ *and* $\gamma \circ \rho$ *are equivalent for* $\mathrm{R}_{a,b}$.

This follows from considering the homotopy $\varphi : \mathrm{I} \times \mathrm{I} \to \mathrm{X}$ defined by $\varphi(t, \xi) = \gamma((1 - \xi)t + \xi\rho(t))$.

Let $\mathrm{E}_{a,b}$ denote the set of equivalence classes for the relation $\mathrm{R}_{a,b}$. Given three points $a, b, c \in \mathrm{X}$, we shall define a mapping $\mathrm{E}_{a,b} \times \mathrm{E}_{b,c} \to \mathrm{E}_{a,c}$, which we shall denote by $(u, v) \mapsto u \cdot v$ (or simply uv). Consider two paths $\gamma_1 \in \Omega_{a,b}$, $\gamma_2 \in \Omega_{b,c}$. From these we construct a path $\gamma \in \Omega_{a,c}$ by putting

(16.27.1.1)
$$\gamma(t) = \begin{cases} \gamma_1(2t) & \text{for } 0 \leqq t \leqq \frac{1}{2}, \\ \gamma_2(2t - 1) & \text{for } \frac{1}{2} \leqq t \leqq 1. \end{cases}$$

We write $\gamma = \gamma_1\gamma_2$ (this is a *juxtaposition* of paths equivalent to γ_1 and γ_2, chosen in a particular way). Let γ_1', γ_2' be paths equivalent, respectively, to γ_1, γ_2. Then $\gamma_1'\gamma_2'$ is equivalent to $\gamma_1\gamma_2$. For we may suppose that the homotopies φ_1 of γ_1 into γ_1' and φ_2 of γ_2 into γ_2' are defined on the same set

$\mathrm{I} \times [\alpha, \beta]$, and then we construct a homotopy φ of $\gamma_1\gamma_2$ into $\gamma_1'\gamma_2'$ by defining

$$\varphi(t, \xi) = \begin{cases} \varphi_1(2t, \xi) & \text{for } 0 \leqq t \leqq \frac{1}{2}, \\ \varphi_2(2t - 1, \xi) & \text{for } \frac{1}{2} \leqq t \leqq 1; \end{cases}$$

the function φ so defined is continuous, because $\varphi_1(1, \xi) = \varphi_2(0, \xi) = b$ for all $\xi \in [\alpha, \beta]$. Consequently the class of $\gamma_1\gamma_2$ in $\mathrm{E}_{a,b}$ depends only on the classes u of γ_1 and v of γ_2, and this class we denote by $u \cdot v$.

For each $a \in \mathrm{X}$ we shall denote by e_a the class in $\mathrm{E}_{a,a}$ of the constant path $t \mapsto a$. Also, given $a, b \in \mathrm{X}$ and a path $\gamma \in \Omega_{a,b}$, the class of the opposite path γ^0 in $\mathrm{E}_{b,a}$ depends only on the class of γ in $\mathrm{E}_{a,b}$. For if γ' is equivalent to γ under a homotopy φ, then γ'^0 is equivalent to γ^0 under the homotopy

$$(t, \xi) \mapsto \varphi(1 - t, \xi).$$

If u is the class of γ, we denote by u^{-1} the class of γ^0. Clearly $(u^{-1})^{-1} = u$.

(16.27.2) (i) *Let a, b, c, d be four points of* X. *If $u \in \mathrm{E}_{a,b}$, $v \in \mathrm{E}_{b,c}$, $w \in \mathrm{E}_{c,d}$, then $(uv)w = u(vw)$* (and therefore we write this product as uvw, without brackets).

(ii) If $a, b \in \mathrm{X}$, *then $e_a u = u$ and $u e_b = u$, for all $u \in \mathrm{E}_{a,b}$.*

(iii) If $a, b \in \mathrm{X}$, *then $uu^{-1} = e_a$ and $u^{-1}u = e_b$, for all $u \in \mathrm{E}_{a,b}$.*

(i) Let $\gamma_1 \in u$, $\gamma_2 \in v$, $\gamma_3 \in w$; put $\gamma = (\gamma_1\gamma_2)\gamma_3$, $\gamma' = \gamma_1(\gamma_2\gamma_3)$. Then we have

$$\gamma(t) = \begin{cases} \gamma_1(4t) & \text{for } 0 \leqq t \leqq \frac{1}{4}, \\ \gamma_2(4t - 1) & \text{for } \frac{1}{4} \leqq t \leqq \frac{1}{2}, \\ \gamma_3(2t - 1) & \text{for } \frac{1}{2} \leqq t \leqq 1, \end{cases}$$

and

$$\gamma'(t) = \begin{cases} \gamma_1(2t) & \text{for } 0 \leqq t \leqq \frac{1}{2}, \\ \gamma_2(4t - 2) & \text{for } \frac{1}{2} \leqq t \leqq \frac{3}{4}, \\ \gamma_3(4t - 3) & \text{for } \frac{3}{4} \leqq t \leqq 1. \end{cases}$$

Now consider the function $\rho : \mathrm{I} \to \mathrm{I}$ defined by

$$\rho(t) = \begin{cases} 2t & \text{for } 0 \leqq t \leqq \frac{1}{4}, \\ t + \frac{1}{4} & \text{for } \frac{1}{4} \leqq t \leqq \frac{1}{2}, \\ \frac{1}{2}(t + 1) & \text{for } \frac{1}{2} \leqq t \leqq 1. \end{cases}$$

Clearly ρ is continuous; $\rho(0) = 0$, $\rho(1) = 1$, and we have

$$\gamma(t) = \gamma'(\rho(t))$$

for all $t \in \mathrm{I}$. Hence the result (16.27.1).

(ii) We shall prove that $ue_b = u$; the proof that $e_a u = u$ is similar. If $\gamma \in u$, a path in the class ue_b is defined by

$$\gamma'(t) = \begin{cases} \gamma(2t) & \text{for } 0 \leqq t \leqq \frac{1}{2}, \\ b & \text{for } \frac{1}{2} \leqq t \leqq 1. \end{cases}$$

Define a homotopy $\varphi : \mathrm{I} \times \mathrm{I} \to \mathrm{X}$ of γ' into γ as follows:

$$\varphi(t, \xi) = \begin{cases} \gamma\left(\dfrac{2t}{1+\xi}\right) & \text{for } 0 \leqq t \leqq \frac{1}{2}(1+\xi), \\ b & \text{for } \frac{1}{2}(1+\xi) \leqq t \leqq 1. \end{cases}$$

(iii) We shall prove that $uu^{-1} = e_a$; the proof of the other relation is similar. Let $\gamma \in u$, then $\gamma' = \gamma\gamma^0$ is defined by

$$\gamma'(t) = \begin{cases} \gamma(2t) & \text{for } 0 \leqq t \leqq \frac{1}{2}, \\ \gamma(2-2t) & \text{for } \frac{1}{2} \leqq t \leqq 1. \end{cases}$$

Define a homotopy $\varphi : \mathrm{I} \times \mathrm{I} \to \mathrm{X}$ of the constant path $t \mapsto a$ into γ' as follows:

$$\varphi(t, \xi) = \begin{cases} \gamma(2t\xi) & \text{for } 0 \leqq t \leqq \frac{1}{2}, \\ \gamma(2\xi - 2t\xi) & \text{for } \frac{1}{2} \leqq t \leqq 1. \end{cases}$$

From **(16.27.2)** we deduce:

(16.27.3) (i) *For each* $a \in \mathrm{X}$, *the mapping* $(u, v) \mapsto uv$ *defines a group structure on* $\mathrm{E}_{a,a}$.

(ii) *If* $a, b \in \mathrm{X}$ *and* $f \in \mathrm{E}_{a,b}$, *then the mapping* $u \mapsto fuf^{-1}$ *is an isomorphism of the group* $\mathrm{E}_{b,b}$ *onto the group* $\mathrm{E}_{a,a}$.

Remarks

(16.27.3.1) (i) With the notation of **(16.27.3(ii))**, if $\gamma \in u$ and $\alpha \in f$, then the loops γ and $\gamma' = (\alpha\gamma)\alpha^0$ are *homotopic* (as loops): for by definition, γ' is given by the equations

$$\gamma'(t) = \begin{cases} \alpha(4t) & \text{for } 0 \leqq t \leqq \frac{1}{4}, \\ \gamma(4t-1) & \text{for } \frac{1}{4} \leqq t \leqq \frac{1}{2}, \\ \alpha(2-2t) & \text{for } \frac{1}{2} \leqq t \leqq 1, \end{cases}$$

and we define a loop homotopy φ of γ' onto γ by

$$\varphi(t, \xi) = \begin{cases} \alpha(\xi + 4(1-\xi)t) & \text{for } 0 \leqq t \leqq \frac{1}{4} \text{ and } 0 \leqq \xi \leqq 1, \\ \gamma(4t-1) & \text{for } \frac{1}{4} \leqq t \leqq \frac{1}{2} \text{ and } 0 \leqq \xi \leqq 1, \\ \alpha(\xi + 2(1-\xi)(1-t)) & \text{for } \frac{1}{2} \leqq t \leqq 1 \text{ and } 0 \leqq \xi \leqq 1. \end{cases}$$

(ii) It follows from **(16.27.2)** that for each $u_0 \in \mathrm{E}_{a,b}$, the mapping $v \mapsto u_0 v$ (resp. $w \mapsto wu_0$) of $\mathrm{E}_{b,c}$ into $\mathrm{E}_{a,c}$ (resp. of $\mathrm{E}_{c,a}$ into $\mathrm{E}_{c,b}$) is *bijective.*

(16.27.4) *For each pair (a, b) of points of* X, *the set* $E_{a,b}$ *is at most denumerable.*

We shall first prove the following lemma, which generalizes (7.4.4):

(16.27.4.1) *Let* E *be a metrizable compact space,* F *a separable metrizable space,* d (resp. d') *a distance defining the topology of* E (resp. F). *For any two mappings* $f, g : E \to F$, *put*

$$\rho(f, g) = \sup_{t \in E} d'(f(t), g(t)).$$

Then ρ *is a distance on the set* $\mathscr{C}_F(E)$ *of continuous mappings of* E *into* F, *with respect to which* $\mathscr{C}_F(E)$ *is a separable metric space.*

The fact that ρ is a distance follows immediately from the fact that $\rho(f, g)$ is *finite* whenever f and g are continuous (3.17.10). For each pair m, n of integers >0, let G_{mn} be the set of functions $f \in \mathscr{C}_F(E)$ such that the relation $d(x, x') \leqq 1/m$ implies that $d'(f(x), f(x')) \leqq 1/n$. Since each $f \in \mathscr{C}_F(E)$ is uniformly continuous (3.16.15), it follows that for each fixed $n > 0$, $\mathscr{C}_F(E)$ is the union of the G_{mn} for $m > 0$. Let $\{a_1, \ldots, a_{p(m)}\}$ be a finite subset of E such that the open balls with centers a_i and radius $1/m$ cover E. Also let $(b_r)_{r \geqq 0}$ be a denumerable dense sequence in F. For each mapping $\varphi : \{1, 2, \ldots, p(m)\} \to \mathbf{N}$, let H_φ be the set of $f \in G_{mn}$ such that $d'(f(a_k), b_{\varphi(k)}) \leqq 1/n$ for $1 \leqq k \leqq p(m)$. From the definition of the b_r, G_{mn} is the union of the H_φ for all $\varphi \in \mathbf{N}^{p(m)}$. Let C_{mn} be the set of $\varphi \in \mathbf{N}^{p(m)}$ such that $H_\varphi \neq \varnothing$, and for each $\varphi \in C_{mn}$ choose an element $g_\varphi \in H_\varphi$; finally, let L_{mn} denote the denumerable set of g_φ for $\varphi \in C_{mn}$. Let $f \in G_{mn}$ and let $\varphi \in C_{mn}$ be such that $f \in H_\varphi$. Then it follows immediately from the definitions that $d'(f(x), g_\varphi(x)) \leqq 4/n$ for all $x \in E$; in other words, $\rho(f, g_\varphi) \leqq 4/n$. Hence the union of the L_{mn} is dense in $\mathscr{C}_F(E)$: for each integer $n > 0$ and each $f \in \mathscr{C}_F(E)$ there exists m such that $f \in G_{mn}$, and we have just seen that the distance from f to L_{mn} is $\leqq 4/n$.

We now apply this lemma to the set $\Omega_{a,b}$, endowed with the distance $\rho(\gamma_1, \gamma_2) = \sup_{t \in I} d(\gamma_1(t), \gamma_2(t))$, where d is a distance defining the topology of X. It follows that there is a sequence (γ_n) which is dense in $\Omega_{a,b}$. On the other hand, it follows from (16.26.4.1) that, for each $\gamma \in \Omega_{a,b}$, there exists $\varepsilon > 0$ such that the relation $\rho(\gamma, \gamma') \leqq \varepsilon$ implies that the paths γ, γ' are equivalent; but since there exists a γ_n such that $\rho(\gamma, \gamma_n) \leqq \varepsilon$, it follows that $E_{a,b}$ is the set of equivalence classes of the γ_n; hence it is at most denumerable.

(16.27.5) The group $E_{a,a}$ is called the *fundamental group of the manifold* X *at the point* a and is denoted by $\pi_1(X, a)$. *Up to isomorphism* it is independent of the point a, and we shall use the notation $\pi_1(X)$ to denote any one of the groups $\pi_1(X, a)$, and refer to $\pi_1(X)$ as the *fundamental group of* X.

A connected differential manifold X is said to be *simply connected* if $\pi_1(X, a)$ is the group of one element for some $a \in X$ (and hence for *all* $a \in X$); or, equivalently, if every loop with origin a is *homotopic to the constant loop* $t \mapsto a$ under a loop homotopy leaving the point a fixed.

(16.27.6) Let Y be another connected differential manifold, $f: X \to Y$ a continuous mapping. For each pair of points a, b in X and each path $\gamma \in \Omega_{a, b}$, $f \circ \gamma$ is a path belonging to $\Omega_{f(a), f(b)}$. If $\gamma, \gamma' \in \Omega_{a, b}$ are equivalent paths, under a homotopy φ, then it is clear that $f \circ \gamma$ and $f \circ \gamma'$ are equivalent under the homotopy $f \circ \varphi$. Hence, as γ runs through a class $u \in E_{a, b}$, the paths $f \circ \gamma$ all belong to the same class $f_*(u) \in E_{f(a), f(b)}$. Moreover, it is easily verified that if $u \in E_{a, b}$ and $v \in E_{b, c}$, then $f_*(uv) = f_*(u)f_*(v)$. In particular, the restriction of f_* to $\pi_1(X, a)$ is a *homomorphism* of this group into the group $\pi_1(Y, f(a))$. Finally, if Z is a third manifold and $g : Y \to Z$ a continuous mapping, we have $(g \circ f)_* = g_* \circ f_*$.

Examples

(16.27.7) A manifold X is said to be *contractible* if there exists a point $a \in X$ and a homotopy φ of the identity mapping 1_X onto the constant mapping $x \mapsto a$ which leaves a fixed: in other words, φ is a continuous mapping of $X \times I$ into X such that $\varphi(a, \xi) = a$ for all $\xi \in I$ and $\varphi(x, 0) = x$, $\varphi(x, 1) = a$ for all $x \in X$.

For each loop $\gamma \in \Omega_{a, a}$, the mapping $(t, \xi) \mapsto \varphi(\gamma(t), \xi)$ is then a homotopy of γ onto the constant loop $t \mapsto a$ which leaves a fixed. Consequently *a contractible manifold is simply-connected.* For example, $\mathbf{R}^n$ is contractible, because the mapping $\varphi(x, \xi) = \xi x$ satisfies the above conditions for $a = 0$.

(16.27.8) *For $n \geqq 2$, the sphere $\mathbf{S}_n$ is simply-connected.*

Let $\gamma : I \to \mathbf{S}_n$ be a loop with origin a. If $\gamma(I) \neq \mathbf{S}_n$, then γ is homotopic in $\mathbf{S}_n$ to the constant loop. For if $b \notin \gamma(I)$, then $\mathbf{S}_n - \{b\}$ is homeomorphic to $\mathbf{R}^n$ (16.8.10), and therefore γ is homotopic to the constant loop already in $\mathbf{S}_n - \{b\}$. Hence we have to show that there exists a loop γ' with origin a which is loop-homotopic to γ and is such that $\gamma'(I) \neq \mathbf{S}_n$. Let d be the distance on $\mathbf{S}_n$ induced by the Euclidean distance on $\mathbf{R}^{n+1}$, and let $\varepsilon \in]0, 1[$ be a number satisfying the condition of **(16.26.4)** for the mapping γ. Then there exists a mapping $\gamma_1 : I \to \mathbf{S}_n$ which is the restriction of a C^∞-mapping defined on a neighborhood of I, such that $d(r(t), \gamma_1(t)) \leqq \frac{1}{8}\varepsilon$ for $t \in I$ **(16.25.5)**. Let $b = \gamma_1(0)$, $c = \gamma_1(1)$, so that b and c belong to the open neighborhood V of a in $\mathbf{S}_n$ consisting of the points x such that $d(x, a) < \frac{1}{2}\varepsilon$, which is homeomorphic to an open ball in $\mathbf{R}^n$ (16.2.3). Let $\delta \in]0, \frac{1}{2}[$ be such that the relation $|t - t'| < \delta$ (for $t, t' \in I$) implies that $d(\gamma(t'), \gamma(t)) < \frac{1}{4}\varepsilon$ **(3.16.5)**. Then there exist continuous mappings $\gamma_0 : [0, \delta] \to \mathrm{V}$ and $\gamma_2 : [1 - \delta, 1] \to \mathrm{V}$ such that $\gamma_0(0) = a$, $\gamma_0(\delta) = b$, $\gamma_2(1 - \delta) = c$, $\gamma_2(1) = a$. Now define γ' on I as follows:

$$\gamma'(t) = \begin{cases} \gamma_0(t) & \text{for} \quad 0 \leqq t \leqq \delta, \\ \gamma_1\left(\dfrac{t - \delta}{1 - 2\delta}\right) & \text{for} \quad \delta \leqq t \leqq 1 - \delta, \\ \gamma_2(t) & \text{for} \quad 1 - \delta \leqq t \leqq 1. \end{cases}$$

Since $|t - t'| < \delta$ implies that $d(\gamma_1(t), \gamma_1(t')) < \frac{1}{2}\varepsilon$, it follows immediately that $d(\gamma(t), \gamma'(t)) \leqq \varepsilon$ for all $t \in \mathrm{I}$. Hence, by (16.26.4), γ and γ' are equivalent in $\Omega_{a,a}$. On the other hand, by virtue of the hypothesis $n \geqq 2$ and Sard's theorem, $\gamma_1(\mathrm{I})$ is nowhere dense in S_n (16.23.2); by construction, this implies that $\gamma'(\mathrm{I}) \neq \mathrm{S}_n$, and the proof is complete.

We shall prove later (Chapter XXIV) that for $n \geqq 1$ *the sphere* S_n *is not contractible* (cf. Section 16.30, Problem 9(b)).

(16.27.9) On the other hand, the circle $\mathbf{U} = \mathbf{S}_1$ is *not* simply-connected. More precisely, $\pi_1(\mathbf{S}_1)$ is isomorphic to $\mathbf{Z}$. In fact, we have already defined in Chapter IX and the Appendix to Chapter IX the *index* $j(0; \gamma)$ of a loop γ in $\mathbf{S}_1$ and shown that it depends only on the class u of γ in $\pi_1(\mathbf{S}_1, a)$, where a is the origin of γ. Moreover, we have shown that if this index is denoted by $i(u)$, then i is a *homomorphism* of $\pi_1(\mathbf{S}_1, a)$ onto $\mathbf{Z}$ (9.8.4). Finally, this homomorphism is *injective*, for if $j(0; \gamma) = 0$ the proof of (Ap.2.8) shows that γ is homotopic to a constant loop.

(16.27.10) *Let* X, Y *be two connected differential manifolds,* $a \in \mathrm{X}$ *and* $b \in \mathrm{Y}$. *Then the group* $\pi_1(\mathrm{X} \times \mathrm{Y}, (a, b))$ *is isomorphic to the direct product* $\pi_1(\mathrm{X}, a) \times \pi_1(\mathrm{Y}, b)$. *In particular, the product of two simply-connected manifolds is simply-connected.*

We define a canonical bijection $f : \Omega_{a,a} \times \Omega_{b,b} \to \Omega_{(a,b),(a,b)}$ as follows: if $\alpha \in \Omega_{a,a}$ and $\beta \in \Omega_{b,b}$, then $f(\alpha, \beta)$ is the loop $t \mapsto (\alpha(t), \beta(t))$. The inverse bijection f^{-1} takes a loop $\gamma \in \Omega_{(a,b),(a,b)}$ to the pair $(\mathrm{pr}_1 \circ \gamma, \mathrm{pr}_2 \circ \gamma)$. Clearly we have $f(\alpha_1\alpha_2, \beta_1\beta_2) = f(\alpha_1, \beta_1)f(\alpha_2, \beta_2)$. Finally, if α, α' are two loops belonging to $\Omega_{a,a}$, which are equivalent under a homotopy $\varphi : \mathrm{I} \times \mathrm{I} \to \mathrm{X}$, and if β, β' are two loops belonging to $\Omega_{b,b}$, which are equivalent under a homotopy $\psi : \mathrm{I} \times \mathrm{I} \to \mathrm{Y}$, then the loops $f(\alpha, \beta)$ and $f(\alpha', \beta')$ are equivalent under the homotopy $(\varphi, \psi) : \mathrm{I} \times \mathrm{I} \to \mathrm{X} \times \mathrm{Y}$. Conversely, if $f(\alpha, \beta)$ and $f(\alpha', \beta')$ are equivalent under a homotopy $\theta : \mathrm{I} \times \mathrm{I} \to \mathrm{X} \times \mathrm{Y}$, then α and α' are equivalent under $\mathrm{pr}_1 \circ \theta$, and β and β' are equivalent under $\mathrm{pr}_2 \circ \theta$. Hence we have a bijection $\bar{f}$ of $\mathrm{E}_{a,a} \times \mathrm{E}_{b,b}$ onto $\mathrm{E}_{(a,b),(a,b)}$ which assigns to each pair of classes (u, v) the class of $f(\alpha, \beta)$ for $\alpha \in u$ and $\beta \in v$; and $\bar{f}$ is a group isomorphism.

PROBLEMS

1. A topological space X is said to be *arcwise-connected* if, for each pair of points a, b in X, there exists a path in X with origin a and endpoint b. A subset A of X is said to be arcwise-connected if the subspace A is arcwise-connected.

(a) An arcwise-connected space is connected. In $\mathbf{R}^2$ the graph A of the function $y = \sin(1/x)$, defined on $]0, +\infty[$, is arcwise-connected, but its closure $\bar{A}$ is connected but not arcwise-connected.
(b) Show that propositions (3.19.3), (3.19.4), and (3.19.7) remain valid when "connected" is replaced by "arcwise-connected" throughout.
(c) The union $C'(x)$ of all the arcwise-connected subsets of X which contain the point $x \in X$ is an arcwise-connected set, called the *arcwise-connected component of x in* X; it is a closed subset of X.
(d) A topological space X is said to be *locally arcwise-connected* if each point of X has a fundamental system of arcwise-connected neighborhoods. Show that this condition is equivalent to the following: the arcwise-connected components of each open subset U of X are open in X. The arcwise-connected components of U are then equal to the connected components of U (in the sense of (3.19)).

2. If X is any topological space, the sets $\Omega_{a,b}$ can be defined as in (16.27) whenever a and b belong to the same arcwise-connected component of X. If X is separable and metrizable, then so is $\Omega_{a,b}$ with respect to the topology defined in (16.27.4.1). Two paths γ_1, $\gamma_2 \in \Omega_{a,b}$ are equivalent with respect to the relation $R_{a,b}$ if and only if they belong to the same arcwise-connected component of $\Omega_{a,b}$. The group $\pi_1(X, a)$ is defined as in (16.27.5).

(a) Generalize the result of (16.27.10), and show that the canonical bijection

$$\Omega_{a,a} \times \Omega_{b,b} \to \Omega_{(a,b),(a,b)}$$

is a homeomorphism.
(b) Show that the mapping $(\gamma_1, \gamma_2) \mapsto \gamma_1\gamma_2$ defined in (16.27.1.1) is a continuous mapping of $\Omega_{a,b} \times \Omega_{b,c}$ into $\Omega_{a,c}$.
(c) For each $\gamma \in \Omega_{a,a}$ and each $\xi \in [0, 1]$, let $\psi(\gamma, \xi)$ denote the loop $\gamma' \in \Omega_{a,a}$ defined by

$$\gamma'(t) = \begin{cases} \gamma\left(\dfrac{2t}{1+\xi}\right) & \text{for } 0 \leq t \leq \tfrac{1}{2}(1+\xi), \\ a & \text{for } \tfrac{1}{2}(1+\xi) \leq t \leq 1. \end{cases}$$

If ε_a is the constant loop equal to a, show that ψ is a homotopy of the mapping $\gamma \mapsto \gamma\varepsilon_a$ of $\Omega_{a,a}$ into $\Omega_{a,a}$, into the identity mapping of $\Omega_{a,a}$.
(d) Let f, g be homotopic mappings of X into Y, and let $a \in X$. Show that there exists an isomorphism u of $\pi_1(Y, f(a))$ onto $\pi_1(Y, g(a))$ such that $g_* = u \circ f_*$. Deduce that if X, Y are arcwise-connected and of the same homotopy type (Section 16.26, Problem 2), then $\pi_1(X)$ and $\pi_1(Y)$ are isomorphic.

3. Deduce from (16.27.10) that the Hopf fibration of S_3 over S_2 (16.14.10) is not trivializable.

4. Let X be a simply-connected, arcwise-connected space. If a, b are any two points of X, show that the set $E_{a,b}$ consists of a single element. (If $\gamma, \gamma' \in \Omega_{a,b}$, show that γ' is equivalent (with respect to $R_{a,b}$) to $(\gamma\gamma^0)\gamma'$.)

5. Let X be a locally compact metrizable space, and let A, B be closed subsets of X such that $X = A \cup B$. Suppose that A, B are each arcwise-connected and simply-connected; suppose also that $A \cap B$ is arcwise-connected and that, for each $x \in A \cap B$, there exists a fundamental system of open neighborhoods V of x for each of which $V \cap B$ is arcwise-connected and simply-connected, and $V \cap A \cap B$ arcwise-connected. Show that X is arcwise-connected and simply-connected. (Let $\gamma : I \to X$ be a loop with origin $a \in A$, and suppose that $\gamma(I)$ meets B. Then the inverse image under γ of $\gamma(I) \cap (X - A)$ is the union of a (finite or infinite) sequence of pairwise disjoint open intervals I_m in I. For each integer n, cover $\gamma(I)$ by a finite number of open subsets $U_{\alpha,n}$ of X, of diameter $\leqq 1/n$, such that $U_{\alpha,n} \cap B$ is arcwise-connected and simply-connected, and $U_{\alpha,n} \cap A \cap B$ arcwise-connected. Then consider the finite set of intervals I_k such that $\gamma(I_k)$ is contained in $U_{\alpha,n}$ but not in $U_{\alpha,n+1}$; finally, use Problem 4.) Hence give another proof of the fact that S_n is simply connected for $n \geqq 2$.

28. COVERING SPACES AND THE FUNDAMENTAL GROUP

(16.28.1) *Let* (X, B, p) *be a covering of a differential manifold* B (16.12.4), f *a continuous mapping of a* connected *topological space* Z *into* B, *and* g_1, $g_2 : Z \to X$ *two continuous liftings* (16.12.1) *of* f. *If there exists* $a \in Z$ *such that* $g_1(a) = g_2(a)$, *then* $g_1 = g_2$.

The set of points $z \in Z$ such that $g_1(z) = g_2(z)$ is nonempty and closed in Z (12.3.5); hence it is enough to show that it is also *open* in Z. Now, if $z_0 \in Z$ is such that $g_1(z_0) = g_2(z_0) = x_0$, there exists an open neighborhood U of x_0 such that $p|U$ is a homeomorphism of U onto the open set $p(U) \subset B$. Since g_1, g_2 are continuous, there exists a neighborhood V of z_0 in Z such that $g_1(V) \subset U$ and $g_2(V) \subset U$. However, for all $z \in V$, we have $p(g_1(z)) = f(z) = p(g_2(z))$; since $p|U$ is injective, it follows that $g_1(z) = g_2(z)$ for all $z \in V$. This completes the proof.

(16.28.2) *Let* (X, B, p) *be a covering of a differential manifold* B.

(i) *For each path* $\beta : I \to B$ *with origin* $b \in B$, *and for each point* $a \in p^{-1}(b)$, *there exists a unique path* $\gamma : I \to X$ *with origin* a *such that* $\beta = p \circ \gamma$ (called the *lifting* of β, with origin a).

(ii) *Let $\beta : \mathrm{I} \to \mathrm{B}$, $\beta' : \mathrm{I} \to \mathrm{B}$ be two paths with origins b, b', respectively, and let $\gamma : \mathrm{I} \to \mathrm{X}$ be the lifting of β with origin a.* If $\varphi : \mathrm{I} \times \mathrm{I} \to \mathrm{B}$ *is a homotopy of β into β', then there exists a unique homotopy ψ of γ into a path $\gamma' : \mathrm{I} \to \mathrm{X}$ such that $\varphi = p \circ \psi$.* (ψ is called the *lifting* of φ such that $\psi(0, 0) = a$.)

(i) The uniqueness of γ follows from (**16.28.1**). To prove the existence of γ, let E denote the set of $s \in \mathrm{I}$ for which there exists a continuous mapping $\gamma_s : [0, s] \to \mathrm{X}$ such that $\gamma_s(0) = a$ and $p(\gamma_s(t)) = \beta(t)$ for $0 \leqq t \leqq s$. This set E does not consist only of 0, because there exists a neighborhood U_0 of a such that $p | \mathrm{U}_0$ is a homeomorphism of U_0 onto the open set $p(\mathrm{U}_0)$ of B; if $s_0 > 0$ is such that $\beta(t) \in p(\mathrm{U}_0)$ for $0 \leqq t \leqq s_0$, put $\gamma_{s0}(t) = q_0(\beta(t))$ for $0 \leqq t \leqq s_0$, where $q_0 : p(\mathrm{U}_0) \to \mathrm{U}_0$ is the inverse of the homeomorphism $p | \mathrm{U}_0$; then $p(\gamma_{s0}(t)) = \beta(t)$ for $0 \leqq t \leqq s_0$, so that $s_0 \in \mathrm{E}$. By virtue of (**16.28.1**), if $s < s'$ in E, the restriction of $\gamma_{s'}$ to $[0, s]$ is equal to γ_s. Consider now the least upper bound λ of E in $[0, 1]$; there exists a continuous mapping γ' of $[0, \lambda[$ into X such that $\gamma'(0) = a$ and $p(\gamma'(t)) = \beta(t)$ for $0 \leqq t < \lambda$. Let V be a connected open neighborhood of $\beta(\lambda)$ in B over which the covering space X is trivializable; then $p^{-1}(\mathrm{V})$ is a (finite or infinite) union of pairwise disjoint connected open sets U_n such that $p | \mathrm{U}_n$ is a homeomorphism onto V for each n. Choose $t_0 < \lambda$ and sufficiently close to λ so that $\beta([t_0, \lambda])$ is contained in V, and let n be such that $\gamma'(t_0) \in \mathrm{U}_n$. Since $\gamma'([t_0, \lambda[)$ is connected and contained in $p^{-1}(\mathrm{V})$, it must be contained in U_n; consequently, if $q_n : \mathrm{V} \to \mathrm{U}_n$ is the inverse of the homeomorphism $p | \mathrm{U}_n$, then $\gamma'(t) = q_n(\beta(t))$ for $t_0 \leqq t < \lambda$. This shows first of all that γ' can be extended by continuity to the point λ, by defining $\gamma'(\lambda) = q_n(\beta(\lambda))$. Second it shows that $\lambda = 1$, otherwise there would exist s_1 such that $\lambda < s_1 < 1$ and such that $\beta([\lambda, s_1]) \subset \mathrm{V}$; we can then extend γ' to γ_{s_1} by putting $\gamma_{s_1}(t) = q_n(\beta(t))$ for $\lambda \leqq t \leqq s_1$, and we shall have $s_1 \in \mathrm{E}$, contrary to the definition of λ. This completes the proof of (i).

(ii) For each $t \in \mathrm{I}$, there exists a unique lifting of the path $\xi \mapsto \varphi(t, \xi)$ with origin $\gamma(t)$. Let $\xi \mapsto \psi_t(\xi)$ denote this lifting. Then we have to prove that the mapping $\psi : (t, \xi) \mapsto \psi_t(\xi)$ is *continuous* on $\mathrm{I} \times \mathrm{I}$. Let d be a distance defining the topology of B. Since $\varphi(\mathrm{I} \times \mathrm{I})$ is compact, the same argument as in (**9.6.3**) shows that there exists a number $\rho > 0$ such that, for each $b \in \varphi(\mathrm{I} \times \mathrm{I})$, the ball $\mathrm{B}(b; \rho)$ is connected and the covering space X is trivializable over $\mathrm{B}(b; \rho)$. Next, there exists $\varepsilon > 0$ such that the relations $|t - t'| \leqq \varepsilon$, $|\xi - \xi'| \leqq \varepsilon$ imply that $|\varphi(t, \xi) - \varphi(t', \xi')| < \frac{1}{2}\rho$. Let $(t_i)_{0 \leqq i \leqq r}$ be an increasing sequence in I such that $t_0 = 0$, $t_r = 1$, and $|t_{i+1} - t_i| \leqq \varepsilon$ for $0 \leqq i \leqq r - 1$. We shall prove by induction on i that ψ is continuous on $\mathrm{I} \times [0, t_{i+1}]$. For this, it is enough to prove that ψ is continuous on $\mathrm{Q}_{ij} = [t_j, t_{j+1}] \times [t_i, t_{i+1}]$ for each $j = 0, 1, \ldots, r - 1$. Put $b_{ij} = \varphi(t_j, t_i)$. Then $p^{-1}(\mathrm{B}(b_{ij}; \rho))$ is the union of a sequence of pairwise disjoint connected open sets (U_n) such that $p | \mathrm{U}_n$ is a homeomorphism of U_n onto $\mathrm{B}(b_{ij}; \rho)$ for each n.

Since by hypothesis $t \mapsto \psi(t, t_i)$ is continuous, the image under this mapping of $[t_j, t_{j+1}]$, being connected and contained in $p^{-1}(\mathrm{B}(b_{ij}; \rho))$ by virtue of the choice of ε, is contained in some U_n; then, for $t_j \leqq t \leqq t_{j+1}$, since $\psi_t(t_i) \in \mathrm{U}_n$ and $\psi_t([t_i, t_{i+1}])$ is contained in $p^{-1}(\mathrm{B}(b_{ij}; \rho))$ (again by the choice of ε), it follows that $\psi_t([t_i, t_{i+1}]) \subset \mathrm{U}_n$, so that finally $\psi(\mathrm{Q}_{ij}) \subset \mathrm{U}_n$; but then, if $q_n : \mathrm{B}(b_{ij}; \rho) \to \mathrm{U}_n$ is the inverse of the homeomorphism $p | \mathrm{U}_n$, we have $\psi(t, \xi) = q_n(\varphi(t, \xi))$ in Q_{ij}, and the proof is complete.

(16.28.3) Let $(\mathrm{X}, \mathrm{B}, p)$ be a covering of a differential manifold B, let $b \in \mathrm{B}$, and let $\mathrm{X}_b = p^{-1}(b)$ be the (discrete) fiber of X over b.

We shall define a *right action* of the fundamental group $\pi_1(\mathrm{B}, b)$ on the fiber $\mathrm{X}_b : (a, u) \mapsto a \cdot u$, as follows: Let $\gamma \in u$ be a *loop* in B with origin b. Then it follows from (16.28.2) that there exists a unique *path* γ_a in X with origin a, such that $p \circ \gamma_a = \gamma$. Further, if γ' is another loop in B with origin b, belonging to the class u, then it is homotopic to γ by a homotopy $\varphi : \mathrm{I} \times \mathrm{I} \to \mathrm{B}$ which leaves b fixed. Now this homotopy has a unique lifting, by virtue of (16.28.2), to a homotopy $\psi : \mathrm{I} \times \mathrm{I} \to \mathrm{X}$ of γ_a into γ'_a, leaving a fixed, and such that $p \circ \psi = \varphi$. In particular, for each $\xi \in \mathrm{I}$, we have $p(\psi(1, \xi)) = \varphi(1, \xi) = b$; in other words, $\xi \mapsto \psi(1, \xi)$ is a *continuous* mapping of I into X_b; but since X_b is a discrete space, this mapping is *constant* (3.19.7), so that $\gamma_a(1) = \gamma'_a(1)$. The point $\gamma_a(1) \in \mathrm{X}_b$ therefore depends only on the class $u \in \pi_1(\mathrm{B}, b)$, and it is this point that is denoted by $a \cdot u$.

We have thus defined an operation of $\pi_1(\mathrm{B}, b)$ on X_b; for we have

$$\text{(16.28.3.1)} \quad \begin{cases} a \cdot e_b = a & \text{for all } a \in \mathrm{X}_b, \\ a \cdot (uv) = (a \cdot u) \cdot v & \text{for } a \in \mathrm{X}_b, \quad u, v \in \pi_1(\mathrm{B}, b) \end{cases}$$

directly from the definitions.

(16.28.4) *Let* X *be a connected differential manifold,* $(\mathrm{Y}, \mathrm{X}, p)$ *a covering of* X, a *a point of* X, *and* $\mathrm{Y}_a = p^{-1}(a)$ *the fiber over* a.

(i) Y *is connected if and only if* $\pi_1(\mathrm{X}, a)$ *acts transitively on* Y_a. *If this condition is satisfied, then for each point* $c \in \mathrm{Y}_a$ *the homomorphism* $p_* : \pi_1(\mathrm{Y}, c) \to \pi_1(\mathrm{X}, a)$ *is injective and its image is the stabilizer* S_c *of* c (for the action of $\pi_1(\mathrm{X}, a)$ on X_a) (12.10).

(ii) *If* Y *is connected and if* p_* *is surjective (hence bijective), then* p *is a diffeomorphism, so that the covering* Y *is trivializable. In particular, every connected covering of a simply-connected manifold is trivializable.*

(iii) *In order that* Y *should be connected and simply-connected, it is necessary and sufficient that the group* $\pi_1(\mathrm{X}, a)$ *should act transitively and freely on* Y_a (which is therefore homeomorphic to $\pi_1(\mathrm{X}, a)$).

(i) If Y is connected, then for any two points c_1, $c_2 \in Y_a$ there exists a path γ in Y with origin c_1 and endpoint c_2. The path $p \circ \gamma$ is a loop with origin a in X, and γ is its unique lifting with origin c_1 (16.28.2); hence $c_2 = c_1 \cdot u$, where u is the class of $p \circ \gamma$, and so $\pi_1(X, a)$ acts transitively on the fiber Y_a. Conversely, if the action is transitive, then any two points c_1, $c_2 \in Y_a$ can be joined by a path in Y, by definition (16.28.3). Next, if b is any point of Y, there exists a path in X joining $p(b)$ to a, and the lifting of this path with origin b (16.28.2) has its end point in Y_a. Hence every point of Y can be joined to a given point $c \in Y_a$ by a path and therefore Y is connected (3.19.3).

Now let γ be a loop in Y with origin c, whose image $p \circ \gamma$ is homotopic to the constant loop with origin a. Then, by (16.28.3), the loop γ, being the lifting of $p \circ \gamma$, is homotopic to the constant loop with origin c. This shows that the homomorphism $p_* : \pi_1(Y, c) \to \pi_1(X, a)$ is *injective*. Its image consists of the classes u of loops with origin a in X whose lifting with origin c is a *loop* with origin c, i.e., the $u \in \pi_1(X, a)$ such that $c \cdot u = c$.

(ii) The hypothesis implies that $c \cdot u = c$ for all $u \in \pi_1(X, a)$, and since $\pi_1(X, a)$ acts transitively on Y_a, we have $Y_a = \{c\}$. Since X is connected, all the fibers are isomorphic to Y_a, hence consist of a single point; in other words, p is bijective, and is therefore a diffeomorphism (because it is a local diffeomorphism). The hypothesis that p_* is surjective is clearly satisfied when X is simply connected, because then $\pi_1(X, a)$ is the group of one element.

(iii) If $\pi_1(Y, c) = \{e\}$, the stabilizer S_c consists only of the identity element, by (i), and since $\pi_1(X, a)$ acts transitively on Y_a, the stabilizer of each point of Y_a consists only of the identity element, so that $\pi_1(X, a)$ acts freely. The converse is an immediate consequence of (i).

Example

(16.28.5) From (16.14.10) we know that $\mathbf{S}_n$ is a two-sheeted covering space of the projective space $\mathbf{P}_n(\mathbf{R})$. Since $\mathbf{S}_n$ is connected and simply-connected for $n \geqq 2$ (16.27.5) it follows that $\pi_1(\mathbf{P}_n(\mathbf{R})) = \mathbf{Z}/2\mathbf{Z}$ for $n \geqq 2$. (For $n = 1$, however, $\mathbf{P}_1(\mathbf{R})$ is diffeomorphic to $\mathbf{S}_1$ (16.11.12) and its fundamental group is therefore isomorphic to $\mathbf{Z}$ (16.27.9).)

(16.28.6) *Let X be a connected differential manifold, (Y, X, p) a covering of X. Then for each connected component Z of Y, (Z, X, p|Z) is a covering of X. In particular, each covering of a connected and simply-connected manifold is trivializable.*

Let $x \in \overline{p(Z)}$ and let U be a connected open neighborhood of x in X such that Y is trivializable over U; in other words, $p^{-1}(U)$ is the union of a (finite or infinite) sequence of pairwise disjoint open sets U_n such that the restriction $p_n : U_n \to U$ of p is a homeomorphism for each n. At least one of the U_n intersects Z, and since U is connected and Z is a connected component of Y, it follows that the U_n which intersect Z are *contained* in Z. This shows already that $x \in p(Z)$; hence $p(Z)$ is both open and closed in X and is therefore equal to X. Moreover, $Z \cap p^{-1}(U)$ is the union of a subsequence of the U_n, which proves that Z is a covering space of X. The last assertion of (16.28.6) follows from the first and from (16.28.4).

We remark that if Y is an *n-sheeted covering* of X, then Y has *at most* n connected components, by virtue of (16.28.6).

Example

(16.28.7) Let X be a *nonorientable* connected differential manifold. Then $\pi_1(X) \neq \{e\}$. For by (16.21.16) there exists an *orientable* two-sheeted covering Y of X. If X were simply-connected then Y would have two connected components Y_1, Y_2, and each of the projections $Y_1 \to X$, $Y_2 \to X$ would be a diffeomorphism; but this is absurd, because Y_1 is orientable and X is not.

(16.28.8) (Monodromy principle) *Let* (Y, X, p) *be a covering of a differential manifold* X, *and let* $f : X' \to X$ *be a* C^∞*-mapping of a connected, simply-connected, differential manifold* X' *into* X. *Let* $a' \in X'$ *and let* $b \in p^{-1}(f(a'))$. *Then there exists a unique* C^∞*-mapping* $g : X' \to Y$ *such that* $g(a') = b$ *and* $p \circ g = f$ (the mapping g is said to be a *lifting of* f).

Consider the *inverse image* Y' under f of the covering Y of X, which is a covering of X' (16.12.8). Let $p' : Y' \to X'$ and $f' : Y' \to Y$ be the canonical projections, and let b' be the point of Y' such that $f'(b') = b$, $p'(b') = a'$. The connected component Y'_0 of Y' containing b' is a connected covering of X' (16.28.6); hence $p' | Y'_0$ is a diffeomorphism of Y'_0 onto X' (16.28.4). If $q' : X' \to Y'_0$ is the inverse diffeomorphism, then the mapping $g = f' \circ q'$ has the required properties, because

$$p \circ g = (p \circ f') \circ q' = (f \circ p') \circ q' = f$$

and $g(a') = f'(b') = b$. The uniqueness of g follows from (16.28.1).

Example

(16.28.9) Let U be a simply-connected open set in $\mathbf{C}^n$ and let f be a holomorphic function on U which does not vanish on U. Then the monodromy principle can be applied by considering f as a mapping of U into $X = \mathbf{C} - \{0\}$, and by taking Y to be the Riemann surface of the logarithmic function (16.12.4.2). It follows (16.8.11) that there exists a holomorphic function g on U such that $f = e^g$ (and g is unique up to a constant integer multiple of $2\pi i$).

PROBLEMS

1. Generalize the definitions and results of (16.28.1) to (16.28.4) by replacing coverings of differential manifolds by C^0-coverings (Section 16.25, Problem 8) of arcwise-connected topological spaces.

2. Let X be an arcwise-connected space, (Y, X, p) a C^0-covering of X and G the group of X-automorphisms of this covering (16.12.1).

   (a) Show that G (considered as a discrete group) acts freely on Y. (Use (16.28.1).)
   (b) If Y is arcwise-connected and G acts transitively on one fiber of Y, then G acts transitively on all fibers of Y. (Join two points of Y by a path.)
   (c) A C^0-covering (Y, X, p) is said to be a *Galois covering* of X if Y is arcwise-connected and G acts transitively on each fiber (which implies that G acts simply transitively on each fiber, by virtue of (a)). Then (Y, X, p) is a principal bundle with the discrete group G as structure group.

   Let $a \in X$ and $b \in p^{-1}(a)$. For each $\sigma \in \pi_1(X, a)$, there exists a unique element $g_\sigma \in G$ such that $g_\sigma(b) = b \cdot \sigma$. Show that the mapping $\varphi_b : \sigma \mapsto g_\sigma$ is a *surjective homomorphism* of $\pi_1(X, a)$ onto the group G^0 opposite to G, and that its kernel is the (normal) subgroup of $\pi_1(X, a)$ which is the image of $\pi_1(Y, b)$ under the homomorphism p_*: in other words, we have an exact sequence

$$1 \to \pi_1(Y, b) \xrightarrow{p_*} \pi_1(X, a) \xrightarrow{\varphi_b} G^0 \to 1.$$

3. Let G be a discrete group acting continuously and freely on a locally compact metrizable space Y; let C be the set of points $(y, z) \in Y \times Y$ such that $z = s \cdot y$ for some (unique) $s \in G$. Put $s = \varphi(y, z)$ for $(y, z) \in C$. Show that the following conditions are equivalent:

   (a) G acts properly on Y (Section 12.10, Problem 1);
   (b) for each $y \in Y$ there exists a neighborhood V of y in Y such that $s \cdot V \cap V = \varnothing$ for all $s \neq e$ in G;
   (c) C is a closed subset of $Y \times Y$ and $\varphi : C \to G$ is continuous.

4. Let Y be an arcwise-connected space, and G a discrete group acting continuously and freely on Y, and satisfying condition (b) of Problem 3. If $p : Y \to Y/G$ is the canonical mapping, show that $(Y, Y/G, p)$ is a Galois C^0-covering whose group of automorphisms is canonically isomorphic to G.

5. Let (Y, X, p) be a Galois C^0-covering of X (Problem 2) and let G be the opposite of its group of automorphisms. For each discrete space F on which G acts on the left, the

fiber bundle $Y \times^G F$ with fiber-type F associated with Y (16.14.7) is a C^0-covering of X. For $Y \times^G F$ to be connected it is necessary and sufficient that G should *act transitively* on F, so that if Γ is the stabilizer of a point of F, then F can be identified with G/Γ; and then $Y \times^G F$ can be identified with Y/Γ, and Y is a Galois C^0-covering of $Y \times^G F$, whose automorphism group is isomorphic to the opposite of Γ.

6. Let X be an arcwise-connected, locally arcwise-connected topological space.

(a) Show that every arcwise-connected component of a C^0-covering (Y, X, p) of X is a C^0-covering of X. (Consider an arcwise-connected open neighborhood of a point of X, over which Y is trivializable.)
(b) Let $a \in X$, and let U be an arcwise-connected open neighborhood of a in X, such that the canonical image of $\pi_1(U, a)$ in $\pi_1(X, a)$ consists only of the identity element. Show that every C^0-covering of X is trivializable over U. (If (Y, X, p) is a C^0-covering, consider a connected component of $p^{-1}(U)$ and show that $\pi_1(U, a)$ acts trivially on the fiber over a of this covering of U.)

7. Let X be an arcwise-connected space, (Y, X, p) a connected C^0-covering of X. Let X' be an arcwise-connected, locally arcwise-connected space and $f: X' \to X$ a continuous mapping. Let $a' \in X'$, $a = f'(a')$, $b \in p^{-1}(a)$. Show that f admits a continuous lifting $g: X' \to Y$ such that $g(a') = b$ if and only if the image of the homomorphism $f_* : \pi_1(X', a') \to \pi_1(X, a)$ is contained in the image of the homomorphism

$$p_* : \pi_1(Y, b) \to \pi_1(X, a).$$

(With the notation of the proof of (16.28.8), apply (16.28.4(ii)) to Y'_0.)

8. (a) Let X be an arcwise-connected, locally arcwise-connected space, and let (Y, X, p), (Y', X, p') be two C^0-coverings of X. If Y is arcwise-connected, show that for every X-morphism $g: Y' \to Y$, (Y', Y, g) is a covering. (If U is an arcwise-connected open subset of X such that Y and Y' are trivializable over U, let V be a connected component of $p^{-1}(U)$, and consider $g^{-1}(V)$.)
(b) Suppose in addition that Y' is arcwise connected. Let $a \in X$, $b \in p^{-1}(a)$, $b' \in p'^{-1}(a)$. Show that there exists an X-morphism $g: Y' \to Y$ such that $g(b') = b$ if and only if the image under p'_* of $\pi_1(Y', b')$ in $\pi_1(X, a)$ is contained in the image under p_* of $\pi_1(Y', b)$ in $\pi_1(X, a)$; the X-morphism g is then unique. (Use Problem 7.)
(c) Deduce that, for each $\sigma \in \pi_1(X, a)$ and $b \in p^{-1}(a)$, there exists an X-*automorphism* of Y mapping b to $b \cdot \sigma$ if and only if σ belongs to the normalizer in $\pi_1(X, a)$ of the stabilizer of b (for the action of $\pi_1(X, a)$ on the fiber over a).
(d) Deduce from (c) that an arcwise-connected covering (Y, X, p) of X is Galois (Problem 2) is and only if, for some point $b \in p^{-1}(a)$, the image under p_* of $\pi_1(Y, b)$ is a normal subgroup of $\pi_1(X, a)$.

9. In $\mathbf{R}^3$, let T be the union of the following sets: the segment with endpoints $(0, -1, 1)$ and $(0, 2, 1)$, the segment with endpoints $(0, 2, 1)$ and $(1, 2, 0)$, the segment with endpoints $(1, 2, 0)$ and $(1, 0, 0)$, and the set of points $(x, \sin(\pi/x), 0)$ for $0 < x \leq \pi$. Let X be the projection of T on $\mathbf{R}^2$ (identified with the plane $z = 0$), and let Y be the union of the sets $T + n\mathbf{e}_3$ with $n \in \mathbf{Z}$. Show that X is arcwise-connected, that Y is a connected C^0-covering of X whose arcwise-connected components are the sets $T + n\mathbf{e}_3$, which are not C^0-coverings of X; and that Y is not trivializable, although X is simply-connected.

10. Let B be a separable, metrizable, locally compact space, B′ a topological space, and (X, B′, π) a C^0-fibration whose fibers are differential manifolds. Let f, g be two continuous mappings of B into B′ and let $\varphi : B \times I \to B'$ be a homotopy of f into g. Suppose that there exists a continuous lifting $h : B \to X$ of f. Show that there then exists a continuous lifting $\psi : B \times I \to X$ of the homotopy φ. Moreover if, for some $b_0 \in B$, $\varphi(b_0, \xi)$ is independent of $\xi \in I$, then ψ can be chosen so that $\psi(b_0, \xi) \in X$ is independent of ξ (*homotopy lifting theorem*: apply Section 16.26, Problem 6 and consider the inverse image $\varphi^*(X)$).

29. THE UNIVERSAL COVERING OF A DIFFERENTIAL MANIFOLD

(16.29.1) *Let X be a connected differential manifold. Then there exists a connected, simply-connected covering* (Z, X, p) *of* X. *If* (Y, X, π) *is a connected covering of* X, b *a point of* Y, $a = \pi(b)$, *and* c *a point of* $p^{-1}(a)$, *then there exists a unique* X*-morphism* $f : Z \to Y$ (16.12.1) *such that* $f(c) = b$, *and* (Z, Y, f) *is a covering of* Y. *In particular, if* Z′ *is a connected, simply-connected covering of* X, *then there exists an* X*-isomorphism of* Z *onto* Z′.

We shall prove the second assertion first (the third will then follow immediately, by reason of (16.28.4)). The existence and uniqueness of the X-morphism f follow from the monodromy principle (16.28.8) applied to $p : Z \to X$ and the covering Y. To show that (Z, Y, f) is a covering, let us first show that f is *surjective*. Let $y \in Y$; because Y is connected, there exists a path $\beta : I \to Y$ from b to y. The path $\pi \circ \beta : I \to X$ from a to $\pi(y)$ then lifts to a path $\gamma : I \to Z$ from c to a point $z \in p^{-1}(\pi(y))$ (16.28.2). It is clear that $f \circ \gamma$ is a path with origin b which lifts $\pi \circ \beta$, hence is equal to β, and therefore $f(z) = y$. Now consider a connected open neighborhood U of $\pi(y)$ such that $\pi^{-1}(U)$ is the union of a sequence of pairwise disjoint open sets V_n, the restriction π_n of π to each V_n being a diffeomorphism of V_n onto U, and such that $p^{-1}(U)$ is the union of a sequence of pairwise disjoint open sets W_m, the restriction p_m of p to each W_m being a diffeomorphism of W_m onto U. Suppose that $y \in V_{n_0}$; if a point $z' \in W_m$ is such that $f(z) \in V_{n_0}$, then $f(W_m) \subset V_{n_0}$ because W_m is connected; but the definition of f then shows that the restriction of f to W_m can be written as $\pi_{n_0}^{-1} \circ p_m$, and is therefore a diffeomorphism of W_m onto V_{n_0}. This proves that (Z, Y, f) is a covering of Y.

We shall now establish the existence of the covering (Z, X, p) with the properties stated, by the method of construction of (16.13.3). Consider a covering (U_α) of X formed of open sets homeomorphic to open balls in $\mathbf{R}^n$, and therefore *simply-connected*. For each index α, fix a point $a_\alpha \in U_\alpha$; for each $x \in U_\alpha$, we shall denote by $c_\alpha(x)$ the element of $E_{a_\alpha, x}$ which is the class of *all* paths from a_α to x which are contained in U_α (by hypothesis, all these paths are equivalent in $\Omega_{a_\alpha, x}$). Let α_0 be one of the indices, and write $a = a_{\alpha_0}$ to simplify the notation. For each $x \in U_\alpha$, the mapping $u \mapsto u \cdot c_\alpha(x)$ is a *bijection* of E_{a, a_α} onto $E_{a, x}$; we shall denote this bijection by $f_\alpha(x)$. Given two indices α, β and $x \in U_\alpha \cap U_\beta$, we put

$$f_{\beta\alpha}(x) = (f_\beta(x))^{-1} \circ f_\alpha(x) : u \mapsto uc_\alpha(x)(c_\beta(x))^{-1},$$

which is a bijection of E_{a, a_α} onto E_{a, a_β}. The mapping $x \mapsto f_{\beta\alpha}(x)$ is constant on each connected component V of $U_\alpha \cap U_\beta$; for if x, y are two points of V, there exists a path $\sigma \in \Omega_{x, y}$ *contained in* V, and if s is the class of this path in $E_{x, y}$, we have $c_\alpha(y) = c_\alpha(x)s$ and $c_\beta(y) = c_\beta(x)s$, whence the assertion follows. Finally, the above definitions show that for any three indices α, β, γ, we have

(16.29.1.1) $$f_{\gamma\alpha}(x) = f_{\gamma\beta}(x) \circ f_{\beta\alpha}(x) \qquad \text{for all} \quad x \in U_\alpha \cap U_\beta \cap U_\gamma.$$

Since the sets E_{a, a_α} are denumerable, they may be considered as *discrete* differential manifolds. Apply the method of (16.13.3) to these manifolds, by defining mappings

$$\psi_{\beta\alpha} : (U_\alpha \cap U_\beta) \times E_{a, a_\alpha} \to (U_\alpha \cap U_\beta) \times E_{a, a_\beta}$$

by the formula $\psi_{\beta\alpha}(x, u) = (x, f_{\beta\alpha}(x)(u))$. Since $x \mapsto f_{\beta\alpha}(x)$ is locally constant, the mapping $(x, u) \mapsto f_{\beta\alpha}(x)(u)$ is of class C^∞, and for each $x \in U_\alpha \cap U_\beta$ we have seen that $f_{\beta\alpha}(x)$ is a bijection of E_{a, a_α} onto E_{a, a_β}; finally, it follows from (16.29.1.1) that the $\psi_{\beta\alpha}$ satisfy the patching condition (16.13.1.1). We have therefore defined a *covering* (Z, X, p) of X, and it remains to show that Z is connected and simply-connected. We shall apply (16.28.4) by showing that on the fiber $p^{-1}(a) = E_{a, a} = \pi_1(X, a)$, the fundamental group $\pi_1(X, a)$ acts by *right multiplication* (for the group structure of $\pi_1(X, a)$); this will complete the proof.

Consider a path $\lambda : [0, 1] \to X$ with origin $\lambda(0) = a$. For each $\xi \in [0, 1]$, let λ_ξ be the path $[0, \xi] \to X$ obtained by restricting λ to $[0, \xi]$, and let u_ξ be the class of λ_ξ in $E_{a, \lambda(\xi)}$. Let μ be the lifting of λ to Z with origin b, the canonical image in $p^{-1}(U_{\alpha_0})$ of the point $(a, e_a) \in U_{\alpha_0} \times E_{a, a}$ (16.28.2). We shall show that, for all $\xi \in [0, 1]$ and *all* α such that $\lambda(\xi) \in U_\alpha$, the point $\mu(\xi)$ is the canonical image (16.13.3) in $p^{-1}(U_\alpha)$ of the point

$$(\lambda(\xi), f_\alpha(\lambda(\xi))^{-1}(u_\xi)) \in U_\alpha \times E_{a, a_\alpha}.$$

Notice first that if this is true for *one* index α such that $\lambda(\xi) \in U_\alpha$, then it is true for all other indices β such that $\lambda(\xi) \in U_\beta$. This follows from the definition of the transition diffeomorphisms given earlier. Let $A \subset [0, 1]$ be the set of points ξ for which the property in question holds; clearly $0 \in A$, and it will suffice to show that A is both open and closed in [0, 1] (3.19.1). Let then $\xi \in \bar{A}$; then there is a neighborhood V of ξ in [0, 1] such that $\lambda(V) \subset U_\alpha$ for some index α; hence $\mu(V)$ is contained in an open set $W_\alpha \subset Z$ such that $p|W_\alpha : W_\alpha \to U_\alpha$ is a diffeomorphism. Let η be a point of $A \cap V$. We have to show that

$$f_\alpha(\lambda(\xi'))^{-1}(u_{\xi'}) = f_\alpha(\lambda(\eta))^{-1}(u_\eta)$$

for all $\xi' \in V$. Now if we suppose for example that $\xi' > \eta$, then the relation $\lambda([\eta, \xi']) \subset U_\alpha$ implies that $c_\alpha(\lambda(\eta))^{-1}c_\alpha(\lambda(\xi'))$ is the class in $E_{\lambda(\eta), \lambda(\xi')}$ of the path $t \mapsto \lambda(t)$ restricted to $[\eta, \xi']$; but, by definition, $u_\eta^{-1}u_{\xi'}$ is also the class of this path, and the assertion is proved.

Suppose now that λ is a *loop* with origin a. If u is its class in $E_{a,a} = \pi_1(X, a)$, then the point $\mu(1)$ is the image in $p^{-1}(U_{\alpha_0})$ of $(a, f_{\alpha_0}(a)^{-1}(u))$; but $c_{\alpha_0}(a) = e_\alpha$ by definition, hence $\mu(1)$ is the image of (a, u), and since $u = e_a \cdot u$, this completes the proof.

The connected and simply-connected covering (Z, X, p), defined up to X-isomorphism, is called the *universal covering* of X.

(16.29.2) We remark that (Z, X, p) is a *principal bundle* with structure group $\pi_1(X, a)$ acting *on the left*: if $\varphi_\alpha : U_\alpha \times E_{a, a_\alpha} \to Z$ is the mapping defined in the above construction, then for $v \in \pi_1(X, a)$ and $(x, u) \in U_\alpha \times E_{a, a_\alpha}$, we have $v \cdot \varphi_\alpha(x, u) = \varphi_\alpha(x, v \cdot u)$.

(16.29.3) It is clear that if X is a real-analytic manifold (resp. a complex manifold), then the construction of **(16.29.1)** defines the universal covering space Z as a real-analytic manifold (resp. a complex manifold), p being an analytic (resp. holomorphic) mapping.

Example

(16.29.4) The Riemann surface Y of the logarithmic function **(16.8.11)** is a universal covering of $\mathbf{C}^* = \mathbf{C} - \{0\}$. For since $\mathbf{C}^*$ is isomorphic to $\mathbf{U} \times \mathbf{R}_+^*$ **(16.8.10)**, its fundamental group is isomorphic to $\mathbf{Z}$ **(16.27.10)**, and it acts simply transitively on the fiber of the point $z = 1$, as we have seen in **(16.12.4)**.

PROBLEMS

1. (a) Let X be an arcwise-connected, locally arcwise-connected topological space. Show that there exists a simply-connected arcwise-connected C^0-covering space Y of X if and only if there exists a covering of X by arcwise-connected open sets U_α such that the canonical homomorphism $\pi_1(U_\alpha, a_\alpha) \to \pi_1(X, a_\alpha)$ is trivial, for some $a_\alpha \in U_\alpha$. (To show that the condition is necessary, consider a simply-connected C^0-covering (Z, X, p) of X,

choose the U_α such that Z is trivializable over U_α, and consider the connected components of $p^{-1}(U_\alpha)$. To show that the condition is sufficient, follow the existence proof in (16.29.1).)

(b) If X satisfies the conditions of (a), show that every C^0-covering space of a C^0-covering space of X is a C^0-covering space of X (reduce to connected covering spaces).

(c) Suppose that X satisfies the conditions of (a), and let Y be the simply-connected covering space of X. Then Y is a Galois C^0-covering space (Section 16.28, Problem 2) whose automorphism group G is isomorphic to the opposite of $\pi_1(X)$. Every connected C^0-covering space of X is of the form $Y \times^G F$ (Section 16.28, Problem 2).

(d) Show that there exists a canonical one–one correspondence between the subgroups of $\pi_1(X)$ and the isomorphism classes of connected C^0-coverings of X. A subgroup Γ of $\pi_1(X)$ corresponds to a connected Galois C^0-covering of X if and only if Γ is normal in $\pi_1(X)$, and the automorphism group of the covering is then isomorphic to the opposite of $\pi_1(X)/\Gamma$ ("Galois theory of coverings").

2. Let X be the union of the two circles $|z-1|^2 = 1$, $|z-2|^2 = 4$ in **C**. Show that X admits a universal C^0-covering space, and describe it explicitly.

Let Y be the compact subspace of **C** which is the union of the circles $|z-(1/n)|^2 = 1/n^2$ for all $n \geq 1$. Show that Y does not admit a universal C^0-covering space (although Y is arcwise-connected and locally arcwise-connected). Describe the nontrivializable connected C^0-coverings of Y.

3. Let K be a Galois extension of degree n of the field **C**(X) of rational functions in one indeterminate with coefficients in **C**. There exists an element $\theta \in K$ which generates K and is a root of an irreducible polynomial of degree n in Y:

$$F(X, Y) = Y^n + a_1(X)Y^{n-1} + \cdots + a_n(X),$$

where the a_j are polynomials in **C**[X]. Let $\Delta(X)$ be the discriminant of F (regarded as a polynomial in Y).

(a) Let $z_1, \ldots, z_N$ be the zeros of Δ in **C**, and let U be the complement in **C** of the set $\{z_1, \ldots, z_N\}$. Let Z be the set of points $(x, y) \in \mathbf{C}^2$ such that $F(x, y) = 0$. Show that $R = \mathrm{pr}_1^{-1}(U) \cap Z$ is a *connected* n-sheeted covering of U. (Use Section 16.12, Problem 2(f). To show that R is connected, consider a connected component of R and the elementary symmetric functions of the second projections of the points where this component meets $\mathrm{pr}_1^{-1}(x)$ for $x \in U$; then use the fact that F is irreducible.)

(b) Let Γ be the Galois group of K over **C**(X). For each $\sigma \in \Gamma$ we have

$$\sigma(\theta) = \sum_{j=0}^{n-1} b_{\sigma,j}(X)\theta^j,$$

where the $b_{\sigma,j}$ are rational functions in X. For each $x \in U$ other than the poles of the $b_{\sigma,j}$ ($\sigma \in \Gamma$, $0 \leq j \leq n-1$) and each $y \in \mathbf{C}$ such that $(x, y) \in R$, put

$$u_\sigma(x, y) = \sum_{j=0}^{n-1} b_{\sigma,j}(x)y^j.$$

Show that the mappings u_σ extend by continuity to the whole of R (9.15.2) and that u_σ is an automorphism of the covering R of U. The mapping $\sigma \mapsto u_\sigma$ is an *isomorphism* of Γ onto the group of automorphisms of R, and R is a Galois covering.

(c) Generalize the results of (a) by replacing K by an arbitrary finite algebraic extension of C(X). (Embed L in a Galois extension K of C(X), use (b) above and Galois theory.)

4. With the notation of Problem 3 (but not assuming K to be Galois over C(X)), suppose that $\partial F/\partial X$ and $\partial F/\partial Y$ do not simultaneously vanish at any point of Z. Then there exists a unique structure of a complex manifold on Z which induces the complex-analytic structure of U.

If $F(X, Y) = Y^2 - X^2 + X^3$, show that F is irreducible over C(X) but that there exists no structure of differential manifold on Z compatible with the topology of Z as a subspace of C^2.

5. Let X be a metrizable, arcwise-connected, locally arcwise-connected space, Y a metrizable space, $p : Y \to X$ a continuous mapping. Suppose that for each $x \in X$ the subspace $p^{-1}(x)$ of Y is *discrete*.

(a) Suppose that, for each $a \in X$, each path $\beta : I \to X$ with origin a (where $I = [0,1]$) and each point $b \in p^{-1}(a)$, there exists a unique path $\gamma : I \to Y$ with origin b such that $\beta = p \circ \gamma$ (cf. (16.28.2)). Show that for each $a \in X$ and each $b \in p^{-1}(a)$, there exists $\delta > 0$ with the following property: for each *loop* $\beta : I \to X$ with origin a, contained in the ball $B(a; \delta)$, the unique path $\gamma : I \to Y$ with origin b which lifts β is a *loop*. (Argue by contradiction, and show that if the result is false there exist two sequences (λ_n), (μ_n) in [0, 1] such that $\lambda_n < \mu_n < \lambda_{n+1}$ and $\lambda_n \to 1$, $\mu_n \to 1$, and a loop $\beta : I \to X$ with origin a having the following properties: The restriction β_n of β to $[\lambda_n, \mu_n]$ is a loop with origin a and of diameter $< 1/n$, and its lifting $\gamma_n : [\lambda_n, \mu_n] \to Y$ with origin b is not a loop. Then define β on $[\mu_n, \lambda_{n+1}]$ so that up to equivalence the restriction of β to $[\mu_n, \lambda_{n+1}]$ is the *opposite* of the loop β_n, and lift this loop to a path in Y with origin $\gamma_n(\mu_n)$ and endpoint b.)

(b) Deduce from (a) that, under the same hypotheses, every continuous mapping $\varphi : I \times I \to X$ such that $\varphi(0, 0) = a$ lifts uniquely to a continuous mapping $\psi : I \times I \to Y$ such that $\psi(0, 0) = b$. (First lift the path $t \mapsto \varphi(0, t)$ to a path $\gamma : t \mapsto \gamma(t)$ such that $\gamma(0) = b$; then lift each path $s \mapsto \varphi(s, t)$ to a path with origin $\gamma(t)$; we have to prove that the function $\psi : I \times I \to Y$ so defined is continuous at each point (s_0, t_0). By virtue of (a), there exists $\alpha > 0$ such that, for each loop in $I \times I$ with origin (s_1, t_0) (where $s_1 \in [0, 1]$) and diameter $\leq \alpha$, the image under φ of this loop lifts to a loop with origin $\psi(s_1, t_0)$ in Y. Prove that ψ is continuous at each point (s, t_0) such that

$$\tfrac{1}{2}k\alpha \leq s \leq \tfrac{1}{2}(k+1)\alpha$$

by induction on k, arguing by contradiction.)

(c) Suppose that X, Y, p satisfy the hypotheses of (a) and also that X admits a simply-connected arcwise-connected C^0-covering space (Problem 1). Show that (Y, X, p) is a covering of X. (Using the result of (b), show that for each of the open sets U_α defined in Problem 1(a), and for each point of $p^{-1}(U_\alpha)$, there exists a unique continuous section of $p^{-1}(U_\alpha)$ passing through this point.)

6. Let X be a simply-connected, arcwise-connected topological space, and let Y be an arcwise-connected, locally arcwise-connected space which admits a simply-connected, arcwise-connected C^0-covering $\tilde{Y}$ (Problem 1). Suppose that there exists a *surjective* continuous mapping $f : X \to Y$ such that for each $y \in Y$ the fiber $f^{-1}(y)$ is connected. Show that Y is simply-connected. (Lift f to a continuous mapping of X into $\tilde{Y}$.)

30. COVERING SPACES OF A LIE GROUP

(16.30.1) Let G be a *connected Lie group* and let $m : (x, y) \mapsto xy$ be the C^∞-mapping of $G \times G$ into G which defines its group structure. Let (Z, G, p) be a *universal covering* (16.29.1) of the manifold underlying G, and let $\tilde{e}$ be a point of $p^{-1}(e)$. We shall show that there exists on Z a unique Lie group structure such that $\tilde{e}$ is the identity element and $p : Z \to G$ a group homomorphism. The space $Z \times Z$ is connected (3.20.16) and simply-connected (16.27.10). Consider the composite mapping

$$Z \times Z \xrightarrow{p \times p} G \times G \xrightarrow{m} G;$$

by virtue of the monodromy principle (16.28.8) this C^∞-mapping lifts uniquely to a C^∞-mapping $\tilde{m} : Z \times Z \to Z$ such that $\tilde{m}(\tilde{e}, \tilde{e}) = \tilde{e}$. We have to show that $\tilde{m}$ defines on Z a group structure, which by construction will be the unique group structure for which p is a homomorphism. First of all, the multiplication $\tilde{m}$ on Z is associative, for the two mappings $(x, y, z) \mapsto \tilde{m}(\tilde{m}(x, y), z)$ and $(x, y, z) \mapsto \tilde{m}(x, \tilde{m}(y, z))$ of $Z \times Z \times Z$ into Z are both liftings of the same mapping $(x, y, z) \mapsto p(x)p(y)p(z)$ and are equal at the point $(\tilde{e}, \tilde{e}, \tilde{e})$; hence they coincide because $Z \times Z \times Z$ is simply-connected (16.27.10). Secondly, $\tilde{e}$ is a neutral element for $\tilde{m}$, because the mappings $x \mapsto \tilde{m}(x, \tilde{e})$, $x \mapsto \tilde{m}(\tilde{e}, x)$ and the identity mapping of Z are three liftings of the same mapping $x \mapsto p(x)$ which take the same value at the point $\tilde{e}$, and therefore coincide for the same reason as before. Finally, to prove the existence of inverses, consider the C^∞-mapping $x \mapsto (p(x))^{-1}$ of Z into G, which lifts to a C^∞-mapping $\tilde{i} : Z \to Z$ such that $\tilde{i}(\tilde{e}) = \tilde{e}$. The three mappings $x \mapsto \tilde{m}(x, \tilde{i}(x))$, $x \mapsto \tilde{m}(\tilde{i}(x), x)$, and $x \mapsto \tilde{e}$ are liftings of the constant mapping $x \mapsto e$ which take the same value at the point $\tilde{e}$, and therefore once again they coincide.

The manifold Z, endowed with the Lie group structure just defined, is called the *universal covering group of* G and is denoted by $\tilde{G}$. From now on we shall write xy in place of $\tilde{m}(x, y)$ $(x, y \in \tilde{G})$.

(16.30.2) Let $x, y \in \tilde{G}$, and let $t \mapsto \gamma(t)$ be a path in $\tilde{G}$, defined on the interval $I = [0, 1]$, with origin $\tilde{e}$ and endpoint y. Then $t \mapsto x\gamma(t)$ is a path in $\tilde{G}$ from x to xy and is the unique lifting with origin x of the path $t \mapsto p(x)p(\gamma(t))$ in G. In particular, suppose that x and y belong to the kernel $p^{-1}(e)$ of the homomorphism p. Then the remark above shows that xy is equal to $x \cdot u_y$, where u_y is the class in $\pi_1(G, e)$ of all the loops with origin e which lift to a path from $\tilde{e}$ to y (16.28.2). If z is another point of $p^{-1}(e)$ we have $xyz = x \cdot u_{yz} = x \cdot (u_y u_z)$ by (16.28.2.1); hence $u_{yz} = u_y u_z$. This proves that $y \mapsto u_y$ is an *isomorphism of the discrete group* $\pi_1(G, e)$ *onto the discrete subgroup* $p^{-1}(e)$ *of* $\tilde{G}$. Furthermore:

(16.30.2.1) *The subgroup* $p^{-1}(e)$ *of* $\tilde{G}$ *is contained in the center of* $\tilde{G}$.

For $p^{-1}(e)$ is a normal subgroup of $\tilde{G}$, and the assertion therefore is a particular case of the following general lemma:

(16.30.2.2) *In a connected topological group* H, *every discrete normal subgroup* D *is contained in the center of* H.

For each $d \in D$, the mapping $x \mapsto xdx^{-1}$ of H into D is continuous and therefore constant (3.19.7); hence its value is $ede^{-1} = d$. Consequently $xd = dx$ for all $x \in H$.

From (16.30.2.1) it follows that the fundamental group $\pi_1(G)$ of a Lie group G is *commutative.*

It follows also that G is isomorphic to $\tilde{G}/p^{-1}(e)$ (16.10.8). Conversely, if L is a connected, simply-connected Lie group, and N is a *discrete normal* subgroup of L (hence contained in the center of L), then L is a universal covering group of L/N (16.14.2).

(16.30.3) *Let* G *be a simply-connected Lie group,* G′ *a connected Lie group,* $\tilde{G}'$ *the universal covering group of* G′, *and* $p : \tilde{G}' \to G'$ *the canonical projection. Then for each Lie group homomorphism* $f : G \to G'$, *there exists a unique Lie group homomorphism* $g : G \to \tilde{G}'$ *such that* $f = p \circ g$.

The existence and uniqueness of a C^∞-mapping $g : G \to \tilde{G}'$ such that $f = p \circ g$ and $g(e) = \tilde{e}'$ (where e, $\tilde{e}'$ are the identity elements of G and $\tilde{G}'$, respectively) follows from the monodromy principle (16.28.8). To show that g is a homomorphism, consider the two mappings $(x, y) \mapsto g(x)g(y)$ and $(x, y) \mapsto g(xy)$ of $G \times G$ into $\tilde{G}'$. Since

$$p(g(x)g(y)) = f(x)f(y) = f(xy) = p(g(xy)),$$

both these mappings are liftings of $(x, y) \mapsto f(xy)$ and take the same value at the point (e, e); hence they are equal (16.28.1).

(16.30.4) Consider now an arbitrary *connected covering space* (Z_1, G, p_1) of a connected Lie group G, and let e_1 be a point of the fiber $p_1^{-1}(e)$. We shall show that there exists a unique structure of Lie group on Z_1 for which e_1 is the identity element and p_1 is a homomorphism. Let q be the unique G-morphism of $\tilde{G}$ onto Z_1 such that $q(\tilde{e}) = e_1$ (16.29.1), which makes $(\tilde{G}, Z_1, q)$ a covering of Z_1. We shall show that the multiplication m_1 on Z_1 is defined as follows: if $x_1 = q(x)$ and $y_1 = q(y)$, then $m_1(x_1, y_1) = q(xy)$. To show that m_1 is well-defined we must show that if $q(x) = q(x')$, then $q(xy) = q(x'y)$ and $q(yx) = q(yx')$. For this, consider the two mappings $z \mapsto q(xz)$ and $z \mapsto q(x'z)$; they are liftings to Z_1 of the same mapping of $\tilde{G}$ into G, because $p_1(q(xz)) = p(xz) = p(x)p(z)$ and $p_1(q(x'z)) = p(x'z) = p(x')p(z)$, and $p(x) = p(x')$ because $q(x) = q(x')$; also, for $z = \tilde{e}$ we have $q(x\tilde{e}) = q(x'\tilde{e})$, and therefore $q(xy) = q(x'y)$ for all $y \in \tilde{G}$ by virtue of (16.28.8). The proof that $q(yx) = q(yx')$ is similar. It follows that m_1 is well-defined and therefore defines a group structure on Z_1 such that q is a homomorphism. Furthermore, because q is a local diffeomorphism, it follows immediately that m_1 is of class C^∞, and that the inverse mapping $x_1 \mapsto x_1^{-1}$ is also of class C^∞.

It is clear that $p_1 : Z_1 \to G$ is a homomorphism for the group structure just defined on Z_1, and this group structure is the *unique* Lie group structure with this property, for which e_1 is the identity element. For if $x_1, y_1 \in Z_1$ and if α, β are paths in Z_1, defined on [0, 1], with origin e_1 and endpoints, respectively, x_1 and y_1, then the hypothesis that p_1 is a homomorphism implies that the path $t \mapsto \alpha(t)\beta(t)$ is a lifting to Z_1 of the path $t \mapsto p_1(\alpha(t))p_1(\beta(t))$ in G, and this lifting has origin e_1, and hence is uniquely determined (16.28.1).

Since q is surjective, the kernel $p_1^{-1}(e)$ is equal to $q(p^{-1}(e))$. Hence, bearing in mind (16.14.2), we see that we obtain (up to isomorphism) *all* the coverings (G_1, G, p_1), where G_1 is a connected Lie group and p_1 a homomorphism, by taking the quotient group $\tilde{G}/D$, where D is a subgroup of $p^{-1}(e)$, and $p_1^{-1}(e)$ is isomorphic to $p^{-1}(e)/D$ (and hence isomorphic to a quotient of $\pi_1(G)$).

(16.30.5) *Let* G *be a connected Lie group and* H *a Lie subgroup of* G *such that* G/H *is simply-connected. Then* H *is connected and* $\pi_1(G)$ *is isomorphic to a quotient of* $\pi_1(H)$.

Let H_0 be the identity component of H, so that H/H_0 is discrete (12.11.2). It follows from (16.14.9) that G/H_0 is then a covering space of G/H; since it is connected (3.19.7) and G/H is simply-connected, it follows that G/H_0 is canonically diffeomorphic to G/H (16.28.4), hence $H = H_0$ and H is connected.

Consider now the universal covering $(\tilde{G}, G, p)$ of G, and the covering $\tilde{H} = p^{-1}(H)$ induced by $\tilde{G}$ (16.12.8). Since $G = \tilde{G}/p^{-1}(e)$ and $H = \tilde{H}/p^{-1}(e)$ it follows, by (16.14.9) applied to the groups $\tilde{G} \supset \tilde{H} \supset p^{-1}(e)$, that G/H is diffeomorphic to $\tilde{G}/\tilde{H}$; consequently $\tilde{G}/\tilde{H}$ is simply-connected, and since $\tilde{G}$ is connected, it follows from the first part of the proof that $\tilde{H}$ is connected; but then it follows from (16.30.4) that $\pi_1(G, e) = p^{-1}(e)$ is isomorphic to a quotient of $\pi_1(H)$.

Examples

(16.30.6) By (16.11.5) and (16.27.8), the spaces $\mathbf{SO}(n+1)/\mathbf{SO}(n)$ are simply-connected for $n \geqq 2$; $\mathbf{SU}(n+1)/\mathbf{SU}(n)$ for $n \geqq 1$; and $\mathbf{U}(n+1, \mathbf{H})/\mathbf{U}(n, \mathbf{H})$

for $n \geqq 0$. Since $\mathbf{SU}(1) = \mathbf{U}(0, \mathbf{H}) = \{e\}$, it follows from (16.30.5) that the Lie groups $\mathbf{SU}(n)$ and $\mathbf{U}(n, \mathbf{H})$ are *simply-connected* for all $n \geqq 0$. Since each matrix $X \in \mathbf{U}(n, \mathbf{C})$ can be written uniquely in the form

$$X = \begin{pmatrix} \det(X) & & & 0 \\ & 1 & & \\ & & \ddots & \\ 0 & & & 1 \end{pmatrix} \cdot Y,$$

where $Y \in \mathbf{SU}(n)$, the *manifold* $\mathbf{U}(n, \mathbf{C})$ is diffeomorphic to the product $\mathbf{S}_1 \times \mathbf{SU}(n)$, hence $\pi_1(\mathbf{U}(n, \mathbf{C})) = \mathbf{Z}$ for $n \geqq 1$ ((16.27.9) and (16.27.10)).

As to the groups $\mathbf{SO}(n)$, observe that the mapping p which associates to each quaternion z of norm 1 the rotation $u_z : x \mapsto zxz^{-1}$ in $\mathbf{R}^3$ (identified with the space of *pure* quaternions) is a surjective homomorphism of the Lie group $\mathbf{U}(1, \mathbf{H})$ (the multiplicative group of quaternions of norm 1) onto $\mathbf{SO}(3)$, with kernel $\{-1, +1\}$. Hence ((16.9.9 (iv)) and (16.14.2)) $\mathbf{U}(1, \mathbf{H})$ is a *two-sheeted covering* of the group $\mathbf{SO}(3)$. Since $\mathbf{U}(1, \mathbf{H})$ is simply-connected, we have $\pi_1(\mathbf{SO}(3)) = \mathbf{Z}/2\mathbf{Z}$. From the remarks above and from (16.30.5), it follows that $\pi_1(\mathbf{SO}(n))$ for $n > 3$ is either $\mathbf{Z}/2\mathbf{Z}$ or trivial. (We shall see later (Chapter XXI) that it is in fact equal to $\mathbf{Z}/2\mathbf{Z}$; cf. Problem 10.) Recall finally that $\mathbf{SO}(2)$ is isomorphic to the multiplicative group $\mathbf{U} = \mathbf{U}(1, \mathbf{C})$ of complex members of absolute value 1, and therefore $\pi_1(\mathbf{SO}(2)) = \mathbf{Z}$ (16.27.9).

(16.30.7) *Let* G, G′ *be two Lie groups such that* G *is connected and simply-connected. Then every local homomorphism* (resp. C^∞ *local homomorphism*) *h from* G *to* G′ *which is defined on a connected open neighborhood* U *of e has a unique extension to a homomorphism of topological groups* (resp. *of Lie groups*) *of* G *into* G′.

Let V be a symmetric connected open neighborhood of e such that $V^2 \subset U$. Then every $x \in G$ can be written as a product $x_1 x_2 \cdots x_n$ with $x_i \in V$ $(1 \leqq i \leqq n)$ (12.8). If $\bar{h}$ is a homomorphism of G into G′ which extends h, then we have $\bar{h}(x) = h(x_1)h(x_2) \cdots h(x_n)$, so that $\bar{h}$ is unique. To prove the existence of $\bar{h}$, we observe first that if we can establish the existence of a homomorphism of abstract groups which extends h, then this homomorphism will automatically be continuous (12.8.4) (resp. of class C^∞ (16.9.7)). We shall first prove the following lemma:

(16.30.7.1) *If* $x_1, \ldots, x_n \in V$ *are such that* $x_1 x_2 \cdots x_n = e$, *then*

$$h(x_1)h(x_2) \cdots h(x_n) = e'.$$

Put $y_0 = e$, $y_j = y_{j-1}x_j$ for $1 \leqq j \leqq n$, so that $y_n = e$. Since $y_j \in y_{j-1}V$ for $1 \leqq j \leqq n$, there exists a path

$$\gamma_j : \left[\frac{j-1}{n}, \frac{j}{n}\right] \to y_{j-1}V$$

from y_{j-1} to y_j. The path $\gamma : I = [0, 1] \to G$, obtained by juxtaposing the γ_j, is therefore a loop with origin e. By hypothesis there exists a homotopy $\varphi : I \times I \to G$ of γ into the constant loop $I \to \{e\}$, leaving the point e fixed. Let d be a left-invariant distance defining the topology of G (12.9.1) and let $\varepsilon > 0$ be a number such that the relation $d(e, z) \leqq \varepsilon$ implies $z \in V$; next let $\delta > 0$ be sufficiently small that the relations $|t - t'| \leqq \delta$ and $|\xi - \xi'| \leqq \delta$ imply that $d(\varphi(t, \xi), \varphi(t', \xi')) \leqq \varepsilon$ (3.16.5). We can always reduce to the case where $1/n \leqq \delta$. For put $t_j = j/n$, so that $\gamma(t_j) = y_j$ $(0 \leqq j \leqq n)$, and let $t'_1, \dots, t'_m$ be points of the interval $[t_{j-1}, t_j]$ such that $|t'_1 - t_{j-1}| \leqq \delta$, $|t_j - t'_m| \leqq \delta$, and $|t'_{k+1} - t'_k| \leqq \delta$ for $1 \leqq k \leqq m - 1$. If we put $z_k = \gamma(t'_k)$, then $z_k \in y_j V$ for $1 \leqq k \leqq m$, and $z_{k+1} \in z_k V$. Hence, by induction on k, we conclude that

$$h(y_j^{-1}z_1)h(z_1^{-1}z_2) \cdots h(z_{k-1}^{-1}z_k) = h(y_j^{-1}z_k)$$

by virtue of the hypothesis on h, and consequently

$$h(y_j^{-1}z_1)h(z_1^{-1}z_2) \cdots h(z_{m-1}^{-1}z_m)h(z_m^{-1}y_{j+1}) = h(y_j^{-1}y_{j+1}) = h(x_{j+1}).$$

Hence the product $h(x_1)h(x_2) \cdots h(x_n)$ is not changed by replacing the sequence $x_1, \dots, x_n$ by $x_1, \dots, x_j, y_j^{-1}z_1, z_1^{-1}z_2, \dots, z_m^{-1}y_{j+1}, x_{j+2}, \dots, x_n$, which proves our assertion.

This being so, put $\varphi(t_j, t_k) = y_{jk}$, so that $y_{j0} = y_j$ and $y_{jn} = e$ for all j. It is enough to prove that

$$h(y_{0k}^{-1}y_{1k})h(y_{1k}^{-1}y_{2k}) \cdots h(y_{n-1,k}^{-1}y_{nk}) = h(y_{0,k+1}^{-1}y_{1,k+1}) \cdots h(y_{n-1,k+1}^{-1}y_{n,k+1})$$

for all k. This will follow by induction on j from the relation

$$(*_j) \qquad h(y_{0,k+1}^{-1}y_{1,k+1}) \cdots h(y_{j-1,k+1}^{-1}y_{j,k+1}) = h(y_{0k}^{-1}y_{1k}) \cdots h(y_{j-1,k}^{-1}y_{jk})h(y_{jk}^{-1}y_{j,k+1})$$

since $y_{nk} = e$. Now, to prove $(*_j)$ by induction, it suffices to prove that

$$h(y_{j,k+1}^{-1}y_{j+1,k+1}) = (h(y_{jk}^{-1}y_{j,k+1}))^{-1}h(y_{jk}^{-1}y_{j+1,k})h(y_{j+1,k}^{-1}y_{j+1,k+1});$$

but we have $y_{j,k+1}^{-1}y_{j+1,k+1} \in V$ and $y_{jk}^{-1}y_{j,k+1} \in V$; hence $y_{jk}^{-1}y_{j+1,k+1} \in U$, which shows that $h(y_{jk}^{-1}y_{j,k+1})h(y_{j,k+1}^{-1}y_{j+1,k+1}) = h(y_{jk}^{-1}y_{j+1,k+1})$. Similarly $h(y_{jk}^{-1}y_{j+1,k})h(y_{j+1,k}^{-1}y_{j+1,k+1}) = h(y_{jk}^{-1}y_{j+1,k+1})$. This completes the proof of (16.30.7.1).

Now each $x \in G$ can be expressed in at least one way as a product $x_1 x_2 \cdots x_n$, with $x_j \in V$ for each j. If also $x = x'_1 x'_2 \cdots x'_m$ with $x'_k \in V$ for each k, then the relation $x_1 \cdots x_n x'^{-1}_m \cdots x'^{-1}_1 = e$ implies, by (16.30.7.1), that $h(x_1)h(x_2) \cdots h(x_n) = h(x'_1)h(x'_2) \cdots h(x'_m)$. We may therefore define $\bar{h}(x)$ to be $h(x_1)h(x_2) \cdots h(x_n)$, since this product depends only on x. If $y = y_1 \cdots y_p$ is another element of G, with $y_i \in V$ for $1 \leqq i \leqq p$, then we have $xy = x_1 \cdots x_n y_1 \cdots y_p$, so that

$$\bar{h}(xy) = h(x_1) \cdots h(x_n)h(y_1) \cdots h(y_p) = \bar{h}(x)\bar{h}(y)$$

and therefore $\bar{h}$ is a homomorphism. It remains to show that $\bar{h}(x) = h(x)$ for $x \in U$. To do this, join x to e by a path $\alpha : I \to U$, then take a sequence of points $e = z_0, z_1, \ldots, z_{n-1}, z_n = x$ in $\alpha(I)$ such that $z_{i-1}^{-1} z_i \in V$ for $1 \leqq i \leqq n$. Since $z_i \in U$ for all i, we have $h(z_1)h(z_1^{-1} z_2) \cdots h(z_{i-1}^{-1} z_i) = h(z_i)$ by induction on i, and hence $h(x) = \bar{h}(x)$ as required.

PROBLEMS

1. Let X be a topological space on which is defined a law of composition $(x, y) \mapsto x * y$, which is a continuous mapping of $X \times X$ into X. If $\gamma_1 : I \to X$, $\gamma_2 : I \to X$ (where $I = [0, 1]$) are two paths, let $\gamma_1 * \gamma_2$ denote the path $t \mapsto \gamma_1(t) * \gamma_2(t)$.

   (a) Show that if γ'_1 is homotopic to γ_1 (resp. γ'_2 homotopic to γ_2) under a homotopy which keeps fixed the endpoints of the paths, then $\gamma'_1 * \gamma_2$ (resp. $\gamma_1 * \gamma'_2$) is homotopic to $\gamma_1 * \gamma_2$ by a homotopy which keeps fixed the endpoints of the paths.
   (b) For each $a \in X$ and each path γ in X, let $a * \gamma$ (resp. $\gamma * a$) denote the path $t \mapsto a * \gamma(t)$ (resp. $t \mapsto \gamma(t) * a$). Let e, a, b be three elements of X such that $e * e = e$, $\gamma_1 \in \Omega_{e,a}$, $\gamma_2 \in \Omega_{e,b}$. Show that $\gamma_1 * \gamma_2$ is homotopic to each of $(\gamma_1 * e)(a * \gamma_2)$ and $(e * \gamma_2)(\gamma_1 * b)$. (Use (16.27.1).)
   (c) Suppose in addition that the two mappings $x \mapsto x * e$ and $x \mapsto e * x$ of X into itself are homotopic to 1_X (16.26.1). Deduce from (b) that the group $\pi_1(X, e)$ is *commutative*. Consider the case in which X is a topological group.

2. With the same hypotheses on X as in Problem 1(c), let (Y, X, p) be an arcwise-connected covering of X, and let $e' \in p^{-1}(e)$.

   (a) Let γ_1, γ_2 be two paths in X with origin e and the same endpoint. Suppose that these paths lift to two paths γ'_1, γ'_2 in Y with origin e' and the same endpoint. Show that, for each path α in X with origin e, the paths $\gamma_1 * \alpha$ and $\gamma_2 * \alpha$ (resp. $\alpha * \gamma_1$ and $\alpha * \gamma_2$) lift to two paths in Y′ with origin e' and the same endpoint. (Argue as in Section 16.27, Problem 4.)
   (b) Suppose in addition that X is locally arcwise-connected. Show that there exists a unique continuous law of composition $(x' y') \mapsto x' * y'$ on Y such that $p(x' * y') = p(x') * p(y')$ and $e' * e' = e'$. In particular, if the law of composition on X makes X a topological group, then the law of composition on Y makes Y a topological group.

If we identify $p^{-1}(e)$ with a quotient of $\pi_1(X, e)$ by means of the action of $\pi_1(X, e)$ on $p^{-1}(e)$, then $p^{-1}(e)$ becomes a subgroup of Y contained in the center of Y. Furthermore, Y is then a Galois covering space of X, whose automorphism group G may be identified with the opposite of the kernel of the homomorphism $p : Y \to X$. If X satisfies the condition of Section **16.29**, Problem 1, we may take Y to be the universal covering of X, and $p^{-1}(e)$ is then isomorphic to $\pi_1(X, e)$.

3. Let I be the interval [0, 1] of **R**, let X be a separable metrizable topological space, and let a be a point of X. For each $n \geqq 0$, let $\mathscr{P}_n(X, a)$ denote the subspace of $\mathscr{C}_X(I^n)$ (**16.27.4.1**) consisting of the continuous mappings of I^n into X which are equal to a on the frontier C_{n-1} of I^n in $\mathbf{R}^n$. (For $n = 0$, we define $I^0 = \{0\}$, and then $\mathscr{P}_0(X, a)$ is canonically identified with X.) We have $\mathscr{P}_1(X, a) = \Omega_{a,a}$ in the notation of (**16.27**).

(a) For $n \geqq 1$, let a_n denote the constant mapping $I^n \to \{a\}$. For each $f \in \mathscr{P}_n(X, a)$, let $\tilde{f} \in \mathscr{P}_1(\mathscr{P}_{n-1}(X, a), a_{n-1})$ be the mapping defined by $\tilde{f}(t)(x_2, \dots, x_n) = f(t, x_2, \dots, x_n)$ for $(t, x_2, \dots, x_n) \in I^n$. Show that $f \mapsto \tilde{f}$ is a homeomorphism of the space $\mathscr{P}_n(X, a)$ onto $\mathscr{P}_1(\mathscr{P}_{n-1}(X, a), a_{n-1})$. Let $\pi_n(X, a)$ denote the fundamental group $\pi_1(\mathscr{P}_{n-1}(X, a), a_{n-1})$, which corresponds canonically to the set of arcwise-connected components of $\mathscr{P}_n(X, a)$. For $n = 1$, the identification of $\mathscr{P}_0(X, a)$ with X identifies $\pi_1(X, a)$ with the fundamental group of X, which justifies the notation. The group $\pi_n(X, a)$ is called the nth *homotopy group* of X at the point a.
(b) Show that the elements of $\pi_n(X, a)$ may be identified with the homotopy classes of continuous mappings of $\mathbf{S}_n$ into X which map $\mathbf{e}_1$ to a (for homotopies leaving the point $\mathbf{e}_1$ fixed).
(c) Show that if $n \geqq 2$, the group $\pi_n(X, a)$ is *commutative*. (Use Problem 1 above and Section **16.27**, Problem 2(c).)
(d) Let $\gamma : I \to X$ be a path in X from a to b, and let $f \in \mathscr{P}_n(X, a)$. Show that there exists a continuous mapping $g : I^n \times I \to X$ such that, if we put $f_t(x_1, \dots, x_n) = g(x_1, \dots, x_n, t)$, then $f_0 = f$ and $f_t \in \mathscr{P}_n(X, \gamma(t))$ for all $t \in I$. (Use Section **16.26**, Problem 3(c).) Moreover, the class of f_1 in $\pi_n(X, b)$ depends only on the class of f in $\pi_n(X, a)$ and the class of γ in $\Omega_{a,b}$ (same method). Deduce that $\pi_n(X, a)$ and $\pi_n(X, b)$ are isomorphic groups. When X is arcwise-connected, we denote by $\pi_n(X)$ any of the groups $\pi_n(X, a)$.

Show that $\pi_n(X) = \{0\}$ if X is contractible. Show also that if X and Y have the same homotopy type (Section **16.26**, Problem 2) and are arcwise-connected, then $\pi_n(X)$ and $\pi_n(Y)$ are isomorphic.

4. With the notation and hypotheses of Problem 3, let A be a closed subset of X containing the point a. Let K_{n-1} be the complement in C_{n-1} of the set of points $(x_1, \dots, x_{n-1}, 0)$ such that $|x_j| < 1$ for $1 \leqq j \leqq n-1$. Let $\mathscr{Q}_n(X, A, a)$ denote the set of continuous mappings $f : I^n \to X$ such that $f(C_{n-1}) \subset A$ and $f(K_{n-1}) = \{a\}$, considered as a subspace of $\mathscr{C}_X(I^n)$.

(a) Suppose that $n \geqq 2$. For each $f \in \mathscr{Q}_n(X, A, a)$, let $\tilde{f} \in \mathscr{P}_1(\mathscr{Q}_{n-1}(X, A, a), a_{n-1})$ be the mapping defined by $\tilde{f}(t)(x_2, \dots, x_n) = f(t, x_2, \dots, x_n)$. Show that $f \mapsto \tilde{f}$ is a homeomorphism of the space $\mathscr{Q}_n(X, A, a)$ onto the space $\mathscr{P}_1(\mathscr{Q}_{n-1}(X, A, a), a_{n-1})$. Let $\pi_n(X, A, a)$ denote the fundamental group $\pi_1(\mathscr{Q}_{n-1}(X, A, a), a_{n-1})$; it is called the nth *homotopy group of* X *modulo* A at the point a. We have $\pi_n(X, \{a\}, a) = \pi_n(X, a)$.

(b) Show that for $n \geqq 3$, the group $\pi_n(X, A, a)$ is isomorphic to

$$\pi_2(\mathscr{Q}_{n-2}(X, A, a), a_{n-2})$$

and hence is *commutative*.
(c) Let $\pi_1(X, A, a)$ denote the set of arcwise-connected components of the space $\mathscr{Q}_1(X, A, a)$ (consisting of the paths in X with both origin and endpoint in A). Define a canonical bijection of $\pi_n(X, A, a)$ onto $\pi_1(\mathscr{P}_{n-1}(X, a), \mathscr{P}_{n-1}(A, a), a_{n-1})$. (Remark that $\mathscr{Q}_n(X, A, a)$ is canonically homeomorphic to $\mathscr{Q}_1(\mathscr{P}_{n-1}(X, a), \mathscr{P}_{n-1}(A, a), a_{n-1}$.)

5. (a) Let X, Y be two separable metrizable spaces, A (resp. B) a closed subset of X (resp. Y), a (resp. b) a point of A (resp. B). For each continuous mapping $f: X \to Y$ such that $f(A) \subset B$ and $f(a) = b$, define canonically a mapping

$$f_* : \pi_n(X, A, a) \to \pi_n(Y, B, b).$$

If $n \geqq 2$, or if $n = 1$ and $A = \{a\}$, $B = \{b\}$, the mapping f_* is a group homomorphism.
(b) Deduce from (a) that if $i: X \to X$ is the identity mapping and $j: A \to X$ the canonical injection, then there exist corresponding canonical mappings

$$i_* : \pi_n(X, a) \to \pi_n(X, A, a) \quad \text{for all} \quad n \geqq 1, \qquad j_* : \pi_n(A, a) \to \pi_n(X, a) \quad \text{for all} \quad n \geqq 0$$

(where $\pi_0(X, a)$ denotes the set of arcwise-connected components of X). Define also a canonical mapping of $\mathscr{Q}_n(X, A, a)$ into $\mathscr{P}_{n-1}(A, a)$ (for $n \geqq 1$) by restricting $f \in \mathscr{Q}_n(X, A, a)$ to I^{n-1} (the set of points $(x_1, \ldots, x_n) \in I^n$ with $x_n = 0$). Show that this mapping induces canonically a mapping $\partial : \pi_n(X, A, a) \to \pi_{n-1}(A, a)$, which is a homomorphism for $n \geqq 2$.
(c) Show that the mappings defined in (b) form an exact sequence

$$\pi_0(X, a) \xleftarrow[j_*]{} \pi_0(A, a) \xleftarrow[\partial]{} \pi_1(X, A, a) \xleftarrow[i_*]{} \pi_1(X, a) \xleftarrow[j_*]{} \pi_1(A, a)$$
$$\longleftarrow \cdots \xleftarrow[i_*]{} \pi_n(X, a) \xleftarrow[j_*]{} \pi_n(A, a) \xleftarrow[\partial]{} \pi_{n+1}(X, A, a) \xleftarrow[i_*]{} \cdots$$

(the *homotopy exact sequence*): For the first three mappings, this means that the image of ∂ in $\pi_0(A, a)$ is the inverse image under j_* of the arcwise-connected component of a in X, and that the image of i_* in $\pi_1(X, A, a)$ is the inverse image under ∂ of the arcwise-connected component of a in A. (Using Problem 4(c), reduce to proving exactness for the first four mappings.)

6. Let B be a separable, metrizable, locally compact space, and (X, B, p) a C^0-fibration whose fibers are differential manifolds.

(a) Let $b \in B$. Since $p(X_b) = \{b\}$, we have a mapping $p_* : \pi_n(X, X_b, a) \to \pi_n(B, b)$ for $a \in X_b$ (Problem 5(a)). Show that p_* is a bijection. (Use the homotopy lifting theorem; cf. Section **16.28**, Problem 10.)
(b) Deduce from (a) and from Problem 5(c) the *homotopy exact sequence of fiber bundles*

$$0 \leftarrow \pi_0(B, b) \leftarrow \pi_0(X, a) \leftarrow \pi_0(X_b, a) \leftarrow \pi_1(B, b) \leftarrow \cdots$$
$$\cdots \leftarrow \pi_n(B, b) \leftarrow \pi_n(X, a) \leftarrow \pi_n(X_b, a) \leftarrow \pi_{n+1}(B, b) \leftarrow \cdots$$

7. Let X be a connected differential manifold, (Y, X, p) a connected covering of X. Show

that for $n \geqq 2$, the groups $\pi_n(X)$ and $\pi_n(Y)$ are isomorphic. (Use the exact homotopy sequence.) In particular, $\pi_n(S_1) = \{0\}$ for $n \geqq 2$.

8. Let G be a connected Lie group, H a connected closed Lie subgroup of G. Show that $\pi_1(G/H)$ is commutative. (Use the exact homotopy sequence.)

9. (a) Let $f: \mathbf{S}_m \to \mathbf{S}_n$ be a continuous mapping, where $m < n$, such that $f(\mathbf{e}_1) = \mathbf{e}_1$. Show that f is homotopic to a C^∞-mapping $g: \mathbf{S}_m \to \mathbf{S}_n$ such that $g(\mathbf{e}_1) = \mathbf{e}_1$, under a homotopy leaving $\mathbf{e}_1$ fixed. (Argue as in (16.27.8).) Deduce that $\pi_m(\mathbf{S}_n) = \{0\}$ for $m < n$.
(b) Identify $\mathbf{S}_{n-1}$ with the intersection of $\mathbf{S}_n$ and the hyperplane $\xi^{n+1} = 0$, and suppose that $n \geqq 2$. Show that every continuous mapping $f: \mathbf{S}_n \to \mathbf{S}_n$ such that $f(\mathbf{e}_1) = \mathbf{e}_1$ is homotopic to a C^∞-mapping $g: \mathbf{S}_n \to \mathbf{S}_n$ such that $g(\mathbf{e}_1) = \mathbf{e}_1$ and such that $g(\mathbf{S}_{n-1})$ contains neither of the points $\pm \mathbf{e}_{n+1}$, under a homotopy leaving $\mathbf{e}_1$ fixed. (Same method.) Deduce that (for $n \geqq 2$) the group $\pi_n(\mathbf{S}_n)$ is isomorphic to $\pi_{n-1}(\mathbf{S}_{n-1})$ and hence to $\mathbf{Z}$. (Observe that if D_+ (resp. D_-) is the hemisphere of $\mathbf{S}_n$ defined by $\xi^{n+1} \geqq 0$ (resp. $\xi^{n+1} \leqq 0$), then $\pi_n(D_+, \mathbf{S}_{n-1}, \mathbf{e}_1)$ (resp. $\pi_n(D_-, \mathbf{S}_{n-1}, \mathbf{e}_1)$) is canonically isomorphic to $\pi_{n-1}(\mathbf{S}_{n-1}, \mathbf{e}_1)$, by means of the homotopy exact sequence.)
(c) If $n \geqq 2$, show that the group $\pi_n(\mathbf{S}_3)$ is canonically isomorphic to $\pi_n(\mathbf{S}_2)$. (Use Problem 7 and the Hopf fibration of $\mathbf{S}_3$ (16.14.10).) In particular, $\pi_3(\mathbf{S}_2)$ is isomorphic to $\mathbf{Z}$.
(d) If $2 \leq p \leq n$ and $k < n - p$, show that the kth homotopy group $\pi_k(\mathbf{S}_{n,p})$ of the Stiefel manifold $\mathbf{S}_{n,p}$ is zero. (Use the fibration of $\mathbf{S}_{n,p}$ defined in (16.14.10), and (a) above.)

10. Show that, for $n \geqq 4$, $\pi_1(\mathbf{SO}(n))$ is isomorphic to $\pi_1(\mathbf{SO}(n-1))$ (and hence isomorphic to $\mathbf{Z}/2\mathbf{Z}$). (Apply the exact homotopy sequence to the principal bundle $\mathbf{SO}(n)$ with structure group $\mathbf{SO}(n-1)$, and use Problem 9(a).)

11. (a) Let G be a simply-connected Lie group, H a connected Lie subgroup of G. Show that the homogeneous space G/H is simply-connected (cf. Section 16.29, Problem 6).
(b) Let G be a connected Lie group, $\tilde{G}$ its universal covering group, $p: \tilde{G} \to G$ the canonical projection. For each connected Lie subgroup H of G, let $\tilde{H} = p^{-1}(H)$ and let $\tilde{H}_0$ be the identity component of $\tilde{H}$. Then there exists a unique C^∞-mapping f of $\tilde{G}/\tilde{H}_0$ onto G/H such that the diagram

$$\begin{array}{ccc} \tilde{G} & \xrightarrow{\;p\;} & G \\ \downarrow & & \downarrow \\ \tilde{G}/\tilde{H}_0 & \xrightarrow[f]{} & G/H \end{array}$$

is commutative, where the vertical arrows are the canonical mappings. Show that $(\tilde{G}/\tilde{H}_0, G/H, f)$ is a universal covering of the manifold G/H, and that $\pi_1(G/H)$ is isomorphic to $\tilde{H}/\tilde{H}_0$.

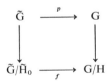

CHAPTER XVII

DIFFERENTIAL CALCULUS ON A DIFFERENTIAL MANIFOLD: I. DISTRIBUTIONS AND DIFFERENTIAL OPERATORS

The previous chapter has provided us with the necessary algebraic and topological foundations for analysis on differential manifolds, which is our goal. The task now is to generalize to the context of differential manifolds the classical notions of differential calculus: derivatives, partial derivatives, differential equations, and partial differential equations. This is neither obvious nor simple, since we no longer have at our disposal the underlying vector space structure which served to define the notion of the derivative of a mapping of an open subset Ω of $\mathbf{R}^n$ into $\mathbf{R}^m$. If we bear in mind the definition of differential manifolds by means of charts, we can of course seek by these means to bring everything back to the classical definitions; but it is essential to verify that in this way we obtain notions which are *intrinsic* to the differential manifold, that is to say, which do not depend on the choice of charts. Now, once we have acquired the notion of the tangent vector space at a point to a manifold (16.5), the only "infinitesimal" notion which intuitively appears to be intrinsic is that of the tangent linear mapping (16.5.3) and, for real-valued functions on a manifold, the notion of the *differential* (16.5.7) which is essentially a particular case of the previous notion. A generalization of the notion of a "partial derivative" appears to be more problematical, because in $\mathbf{R}^n$ this notion is tied to the choice of a particular basis of this vector space. However, if we observe that, for a mapping f of $\Omega \subset \mathbf{R}^n$ into $\mathbf{R}^m$, the partial derivative $D_j f(x)$ $(1 \leqq j \leqq n)$ is just the value $Df(x) \cdot \mathbf{e}_j$ of the total derivative $Df(x)$ at the particular vector $\mathbf{e}_j$, we are naturally led to define the "derivative of a real-valued function f in the direction of a tangent vector $\mathbf{h}_x$ at the point x" to be the value $\theta_{\mathbf{h}_x} \cdot f = \langle df(x), \mathbf{h}_x \rangle$ of the differential $df(x) = d_x f$ at the point x, for the vector $\mathbf{h}_x$.† From here we pass to the generalization of the

† Some authors *define* tangent vectors at a point by identifying them with these operators. We have preferred to give a more "geometrical" definition, which remains close to intuition and distinguishes carefully this intuitive aspect from the operator aspect.

partial derivative, not any more at a point, but as a function of the point, by taking for each point x a tangent vector $\mathbf{h}_x$ depending on x, that is to say a *vector field* X on the manifold (16.15.4); the function $x \mapsto (\theta_X \cdot f)(x) = \theta_{X(x)} \cdot f$ is then the generalization of the notion of a first-order partial derivative.

The difficulties appear to be much more serious when it comes to the generalization of higher-order partial derivatives, where our "geometrical" intuition leads us badly astray. We arrive in fact at the right definition by an unexpected detour: a linear combination $f \mapsto \sum_{|\alpha| \leqq m} c_\alpha \mathrm{D}^\alpha f(x)$ of partial derivatives of order $\leqq m$ at a point $x \in \Omega \subset \mathbf{R}^n$ can be characterized as a *linear form* on the vector space of m times continuously differentiable functions on Ω, having the following two properties: (i) it takes the same value for two functions which are equal in some neighborhood of x (in other words, it is a "local" operator); (ii) it is *continuous* for a topology in which two functions are "close" to each other if all their partial derivatives of order $\leqq m$ are close to each other in the sense of the topology of uniform convergence on compact sets (defined in (12.14.6)). It is clear that we cannot *define* partial derivatives (for a function on $\Omega \subset \mathbf{R}^n$) in this way without running a vicious circle; but it is perfectly possible to transport this definition to a differential manifold M by means of a chart: for it can be verified that, although the partial derivatives of the local expression (16.3) of a function defined in a neighborhood of a point $x \in$ M depend on the choice of chart, nevertheless the notion of an "m times continuously differentiable function" and the topology envisaged above *do not depend on the particular chart chosen* (17.2).

The notion at which we arrive in this way is that of a *point-distribution*, which is a particular case of the notions of *distribution* and *current* on a differential manifold. These notions, which embrace simultaneously the two basic concepts of infinitesimal calculus—derivative at a point, and integral—have become fundamental in contemporary analysis, and we therefore begin this chapter with an elementary account of them (Sections 17.3–17.12). In this account we have emphasized the "operator" aspect of distributions, much more than the "generalized function" aspect which some authors put in the foreground. The latter aspect in fact acquires a meaning only when one has available a privileged measure (17.5.3) on the manifold under consideration, and to give it the leading place obscures the fundamental role played by distributions in the theory of linear functional equations (Chapter XXIII). For distributions on a Lie group, the group structure is reflected by the operation of *convolution* of distributions (17.11), which generalizes convolution of measures (14.5). The fundamental importance of this operation will appear clearly only when we come to Fourier analysis (Chapter XXII), but already in this chapter its utilization for the regularization of distributions renders it very valuable. Furthermore, convolution of point-distributions lies at the base of the infinitesimal study of Lie groups (Chapter XIX).

Just as the notion of a vector field generalizes that of first-order partial differentiation, so the notion which generalizes partial differentiation of arbitrary order is that of a *field of point-distributions* (which transforms real-valued functions into real-valued functions). However, this notion is not general enough for all applications, because it is necessary to be able to "differentiate" not only real-valued functions but also vector-valued functions, and more generally (in conformity with the very nature of differential manifolds) functions which, at each point x of a manifold M. take their values in a vector space *depending on* x—in other words, sections of a vector bundle E over M. Guided by the notion of a point-distribution, we arrive thus at the general notion of a *differential operator on a vector bundle* E, which transforms (differentiable) sections of E into sections of another vector bundle F (which may or may not be the same as E) over M (17.13). It is this notion which will enable us to define intrinsically the concept of a *linear partial differential equation* on a manifold (Chapter XXIII). The rest of the chapter is devoted to the study of certain first-order differential operators which are essential for the following chapters.

The fundamental property of symmetry of the (total) second derivative (8.12.2) is reflected, in analysis on manifolds, by the existence of two first-order differential operators which depend only on the structure of the differential manifold: *exterior differentiation of differential forms* (17.15), which transforms a p-form into a $(p+1)$-form; and the *Lie derivative* $\mathbf{Z} \mapsto \theta_X \cdot \mathbf{Z}$ which, for a given vector field X, transforms a tensor field $\mathbf{Z}$ of type (r, s) into a tensor field of the same type (17.14). It should be emphasized that the value of the Lie derivative $\theta_X \cdot \mathbf{Z}$ at a point x depends not only on the value $X(x)$ of the vector field X at the point x, but also on its values in a neighborhood of x. It is not possible to define intrinsically, for a tensor field $\mathbf{Z}$ which is not a scalar function, the notion of "derivative at a point x in the direction of a tangent vector $\mathbf{h}_x$," at any rate in terms of the manifold structure alone. The reason for this is that the manifold structure provides no *intrinsic* way of comparing tangent vectors at two distinct points, strange though this may appear to our "intuition." In order to be able to make such comparisons, it is necessary to endow the manifold with an additional structure defined by what is called a *linear connection*.† This notion is introduced in Section 17.16 for an arbitrary vector bundle; its relation with an analogous notion for principal bundles will be brought out in Chapter XX, where we shall also study in detail the most important type of linear connection, the *Levi–Civita connection* on a Riemannian manifold. In this chapter we shall only show how the

† It appears intuitively obvious that we should be able to compare tangent vectors at different points, precisely because we think always of a manifold as embedded in some $\mathbf{R}^n$, which carries a *canonical* linear connection so "natural" that we take it for granted.

presence of a linear connection allows us to define the notions of *covariant derivative of a tensor field at a point x in the direction of a tangent vector* $\mathbf{h}_x$ (which generalizes the notion of the derivative of a scalar function at x in the direction of $\mathbf{h}_x$) (17.17), and of *covariant exterior differential of a differential form with values in a vector bundle* (17.19), which generalizes the notion of the exterior differential of a scalar-valued differential form, and leads (17.20) to the notions of curvature and torsion of a linear connection.

1. THE SPACES $\mathscr{E}^{(r)}(\mathrm{U})$ (U OPEN IN $\mathbf{R}^n$)

Recall (9.1) that, for each "multi-index" $\nu = (\nu_1, \nu_2, \ldots, \nu_n) \in \mathbf{N}^n$ we put $|\nu| = \sum_{j=1}^{n} \nu_j$ (the *total degree* of ν) and $\nu! = \nu_1!\,\nu_2! \cdots \nu_n!$. If $\nu = (\nu_j)$ and $\nu' = (\nu'_j)$ are multi-indices, we put $\nu + \nu' = (\nu_j + \nu'_j)$, and the relation $\nu \leqq \nu'$ means that $\nu_j \leqq \nu'_j$ for all j; in which case we define $\nu' - \nu$ to be the multi-index $(\nu'_j - \nu_j)$. For each vector $\mathbf{x} = (x_j) \in \mathbf{C}^n$, put $\mathbf{x}^\nu = x_1^{\nu_1} x_2^{\nu_2} \cdots x_n^{\nu_n} \in \mathbf{C}$. Finally let D^ν denote the partial differentiation operator $\mathrm{D}_1^{\nu_1}\mathrm{D}_2^{\nu_2} \cdots \mathrm{D}_n^{\nu_n}$ (D^0 being the identity mapping).

Let U be an open subset of $\mathbf{R}^n$. For each integer $r > 0$, we denote by $\mathscr{E}_{\mathbf{C}}^{(r)}(\mathrm{U})$ or $\mathscr{E}^{(r)}(\mathrm{U})$ the complex vector space of all C^r-mappings of U into $\mathbf{C}$. The intersection $\mathscr{E}_{\mathbf{C}}(\mathrm{U})$ or $\mathscr{E}(\mathrm{U})$ of the decreasing sequence of spaces $\mathscr{E}^{(r)}(\mathrm{U})$ is therefore the space of all C^∞-mappings of U into $\mathbf{C}$.

We shall show that $\mathscr{E}(\mathrm{U})$ (resp. $\mathscr{E}^{(r)}(\mathrm{U})$) can be endowed with the structure of a Hausdorff locally convex topological vector space, defined by a sequence of seminorms (hence *metrizable* (12.14.5)), and having the following property:

(*) *A sequence* (f_k) *of functions belonging to* $\mathscr{E}(\mathrm{U})$ (resp. $\mathscr{E}^{(r)}(\mathrm{U})$) *converges to* 0 *if and only if, for each compact subset* K *of* U *and each multi-index* ν (resp. *each multi-index* ν *such that* $|\nu| \leqq r$), *the sequence of restrictions of the* $\mathrm{D}^\nu f_k$ *to* K *converges uniformly to* 0.

It follows immediately from (3.13.14), applied to the identity mapping, that if such a topology exists it is *unique*. As to its existence, an increasing sequence (K_m) of compact subsets of U will be said to be *fundamental* if U is the union of the K_m, and each K_m is contained in the interior of K_{m+1}. Such sequences exist (3.18.3). For such a sequence (K_m), for each pair of integers $s \geqq 0$ and $m > 0$ and each function $f \in \mathscr{E}^{(r)}(\mathrm{U})$, where $r \geqq s$, put

$$p_{s,m}(f) = \sup_{x \in \mathrm{K}_m,\, |\nu| \leqq s} |\mathrm{D}^\nu f(x)|. \tag{17.1.1}$$

It is clear that for $s \leqq r$ each of the $p_{s,m}$ is a *seminorm* (12.14) on $\mathscr{E}^{(r)}(\mathrm{U})$ and that $p_{s,m} \leqq p_{r,m}$. The topology on $\mathscr{E}(\mathrm{U})$ (resp. $\mathscr{E}^{(r)}(\mathrm{U})$) defined by the seminorms $p_{s,m}$ for $m > 0$ and $s \geqq 0$ (resp. $0 \leqq s \leqq r$) is Hausdorff, for if $p_{0,m}(f) = 0$ for all m, then f vanishes on each K_m and hence on U. This topology satisfies the condition (∗) (because each compact $\mathrm{K} \subset \mathrm{U}$ is contained in some K_m, by virtue of the Borel–Lebesgue axiom), and hence it is independent of the fundamental sequence (K_m) chosen. We remark also that the topology on $\mathscr{E}^{(r)}(\mathrm{U})$ is defined already by the seminorms $p_{r,m}$ $(m > 0)$.

A subset H of $\mathscr{E}(\mathrm{U})$ will be said to be *bounded* if each of the seminorms $p_{s,m}$ is bounded in H. This property depends only on the topology of $\mathscr{E}(\mathrm{U})$ (12.14.12).

(17.1.2) (i) *The spaces $\mathscr{E}^{(r)}(\mathrm{U})$ and $\mathscr{E}(\mathrm{U})$ are separable Fréchet spaces. More precisely, there exists a sequence of functions in $\mathscr{E}(\mathrm{U})$, with compact supports contained in* U, *which is dense in each of the spaces $\mathscr{E}^{(r)}(\mathrm{U})$ and $\mathscr{E}(\mathrm{U})$.*

(ii) *Every bounded subset in $\mathscr{E}(\mathrm{U})$ is relatively compact in $\mathscr{E}(\mathrm{U})$.*

(i) Let (f_p) be a Cauchy sequence in $\mathscr{E}(\mathrm{U})$ (resp. $\mathscr{E}^{(r)}(\mathrm{U})$). It follows from the definition of the seminorms $p_{s,m}$ that for each multi-index ν (resp. each multi-index ν such that $|\nu| \leqq r$) there exists a continuous mapping $f^{(\nu)} : \mathrm{U} \to \mathbf{C}$ such that the sequence $(\mathrm{D}^\nu f_p)$ converges uniformly to $f^{(\nu)}$ on each K_m (7.2.1). Putting $f = f^{(0)}$, it follows from (8.6.3) that f is indefinitely differentiable (resp. of class C^r) and that $f^{(\nu)} = \mathrm{D}^\nu f$ for all ν (resp. for $|\nu| \leqq r$). Consequently f is the limit of the sequence (f_p) in $\mathscr{E}(\mathrm{U})$ (resp. $\mathscr{E}^{(r)}(\mathrm{U})$), and therefore $\mathscr{E}(\mathrm{U})$ (resp. $\mathscr{E}^{(r)}(\mathrm{U})$) is complete.

To prove the second assertion we remark that by virtue of (12.14.6.2) there exists a sequence (u_q) of continuous functions with compact support which form a dense set in $\mathscr{E}^{(0)}(\mathrm{U}) = \mathscr{C}_{\mathbf{C}}(\mathrm{U})$. Let λ be Lebesgue measure on $\mathbf{R}^n$, and consider the function g of class C^∞ defined in (16.4.1.4), with support I^n. For each $k > 0$, put $g_k(x) = k^n g(kx)$, so that $\int g_k(x)\, d\lambda(x) = 1$. The sequence (g_k) is said to be a *regularizing sequence.*

Since the support of g_k is $k^{-1}\mathrm{I}^n$, for fixed q the support of each of the functions $v_{kq} = g_k * u_q$ is compact and contained in U for all sufficiently large k ((3.18.2) and (14.5.4)). Let us show that the functions v_{kq} with support contained in U form a dense set in $\mathscr{E}(\mathrm{U})$ (resp. in each $\mathscr{E}^{(r)}(\mathrm{U})$). Since $v_{kq}(x) = \int g_k(x-y)u_q(y)\, d\lambda(y)$, it follows from (13.8.6) that v_{kq} is of class C^∞ and that, for each multi-index ν,

$$\mathrm{D}^\nu v_{kq}(x) = \int \mathrm{D}^\nu g_k(x-y)u_q(y)\, d\lambda(y). \tag{17.1.2.1}$$

Fix a real number $\varepsilon > 0$ and two integers m and r. We shall show that for each $f \in \mathscr{E}(U)$ it is possible to find a v_{kq} such that $|D^\nu(f - v_{kq})(x)| \leqq \varepsilon$ for all $x \in K_m$ and all ν such that $|\nu| \leqq r$; this will prove the assertion above. Let F be a C^∞-function which is equal to f on K_{m+1} and has support contained in K_{m+2} (16.4.3). By (13.8.6) the function $f_k = F * g_k$ is of class C^∞, and we have

(17.1.2.2)
$$D^\nu f_k(x) = \int D^\nu g_k(x - y) F(y)\, d\lambda(y) = \int D^\nu F(x - y) g_k(y)\, d\lambda(y).$$

It follows therefore from (14.11.1) that for sufficiently large k we have $|D^\nu(f_k - f)(x)| \leqq \frac{1}{2}\varepsilon$ for all $x \in K_m$ and $|\nu| \leqq r$. On the other hand, if k is sufficiently large, then for all $x \in K_m$ we have, by (3.18.2) and (14.5.4)

$$|D^\nu(f_k - v_{kq})(x)| = \left| \int_{K_{m+1}} D^\nu g_k(x - y)(F(y) - u_q(y))\, d\lambda(y) \right| \leqq N_1(D^\nu g_k) \sup_{y \in K_{m+1}} |F(y) - u_q(y)|.$$

Now, for a fixed and suitably large k, we can by hypothesis find a large q so that the right-hand side of this inequality is $\leqq \frac{1}{2}\varepsilon$ for $|\nu| \leqq r$. This completes the proof of (i) for $\mathscr{E}(U)$; the proof for $\mathscr{E}^{(r)}(U)$ is similar.

(ii) If a sequence (f_k) is bounded in $\mathscr{E}(U)$, then each of the sequences $(D_i f_k)$ $(1 \leqq i \leqq n)$ is uniformly bounded in each K_m, hence it follows from (7.5.1) and the mean value theorem that (f_k) is equicontinuous. By definition of the bounded sequences in $\mathscr{E}(U)$, it follows that for each multi-index ν the sequence $(D^\nu f_k)$ is equicontinuous. By applying Ascoli's theorem (7.5.7) and the diagonal procedure (cf. 12.5.9) we can therefore find a subsequence (f_{k_q}) such that *each* of the sequences $(D^\nu f_{k_q})$ converges uniformly on *each* of the K_m. In other words, the sequence (f_{k_q}) converges in $\mathscr{E}(U)$, and this proves (ii).

(17.1.3) *For each multi-index ν, the linear mapping $f \mapsto D^\nu f$ of $\mathscr{E}(U)$ into $\mathscr{E}(U)$ is continuous.*

For $p_{s,m}(D^\nu f) \leqq p_{s+|\nu|,m}(f)$, from which the assertion follows (12.14.11).

(17.1.4) *For each function $g \in \mathscr{E}(U)$ (resp. $g \in \mathscr{E}^{(r)}(U)$), the linear mapping $f \mapsto fg$ of $\mathscr{E}(U)$ (resp. $\mathscr{E}^{(r)}(U)$) into itself is continuous.*

For each pair of integers s, m (resp. $s \leqq r$ and m), let $a_{s,m}$ be the greatest of the least upper bounds of the $|D^\nu g|$ on K_m for $|\nu| \leqq s$. By Leibniz' formula

(8.13.2), there exists a number $c_{s,m}$ independent of f and g such that $p_{s,m}(fg) \leq c_{s,m} a_{s,m} p_{s,m}(f)$. Hence the result (12.14.11).

(17.1.5) *Let φ be a mapping of class* C^∞ (resp. C^r) *of an open set* $V \subset \mathbf{R}^n$ *into* U. *Then the linear mapping* $f \mapsto f \circ \varphi$ *of* $\mathscr{E}(U)$ *into* $\mathscr{E}(V)$ (resp. *of* $\mathscr{E}^{(r)}(U)$ *into* $\mathscr{E}^{(r)}(V)$) *is continuous.*

Put $\varphi = (\varphi_1, \varphi_2, \ldots, \varphi_n)$, where the φ_j are scalar-valued functions. Let (K'_m) be a fundamental sequence of compact subsets of V used to define seminorms $p'_{s,m}$ on $\mathscr{E}(V)$ as in (17.1.1). For each pair of integers s, m, let $a'_{s,m}$ be the greatest of the least upper bounds of the functions $\sup(1, |D^\nu \varphi_j|)$ on K'_m for $|\nu| \leq s$ and $1 \leq j \leq n$. Finally let q be an integer such that $\varphi(K'_m) \subset K_q$. By repeated application of the formula for the partial derivatives of a composite function (8.10.1) we obtain, for each $f \in \mathscr{E}(U)$

$$p_{s,m}(f \circ \varphi) \leq c'_{s,m} a'^{s}_{s,m} p_{s,q}(f),$$

where $c'_{s,m}$ is a constant independent of f and φ. This proves the proposition.

2. SPACES OF C^∞- (resp. C^r-) SECTIONS OF VECTOR BUNDLES

Let X be a pure differential manifold of dimension n and let (E, X, π) be a complex vector bundle of rank N (16.15) over X. If U is an open subset of X, we recall (16.15) that the complex vector space of C^∞-sections of E over U is denoted by $\Gamma(U, E)$. Likewise we shall denote by $\Gamma^{(r)}(U, E)$ the vector space of C^r-sections of E over U. We propose to generalise the results of (17.1) to these spaces: $\mathscr{E}(U)$ is the space $\Gamma(U, E)$ when $X = \mathbf{R}^n$ and $E = X \times \mathbf{C}$ is the trivial complex line bundle over X. Since in general we cannot attach a meaning to the notion of partial derivatives of a section of E, we shall reformulate the problem as follows: we have to show that the space $\Gamma(U, E)$ (resp. $\Gamma^{(r)}(U, E)$) can be endowed with the structure of a Hausdorff locally convex topological space, defined by a sequence of seminorms and having the following property:

(**) *A sequence* $(\mathbf{u}_k)$ *of sections of* $\Gamma(U, E)$ (resp. $\Gamma^{(r)}(U, E)$) *converges to* 0 *if and only if, for each chart* (V, φ, n) *of* X *over which* E *is trivializable, each diffeomorphism*

$$z \mapsto (\varphi(\pi(z)), v_1(z), \ldots, v_N(z))$$

of $\pi^{-1}(V)$ *onto* $\varphi(V) \times \mathbf{C}^N$, *where the* v_j *are linear on each fiber* $\pi^{-1}(x)$, *each compact* $K \subset \varphi(V)$ *and each multi-index* ν (resp. *such that* $|\nu| \leq r$), *the sequence*

$((\mathrm{D}^\nu w_{jk})|\mathrm{K})_{k\geqq 1}$ converges uniformly to 0 for $1 \leqq j \leqq \mathrm{N}$, where $w_{jk}(t) = v_j(\mathbf{u}_k(\varphi^{-1}(t)))$ for $t \in \varphi(\mathrm{V})$.

Here again, by virtue of (3.13.14), the topology is necessarily unique. To establish existence, consider an at most denumerable family of charts $(\mathrm{V}_\alpha, \varphi_\alpha, n)$ of U such that the V_α form a locally finite open covering of U, and such that E is trivializable over each V_α (12.6.1). For each α, let

$$z \mapsto (\varphi_\alpha(\pi(z)), v_{1\alpha}(z), \ldots, v_{\mathrm{N}\alpha}(z))$$

be a diffeomorphism of $\pi^{-1}(\mathrm{V}_\alpha)$ onto $\varphi(\mathrm{V}_\alpha) \times \mathbf{C}^\mathrm{N}$, the $v_{j\alpha}$ being linear on each fiber $\pi^{-1}(x)$. Let $(\mathrm{K}_{m\alpha})_{m\geqq 1}$ be a fundamental sequence of compact subsets of $\varphi_\alpha(\mathrm{V}_\alpha)$, and let $p'_{s,m,\alpha}$ be the corresponding seminorms (17.1.1) on $\mathscr{E}(\varphi_\alpha(\mathrm{V}_\alpha))$ (resp. $\mathscr{E}^{(r)}(\varphi_\alpha(\mathrm{V}_\alpha))$). For each section $\mathbf{u} \in \Gamma(\mathrm{U}, \mathrm{E})$ (resp. $\mathbf{u} \in \Gamma^{(r)}(\mathrm{U}, \mathrm{E})$) we define

$$p_{s,m,\alpha}(\mathbf{u}) = \sum_{j=1}^{\mathrm{N}} p'_{s,m,\alpha}(v_{j\alpha} \circ \mathbf{u}_\alpha \circ \varphi_\alpha^{-1}), \tag{17.2.1}$$

where $\mathbf{u}_\alpha$ is the restriction of $\mathbf{u}$ to V_α. It is clear that the $p_{s,m,\alpha}$ are seminorms, and that if $p_{0,m,\alpha}(\mathbf{u}) = 0$ for all m and all α, then $\mathbf{u}(x) = 0$ in each V_α and therefore $\mathbf{u} = 0$. To show that the topology defined by these seminorms satisfies the condition (**), we have to show that if $(\mathbf{u}_k)$ tends to 0 in the topology defined by the seminorms $p_{s,m,\alpha}$, then the $\mathrm{D}^\nu w_{jk}$ converge uniformly to 0 on K (in the notation of (**)). The compact set $\varphi^{-1}(\mathrm{K})$ meets only a finite number of the open sets V_α; by applying the Borel–Lebesgue axiom to the union of these V_α, it follows that there exists an integer m and indices α_h $(1 \leqq h \leqq q)$ such that the $\varphi_{\alpha_h}^{-1}(\mathrm{K}_{m\alpha_h})$ cover $\varphi^{-1}(\mathrm{K})$. If w_{hjk} is the restriction of w_{jk} to $\varphi(\mathrm{V} \cap \mathrm{V}_{\alpha_h})$, then clearly it is enough to show that the restrictions of the $\mathrm{D}^\nu w_{hjk}$ to $\mathrm{K} \cap \varphi(\varphi_{\alpha_h}^{-1}(\mathrm{K}_{m\alpha_h}))$ converge uniformly to 0. Now, for each $z \in \pi^{-1}(\mathrm{V} \cap \mathrm{V}_{\alpha_h})$, we can write

$$v_j(z) = \sum_{l=1}^{\mathrm{N}} c_{ljh}(\pi(z)) v_{l\alpha_h}(z),$$

where the c_{ljh} are C^∞-functions on $\mathrm{V} \cap \mathrm{V}_{\alpha_h}$. Let

$$\psi_h : \varphi(\mathrm{V} \cap \mathrm{V}_{\alpha_h}) \to \varphi_{\alpha_h}(\mathrm{V} \cap \mathrm{V}_{\alpha_h})$$

be the transition diffeomorphism. If we put

$$u_{hlk} = v_{l\alpha_h} \circ (\mathbf{u}_k | \mathrm{V}_{\alpha_h}) \circ \varphi_{\alpha_h}^{-1},$$

then we have

$$w_{hjk}(t) = \sum_{l=1}^{\mathrm{N}} c_{ljh}(\varphi^{-1}(t)) u_{hlk}(\psi_h(t)),$$

and the assertion is now a consequence of (17.1.4) and (17.1.5), because each sequence $(u_{hlk})_{k\geqq 0}$ converges to 0 in the space $\mathscr{E}(\varphi_{\alpha_h}(\mathrm{V}_{\alpha_h}))$ (resp. $\mathscr{E}^{(r)}(\varphi_{\alpha_h}(\mathrm{V}_{\alpha_h}))$).

(17.2.1.1) Equivalently, we may say that a sequence $(\mathbf{u}_k)$ converges to 0 in $\Gamma(\mathrm{U}, \mathrm{E})$ if and only if, for each α, the sequence of restrictions $\mathbf{u}_k | \mathrm{V}_\alpha$ converges to 0 in $\Gamma(\mathrm{V}_\alpha, \mathrm{E})$; and $\Gamma(\mathrm{V}_\alpha, \mathrm{E})$ is isomorphic to $(\mathscr{E}(\varphi_\alpha(\mathrm{V}_\alpha)))^{\mathrm{N}}$, so that we are brought back to the spaces $\mathscr{E}(\mathrm{U})$.

(17.2.2) *The spaces* $\Gamma(\mathrm{U}, \mathrm{E})$ *and* $\Gamma^{(r)}(\mathrm{U}, \mathrm{E})$ *are separable Fréchet spaces. More precisely, there exists a sequence of* C^∞*-sections of* E *over* U, *with compact support contained in* U, *which is dense in each of the spaces* $\Gamma^{(r)}(\mathrm{U}, \mathrm{E})$ *and* $\Gamma(\mathrm{U}, \mathrm{E})$. *Every bounded subset of* $\Gamma(\mathrm{U}, \mathrm{E})$ *is relatively compact in* $\Gamma(\mathrm{U}, \mathrm{E})$.

Let $(\mathbf{u}_k)$ be a Cauchy sequence in $\Gamma(\mathrm{U}, \mathrm{E})$. By virtue of (17.1.2), there exists for each α a section $\mathbf{f}_\alpha \in \Gamma(\mathrm{V}_\alpha, \mathrm{E})$ which is the limit of the sequence $(\mathbf{u}_k | \mathrm{V}_\alpha)$, and for any two indices α, β the restrictions of $\mathbf{f}_\alpha$ and $\mathbf{f}_\beta$ to $\mathrm{V}_\alpha \cap \mathrm{V}_\beta$ coincide, so that the $\mathbf{f}_\alpha$ are restrictions of a section $\mathbf{f} \in \Gamma(\mathrm{U}, \mathrm{E})$. By the definition of the topology of $\Gamma(\mathrm{U}, \mathrm{E})$, the sequence $(\mathbf{u}_k)$ converges to $\mathbf{f}$. In the same way we may show that $\Gamma^{(r)}(\mathrm{U}, \mathrm{E})$ is complete.

To prove the second assertion, we remark that it follows from (17.1.2) that for each α there exists in $\Gamma(\mathrm{V}_\alpha, \mathrm{E})$ a sequence $(\mathbf{u}_{k\alpha})$ of sections with compact support contained in V_α, which is dense in $\Gamma(\mathrm{V}_\alpha, \mathrm{E})$ and in each $\Gamma^{(r)}(\mathrm{V}_\alpha, \mathrm{E})$. If (g_α) is a C^∞-partition of unity subordinate to the covering (V_α), then the sections which are linear combinations of the $g_\alpha \mathbf{u}_{k\alpha}$ are dense in $\Gamma(\mathrm{U}, \mathrm{E})$ and in each $\Gamma^{(r)}(\mathrm{U}, \mathrm{E})$. For each compact $\mathrm{K} \subset \mathrm{U}$ meets only a finite number of sets $\mathrm{Supp}(g_\alpha)$, say those with indices α_h $(1 \leqq h \leqq q)$; for each section $\mathbf{u} \in \Gamma(\mathrm{U}, \mathrm{E})$, the restriction of $\mathbf{u}$ to K therefore coincides with the restriction to K of $\sum_{h=1}^{q} g_{\alpha_h} \mathbf{u}$, and by virtue of (17.1.2) and (17.1.4), each section $g_{\alpha_h} \mathbf{u}$ may be approximated arbitrarily closely by the $g_{\alpha_h} \mathbf{u}_{k\alpha_h}$; whence the assertion follows. Finally, if $(\mathbf{u}_k)$ is a bounded sequence in $\Gamma(\mathrm{U}, \mathrm{E})$, then it follows from (17.1.2) that for each α there exists a convergent subsequence of the sequence $(\mathbf{u}_k | \mathrm{V}_\alpha)$ in $\Gamma(\mathrm{V}_\alpha, \mathrm{E})$. By applying the diagonal procedure, we obtain a convergent subsequence of the sequence $(\mathbf{u}_k)$.

Remark

(17.2.2.1) The last part of the proof may be used, more generally, to show that a vector subspace H of $\Gamma(\mathrm{U}, \mathrm{E})$ is dense in $\Gamma(\mathrm{U}, \mathrm{E})$ provided that, for each $\mathbf{v} \in \mathrm{H}$, the sections $g_\alpha \mathbf{v}$ also belong to H; it is enough to verify that, for each α, the restrictions to V_α of the sections belonging to H form a dense set in $\Gamma(\mathrm{V}_\alpha, \mathrm{E})$.

(17.2.3) If F is another complex vector bundle over X, $\mathbf{u}_0$ an element of $\Gamma(\mathrm{U}, \mathrm{E})$ and $\mathbf{v}_0$ an element of $\Gamma(\mathrm{U}, \mathrm{F})$, then the linear mappings $\mathbf{u} \mapsto \mathbf{u} \otimes \mathbf{v}_0$

and $\mathbf{v} \mapsto \mathbf{u}_0 \otimes \mathbf{v}$ of $\Gamma(U, E)$ and $\Gamma(U, F)$, respectively, into $\Gamma(U, E \otimes F)$ are continuous. Likewise, if $\mathbf{w}_0^* \in \Gamma\left(U, \overset{p}{\bigwedge} E^*\right)$, then the linear mappings $\mathbf{u} \mapsto i(\mathbf{u})\mathbf{w}_0^*$ and $\mathbf{w}^* \mapsto i(\mathbf{u}_0)\mathbf{w}^*$ of $\Gamma(U, E)$ and $\Gamma\left(U, \overset{p}{\bigwedge} E^*\right)$, respectively, into $\Gamma\left(U, \overset{p-1}{\bigwedge} E^*\right)$ are continuous. The same is true of the mappings $\mathbf{s} \mapsto \mathbf{s} \wedge \mathbf{t}_0$ and $\mathbf{t} \mapsto \mathbf{s}_0 \wedge \mathbf{t}$ of $\Gamma\left(U, \overset{p}{\bigwedge} E\right)$ and $\Gamma\left(U, \overset{q}{\bigwedge} E\right)$ into $\Gamma\left(U, \overset{p+q}{\bigwedge} E\right)$.

All these propositions follow from (17.1.4) by virtue of the definitions of (16.18), by reducing to the case where E is *trivial*, as we may assume by use of the general principle stated in (17.2.1.1).

PROBLEMS

1. Let M, N be two differential manifolds. For each integer $r \geq 1$, let $\mathscr{E}^{(r)}(M; N)$ denote the set of all C^r-mappings of M into N.

   (a) Let $f \in \mathscr{E}^{(r)}(M; N)$. Then there exists a denumerable locally finite covering of M by compact sets K_α, such that each K_α (resp. each $f(K_\alpha)$) is contained in the domain of definition of a chart $(U_\alpha, \varphi_\alpha, m_\alpha)$ of M (resp. a chart $(V_\alpha, \psi_\alpha, n_\alpha)$ of N). For each $\varepsilon > 0$ and each α, let $W(f, K_\alpha, \varphi_\alpha, \psi_\alpha, \varepsilon)$ denote the set of all C^r-mappings $g : M \to N$ such that $g(K_\alpha) \subset V_\alpha$ and $\|D^\nu(\psi_\alpha \circ (f|K_\alpha) \circ \varphi_\alpha^{-1})(x) - D^\nu(\psi_\alpha \circ (g|K_\alpha) \circ \varphi_\alpha^{-1})(x)\| \leq \varepsilon$ for all $x \in \varphi_\alpha(K_\alpha)$ and $|\nu| \leq r$. Show that there exists on $\mathscr{E}^{(r)}(M; N)$ a unique topology $\mathscr{T}_r'$ (called the *coarse* C^r-topology) such that, for each $f \in \mathscr{E}^{(r)}(M; N)$, the finite intersections of the sets $W(f, K_\alpha, \varphi_\alpha, \psi_\alpha, 1/n)$ (where the $K_\alpha, \varphi_\alpha, \psi_\alpha$ are fixed and n is variable) form a fundamental system of neighborhoods of f (cf. Section 12.3, Problem 3). Show that the topology so defined is independent of the choice (for each $f \in \mathscr{E}^{(r)}(M; N)$) of the families (K_α), (φ_α), (ψ_α) satisfying the stated conditions.

   (b) If N is a vector bundle over M, show that $\mathscr{T}_r'$ induces on $\Gamma^{(r)}(M; N)$ the topology defined in (17.2).

   (c) If N is a submanifold of a differential manifold P, then $\mathscr{E}^{(r)}(M; N)$ is a subset of $\mathscr{E}^{(r)}(M; P)$. Show that the topology induced on $\mathscr{E}^{(r)}(M; N)$ by the coarse C^r-topology of $\mathscr{E}^{(r)}(M; P)$ is the coarse C^r-topology.

   (d) Let M, N, P be three differential manifolds. Show that the mapping $(f, g) \mapsto g \circ f$ of $\mathscr{E}^{(r)}(M; N) \times \mathscr{E}^{(r)}(N; P)$ into $\mathscr{E}^{(r)}(M; P)$ is continuous with respect to the coarse C^r-topologies on these three spaces.

2. (a) With the notation of Problem 1, for each family (δ_α) of numbers $\delta_\alpha > 0$, let $W(f, (\delta_\alpha))$ denote the intersection of the sets $W(f, K_\alpha, \varphi_\alpha, \psi_\alpha, \delta_\alpha)$ for all α. Show that there exists on $\mathscr{E}^{(r)}(M; N)$ a unique topology $\mathscr{T}_r''$ (called the *fine* C^r-topology) such that, for each $f \in \mathscr{E}^{(r)}(M; N)$, the sets $W(f, (\delta_\alpha))$ (where the $K_\alpha, \varphi_\alpha, \psi_\alpha$ are fixed and the δ_α variable) form a fundamental system of neighborhoods of f. Show that the topology so defined is independent of the choice (for each $f \in \mathscr{E}^{(r)}(M; N)$) of the families (K_α), (φ_α), (ψ_α) satisfying the stated conditions.

(b) If N is a submanifold of a differential manifold P, then $\mathscr{E}^{(r)}(M; N)$ is a subset of $\mathscr{E}^{(r)}(M; P)$. Show that the topology induced on $\mathscr{E}^{(r)}(M; N)$ by the fine C^r-topology on $\mathscr{E}^{(r)}(M; P)$ is the fine C^r-topology.
(c) Give an example of an element $f \in \mathscr{E}^{(r)}(M; N)$ for which there does not exist a denumerable fundamental system of neighborhoods of f for the fine C^1-topology (take $M = N = \mathbf{R}$).
(d) Suppose that $f_0 \in \mathscr{E}^{(r)}(M; N)$ is a homeomorphism of M onto $f_0(M)$ and that f_0 is of rank equal to $\dim_x(M)$ at each point $x \in M$. If P is another differential manifold, show that for each $g_0 \in \mathscr{E}^{(r)}(N; P)$ the mapping $(f, g) \mapsto g \circ f$ of $\mathscr{E}^{(r)}(M; N) \times \mathscr{E}^{(r)}(N; P)$ into $\mathscr{E}^{(r)}(M; P)$ is continuous at the point (f_0, g_0) with respect to the fine C^r-topologies.
(e) With the notation of (d), give an example in which f_0 is of rank equal to $\dim_x(M)$ at each point x of M and is injective, but is not a homeomorphism of M onto $f_0(M)$, and the mapping $g \mapsto g \circ f_0$ fails to be continuous at some point g_0 with respect to the fine C^r-topologies (cf. (16.8.5)).
(f) Suppose that $f_0 \in \mathscr{E}^{(r)}(M; N)$ is of rank equal to $\dim_x(M)$ at each point $x \in M$. Show that there exists a neighborhood V_0 of f_0 in $\mathscr{E}^{(r)}(M; N)$ for the fine C^r-topology such that each $g \in V_0$ is also of rank equal to $\dim_x(M)$ at each $x \in M$. If moreover f_0 is injective, show that there exists a neighborhood $V_1 = W(f_0, (\delta_\alpha))$ of f_0 contained in V_0 such that the restriction of each $g \in V_1$ to each K_α is injective (argue by contradiction). Give an example in which f_0 is injective and such that there exist noninjective mappings of M into N in each neighborhood of f_0 for the fine C^r-topology (cf. (16.8.5)). If, however, f_0 is a homeomorphism of M onto $f_0(M)$, then there exists a neighborhood V_2 of f_0 contained in V_1 (for the fine C^r-topology) such that each $g \in V_2$ is a homeomorphism of M onto $g(M)$. (Consider first a covering of M by compact sets L_α such that $L_\alpha \subset \mathring{K}_\alpha$ for all α. If d is a distance defining the topology of N, show that $d(f_0(L_\alpha), f_0(M - \mathring{K}_\alpha)) > 0$, and deduce that for a suitable choice of the family (δ'_α), the functions $g \in W(f_0, (\delta'_\alpha)) \subset V_1$ are injective. Put $L(f_0) = \overline{f_0(M)} - f_0(M)$. By considering sequences (x_n) in M with no cluster values, show that if $g \in V_1$ and if the sequence $g(x_n)$ converges, then its limit must belong to $L(f_0)$. Finally remark that the distance from $f_0(K_\alpha)$ to $L(f_0)$ is >0, and use this remark to construct a neighborhood $V_2 \subset W(f_0, (\delta'_\alpha))$ such that $g(M) \cap L(f_0) = \varnothing$ for all $g \in V_2$.) If moreover $f_0(M)$ is closed, then so is $g(M)$ for each $g \in V_2$. Finally, if M and N are connected and if $f_0(M) = N$, then $g(M) = N$ for all $g \in V_2$.

3. (a) Let U be an open subset of $\mathbf{R}^n$, A a compact subset of U, and V a neighborhood of A such that $\bar{V} \subset U$. For each C^r-function f on U and each $\varepsilon > 0$, show that there exists a C^r-mapping $f_1 : U \to \mathbf{R}$ which coincides with f on $U - V$ and is of class C^∞ on a neighborhood of A, and such that $|D^\nu f_1(x) - D^\nu f(x)| \leq \varepsilon$ for all $x \in U$ and all ν such that $|\nu| \leq r$. (Let g be a C^r-function equal to f on a compact neighborhood of A contained in V, and zero outside this neighborhood; take a suitable regularization h of g and consider the function $f_1 = f + (h - g)$.)
(b) Let M, N be two differential manifolds. Show that for each function $f_0 \in \mathscr{E}^{(r)}(M; N)$ and each neighborhood $W(f_0, (\delta_\alpha))$ of f_0 in the fine C^r-topology (Problem 2), there exist C^∞-mappings of M into N in this neighborhood (proceed by induction as in (16.12.11), using (a) above).

4. Show that in $\mathscr{E}^{(r)}(M; N)$ the set of mappings which are transversal over a submanifold Z of N is a dense open set in the fine C^r-topology (cf. Section 16.25, Problem 16).

3. CURRENTS AND DISTRIBUTIONS

(17.3.1) Let X be a pure differential manifold of dimension n, and consider the complex vector bundles $\left(\bigwedge^p \mathrm{T}(\mathrm{X})^*\right)_{(\mathbf{C})}$ on X, whose global sections are the complex-valued *differential p-forms* on X (16.20.1), for $0 \leqq p \leqq n$. For $p=0$, these are by definition the complex-valued *functions*. For brevity we shall denote by $\mathscr{E}_p^{(r)}(\mathrm{X})$ (resp. $\mathscr{E}_p(\mathrm{X})$) the Fréchet space $\Gamma^{(r)}\left(\mathrm{X}, \left(\bigwedge^p \mathrm{T}(\mathrm{X})^*\right)_{(\mathbf{C})}\right)$ $\left(\text{resp. } \Gamma\left(\mathrm{X}, \left(\bigwedge^p \mathrm{T}(\mathrm{X})^*\right)_{(\mathbf{C})}\right)\right)$ of complex differential p-forms of class C^r (resp. C^∞) on X. For each *compact* subset K of X we denote by $\mathscr{D}_p^{(r)}(\mathrm{X}; \mathrm{K})$ (resp. $\mathscr{D}_p(\mathrm{X}; \mathrm{K})$) the vector subspace consisting of the complex differential p-forms of class C^r (resp. C^∞) *with support contained in* K; this is clearly a *closed* subspace of $\mathscr{E}_p^{(r)}(\mathrm{X})$ (resp. $\mathscr{E}_p(\mathrm{X})$). We denote by $\mathscr{D}_p^{(r)}(\mathrm{X})$, (resp. $\mathscr{D}_p(\mathrm{X})$) the *union* of the subspaces $\mathscr{D}_p^{(r)}(\mathrm{X}; \mathrm{K})$ (resp. $\mathscr{D}_p(\mathrm{X}; \mathrm{K})$) as K runs through all the compact subsets of X, i.e., the space of all complex differential p-forms of class C^r (resp. C^∞) *with compact support*. When $p=0$ we shall drop p from these notations.

We can now proceed exactly as in the definition of a measure (13.1), except that the Banach spaces $\mathscr{K}(\mathrm{X}; \mathrm{K})$ are now replaced by Fréchet spaces. A *p-current* (or a *complex p-current*) (or a *current of dimension p*) on X is by definition a *linear form* T on $\mathscr{D}_p(\mathrm{X})$ whose restriction to each Fréchet space $\mathscr{D}_p(\mathrm{X}; \mathrm{K})$ is continuous; in other words (3.13.14), in order to verify that a linear form T on $\mathscr{D}_p(\mathrm{X})$ is a p-current, it must be shown that for each sequence (ω_k) of C^∞ differential p-forms, with supports contained in *the same compact set* K, and which converges to 0 in $\mathscr{E}_p(\mathrm{X})$, the sequence $(\mathrm{T}(\omega_k))$ tends to 0 in $\mathbf{C}$.

A 0-current on X is called a *distribution*.

With the notation of (17.2.1), T is a p-current if and only if, for each compact subset K of X, there exist integers s, m and a finite number of indices $\alpha_1, \ldots, \alpha_r$, together with a constant $a_\mathrm{K} \geqq 0$, such that, for each C^∞ p-form ω with support contained in K, we have

(17.3.1.1) $$|\mathrm{T}(\omega)| \leqq a_\mathrm{K} \cdot \sup_i p_{s, m, \alpha_i}(\omega).$$

(17.3.2) Suppose that a p-current T is such that, for each compact subset K of X, the restriction of T to $\mathscr{D}_p(\mathrm{X}; \mathrm{K})$ is *continuous with respect to the topology induced by that of* $\mathscr{D}_p^{(r)}(\mathrm{X}; \mathrm{K})$. In that case T is said to be a p-current of *order* $\leqq r$. The *order* of a current is the smallest integer r with this property (when such integers exist). If there exists no integer r with this property, then T is said to be a current *of infinite order*. It is clear that if T is the restriction to

$\mathscr{D}_p(X)$ of a linear form T′ on $\mathscr{D}_p^{(r)}(X)$ whose restriction to each $\mathscr{D}_p^{(r)}(X; K)$ is continuous, then T is of order $\leqq r$. Conversely, we shall show that if T is of order $\leqq r$, then it is the restriction of such a linear form T′, which moreover is *unique*. Let K be a compact subset of X and let K′ be a compact neighborhood of K in X. Then there exists a C^∞-function h which is equal to 1 on a compact neighborhood of K and which is 0 on X − K′ (16.4.2). For each p-form $\beta \in \mathscr{D}_p^{(r)}(X; K)$, there exists a sequence α_k in $\mathscr{D}_p(X)$ which converges to β with respect to the topology of $\mathscr{E}_p^{(r)}(X)$ (17.2.2). The sequence $(h\alpha_k)$, which belongs to $\mathscr{D}_p(X; K')$, therefore also converges to β in $\mathscr{D}_p^{(r)}(X; K')$ by (17.1.4). In other words, the closure of $\mathscr{D}_p(X; K')$ in $\mathscr{E}_p^{(r)}(X)$ contains $\mathscr{D}_p^{(r)}(X; K)$ and is contained in $\mathscr{D}_p^{(r)}(X; K')$. The existence and uniqueness of T′ now follow immediately by applying (12.9.4). We shall often write T in place of T′.

Examples of currents

(17.3.3) Let x be a point of X, and let $\mathbf{z}_x$ be a tangent p-vector at x (i.e., an element of $\bigwedge^p T_x(X)$). Then the mapping $\alpha \mapsto \langle \mathbf{z}_x, \alpha(x) \rangle$ is a continuous linear form on $\mathscr{E}_p^{(0)}(X)$, hence is a p-current of order 0; it is called the *Dirac p-current* defined by $\mathbf{z}_x$, and is sometimes denoted by $\varepsilon_{\mathbf{z}_x}$. When $p = 0$, we obtain the scalar multiples of the Dirac measure (13.1.3).

(17.3.4) It follows from (17.3.2) that *distributions of order* 0 on X are continuous linear forms on each of the Banach spaces $\mathscr{D}_0^{(0)}(X; K) = \mathscr{K}(X; K)$, and are therefore precisely the (complex) *measures* on X.

(17.3.5) Let T be a p-current and ω a C^∞ differential q-form (i.e., an element of $\mathscr{E}_q(X)$), with $q \leqq p$. For each $(p-q)$-form $\beta \in \mathscr{D}_{p-q}(X; K)$, we have $\omega \wedge \beta \in \mathscr{D}_p(X; K)$, and it follows therefore from (17.2.3) that the linear form $\beta \mapsto T(\omega \wedge \beta)$ is a $(p-q)$-current, which is denoted by $T \wedge \omega$. If T is of order $\leqq r$, then so is $T \wedge \omega$, and in this case we can also define $T \wedge \omega$ when ω is a differential q-form of class C^r. When $q = 0$, so that ω is a complex-valued function g, we write $T \cdot g$ or $g \cdot T$ in place of $T \wedge \omega$; if T is a measure, this definition agrees with that of (13.1.5), because of the fact that the closure of $\mathscr{D}(X; K')$ in $\mathscr{C}_{\mathbf{C}}(X)$ contains $\mathscr{K}(X; K)$ for each compact neighborhood K′ of K.

(17.3.6) Suppose that $p > 0$, and let Y be a C^∞ vector field on X. For each p-form $\alpha \in \mathscr{D}_p(X; K)$, we have $i_Y \cdot \alpha \in \mathscr{D}_{p-1}(X; K)$ (16.18.4). For each $(p-1)$-current T, it therefore follows again from (17.2.3) that the linear form $\alpha \mapsto T(i_Y \cdot \alpha)$ is a p-current, which is of order $\leqq r$ if T is of order $\leqq r$, and which is denoted by ${}^t i_Y \cdot T$.

(17.3.7) If X, Y are two locally compact metrizable topological spaces, a continuous mapping $u : X \to Y$ is said to be *proper* if for each compact subset K of Y, the inverse image $u^{-1}(K)$ is a compact subset of X. It then follows that if F is any *closed* subset of X, its image $u(F)$ is closed in Y. To see this, let (y_n) be a sequence of points in $u(F)$ converging to a point $y \in Y$. Then the set K consisting of the y_n and y is compact, and therefore $F \cap u^{-1}(K)$ is compact; choose for each n a point $x_n \in F \cap u^{-1}(K)$ such that $u(x_n) = y_n$, then the sequence (x_n) has a subsequence (x_{n_k}) converging to a point $x \in F$. Since u is continuous, it follows that $u(x) = y$, that is to say $y \in u(F)$.

Now let X, X′ be differential manifolds and $u : X \to X'$ a mapping of class C^r, where $r \geqq 1$. If α' is any p-form on X′ of class C^s $(s \geqq 0)$, then by the formula (16.20.9.3) the inverse image ${}^t u(\alpha')$ is defined and is a p-form on X of class $C^{\inf(r-1, s)}$; moreover it is clear that $\operatorname{Supp}({}^t u(\alpha'))$ is contained in $u^{-1}(\operatorname{Supp}(\alpha'))$. If we suppose that the mapping u is *proper*, it follows that for each compact subset K of X′ the mapping $\alpha' \mapsto {}^t u(\alpha')$ is a linear mapping of $\mathscr{D}_p^{(r-1)}(X'; K)$ into $\mathscr{D}_p^{(r-1)}(X; u^{-1}(K))$. Furthermore, this mapping is continuous; this follows immediately from (17.2), the local expression of ${}^t u(\alpha')$ (16.20.9.2), and (17.1.4). Hence, for each p-current T of order $\leqq r-1$ on X, the linear form $\alpha' \mapsto T({}^t u(\alpha'))$ on $\mathscr{D}_p^{(r-1)}(X')$ is a p-current of order $\leqq r-1$; it is denoted by $u(T)$ and is called the *image* of T by u. If v is a proper mapping of class C^r of X′ into another differential manifold X″, then $v \circ u$ is proper of class C^r, and we have $(v \circ u)(T) = v(u(T))$ for each p-current T of order $\leqq r-1$ on X. If u is a diffeomorphism, then $u(T)$ is defined for every current T on X and has the same order as T. If T is a distribution on X, then $u(T)$ is the distribution on X′ such that $u(T)(g) = T(g \circ u)$ for each function $g \in \mathscr{D}^{(r-1)}(X')$. But this formula makes sense also for functions $g \in \mathscr{D}^{(r)}(X')$; hence $u(T)$ is defined also for distributions of order r. When u is a homeomorphism and T is a measure, we recover the definition of (13.1.6).

If G is a Lie group which acts differentiably on X on the left (resp. on the right), the image of a current T under the diffeomorphism $x \mapsto s \cdot x$ (resp. $x \mapsto x \cdot s$) will be denoted by $\gamma(s)T$ (resp. $\delta(s^{-1})T$. When X = G, so that G is acting on itself by left translation (resp. right translation), and T is a measure, then $\gamma(s)T$ (resp. $\delta(s)T$) coincides with the measure so denoted in (14.1.2).

If X is a differential manifold, Y a closed submanifold of X, then the canonical immersion $j : Y \to X$ is proper; hence, for each p-current T on Y, the image $j(T)$ is defined and is a p-current on X. By considering the local expressions it is immediately verified that $j(T)$ has the same order as T. For measures, this notion agrees with that defined in (13.1.7).

(17.3.8) The set of all p-currents on X forms a vector space, which we denote by $\mathscr{D}'_p(X)$. The subspace of currents of order r is denoted by $\mathscr{D}_p'^{(r)}(X)$. When $p = 0$ we suppress p from the notation, so that $\mathscr{D}'(X)$ denotes the space of distributions on X, and $\mathscr{D}'^{(r)}(X)$ the subspace of distributions of order r. If α is a differential p-form and T a p-current, we shall often write $\langle T, \alpha \rangle$ or $\langle \alpha, T \rangle$ in place of $T(\alpha)$. For example, for a Dirac p-current (17.3.3), we have $\langle \varepsilon_{\mathbf{z}_x}, \alpha \rangle = \langle \mathbf{z}_x, \alpha(x) \rangle$.

If T is a distribution, we write $\langle T, f \rangle$ or $\langle f, T \rangle$ or even

(17.3.8.1)
$$\int f(x)\, dT(x)$$

in place of $T(f)$, for a function $f \in \mathscr{D}(X)$, whenever there is no risk of ambiguity.

4. LOCAL DEFINITION OF A CURRENT. SUPPORT OF A CURRENT

(17.4.1) Let U be an *open* subset of a differential manifold X. For each compact subset K of U, it is clear that the mapping $\alpha \mapsto \alpha | U$ is an *isomorphism* of $\mathscr{D}_p(X; K)$ onto $\mathscr{D}_p(U; K)$ (resp. of $\mathscr{D}_p^{(r)}(X; K)$ onto $\mathscr{D}_p^{(r)}(U; K)$); the inverse isomorphism sends a differential p-form $\beta \in \mathscr{D}_p(U; K)$ to the p-form β^U which is equal to β in U and is zero in $X - U$. (By abuse of notation we shall often write α in place of $\alpha | U$ when α is a differential p-form on X with support contained in U.)

For each p-current T on X, the mapping $\beta \mapsto T(\beta^U)$ is therefore a p-current on U; it is said to be the p-current *induced* by T on U, or the *restriction* of T to U, and is denoted by T_U. The order of T_U is at most equal to that of T, but it may be strictly less than that of T. It should be noted that a current on U is not necessarily the restriction of a current on X (Section 17.5, Problem 2), and that when it is the restriction of a current on X, the latter need not be unique.

However, there is the following result, which generalizes (13.1.9):

(17.4.2) *Let $(U_\lambda)_{\lambda \in L}$ be an open covering of* X. *For each $\lambda \in L$ let T_λ be a p-current on U_λ, such that for each pair of indices λ, μ the restrictions of T_λ and T_μ to $U_\lambda \cap U_\mu$ are equal. Then there exists a unique p-current* T *on* X *whose restriction to U_λ is equal to T_λ for each $\lambda \in L$.*

We shall not write out the proof in full detail; it is based, step by step, on the proof of (13.1.9), with the obvious modifications. We begin by writing a p-form $\alpha \in \mathscr{D}_p(X)$ as $\sum_i \alpha_i$, where $\alpha_i = h_i \alpha$ and $\operatorname{Supp}(h_i) \subset U_{\lambda_i}$ for a suitable λ_i; for this it is sufficient to invoke (16.4.2) instead of (12.6.4). It follows that T is unique and necessarily given by $T(\alpha) = \sum_i T_{\lambda_i}(\alpha_i)$, and the proof of the fact that this formula does define a linear form on $\mathscr{D}_p(X)$ (in other words, that the number $T(\alpha)$ does not depend on the particular decomposition of α chosen) goes over without change. It remains to show that if a sequence (α_k) tends to 0 in $\mathscr{D}_p(X; K)$, then $T(\alpha_k) \to 0$. We may take the same finite sequence (h_i) for all the α_k, and each of the sequences $(T_{\lambda_i}(h_i \alpha_k))_{k \geq 1}$ then tends to zero by virtue of (17.1.4); this proves our assertion.

(17.4.3) It follows in particular from (17.4.2) that if the restriction of a current T to each member of a family of open sets U_λ is zero, then the restriction of T to the union of the U_λ is also zero. Hence there is a *largest* open subset V of X such that the restriction of T to V is zero; the complement $S = \complement V$ is called the *support* of T, and is denoted by Supp(T). A point $x \in X$ belongs to the support of a p-current T if and only if, for each neighborhood V of x, there exists a p-form $\alpha \in \mathscr{D}_p(X)$ with support contained in V and such that $T(\alpha) \neq 0$. If T_1, T_2 are two p-currents, it is clear that

$$\mathrm{Supp}(T_1 + T_2) \subset \mathrm{Supp}(T_1) \cup \mathrm{Supp}(T_2),$$

and that if $\omega \in \mathscr{E}_q(X)$ is a q-form with $q \leq p$, then

$$\mathrm{Supp}(T \wedge \omega) \subset \mathrm{Supp}(T) \cap \mathrm{Supp}(\omega).$$

If $\pi: X \to X'$ is a proper mapping of class C^r, then for any current T on X of order $\leq r - 1$, we have $\mathrm{Supp}(\pi(T)) \subset \pi(\mathrm{Supp}(T))$. If π is a diffeomorphism of X onto X′, then $\mathrm{Supp}(\pi(T)) = \pi(\mathrm{Supp}(T))$. If Y is a closed submanifold of X, and $j: Y \to X$ the canonical immersion, then $\mathrm{Supp}(j(T)) = \mathrm{Supp}(T)$ for any current T on Y. When $p = 0$, we recover the definition of the support of a measure on X (13.19), by virtue of the fact that each space $\mathscr{K}(X; K)$ is contained in the closure of $\mathscr{D}(X; K')$ in $\mathscr{K}(X; K')$, whenever K′ is a compact neighborhood of the compact set K.

(17.4.4) Let X, X′ be two pure differential manifolds of the same dimension n, and let $\pi : X' \to X$ be a local diffeomorphism (16.5.6). Then for each current T on X there exists a unique current T′ on X′ with the following property: for each open subset U′ of X′ such that the restriction $\pi_{U'} : U' \to \pi(U')$ is a diffeomorphism, we have $\pi_{U'}(T'_{U'}) = T_{\pi(U')}$. For there exists an open covering (U'_λ) of X′ such that each of the restrictions $\pi_{U'_\lambda}$ is a diffeomorphism; if $\pi^{-1}_{U'_\lambda}$ is the inverse diffeomorphism, put $T'_\lambda = \pi^{-1}_{U'_\lambda}(T_{\pi(U'_\lambda)})$. For any two indices λ, μ, the mappings $\pi_{U'_\lambda}$ and $\pi_{U'_\mu}$ agree by definition on $U'_\lambda \cap U'_\mu$, hence $\pi^{-1}_{U'_\lambda}$ and $\pi^{-1}_{U'_\mu}$ agree on $\pi(U'_\lambda) \cap \pi(U'_\mu)$. This implies that T'_λ and T'_μ have the same restriction to $U'_\lambda \cap U'_\mu$, and the existence and uniqueness of T′ therefore follows from (17.4.2). The current T′ is called the *inverse image* of T by π, and is denoted by ${}^t\pi(T)$.

For example, if X′ is a universal covering of X (16.29), then it follows from the definitions that the fundamental group of X leaves *invariant* the inverse image on X′ of every current on X. Conversely, every current T′ on X′ having this property is the inverse image of a current T on X. To see this, we take a covering (U_λ) of X by connected open sets over which X′ is trivializable, and we define T_U to be the image by the canonical projection π of the restriction of T′ to any one of the connected components of $\pi^{-1}(U_\lambda)$.

More particularly, taking $X' = \mathbf{R}^n$ and $X = \mathbf{T}^n$, the inverse images on $\mathbf{R}^n$ of currents on $\mathbf{T}^n$ are precisely those which are invariant under the group $\mathbf{Z}^n$ (acting on $\mathbf{R}^n$ by translations); such currents are said to be *periodic* with $\mathbf{Z}^n$ as group of periods.

(17.4.5) Let A be the support of a p-current T. We shall show that it is possible to attach a meaning to $T(\alpha)$ for *some* p-forms $\alpha \in \mathscr{E}_p(X)$ which are not compactly supported: it is enough that $A \cap \operatorname{Supp}(\alpha)$ should be *compact* (which will always be the case if $A = \operatorname{Supp}(T)$ is compact). For if $h : X \to [0, 1]$ is a C^∞-mapping which is equal to 1 on some compact neighborhood of $A \cap \operatorname{Supp}(\alpha)$, and is compactly supported (16.4.2), then $T(h\alpha)$ is defined because $h\alpha$ has compact support. Furthermore, if h_1 is another function having the same properties as h, we have $T(h_1\alpha) = T(h\alpha)$, because there exists an open neighborhood V of $A \cap \operatorname{Supp}(\alpha)$ on which $h(x) = h_1(x)$, and the support of $(h - h_1)\alpha$ is contained in $\operatorname{Supp}(\alpha) \cap \complement V$, and therefore does not intersect A. We may therefore define $T(\alpha)$ to be $T(h\alpha)$ for any function h with the properties stated above. It is immediate that the set of p-forms $\alpha \in \mathscr{E}_p(X)$ such that $A \cap \operatorname{Supp}(\alpha)$ is compact is a vector subspace of $\mathscr{E}_p(X)$, and that $\alpha \mapsto T(\alpha)$ is a *linear form* on this vector space. Next, consider a sequence (α_n) of p-forms in $\mathscr{E}_p(X)$, such that (i) all the sets $A \cap \operatorname{Supp}(\alpha_n)$ are contained in a *fixed* compact set K; (ii) the sequence (α_n) tends to 0 in $\mathscr{E}_p(X)$. Then $T(\alpha_n) \to 0$. For if we choose as above a function h which is equal to 1 on some compact neighborhood of K, then the sequence $(h\alpha_n)$ tends to 0 in $\mathscr{D}_p(X; V)$ (17.1.4); since $T(\alpha_n) = T(h\alpha_n)$ for each n, our assertion follows immediately.

5. CURRENTS ON AN ORIENTED MANIFOLD. DISTRIBUTIONS ON $\mathbf{R}^n$

(17.5.1) Consider now an *oriented* pure differential manifold X of dimension n. We have defined in (16.24.2) the notion of an *integrable* differential n-form υ on X, and its integral, denoted by $\int \upsilon$ or $\int_X \upsilon$. Now consider a *locally integrable* differential $(n-p)$-form β, where $0 \leq p \leq n$; for each p-form $\alpha \in \mathscr{D}_p^{(0)}(X; K)$, the n-form $\beta \wedge \alpha$ is locally integrable and has support contained in K, hence is integrable. We shall show that the linear form $\alpha \mapsto \int \beta \wedge \alpha$ on $\mathscr{D}_p^{(0)}(X)$ is a p-current (or order 0). The proof reduces immediately to the situation where X is an open set U in $\mathbf{R}^n$, and then we have

$$\int \beta \wedge \alpha = \sum_H \pm \int b_H(x) a_{I-H}(x)\, d\xi^1\, d\xi^2 \cdots d\xi^n,$$

where the b_{H} (resp. the $a_{\mathrm{I-H}}$) are the coefficients of β (resp. α) relative to the canonical basis of the $\mathscr{C}(\mathrm{U})$-module $\mathscr{E}_{n-p}^{(0)}(\mathrm{U})$ (resp. $\mathscr{E}_{p}^{(0)}(\mathrm{U})$), and in the summation $\mathrm{I} = \{1, 2, \ldots, n\}$ and H runs through all subsets of $n - p$ elements of I. Then we have to show that each of the linear mappings $a_{\mathrm{I-H}} \mapsto \int b_{\mathrm{H}}(x) a_{\mathrm{I-H}}(x)\, dx$ is continuous on each of the Banach spaces $\mathscr{K}(\mathrm{U}; \mathrm{K})$, where K is any compact subset of U; and this follows from (13.13) because each of the functions b_{H} is locally integrable.

Let T_β be the p-current so defined. If we denote by $\mathscr{E}_{n-p,\,\mathrm{loc}}(\mathrm{X})$ the vector space of locally integrable differential $(n-p)$-forms on X, then we have a linear mapping $\beta \mapsto \mathrm{T}_\beta$ of $\mathscr{E}_{n-p,\,\mathrm{loc}}(\mathrm{X})$ into $\mathscr{D}_p'^{(0)}(\mathrm{X})$. From (13.14.4) it follows immediately that the kernel of this mapping is the subspace of *negligible* $(n-p)$-forms. Since the support of a Lebesgue measure on X is the whole of X, the restriction of the mapping $\beta \mapsto \mathrm{T}_\beta$ to the space $\mathscr{E}_{n-p}^{(0)}(\mathrm{X})$ of *continuous* differential $(n-p)$-forms is *injective*, so that such a form may be identified with a p-current of order 0. Under this identification, the notions of *support* are the same for the continuous $(n-p)$-form β and the p-current T_β with which we have identified it. For, by reducing as above to the case where X is an open subset U of $\mathbf{R}^n$, if $x_0 \in \mathrm{Supp}(\beta)$, then there is an index H such that $b_{\mathrm{H}}(x_0) \neq 0$; we can then choose $a_{\mathrm{I-H}}$ such that the integral

$$\int b_{\mathrm{H}}(x) a_{\mathrm{I-H}}(x)\, d\xi^1\, d\xi^2 \cdots d\xi^n$$

is $\neq 0$, and such that $\mathrm{Supp}(a_{\mathrm{I-H}})$ is contained in an arbitrarily small neighborhood V of x_0; defining $a_{\mathrm{I-H'}}$ to be 0 for $\mathrm{H}' \neq \mathrm{H}$, we obtain a form α with support contained in V and such that $\int \beta \wedge \alpha \neq 0$, which proves the assertion.

In particular, for each locally integrable n-form υ, the mapping $f \mapsto \int f\upsilon$ is a *measure* T_υ on X, which is positive if and only if $\upsilon(x) \geqq 0$ almost everywhere (relative to the orientation of X) (13.15.3).

(17.5.1.1) Again, if f is any locally integrable complex *function* on X, then the mapping $\upsilon \mapsto \int f\upsilon$ is an *n-current* T_f on X, of order 0. If U is any open subset of X, the n-current $\mathrm{T}_{\varphi_{\mathrm{U}}}$ on X, where φ_{U} is the characteristic function of U, is called an *open n-chain element* on X, and linear combinations of open n-chain elements are called *open n-chains* on X. By abuse of notation, we shall often write U in place of $\mathrm{T}_{\varphi_{\mathrm{U}}}$, and $\sum_j \lambda_j \mathrm{U}_j$ in place of $\sum_j \lambda_j \mathrm{T}_{\varphi_{\mathrm{U}_j}}$.

(17.5.2) We retain the notation and assumptions of (17.5.1). Let γ be a continuous differential q-form, where $q \leqq p$; then the $(n-p+q)$-form $\beta \wedge \gamma$ is locally integrable, and it follows immediately from the definitions that

(17.5.2.1) $$\mathrm{T}_{\beta \wedge \gamma} = \mathrm{T}_\beta \wedge \gamma.$$

The left-hand side of this formula is meaningful if we suppose only that γ is *measurable and locally bounded*, or, on the other hand, that β is measurable and locally bounded and that γ is locally integrable. Under these conditions we define the right-hand side of (17.5.2.1) by this formula, and then we have

(17.5.2.2) $$T_\gamma \wedge \beta = (-1)^{q(n-p)} T_\beta \wedge \gamma.$$

Next, let $\pi : X \to X'$ be an orientation-preserving diffeomorphism. Then the p-current $\pi(T_\beta)$ is the linear form $\alpha' \mapsto \int_X \beta \wedge {}^t\pi(\alpha')$ on $\mathscr{D}_p^{(0)}(X')$. By (16.24.5.1) we have

$$\int_X \beta \wedge {}^t\pi(\alpha') = \int_X {}^t({}^t\pi^{-1}(\beta)) \wedge {}^t\pi(\alpha') = \int_{X'} {}^t\pi^{-1}(\beta) \wedge \alpha'$$

and therefore

(17.5.2.3) $$\pi(T_\beta) = T_{{}^t\pi^{-1}(\beta)}.$$

If β is a locally integrable differential $(m - p)$-form on an *oriented submanifold* Y of X, of dimension m, then clearly we can form the image $j(T_\beta)$ of T_β under the canonical immersion $j : Y \to X$, and $j(T_\beta)$ is a p-current. However, when $m < n$, this current is *not* in general expressible in the form T_γ for some locally integrable $(n - p)$-form γ on X, even if β is of class C^∞. For example, if $p = 0$, the support of the measure $j(T_\beta)$ is the submanifold Y, which is *negligible* with respect to any Lebesgue measure on X, so that a measure of the form T_γ with support Y is necessarily zero, whereas $j(T_\beta) \neq 0$ in general (cf. (17.10.7)).

Finally, let $\pi : X \to X'$ be a *proper* mapping of class C^r with $r \geqq 1$ (17.3.7), the manifold X' being not necessarily orientable. Then for each locally integrable complex-valued function f on X, $\pi(T_f)$ is an n-current of order 0 on X' (which therefore vanishes if $\dim X' < n$), defined by the formula

(17.5.2.4) $$\langle \pi(T_f), \alpha' \rangle = \int_X f \cdot {}^t\pi(\alpha'),$$

where α' is any continuous, compactly-supported differential n-form on X'. In particular, if we take f to be φ_X, the constant function equal to 1 at all points of X, then $\pi(T_{\varphi_X})$ is called the *n-chain element without boundary* on X (cf. (17.15.5)) defined by the proper mapping π, and we write $\int_\pi \alpha'$ in place of $\int_X {}^t\pi(\alpha')$. When $n = 1$, this is an integral along a particular type of "unending path" (16.27). If X is a *closed submanifold* of X' and π is the canonical injection, we shall sometimes write X in place of $\pi(\mathrm{T}_{\varphi_X})$, and $\int_X \alpha'$ in place of $\int_X {}^t\pi(\alpha')$; but it must be borne in mind that this number depends not only on the manifold X but also on its *orientation*, and changes sign when the orientation is reversed.

(17.5.3) If we fix a C^∞ differential n-form υ_0 belonging to the orientation of X, then every differential n-form on X can be written uniquely as $f\upsilon_0$, where f is a complex-valued function on X. The form $f\upsilon_0$ is locally integrable if and only if f is locally integrable; in other words, the mapping $f \mapsto f\upsilon_0$ is a linear bijection of the space $\mathscr{L}_{\mathrm{loc}}(\mathrm{X})$ of locally integrable complex-valued functions on X onto the space $\mathscr{E}_{n,\,\mathrm{loc}}(\mathrm{X})$. We shall write T_f in place of $\mathrm{T}_{f\upsilon_0}$ and *identify* the *function* f with the corresponding *distribution* T_f.

(17.5.3.1) The choice of υ_0 allows us to *identify* n-currents and distributions (i.e., 0-currents), because $g \mapsto g\upsilon_0$ is an isomorphism of the Fréchet space $\mathscr{D}(\mathrm{X};\mathrm{K}) = \mathscr{D}_0(\mathrm{X};\mathrm{K})$ onto $\mathscr{D}_n(\mathrm{X};\mathrm{K})$, for each compact subset K of X; this follows immediately from (17.1.4). Hence every n-current is uniquely expressible as $g\upsilon_0 \mapsto \mathrm{T}(g)$, where T is a distribution. We shall denote this n-current by $\mathrm{T}_{|\upsilon_0}$ when it is necessary to avoid ambiguity; in this notation, we have $(\mathrm{T}_f)_{|\upsilon_0} = \mathrm{T}_{f\upsilon_0}$ for all $f \in \mathscr{L}_{\mathrm{loc}}(\mathrm{X})$.

(17.5.3.2) In future we shall make these identifications only when X is an *open* subset U of $\mathbf{R}^n$, endowed with the canonical orientation and the canonical n-form $\upsilon_0 = d\xi^1 \wedge d\xi^2 \wedge \cdots \wedge d\xi^n$ restricted to U. Then no risk of confusion arises except as regards the *image* of a current under a diffeomorphism π of U onto an open subset U′ of $\mathbf{R}^n$. If T is a distribution on U and $\pi(\mathrm{T})$ its image on U′, then the image $\pi(\mathrm{T}_{|\upsilon_0})$ is given by

(17.5.3.3)
$$\pi(\mathrm{T}_{|\upsilon_0}) = \mathrm{J}(\pi^{-1})(\pi(\mathrm{T})_{|\upsilon_0}),$$

where $\mathrm{J}(\pi^{-1})$ is the *Jacobian* of the inverse diffeomorphism π^{-1}, by virtue of (16.20.9.4). In particular, for each *function* $f \in \mathscr{L}_{\mathrm{loc}}(\mathrm{U})$ we have

(17.5.3.4)
$$\pi(\mathrm{T}_f) = \mathrm{T}_{f'},$$

where f' is the mapping $x' \mapsto f(\pi^{-1}(x'))\mathrm{J}(\pi^{-1})(x')$.

(17.5.3.5) It is clear that the kernel of the linear mapping $f \mapsto \mathrm{T}_f$ of $\mathscr{L}_{\mathrm{loc}}(\mathrm{U})$ into $\mathscr{D}'(\mathrm{U})$ is formed by the *negligible* complex-valued functions on U (relative to Lebesgue measure). Passing to the quotient, it follows that the space $\mathrm{L}_{\mathrm{loc}}(\mathrm{U})$ of *classes* of locally integrable functions on U (13.13.4) may be canonically identified with a *subspace* of the space of distributions on U. *A fortiori*, we may identify the spaces $L^1(U)$, $L^2(U)$, and $L^\infty(U)$ with subspaces of $\mathscr{D}'(U)$. See (17.8) for questions relating to the topologies of these spaces.

(17.5.4) Questions of a *local* nature concerning currents reduce to the case where the manifold X under consideration is an open subset U of $\mathbf{R}^n$, and it is this case that we shall be mainly concerned with in the remainder of this chapter. Each of the spaces $\mathscr{E}_p^{(r)}(U)$ (resp. $\mathscr{E}_p(U)$) is then an $\mathscr{E}^{(r)}(U)$-module (resp. an $\mathscr{E}(U)$-module) which is *free* of rank $\binom{n}{p}$. Relative to the canonical basis for the p-forms:

$$\zeta_H = d\xi^{i_1} \wedge d\xi^{i_2} \wedge \cdots \wedge d\xi^{i_p}$$

(where H is the set of integers $i_1 < i_2 < \cdots < i_p$), each p-current is of the form

$$\sum_H f_H \zeta_H \mapsto \sum_H \langle T_H, f_H \rangle,$$

where the T_H are *distributions* on U. Hence the study of currents on an open subset U of $\mathbf{R}^n$ reduces to the study of distributions on U, and it is therefore the latter that we shall mainly consider.

(17.5.5) The only concrete examples of distributions that we have given so far have been distributions of order 0 (i.e., *measures*). We shall now show that there exist on $U \subset \mathbf{R}^n$ distributions *of all orders*. For this we observe that, for each compact $K \subset U$ and each multi-index ν, the mapping $f \mapsto D^\nu f$ is a *continuous* linear mapping of $\mathscr{D}(U; K)$ into itself, by virtue of (17.1.3) and the fact that $\operatorname{Supp}(D^\nu f) \subset \operatorname{Supp}(f)$. Consequently, if $T \in \mathscr{D}'(U)$ is any distribution on U, the linear form $f \mapsto T(D^\nu f)$ on $\mathscr{D}(U)$ is a distribution, which we denote by $(-1)^{|\nu|} D^\nu T$; the distribution $D^\nu T$ is called the *derivative of multi-index* ν of the distribution T. For each function $f \in \mathscr{D}(U)$ we have

(17.5.5.1) $$\langle D^\nu T, f \rangle = (-1)^{|\nu|} \langle T, D^\nu f \rangle,$$

from which it follows that

(17.5.5.2) $$D^\nu(D^{\nu'} T) = D^{\nu+\nu'} T$$

for any two multi-indices ν, ν'.

We shall also write $\partial^\nu T / \partial x_1^{\nu_1} \partial x_2^{\nu_2} \cdots \partial x_n^{\nu_n}$ in place of $D^\nu T$. Clearly we have

(17.5.5.3) $$\operatorname{Supp}(D^\nu T) \subset \operatorname{Supp}(T).$$

In the particular case of a distribution of the form T_g, where $g \in \mathscr{E}^{(r)}(U)$ and $|\nu| \leqq r$ (with $r \geqq 1$), we have

(17.5.5.4) $$D^{\nu}T_g = T_{D^{\nu}g}$$

(which justifies the factor $(-1)^{|\nu|}$ introduced in the definition of $D^{\nu}T$). By virtue of (17.5.5.2), it is enough to establish the formula (17.5.5.4) for $|\nu| = 1$. By definition, if $f \in \mathscr{D}(U)$, we have

$$\langle T_g, D_i f\rangle = \iint \cdots \int_{\mathbf{R}^n} g(x) D_i f(x)\, d\xi^1 \cdots d\xi^n$$
$$= \int \cdots \int_{\mathbf{R}^{n-1}} d\xi^1 \cdots d\xi^{i-1}\, d\xi^{i+1} \cdots d\xi^n \int_{\mathbf{R}} g(x) D_i f(x)\, d\xi^i$$

(the integrals being extended to the whole of $\mathbf{R}^n$ by extending the function $gD_i f$ by zero outside $\mathrm{Supp}(f)$); but because $D_i f$ is compactly supported, we obtain by integrating by parts (8.7.5)

$$\int_{\mathbf{R}} g(x) D_i f(x)\, d\xi^i = -\int_{\mathbf{R}} f(x) D_i g(x)\, d\xi^i,$$

from which it immediately follows that $\langle T_g, D_i f\rangle = -\langle T_{D_i g}, f\rangle$.

We remark also that for each function $g \in \mathscr{E}^{(1)}(U)$ and each distribution $T \in \mathscr{D}'(U)$ we have

(17.5.5.5) $$D_i(g \cdot T) = (D_i g) \cdot T + g \cdot D_i T;$$

this follows immediately from the definition (17.5.5.1) and the rule for differentiating a product.

From the definition of the topology of $\mathscr{E}(U)$ (resp. $\mathscr{E}^{(r)}(U)$) by seminorms (17.1.1), it follows that if T is a distribution of order $\leqq q$, then $D^{\nu}T$ is of order *at most* $q + r$, where $r = \sup_{1 \leqq i \leqq n} \nu_i$. We shall see that it can be exactly of this order.

Examples

(17.5.6) The derivatives of the Dirac measure ε_a at a point $a = (a_i) \in \mathbf{R}^n$ are given by

(17.5.6.1) $$\langle D^{\nu}\varepsilon_a, f\rangle = (-1)^{|\nu|} D^{\nu}f(a)$$

for $f \in \mathscr{D}(\mathbf{R}^n)$. This distribution is of order *exactly* $r = \sup_{1 \leqq i \leqq n} \nu_i$. To see this, let j be an index such that $\nu_j = r$, and let $f \in \mathscr{D}(\mathbf{R}^n)$ be a function of the form

$$f(\xi_1, \xi_2, \ldots, \xi_n) = g(\xi_j) \prod_{i \neq j} (\xi_i - a_i)^{\nu_i} u(\xi_1, \ldots, \xi_n),$$

where $u \in \mathscr{D}(\mathbf{R}^n)$ is equal to 1 throughout some neighborhood of a, and $g \in \mathscr{E}(\mathbf{R})$. Now replace g by a sequence of functions $g_k \in \mathscr{E}(\mathbf{R})$ such that the functions $\mathrm{D}^s g_k$ are uniformly bounded on a compact neighborhood of a_j for $s < r$, and such that $\mathrm{D}^r g_k(a_j)$ tends to $+\infty$ as $k \to \infty$. In other words, we may assume that $n = 1$ and, by translation, that $a_j = 0$. If $h_0 \in \mathscr{D}(\mathbf{R})$ is the function defined in (16.4.1.4), such that $h_0(0) \neq 0$, then we may take

$$g_k(x) = k \int_0^x \frac{(x-t)^{r-1}}{(r-1)!} h_0(kt)\, dt$$

(cf. (8.14.2)). We have then

$$\mathrm{D}^s g_k(x) = k \int_0^x \frac{(x-t)^{r-s-1}}{(r-s-1)!} h_0(kt)\, dt$$

for $s < r$; these functions are uniformly bounded for $|x| \leqq 1$ by the number

$$k \int_0^1 h_0(kt)\, dt = \int_0^1 h_0(u)\, du$$

independent of k, whereas $\mathrm{D}^r g_k(0) = k h_0(0)$.

This also enables us to give an example of a distribution of *infinite order* on $\mathbf{R}$, namely the linear form $f \mapsto \sum_{k=0}^{\infty} \mathrm{D}^k f(k)$ on $\mathscr{D}(\mathbf{R})$.

(17.5.7) The *Heaviside function* is by definition the characteristic function of the interval $[0, +\infty[$ of $\mathbf{R}$ and is denoted by Y or Y_1. For $f \in \mathscr{D}(\mathbf{R})$ we have

$$\langle \mathrm{DT}_\mathrm{Y}, f \rangle = -\int_{-\infty}^{+\infty} f'(x)\mathrm{Y}(x)\, dx = -\int_0^{\infty} f'(x)\, dx = f(0)$$

or, identifying Y and T_Y,

$$\mathrm{DY} = \varepsilon_0, \tag{17.5.7.1}$$

the Dirac measure at the origin. Here the order of T_Y is equal to that of its derivative.

(17.5.8) The function g which is equal to $\log x$ for $x > 0$ and is zero for $x \leqq 0$ is locally integrable on $\mathbf{R}$ (because in a neighborhood of 0 we have

$|\log x| < 1/x^\mu$ for $0 < \mu < 1$). We have

$$\langle \mathrm{DT}_g, f\rangle = -\int_0^\infty f'(x)\log x\,dx$$

for $f \in \mathscr{D}(\mathbf{R})$. If $\alpha > 0$, we obtain, on integrating by parts,

$$-\int_\alpha^\infty f'(x)\log x\,dx = (f(\alpha) - f(0))\log\alpha + \int_\alpha^\infty \frac{f(x)-f(0)}{x}\,dx.$$

Since f is continuously differentiable, we may write

$$f(x) - f(0) = xf_1(x),$$

where f_1 is continuous on $\mathbf{R}$. Since $\alpha\log\alpha \to 0$ as $\alpha \to 0$, we obtain therefore

(17.5.8.1) $$\langle \mathrm{DT}_g, f\rangle = \int_0^\infty \frac{f(x)-f(0)}{x}\,dx.$$

The restriction of DT_g to the open interval $\mathrm{U} =]0, +\infty[$ coincides with $\mathrm{T}_{\mathrm{D}g}$ and therefore may be identified with the function $1/x$ on U. It can be shown that the order of DT_g is 1 (Problem 2), whereas its restriction to U is of order 0.

(17.5.9) Consider the function $g(x) = x\sin(1/x)$. Here the function $\mathrm{D}g$, defined for $x \neq 0$, is not integrable on any neighborhood of the origin, but the limits of $\int_{-\infty}^{-\alpha} g'(x)\,dx$ and $\int_\alpha^\infty g'(x)\,dx$ as $\alpha \to 0$ exist, and are denoted by $\int_{-\infty}^0 g'(x)\,dx$ and $\int_0^\infty g'(x)\,dx$ (they are *improper* integrals). The same argument as in (17.5.8) this time gives

(17.5.9.1) $$\langle \mathrm{DT}_g, f\rangle = \int_{-\infty}^0 f(x)g'(x)\,dx + \int_0^\infty f(x)g'(x)\,dx$$

(improper integrals), but it can be shown that the distribution DT_g is again of order 1 (Problem 4).

(17.5.10) Let U be an open subset of $\mathbf{R}^n$, and let $\mathrm{F} \in \mathscr{E}(\mathrm{U})$ be a function satisfying the conditions of (16.24.11). With the notation used there, consider the (positive) measure on the oriented manifold E_u defined by the positive $(n-1)$-form σ_u (16.24.2), and let μ_u be the image of this measure under the canonical immersion $\mathrm{E}_u \to \mathrm{U}$ (13.1.7). The elementary version of Stokes' formula (16.24.11.1) can then be interpreted as giving the *derivative* of the characteristic function $\varphi_{\mathrm{U}_{a,b}}$ (identified with a distribution on U):

(17.5.10.1) $$\mathrm{D}_1\varphi_{\mathrm{U}_{a,b}} = -((\mathrm{D}_1\mathrm{F})\,|\,\mathrm{E}_b)\cdot\mu_b + ((\mathrm{D}_1\mathrm{F})\,|\,\mathrm{E}_a)\cdot\mu_a.$$

(17.5.11) In future, whenever we identify a function $f \in \mathscr{L}_{\text{loc}}(\mathrm{U})$ with the distribution T_f, we shall interpret the expressions $\mathrm{D}_j f\ (1 \leqq j \leqq n)$ as meaning the *distributions* $\mathrm{D}_j\mathrm{T}_f = \partial\mathrm{T}_f/\partial x_j$, *except* when f is *continuously differentiable*, in which case, as we have seen, the distribution $\mathrm{D}_j\mathrm{T}_f$ can be identified with the *continuous function* $\mathrm{D}_j f$. Whenever we say that a function $f \in \mathscr{L}_{\text{loc}}(\mathrm{U})$ has a derivative which "is a function" $\mathrm{D}_j f \in \mathscr{L}_{\text{loc}}(\mathrm{U})$, we shall mean that there exists a locally integrable function g such that $\mathrm{D}_j\mathrm{T}_f = \mathrm{T}_g$, and we shall denote this *function* by $\mathrm{D}_j f$; but, unless the contrary is explicitly stated, we shall never attempt to define this function g in terms of differences $f(x+h) - f(x)$. In any case, the function g is defined only *to within a negligible function*, and it is only when there exists a continuous function (necessarily unique) in the class of g that it is appropriate to choose this function as a privileged representative of the class.

(17.5.12) Consider in particular the case $n = 1$, and let f be a locally integrable complex-valued function on $\mathbf{R}$. For each compact interval $[a, b] \subset \mathbf{R}$, the integral $\int_{[a,b]} f\,d\lambda$ (which by abuse of notation we shall write as $\int_a^b f(t)\,dt$ (13.9.16)) has a meaning. Put

(17.5.12.1)
$$\mathrm{F}(x) = \begin{cases} \displaystyle\int_0^x f(t)\,dt & \text{if } x \geqq 0, \\ \displaystyle -\int_x^0 f(t)\,dt & \text{if } x < 0. \end{cases}$$

Then the function F is *continuous* on $\mathbf{R}$, by the dominated convergence theorem (13.8.4) applied to the sequence of functions $f\varphi_{[x, x_k]}$ (resp. $f\varphi_{[x_k, x]}$), where (x_k) is a decreasing (resp. increasing) sequence with x as limit. We shall show that

(17.5.12.2)
$$\mathrm{DT}_\mathrm{F} = \mathrm{T}_f$$

(or $\mathrm{DF} = f$, with the conventions described above). To prove this, we must show that, for all $u \in \mathscr{D}(\mathbf{R})$,

(17.5.12.3)
$$\int_{-\infty}^{+\infty} \mathrm{F}(t)u'(t)\,dt = -\int_{-\infty}^{+\infty} f(t)u(t)\,dt.$$

Since u has compact support, contained in some interval $[a, b]$, the left-hand side of (17.5.12.3) does not change when $\mathrm{F}(t)$ is replaced by $\mathrm{F}(t) - \mathrm{F}(a)$, and therefore is equal to

(17.5.12.4)
$$\int_a^b u'(t)\,dt \int_a^t f(s)\,ds.$$

However, since the function $f \otimes u'$ is locally integrable with respect to the Lebesgue measure $\lambda \otimes \lambda$ on $\mathbf{R}^2$ (13.21.16), the same is true of its product by the characteristic function of the closed set $\{(s, t) : s \leqq t\}$. Applying the Lebesgue–Fubini theorem (13.21.7) to this product, we see that (17.5.12.4) is equal to

$$\int_a^b f(s)\,ds \int_s^b u'(t)\,dt = -\int_a^b f(s)u(s)\,ds,$$

and (17.5.12.3) now follows. We may therefore say that the distribution $\mathrm{T_F}$ is a *primitive* of T_f. Every other primitive is of the form $\mathrm{T}_{\mathrm{F}+c}$, where c is a constant. For in order that a function $v \in \mathscr{D}(\mathbf{R})$ should have a primitive belonging to $\mathscr{D}(\mathbf{R})$, it is necessary and sufficient that $\int_{-\infty}^{+\infty} v(t)\,dt = 0$; the functions satisfying this relation form a hyperplane H in the vector space $\mathscr{D}(\mathbf{R})$ (A.4.15), and by definition (17.5.5) the distributions T such that $\mathrm{DT} = 0$ are those which *vanish on* H; hence (A.4.15) for such a distribution T, there exists a constant c such that $\mathrm{T}(v) = c\int_{-\infty}^{+\infty} v(t)\,dt$ for all $v \in \mathscr{D}(\mathbf{R})$, in other words $\mathrm{T} = \mathrm{T}_c$.

Remark

(17.5.13) Suppose that, for two indices j, k between 1 and n, the continuous function f on $\mathrm{U} \subset \mathbf{R}^n$ is such that the four derivatives $\mathrm{D}_j f$, $\mathrm{D}_k f$, $\mathrm{D}_j(\mathrm{D}_k f)$, and $\mathrm{D}_k(\mathrm{D}_j f)$ (*in the sense of* (8.9)) exist and are *continuous on* U. Then we have $\mathrm{D}_j(\mathrm{D}_k f) = \mathrm{D}_k(\mathrm{D}_j f)$, because the relation $\mathrm{D}_j(\mathrm{D}_k \mathrm{T}) = \mathrm{D}_k(\mathrm{D}_j \mathrm{T})$ is true for all distributions T, and in particular for T_f; the remarks of (17.5.11) therefore apply.

(17.5.14) For each distribution $\mathrm{T} \in \mathscr{D}'(\mathrm{U})$, there exists a largest open subset V of U such that the restriction of T to V is a function belonging to $\mathscr{E}(\mathrm{V})$. The complement of V in U is called the *singular support* of T and is denoted by Supp sing(T). In examples (17.5.6)–(17.5.9), the singular support of the distributions considered consists of a single point. Clearly we have

$$\mathrm{Supp\ sing}(\mathrm{T}) \supset \mathrm{Supp\ sing}(\mathrm{D}^{\nu}\mathrm{T}), \qquad \mathrm{Supp\ sing}(\mathrm{T}) \supset \mathrm{Supp\ sing}(g \cdot \mathrm{T})$$

for $g \in \mathscr{E}(\mathrm{U})$.

PROBLEMS

1. Let K be a compact interval in **R** such that $0 \in \mathring{K}$, and let $\mathscr{H}$ be the hyperplane in $\mathscr{D}(\mathbf{R}; K)$ consisting of the functions which vanish at 0.

(a) For each $f \in \mathscr{H}$ let $\tilde{f}$ be the function $t \mapsto f(t)/t$. Then the mapping $f \mapsto \tilde{f}$ is an isomorphism of the subspace $\mathscr{H}$ onto $\mathscr{D}(\mathbf{R}; K)$. (Argue as in **(16.8.9.1)**.)
(b) Deduce from (a) that for each distribution T on **R** there exists a distribution S such that $1_{\mathbf{R}} \cdot S = T$, where $1_{\mathbf{R}}$ is the identity mapping $t \mapsto t$. If T is of order $\leqq r$, then S is of order $\leqq r + 1$.

2. Let μ be a *positive* measure on the open interval $]0, \infty[$ in **R**. Show that μ is the restriction of a distribution on **R** if and only if there exists an integer $k \geqq 0$ and a constant $c > 0$ such that $\mu([\varepsilon, 1]) \leqq c \cdot \varepsilon^{-k}$ for $\varepsilon \in]0, 1[$, and that if this condition is satisfied there exists a distribution of order $\leqq k + 1$ on **R** whose restriction to $]0, +\infty[$ is equal to μ. (To show that the condition is necessary, apply **(17.3.1.1)** and consider a function $f \geqq 0$ belonging to $\mathscr{D}^{(k)}(\mathbf{R})$ which is equal to 0 for $t \leqq \frac{1}{2}\varepsilon$ and to 1 for $t \in [\varepsilon, 1]$. To show that the condition is sufficient, show that the measure with base μ and density t^{k+1} on $]0, +\infty[$ is bounded, and apply the result of Problem 1.) If h is the least integer k satisfying the above condition, no distribution which extends μ is of order $< h$.

3. Let $K \subset \mathbf{R}$ be a compact interval, not consisting of a single point.

(a) Show that the mapping $f \mapsto Df$ is an isomorphism of $\mathscr{D}(\mathbf{R}; K)$ onto the subspace $\mathscr{H}$ of $\mathscr{D}(\mathbf{R}; K)$ consisting of the functions f such that $\int_{-\infty}^{+\infty} f(t)\,dt = 0$.
(b) Deduce from (a) that for each distribution T on **R** there exists a distribution S such that DS = T (i.e., a "primitive" of T).
(c) For each positive measure μ on **R**, show that if θ is the increasing function, continuous on the right, which corresponds to μ (and which is defined only up to a constant; cf. Section **13.18**, Problem 6), then $DT_\theta = \mu$.

4. Let f be a locally integrable function with respect to Lebesgue measure on $]0, a]$, where $a > 0$. Suppose that the integral $\int_0^a f(t)\,dt$ is *convergent*, i.e., that the limit of $\int_\varepsilon^a f(t)\,dt$ exists and is finite as $\varepsilon \to 0$. Show that, for each function $g \in \mathscr{D}(\mathbf{R})$, the integral $\int_0^a f(t)g(t)\,dt$ converges, and that $T : g \mapsto \int_0^a f(t)g(t)\,dt$ is a distribution of order $\leqq 1$ on **R**. This distribution T is of order 0 if and only if f is integrable over $]0, a]$. (If (t_n) is a decreasing sequence with limit 0, apply the formulas **(13.20.1)** and **(13.20.3)** on each of the intervals $]t_n, t_{n-1}[$.) If f is of class C^∞, then the singular support of T consists only of the point 0.

5. If a distribution $T \in \mathscr{D}'(\mathbf{R})$ is such that all the derivatives D^kT are measures, then $T = T_f$ for some $f \in \mathscr{E}(\mathbf{R})$. (Remark that if $D^{k+2}T$ is a measure, then $D^kT = T_g$, where g is a continuous function, by using Problem 3(c).)

6. Let $\pi : X \to X'$, $\pi' : X' \to X''$ be two proper mappings of class C^∞. Then for each current T on X we have $\pi'(\pi(T)) = (\pi' \circ \pi)(T)$.

7. Let $\pi : X \to X'$ be a proper mapping of class C^∞. If T is a p-current on X and α' is a C^∞ q-form on X′, show that $\pi(T \wedge {}^t\pi(\alpha')) = \pi(T) \wedge \alpha'$.

8. (a) Let $\pi : X \to X'$ be a C^∞-mapping, and let T′ be a p-current on X′ with the following property; if $A' = \text{Supp}(T')$, every point $x \in \pi^{-1}(A')$ has an open neighborhood V_x in X such that $\pi | V_x$ is a diffeomorphism onto an open subset of X′. Let p_x be the inverse of this diffeomorphism, and consider the p-current $p_x(T'_{\pi(V_x)})$ on V_x for each $x \in \pi^{-1}(A')$. Show that these p-currents are the restrictions of a p-current T on X with support contained in $\pi^{-1}(A)$. This p-current T is called the *inverse image* of T′ by π and is denoted by ${}^t\pi(T')$.
(b) If X, X′ are oriented manifolds of dimension n and if β' is a locally integrable differential $(n-p)$-form on X′, whose support A′ satisfies the condition in (a), show that if we put $\beta = {}^t\pi(\beta')$, then ${}^t\pi(T_{\beta'})$ coincides with T_β in a neighborhood of each point of $\pi^{-1}(A')$ at which π preserves the orientation, and with $-T_\beta$ in a neighborhood of each point of $\pi^{-1}(A')$ at which π reverses the orientation.
(c) Take $X = X' = \mathbf{R}$, and π to be the mapping $t \mapsto t^2 - a^2$, where $a > 0$. Then the inverse image by π of the Dirac measure ε_0 is $\varepsilon_a + \varepsilon_{-a}$. The inverse image by π of the Dirac 1-current $\varepsilon_{E(0)}$ (in the notation of (18.1)) is the 1-current $(1/2a)(\varepsilon_{E(a)} - \varepsilon_{-E(-a)})$.

9. Let X, Y be two oriented pure differential manifolds, of dimensions, respectively, n and m; let $\pi : X \to Y$ be a *submersion* and α a C^∞ differential $(n-m+k)$-form on X (where $k \leqq m$) with compact support. Show that the image $\pi(T_\alpha)$ of the $(m-k)$-current T_α is equal to $T_{\alpha^\flat}$, where $\alpha^\flat$ is the integral of α along the fibers of π (Section 16.24, Problem 11). (Reduce to the case of (16.7.4).)
For each p-current S on Y (with $p \leqq m$), we define the *inverse image* of S by π to be the $(n-m+p)$-current ${}^t\pi(S)$ such that

$$\langle {}^t\pi(S), \alpha \rangle = \langle S, \alpha^\flat \rangle$$

for every compactly supported C^∞ differential $(n-m+p)$-form α on X. If $S = T_{\beta'}$, where β' is a differential $(m-p)$-form on Y, then ${}^t\pi(T_{\beta'}) = T_\beta$, where $\beta = {}^t\pi(\beta')$.
If γ' is any C^∞ differential q-form on Y, then ${}^t\pi(S \wedge \gamma') = {}^t\pi(S) \wedge {}^t\pi(\gamma')$.
In particular, for each point $y \in Y$, ${}^t\pi(\varepsilon_y)$ is the closed $(n-m)$-chain element $\pi^{-1}(y)$, endowed with the orientation induced by π from the orientations of X and Y.

10. Let G be a Lie group of dimension n, and let v_0 be a left-invariant C^∞ differential n-form on G (19.16.4), so that the linear form $f \mapsto \int f v_0$ is a Haar measure β_0 on G (16.24.2).
(a) The image of β_0 under the diagonal mapping $x \mapsto (x, x)$ of G into $G \times G$ is called the (left) *trace measure* on $G \times G$ corresponding to v_0, and is denoted by tr. For each function $f \in \mathscr{K}(G \times G)$ we have therefore $\text{tr}(f) = \int f(x, x)\, d\beta_0(x)$.
(b) If $G = \mathbf{R}$, show that the measure tr on $\mathbf{R}^2$ is equal to $D_2 T_{\varphi_U}$, where U is the set of points $(\xi_1, \xi_2) \in \mathbf{R}^2$ such that $\xi_1 \leqq \xi_2$.
(c) Let $\pi : G \times G \to G$ be the mapping $(x, y) \mapsto xy^{-1}$, which is a submersion. For

each $z \in G$, $\pi^{-1}(z)$ is the left coset $(z, e)D$, where D is the diagonal subgroup of $G \times G$; so we may identify G with $(G \times G)/D$ and π with the canonical mapping

$$G \times G \to (G \times G)/D.$$

Let $v = {}^t\mathrm{pr}_1(v_0) \wedge {}^t\mathrm{pr}_2(v_0)$, which is the left-invariant $2n$-form on $G \times G$ corresponding to the Haar measure $\beta = \beta_0 \otimes \beta_0$. For each differential $2n$-form fv on $G \times G$, where $f \in \mathscr{D}(G \times G)$, show that $(fv)^\flat = f^\flat v_0$, where $f^\flat$ is the function on G (identified with $(G \times G)/D$) defined in Section 14.4, Problem 2, such that $f^\flat(z) = \int_G f(zw, w)\, d\beta_0(w)$. Deduce that, for each n-current T on G, we have $\langle {}^t\pi(T), fv\rangle = \langle T, f^\flat v_0\rangle$. If the Dirac measure ε_e on G is identified with an n-current, then ${}^t\pi(\varepsilon_e)$ is identified with the measure tr defined in (a).

6. REAL DISTRIBUTIONS. POSITIVE DISTRIBUTIONS

(17.6.1) Let X be a differential manifold. For each p the vector bundle $\bigwedge^p T(X)^*$ may be identified with a subbundle of the *real* vector bundle $\left(\bigwedge^p T(X)^*\right)_{(\mathbf{C})}$, which is the direct sum $\bigwedge^p T(X)^* \oplus i \bigwedge^p T(X)^*$ (16.18.5). It follows that $\mathscr{E}_p(X)$ (resp. $\mathscr{E}_p^{(r)}(X)$) is equal to $\mathscr{E}_{p,\mathbf{R}}(X) + i\mathscr{E}_{p,\mathbf{R}}(X)$ (resp. $\mathscr{E}_{p,\mathbf{R}}^{(r)}(X) + i\mathscr{E}_{p,\mathbf{R}}^{(r)}(X)$), where $\mathscr{E}_{p,\mathbf{R}}(X)$ (resp. $\mathscr{E}_{p,\mathbf{R}}^{(r)}(X)$) is the space of *real* differential p-forms of class C^∞ (resp. C^r) on X. A p-current T on X is then said to be *real* if its restriction to $\mathscr{E}_{p,\mathbf{R}}(X)$ is real-valued, and a real p-current is often identified with its restriction to $\mathscr{E}_{p,\mathbf{R}}(X)$. As in (13.2), for each current T the *conjugate* current $\overline{T}$ is defined; the real currents $\mathscr{R}T = \frac{1}{2}(T + \overline{T})$ and $\mathscr{I}T = (1/2i)(T - \overline{T})$ are called, respectively, the *real* and *imaginary* parts of T.

(17.6.2) *Every positive distribution* T *on a differential manifold* X (i.e., *every distribution* T *such that* $T(f) \geqq 0$ *for all* $f \geqq 0$ *in* $\mathscr{D}(X)$) *is a positive measure on* X.

We may assume that X is an open subset U of $\mathbf{R}^n$, by virtue of (13.1.9). Let K be a compact subset of U, and let $h : U \to [0, 1]$ be a C^∞-mapping with compact support and equal to 1 on K. For each real-valued function $f \in \mathscr{D}(U; K)$, we have

$$-\|f\|\, h \leqq f \leqq \|f\|\, h$$

and therefore $-\|f\|\, T(h) \leqq T(f) \leqq \|f\|\, T(h)$. Hence, if now

$$f = f_1 + if_2 \in \mathscr{D}(U; K)$$

with f_1, f_2 real-valued, we have $|T(f)| \leqq 2T(h)\, \|f\|$. This shows that T is a distribution of order 0, hence a *measure*. Moreover, with the same notation as in the proof of (17.1.2), for each compact neighborhood K′ of K contained in U, and for each function $f \in \mathscr{K}(U; K)$, the functions $g_k * f$ belong to $\mathscr{D}(U; K')$ and converge uniformly to f on K′ as $k \to \infty$, hence $T(f) = \lim_{k\to\infty} T(g_k * f)$; but if $f \geqq 0$, then $g_k * f \geqq 0$, hence $T(g_k * f) \geqq 0$ by hypothesis, and so finally $T(f) \geqq 0$. Hence T is a *positive* measure.

7. DISTRIBUTIONS WITH COMPACT SUPPORT. POINT-DISTRIBUTIONS

(17.7.1) A distribution whose support is *compact* is of *finite order*. To prove this we need only consider the case of a distribution T on an open subset U of $\mathbf{R}^n$. Let A be the (compact) support of T and let V be a compact neighborhood of A contained in U. If $h : U \to [0, 1]$ is a C^∞-mapping which is equal to 1 on V and whose support B is compact, then we have $T(f) = T(hf)$ for all $f \in \mathscr{D}(U)$. Define the seminorms (17.1.1) on $\mathscr{E}(U)$ by taking a fundamental sequence (K_m) of compact subsets of U, and choose m large enough so that $B = \text{Supp}(h) \subset K_m$. Then, for each function $f \in \mathscr{D}(U)$, the function hf belongs to $\mathscr{D}(U; K_m)$ and therefore, by virtue of (12.14.11) and the definition of a distribution, there exists an integer r and a constant $a > 0$ such that, for all $f \in \mathscr{D}(U)$,

$$|T(f)| \leqq a \cdot p_{r,m}(hf) \leqq ac \cdot p_{r,m}(f), \tag{17.7.1.1}$$

where c is a constant independent of f (17.1.4). This proves that T is of order $\leqq r$. Also, for each function $f \in \mathscr{E}^{(r)}(U)$, we can define $T(f)$ to be equal to $T(hf)$, because $hf \in \mathscr{E}^{(r)}(U)$. The inequality (17.7.1.1) shows that the linear form T thus extended is *continuous* on $\mathscr{E}^{(r)}(U)$. In the same way we may show that, if a p-current T has compact support, then its value $\langle T, \alpha \rangle$ can be defined for *all* p-forms $\alpha \in \mathscr{E}_p(X)$. If T has order $\leqq r - 1$, this allows us to define the *image* $\pi(T)$ of T by an *arbitrary* mapping π of class C^r (where $r \geqq 1$) from X to X′. For $T({}^t\pi(\alpha'))$ is defined for all differential p-forms $\alpha' \in \mathscr{E}_p^{(r-1)}(X')$; moreover, for each compact subset K′ of X′, the linear mapping $\alpha' \mapsto h \cdot {}^t\pi(\alpha')$ of $\mathscr{D}_p^{(r-1)}(X'; K')$ into $\mathscr{D}_p^{(r-1)}(X; B)$ is continuous, by virtue of (17.2), the local expression of ${}^t\pi(\alpha')$, and (17.1.4). Hence the assertion follows. The relation $\text{Supp}(\pi(T)) \subset \pi(\text{Supp}(T))$ is still valid.

We denote by $\mathscr{E}'(U)$ the subspace of $\mathscr{D}'(U)$ consisting of the distributions with compact support.

(17.7.2) *Let* $T \in \mathscr{E}'(U)$ *be a distribution of order* r *with compact support* $A \subset \mathbf{R}^n$. *Then* $T(f) = 0$ *for all functions* $f \in \mathscr{E}(U)$ *whose derivatives of order* $\leqq r$ *vanish at all points of* A.

For each $\varepsilon > 0$, let A_ε be the compact subset of $\mathbf{R}^n$ consisting of the points whose distance from A (relative to the norm $\|x\| = \sup|\xi_i|$ on $\mathbf{R}^n$) is $\leqq \varepsilon$. Then there exists $\varepsilon_0 > 0$ such that $A_\varepsilon \subset U$ for all $\varepsilon \leqq \varepsilon_0$ (3.18.2). Since the derivatives of order $r + 1$ of f are bounded in absolute value on A_{ε_0} by a number independent of ε, it follows from the hypothesis on f and from Taylor's formula (8.14.3) that, for $\varepsilon \leqq \frac{1}{3}\varepsilon_0$ and any $x \in A_\varepsilon$, we have $|f(x)| \leqq b \cdot \varepsilon^{r+1}$, where b is a constant independent of ε; for there exists a vector t with norm $\leqq 2\varepsilon$ such that $x - t \in A$, and the derivatives of order $\leqq r$ of f are zero at the point $x - t$. The same reasoning holds for each derivative $D^\nu f$ such that $|\nu| \leqq r$, and we may therefore suppose that the constant b is such that

(17.7.2.1) $$\sup_{x \in A_\varepsilon} |D^\nu f(x)| \leqq b \cdot \varepsilon^{r+1-|\nu|}$$

for all $|\nu| \leqq r$ and all $\varepsilon \leqq \frac{1}{3}\varepsilon_0$.

Next, consider the functions g_k introduced in (17.1.2), and put

$$u_k = g_k * \varphi_{A_{2/k}}$$

for each integer $k \geqq 3\varepsilon_0^{-1}$. Since the support of g_k is contained in the ball with center 0 and radius $1/k$, it follows immediately (14.5.4) that $u_k(x) = 1$ for $x \in A_{1/k}$ and that $\operatorname{Supp}(u_k) \subset A_{3/k}$. Moreover, by (13.8.6), for all $x \in \mathbf{R}^n$ and all ν we have

$$D^\nu u_k(x) = k^{|\nu|+n} \int_{A_{2/k}} D^\nu g(k(x - t))\, d\lambda(t),$$

and therefore

$$|D^\nu u_k(x)| \leqq k^{|\nu|+n} \int_{\mathbf{R}^n} |D^\nu g(kt)|\, d\lambda(t) = k^{|\nu|} \int_{\mathbf{R}^n} |D^\nu g(t)|\, d\lambda(t);$$

in other words

(17.7.2.2) $$\sup_{x \in \mathbf{R}^n} |D^\nu u_k(x)| \leqq c \cdot k^{|\nu|}$$

for all ν such that $|\nu| \leqq r$, where c is a constant independent of k.

This being so, since $u_k(x) = 1$ in some neighborhood of A, we have $T(f) = T(u_k f)$ for all k, and by hypothesis we have

(17.7.2.3) $$|T(u_k f)| \leqq a \cdot p_{r,m}(u_k f)$$

for some constant a independent of k. However, the definition (17.1.1) and Leibniz' rule (8.13.2) show that there exists a constant M such that

$$p_{r,m}(u_k f) \leqq M \cdot \sup_{|\rho|+|\sigma| \leqq r,\, x \in A_{3/k}} |D^\rho u_k(x)| \cdot |D^\sigma f(x)|$$

for all k. Using these inequalities, (17.7.2.1), and (17.7.2.2), we obtain

$$p_{r,m}(u_k f) \leqq \mathrm{C}/k,$$

where C is a constant. Hence, by (17.7.2.3), we have

$$\mathrm{T}(f) = \lim_{k \to \infty} \mathrm{T}(u_k f) = 0. \qquad \text{Q.E.D.}$$

We remark that the condition $f(x) = 0$ for all $x \in \mathrm{A}$ does not ensure that $\mathrm{T}(f) = 0$, as is shown by the example in which $n = 1, f(x) = x$ and $\mathrm{T} = \mathrm{D}\varepsilon_0$.

From (17.7.2) we obtain the following corollary:

(17.7.3) *Every distribution* $\mathrm{T} \in \mathscr{E}'(\mathrm{U})$ *whose support consists of a single point* a *is a linear combination of a finite number of derivatives* $\mathrm{D}^\nu \varepsilon_a$ *of the Dirac measure at the point* a.

Let r be the order of T. For each $f \in \mathscr{E}(\mathrm{U})$, we can write

$$f(x) = \sum_{|\nu| \leqq r} \frac{1}{\nu!} \mathrm{D}^\nu f(a)(x-a)^\nu + g(x),$$

where all the derivatives of g of order $\leqq r$ vanish at the point a. By virtue of (17.7.2), we have $\mathrm{T}(g) = 0$ and therefore, if c_ν is the value of T on the function $(x-a)^\nu/\nu!$, we obtain

$$\mathrm{T}(f) = \sum_{|\nu| \leqq r} c_\nu \mathrm{D}^\nu f(a).$$

PROBLEMS

1. Let $\mathrm{T} \in \mathscr{E}'(\mathbf{R}^n)$ be a distribution with compact support K, of order $\leqq m$. For each open neighborhood U of K, show that T is a sum of derivatives of order $\leqq m$ of a finite number of measures with supports contained in U. (Let N be the number of multi-indices ν such that $|\nu| \leqq m$. Also let V be a relatively compact open neighborhood of K such that $\bar{\mathrm{V}} \subset \mathrm{U}$. Show that the mapping $\pi : f \mapsto (\mathrm{D}^\nu f)_{|\nu| \leqq m}$ is an isomorphism of the Banach space $\mathscr{D}^{(m)}(\mathbf{R}^n; \bar{\mathrm{V}})$ onto a closed subspace F of the Banach space $(\mathscr{C}(\bar{\mathrm{V}}))^{\mathrm{N}}$, and hence that $\mathrm{T} \circ \pi^{-1}$ is a continuous linear form on F. Use the Hahn–Banach theorem (Section 12.15, Problem 4) to extend this linear form to a continuous linear form on $(\mathscr{C}(\bar{\mathrm{V}}))^{\mathrm{N}}$.)

2. (a) For each function $f \in \mathscr{D}(\mathbf{R})$, show that the limit

$$\mathrm{T}(f) = \lim_{n \to \infty} \left(\left(\sum_{k=1}^{n} f(1/k) \right) - nf(0) - (\log n) \cdot f'(0) \right)$$

exists and is finite. T is a distribution on $\mathbf{R}$ of order 2, whose support A consists of the points $1/n$ $(n \geqq 1)$ and 0. If $f|A = 0$, then $T(f) = 0$.
(b) Show that T is not expressible as a finite sum of derivatives of order $\leqq 2$ of measures *with support contained in* A.
(c) For each integer $k \geqq 1$, let $f_k \in \mathscr{D}(\mathbf{R})$ be a function which takes its values in the interval $[0, k^{-1/2}]$, is equal to $k^{-1/2}$ for $t \geqq 1/k$, and is zero for $t \leqq 1/(k+1)$. All the derivatives of f_k are zero at the points of A, and the sequence (f_k) converges uniformly to 0; but the sequence $T(f_k)$ tends to $+\infty$. Hence, in order that $T(f_k)$ should tend to 0, it is not sufficient that (f_k) and each of the sequences $(D^\nu f_k)$ should tend to 0 uniformly *on* A.

8. THE WEAK TOPOLOGY ON SPACES OF DISTRIBUTIONS

The space $\mathscr{D}'_p(X)$ of p-currents on a differential manifold X of dimension n is a vector space of linear forms on $\mathscr{D}_p(X)$, and therefore can naturally be equipped with the *weak topology* (12.15.2) defined by the seminorms

(17.8.1)
$$T \mapsto |\langle T, \alpha \rangle|,$$

where α runs through $\mathscr{D}_p(X)$. Whenever we use topological notions in the spaces $\mathscr{D}'_p(X)$, it is always the weak topology that is meant, unless the contrary is expressly stated.

Example

(17.8.1.1) Let (g_k) be a *regularizing sequence* of functions belonging to $\mathscr{D}(\mathbf{R}^n)$, and identify the g_k with distributions. Then the sequence (g_k) *converges weakly* to the Dirac measure ε_0 at the origin.

It follows directly from the definition of the weak topology that the linear mappings $T \mapsto T \wedge \omega$, $T \mapsto \pi(T)$, and $T \mapsto j(T)$ defined in (17.3.5) and (17.3.7) are *continuous.*

Likewise, it follows from the definition (17.5.5.1) that the differentiations $T \mapsto D^\nu T$ on $\mathscr{D}'(U)$ (where U is open in $\mathbf{R}^n$) are *continuous.*

Remark

(17.8.2) The derivatives $D_i T$ of a distribution can also be defined as limits in the space $\mathscr{D}'(U)$: if $(\mathbf{e}_j)$ is the canonical basis of $\mathbf{R}^n$, then we have

(17.8.2.1)
$$D_i(T) = \lim_{t \to 0, t \neq 0} (\gamma(t\mathbf{e}_i)T - T)/t.$$

For if $f \in \mathscr{D}(\mathrm{U})$, the function $(\gamma(t\mathbf{e}_i)f - f)/t$ is the mapping

$$\mathbf{x} \mapsto (f(\mathbf{x} - t\mathbf{e}_i) - f(\mathbf{x}))/t,$$

and Taylor's formula applied to f and its derivatives shows that, as $t \to 0$, the function $(\gamma(t\mathbf{e}_i)f - f)/t$ tends to $-\mathrm{D}_i f$ in the space $\mathscr{D}(\mathrm{U}; \mathrm{K})$, where K is any compact neighborhood of the support of f.

(17.8.3) *Let Z be a metric space, A a subset of Z, and let $z \mapsto \mathrm{T}_z$ be a mapping of A into $\mathscr{D}'_p(\mathrm{X})$, and z_0 a point of $\bar{\mathrm{A}}$. Suppose that, for each p-form $\alpha \in \mathscr{D}_p(\mathrm{X})$, the function $z \mapsto \langle \mathrm{T}_z, \alpha \rangle$ tends to a limit as z tends to z_0 whilst remaining in A. Then the linear form $\alpha \mapsto \lim_{z \to z_0, z \in \mathrm{A}} \langle \mathrm{T}_z, \alpha \rangle$ is a p-current, equal to $\lim_{z \to z_0, z \in \mathrm{A}} \mathrm{T}_z$ in $\mathscr{D}'_p(\mathrm{X})$. In particular, if a sequence (T_k) of p-currents on X is such that, for each p-form $\alpha \in \mathscr{D}_p(\mathrm{X})$, the sequence $\langle \mathrm{T}_k, \alpha \rangle$ converges in* **C**, *then the sequence (T_k) has a limit in $\mathscr{D}'_p(\mathrm{X})$.*

This is an immediate consequence of the Banach–Steinhaus theorem (12.16.5) applied to each of the Fréchet spaces $\mathscr{D}_p(\mathrm{X}; \mathrm{K})$.

Likewise, if Z is an open interval in **R** (resp. an open set in **C**), and if $z \mapsto \mathrm{T}_z$ is *weakly differentiable* (resp. *weakly analytic*) in Z, then the *weak derivative* T'_z is a p-current (12.16.6).

Finally, suppose that Z is locally compact, and let μ be a positive measure on Z. If for each form $\alpha \in \mathscr{D}_p(\mathrm{X})$ the function $z \mapsto \langle \mathrm{T}_z, \alpha \rangle$ is μ-integrable, then by applying (13.10.4) to each of the Fréchet spaces $\mathscr{D}_p(\mathrm{X}; \mathrm{K})$ it follows that the mapping $\alpha \mapsto \int \langle \mathrm{T}_z, \alpha \rangle \, d\mu(z)$ is a p-current, which is denoted by $\int \mathrm{T}_z \, d\mu(z)$ and is called the *weak integral* of $z \mapsto \mathrm{T}_z$ with respect to μ.

(17.8.4) For each integer r, the topology induced on $\mathscr{D}'^{(r)}_p(\mathrm{X})$ by the weak topology of $\mathscr{D}'_p(\mathrm{X})$ is *coarser* than the weak topology of $\mathscr{D}'^{(r)}_p(\mathrm{X})$ considered as a space of linear forms on $\mathscr{D}^{(r)}_p(\mathrm{X})$, because $\mathscr{D}_p(\mathrm{X}) \subset \mathscr{D}^{(r)}_p(\mathrm{X})$. In particular, on the space $\mathrm{M}(\mathrm{X}) = \mathscr{D}'^{(0)}(\mathrm{X})$ of *measures* on X, the topology induced by the weak topology of $\mathscr{D}'(\mathrm{X})$ is *coarser than the vague topology*. More particularly, since the spaces $\mathrm{L}^1(\mathrm{U})$, $\mathrm{L}^2(\mathrm{U})$, and $\mathrm{L}^\infty(\mathrm{U})$ (where U is an open set in $\mathbf{R}^n$) are identified (algebraically) with subspaces of $\mathscr{D}'(\mathrm{U})$, and since the norm topologies on these spaces are in all cases finer than the vague topology, they are *a fortiori* finer than the topologies induced by the weak topology of $\mathscr{D}'(\mathrm{U})$.

In fact, the topologies induced by the weak topology of $\mathscr{D}'(\mathrm{U})$ on these spaces are *strictly coarser* than the other topologies mentioned above. For example, if (g_k) is the "regularizing sequence" of functions introduced in (17.1.2), which converges weakly to ε_0, then it follows from the continuity of differentiation relative to the weak topology that each of the sequences $(D^\nu g_k)$ converges weakly to $D^\nu \varepsilon_0$, which is not a measure for $\nu \neq 0$. We shall see later (17.12.3) that in fact *every* distribution on U is the weak limit of a sequence of functions belonging to $\mathscr{D}(U)$.

(17.8.5) *Let* H *be a bounded subset of* $\mathscr{D}'_p(X)$ (12.15).

(i) *The weak closure of* H *in* $\mathscr{D}'_p(X)$ *is compact and metrizable relative to the weak topology.*

(ii) *If* U *is open in* $\mathbf{R}^n$ *and* H *a bounded subset of* $\mathscr{D}'(U)$, *then* (with the notation of (17.1)) *for each integer m there exists an integer r and a number* $c_{r,m}$ *such that* $|T(f)| \leq c_{r,m} p_{r,m}(f)$ *for all* $f \in \mathscr{D}(U; K_m)$ *and all* $T \in H$.

Assertion (ii) follows from the Banach–Steinhaus theorem (12.16.4) applied to the Fréchet spaces $\mathscr{D}_p(X; K_m)$. Assertion (i) then follows by using (17.2.2), by an argument analogous to that of (13.4.2), which we shall not repeat here. In particular, the restrictions of the distributions $T \in H$ to a relatively compact open set have *bounded orders.*

(17.8.6) *If a sequence* (T_k) *of currents belonging to* $\mathscr{D}'_p(X)$ *converges weakly to* T, *then for each compact subset* K *of* X *and each bounded subset* B *of* $\mathscr{D}_p(X; K)$ (17.1), *the sequence* $(\langle T_k, \alpha \rangle)$ *converges to* $\langle T, \alpha \rangle$ *uniformly on* B.

We reduce to the case where X is an open subset of $\mathbf{R}^n$, and then the proposition is an immediate consequence of (17.8.5(ii)).

(17.8.7) *Let* (T_k) *be a sequence of p-currents on* X, *and* (α_k) *a sequence of q-forms belonging to* $\mathscr{E}_q(X)$, *where* $q \leq p$. *Then the sequence* $(T_k \wedge \alpha_k)$ *converges to zero in* $\mathscr{D}'_{p-q}(X)$ *in each of the following two cases*:

(1) *The sequence* (α_k) *is bounded in* $\mathscr{E}_q(X)$ (17.1) *and the sequence* (T_k) *converges to* 0 *in* $\mathscr{D}'_p(X)$.

(2) *The sequence* (α_k) *converges to* 0 *in the Fréchet space* $\mathscr{E}_q(X)$ *and the sequence* (T_k) *is bounded in* $\mathscr{D}'_p(X)$.

We have to show that, for each $(p-q)$-form $\beta \in \mathscr{D}_{p-q}(X)$, the sequence $(\langle T_k, \alpha_k \wedge \beta \rangle)$ tends to 0. The question reduces immediately to the case where X is an open subset U of $\mathbf{R}^n$, and $p = q = 0$. Let K be the support of β. In case (1), the sequence $(\alpha_k \beta)$ is bounded in $\mathscr{D}(U; K)$ by virtue of (17.1.4), hence the result follows from (17.8.6). In case (2), the sequence $(\alpha_k \beta)$ converges to 0 in $\mathscr{D}(U; K)$ by virtue of (17.1.4), and the result follows from (17.8.5(ii)).

Example

(17.8.8) Let (λ_n) be an increasing sequence of strictly positive real numbers, and let (c_n) be a sequence of complex numbers; suppose that there exist two numbers $\rho, \sigma > 0$ such that $\lambda_n \geqq n^\rho$ and $|c_n| \leqq n^\sigma$ for all $n \geqq 1$. Then the series

(17.8.8.1)
$$\sum_{n=0}^{\infty} c_n e^{i\lambda_n x}$$

is *convergent in* $\mathscr{D}'(\mathbf{R})$. There exists an integer $k > 0$ such that $|c_n/\lambda_n^k| \leqq 1/n^2$, namely any integer k such that $k\rho - \sigma \geqq 2$. Then it is enough to show that the series

(17.8.8.2)
$$\sum_{n=0}^{\infty} \lambda_n^{-k} c_n e^{i\lambda_n x}$$

converges in $\mathscr{D}'(\mathrm{R})$, because if T is its sum, then the series (17.8.8.1) will converge to $\mathrm{D}^k\mathrm{T}$, by virtue of the continuity of differentiation; but the series of continuous functions (17.8.8.2) is normally convergent in $\mathbf{R}$, hence converges also in $\mathscr{D}'(\mathrm{R})$ if its terms are regarded as distributions (17.8.4).

PROBLEMS

1. Let $\mathbf{z}_x$ be a nonzero tangent n-vector at a point $x \in \mathbf{R}^n$, and let (V_k) be a fundamental system of bounded open neighborhoods of x in $\mathbf{R}^n$. Show that the sequence of n-currents (open n-chain elements (17.5.1.1)) $(\lambda_n(\mathrm{V}_k)^{-1}\mathrm{V}_k)$ tends to the limit $c\mathbf{z}_x$, where $c^{-1} = \langle \mathbf{z}_x, v_0(x)\rangle$, v_0 being the canonical n-form on $\mathbf{R}^n$.

2. Let (V_k) be a fundamental system of bounded open neighborhoods of the origin in $\mathbf{R}^{n-1}$. Let σ_k be the locally integrable differential $(n-1)$-form on $\mathbf{R}^n = \mathbf{R} \times \mathbf{R}^{n-1}$ which is equal to $\lambda_{n-1}(\mathrm{V}_k)^{-1}\, d\xi^2 \wedge \ldots \wedge d\xi^n$ on $\mathbf{R} \times \mathrm{V}_k$ and is 0 elsewhere. Show that, as $k \to +\infty$, the sequence of 1-currents T_{σ_k} (17.5.1) tends to the 1-chain element without boundary $\mathbf{R} \times \{0\}$ (17.5.2), $\mathbf{R}$ being canonically oriented.

3. Let T be a distribution on $\mathbf{R}$. For each $h \neq 0$ in $\mathbf{R}$, put $\Delta_h \mathrm{T} = \gamma(h)\mathrm{T} - \mathrm{T}$, and $\Delta_h^p\mathrm{T} = \Delta_h(\Delta_h^{p-1}\mathrm{T})$ for all integers $p > 1$.

   (a) Show that as $h \to 0$ the distribution $(1/h^p)\Delta_h^p\mathrm{T}$ tends to $\mathrm{D}^p\mathrm{T}$.
   (b) Let f be a continuous function on $\mathbf{R}$. The function f is said to be *completely monotone* if, for each integer $p \geqq 1$, $\Delta^p f(x; h, h, \ldots, h)$ has the same sign as h^p for all $h \neq 0$ (Section 8.12, Problem 4). Show that f is then analytic. (Use (a), Problem 5 of Section 17.5, and Problem 7(c) of Section 9.9.) (*S. Bernstein's theorem*.)

(c) Suppose that, for some integer $p \geqq 1$, $\Delta^p f(x; h, h, \ldots, h)/h^p \to 0$ as $h \to 0$, for all $x \in \mathbf{R}$. Show that there exists a dense open subset U of **R** such that, on each connected component of U, f is equal to a polynomial of degree $\leqq p-1$. (Use (a), and observe that if a sequence (f_n) of continuous functions converges pointwise to 0, then $\sup_n (|f_n|)$ is finite at each point; then apply (12.16.2).) If also there exists a Lebesgue-integrable function $g \geqq 0$ such that $|\Delta^p f(x; h, h, \ldots, h)/h^p| \leqq g(x)$ for all $h \neq 0$, then f is a polynomial of degree at most $p-1$.

4. (a) Show that a distribution on $\mathbf{R}^n$ which is *invariant* under all translations of $\mathbf{R}^n$ (i.e., is equal to its image under each translation) is a *constant* function (or, more accurately, a distribution T_c, where c is a constant function on $\mathbf{R}^n$). (Use (17.8.2.1).) What are the translation-invariant n-currents on $\mathbf{R}^n$?

(b) A 1-current on $]0, +\infty[$ which is invariant under all homotheties $h_\lambda : x \mapsto \lambda x$ (where $\lambda > 0$) is of the form T_{cv_0}, where c is a constant. A distribution on $]0, +\infty[$ invariant under all h_λ is of the form T_f, where $f(x) = cx^{-1}$, c being a constant. (Use the isomorphism $x \mapsto \log x$ of $]0, +\infty[$ onto **R**, and (a) above.)

5. (a) Let f be a holomorphic function on an open set $U \subset \mathbf{C}$, and let $z_0 \in U$. Let $S \subset U$ be a closed annulus with center z_0, defined by $r_1 \leqq |z - z_0| \leqq r_2$. Also let u be a C^∞-function on **R**, everywhere $\geqq 0$, with support contained in $[r_1, r_2]$, and such that $\int u(t)\, dt = 1$. Show that

$$f(z_0) = \frac{1}{2\pi} \iint_S f(x + iy) \frac{u(r)}{r}\, dx\, dy,$$

where $r = |z - z_0| = ((x - x_0)^2 + (y - y_0)^2)^{1/2}$.

(b) Let U be an open disk in **C** with center $x_0 \in \mathbf{R}$, let U_+ be the intersection of U with the half-plane $\mathscr{I} z > 0$, and let f be a holomorphic function on U_+, regarded as a distribution belonging to $\mathscr{D}'(U_+)$. Show that f is the restriction to U_+ of a distribution belonging to $\mathscr{D}'(U)$ if and only if, for each compact interval $K \subset \mathbf{R} \cap U$ with center x_0, there exists an integer $k > 0$ and an interval $J =]0, c[$ in **R**, such that

$$\sup_{x \in K,\, y \in J} |y^k f(x + iy)| < +\infty.$$

(To show that the condition is sufficient, consider an iterated primitive of f in U_+. To show that it is necessary, observe that the hypothesis implies that, if J is any interval $]0, c[$ in **R** such that $K \times \bar{J} \subset U_+ \cup \mathbf{R}$, there exists a constant $A > 0$ and a multi-index $\alpha = (\alpha_1, \alpha_2)$ such that

(1) $$\left| \iint_{K \times \bar{J}} f(x + iy) v(x, y)\, dx\, dy \right| \leqq A \cdot \sup_{(x, y) \in K \times \bar{J}} |D^\alpha v(x, y)|$$

for all $v \in \mathscr{D}(U)$. Let $z_0 = x_0 + iy_0 \in U_+$, and let S be a closed annulus with center z_0 contained in U_+. For each $z = x + iy \in U_+$ such that the line passing through z and z_0 meets the real axis at a point $t \in K$, let S_z be the annulus with center z which is the image of S under the homothety with center t and ratio y/y_0. Then by (a) above we have

(2) $$f(z) = \frac{1}{2\pi} \iint_{S_z} f(\xi + i\eta) u_z(\xi, \eta)\, d\xi\, d\eta,$$

where

$$u_z(\xi, \eta) = \frac{u(y_0\, y^{-1} |\xi + i\eta - z|)}{y_0\, y^{-1} |\xi + i\eta - z|}.$$

Then use the inequality (1).)

(c) Show that the extension property in (b) is equivalent to the following: in every open interval $\mathrm{I} \subset \mathrm{U} \cap \mathbf{R}$ with center x_0, the functions $x \mapsto f(x+iy)$ converge to a distribution in $\mathscr{D}'(\mathrm{I})$ as y tends to 0 through positive values. (For the necessity of the condition, use Section 17.12, Problem 6.)

(d) Show that, for each compact interval K in $\mathbf{R}$ and each function $f \in \mathscr{E}(\mathbf{R})$, there exists a sequence (f_n) of polynomials such that, for each integer $p \geqq 0$, the restrictions to K of the functions $\mathrm{D}^p f_n$ converge uniformly on K to $\mathrm{D}^p f$ (cf. (14.11.3)). If $\mathrm{T} \in \mathscr{E}'(\mathbf{R})$ has support contained in K, show that the function u defined by $u(z) = \int (x-z)^{-1}\, d\mathrm{T}(x)$ for $z \in \mathbf{C} - \mathrm{K}$ is analytic on this open set and that for each polynomial f, we have

$$\langle \mathrm{T}, f \rangle = \frac{1}{2\pi i} \int_\gamma u(z) f(z)\, dz,$$

where γ is a suitably chosen circuit in $\mathbf{C} - \mathrm{K}$. Conclude that T is the limit in $\mathscr{E}'(\mathbf{R})$ of a family of functions of the form $x \mapsto \mathrm{F}(x+iy)$, where F is holomorphic in the half-plane $\mathscr{I} z > 0$, and y tends to 0 through positive values.

9. EXAMPLE: FINITE PARTS OF DIVERGENT INTEGRALS

(17.9.1) Let X be an oriented pure differential manifold of dimension n, and let F be a *real*-valued continuous function on X. Suppose that the open set $\mathrm{U}_0 = \{x \in \mathrm{X} : \mathrm{F}(x) > 0\}$ is not empty and that the *frontier* P of U_0 (where $\mathrm{F}(x) = 0$) is *negligible* with respect to Lebesgue measure on X. The function $\mathrm{F}^{-1}\varphi_{\mathrm{U}_0}$, which is equal to 0 in $\mathrm{X} - \mathrm{U}_0$ and coincides with F^{-1} on U_0, is not in general locally integrable in a neighborhood of P, because F^{-1} is not in general bounded in such a neighborhood. If v_0 is a differential n-form on X belonging to the orientation of X, then the mapping $f \mapsto \int_{\mathrm{U}_0} \mathrm{F}^{-1} f v_0$ is a *measure on* $\mathrm{X} - \mathrm{P}$, zero on $\mathrm{X} - (\mathrm{P} \cup \mathrm{U}_0)$, which in general cannot be extended to a *measure on* X. In this section we shall indicate methods of wide applicability of constructing an extension which is a *distribution on* X (more precisely, this distribution will extend to $\mathscr{D}(\mathrm{X})$ the *restriction* of the measure $f \mapsto \int_{\mathrm{U}_0} \mathrm{F}^{-1} f v_0$ to $\mathscr{D}(\mathrm{X} - \mathrm{P})$).

A first method consists of considering the integral $\int_{\mathrm{U}_0} \mathrm{F}^\zeta f v_0$ (where as usual t^ζ means $e^{\zeta \log t}$ for real numbers $t > 0$), which is an *analytic* function of ζ in the half-plane $\mathrm{E}_0 : \mathscr{R}\zeta > 0$ in $\mathbf{C}$, for all functions $f \in \mathscr{K}(\mathrm{X})$ (13.8.6). In other words, on restricting to $\mathscr{D}(\mathrm{X})$, we obtain a distribution $\mathrm{T}_\zeta : f \mapsto \int_{\mathrm{U}_0} \mathrm{F}^\zeta f v_0$

on X, which is a *weakly analytic* function (12.16.6.1) of ζ on E_0, and which we seek to *analytically continue* to a larger open subset of **C**. It may happen that such an analytic continuation exists on an open set containing the point $\zeta = -1$, for each $f \in \mathscr{K}(X)$, in which case its value at $\zeta = -1$ will still be a *measure* on X. Another case which occurs frequently (see the examples below) is that in which an analytic continuation $\zeta \mapsto T_\zeta^{(\alpha)}$ exists on $E_\alpha - \{-1\}$, where E_α is a half-plane $\mathscr{R}\zeta > \alpha$, with $\alpha < -1$, and $T_\zeta^{(\alpha)}$ has a *simple pole* at the point -1; in other words, for each function $f \in \mathscr{D}(X)$, we have

$$T_\zeta^{(\alpha)}(f) = \frac{A(f)}{\zeta + 1} + B_\zeta(f),$$

where A is a linear form on $\mathscr{D}(X)$ and, for all ζ in a neighborhood $V \subset E_\alpha$ of -1, B_ζ is a *distribution* on X and the function $\zeta \mapsto B_\zeta$ is *weakly analytic* in V. It follows then (12.16.6.1) that A is a distribution. Moreover, if $\text{Supp}(f) \cap P = \varnothing$, then we have $A(f) = 0$; for $\zeta \mapsto T_\zeta(f)$ is then an *entire* function of ζ, hence coincides with $\zeta \mapsto T_\zeta^{(\alpha)}(f)$ in $E_\alpha - \{-1\}$; consequently, as $\zeta \to -1$, $T_\zeta(f)$ and $B_\zeta(f)$ both tend to finite limits, whence the assertion follows. Hence we have $\text{Supp}(A) \subset P$. Consequently B_{-1} is an *extension to* $\mathscr{D}(X)$ of the distribution $f \mapsto \int_{U_0} F^{-1} f v_0$ defined on $\mathscr{D}(X - P)$, and is called the *finite part* of this integral. (Of course, there are infinitely many such extensions, obtained by adding to B_{-1} any distribution with support P.)

Examples

(17.9.2) Take $X = \mathbf{R}^n$ $(n \geqq 1)$, and let F be the function $r(x) = \left(\sum_{j=1}^{n} (\xi^j)^2\right)^{1/2}$, so that $U_0 = \mathbf{R}^n - \{0\}$ and $P = \{0\}$. Take v_0 to be the canonical n-form $d\xi^1 \wedge d\xi^2 \wedge \cdots \wedge d\xi^n$.

If σ is the "solid angle" differential $(n-1)$-form on the unit sphere $\mathbf{S}_{n-1}$, then by (16.24.9) we may write

(17.9.2.1) $$T_\zeta(f) = \int_{\mathbf{R}^n - \{0\}} r^\zeta f v_0 = \int_0^{+\infty} \rho^{\zeta+n-1}\, d\rho \int_{\mathbf{S}_{n-1}} f(\rho z)\sigma(z)$$

for $\mathscr{R}\zeta > 0$ and any $f \in \mathscr{K}(\mathbf{R}^n)$, and the function

(17.9.2.2) $$M_f(\rho) = \int_{\mathbf{S}_{n-1}} f(\rho z)\sigma(z)$$

is continuous for $\rho \geqq 0$, compactly supported, and of class C^∞ on $]0, +\infty[$ (13.8.6). Hence we see already that, for $f \in \mathscr{K}(\mathbf{R}^n)$, the function $r^\zeta f$ is integrable not only for $\mathscr{R}\zeta > 0$, but for $\mathscr{R}\zeta > -n$ (13.21.10), and T_ζ is therefore a *measure on* $\mathbf{R}^n$ for all ζ in this half-plane. For each integer $m > 0$ and for $f \in \mathscr{D}(\mathbf{R}^n)$, we shall define an analytic continuation $\zeta \mapsto T_\zeta^{(m)}(f)$ of the function $\zeta \mapsto T_\zeta(f)$, as follows.

Replace f in (17.9.2.2) by its Taylor series up to order $2m$ (8.14.3):

(17.9.2.3) $$f(\rho z) = \sum_{0 \leqq |\nu| \leqq 2m} \frac{1}{\nu!} \rho^{|\nu|} z^\nu D^\nu f(0) + \rho^{2m} g(\rho, z),$$

where $(\rho, z) \mapsto g(\rho, z)$ is continuous on $[0, +\infty[\times \mathbf{S}_{n-1}$ and has support contained in a set of the form $[0, \rho_0[\times \mathbf{S}_{n-1}$. Hence, by splitting the integral into two parts, we have *for* $\mathscr{R}\zeta > -n$

(17.9.2.4) $$\begin{aligned} T_\zeta(f) = &\int_0^1 \rho^{\zeta+2m+n-1}\, d\rho \int_{\mathbf{S}_{n-1}} g(\rho, z)\sigma(z) \\ &+ \sum_{0 \leqq |\nu| \leqq 2m} \frac{c_\nu}{\zeta + |\nu| + n} D^\nu f(0) \\ &+ \int_1^{+\infty} \rho^{\zeta+n-1} M_f(\rho)\, d\rho, \end{aligned}$$

where

(17.9.2.5) $$c_\nu = \int_{\mathbf{S}_{n-1}} \frac{1}{\nu!} z^\nu \sigma(z).$$

Now, on the right-hand side of (17.9.2.4), the last integral is an *entire* function of ζ (13.8.6); the first is an analytic function of ζ in the half-plane $\mathscr{R}\zeta > -n - 2m$; and therefore the right-hand side of (17.9.2.4) is a *meromorphic function* of ζ in the half-plane $\mathscr{R}\zeta > -n - 2m$, having *at most simple poles* at the points of the form $-n - k$, where $0 \leqq k \leqq 2m$. It is this function which is the desired analytic continuation $T_\zeta^{(m)}(f)$. Since $T_\zeta^{(m+1)}(f)$ and $T_\zeta^{(m)}(f)$ coincide with $T_\zeta(f)$ for $\mathscr{R}\zeta > -n$, they coincide throughout the domain of definition of $T_\zeta^{(m)}(f)$ (9.4.2). We shall therefore denote by $T_\zeta(f)$ the function, *meromorphic in the whole complex plane* $\mathbf{C}$, which coincides with each $T_\zeta^{(m)}(f)$ in the domain of definition of the latter. We shall now determine the *residues* of $T_\zeta(f)$ at its poles.

Notice first that the symmetry $z \mapsto -z$ multiplies the form σ by $(-1)^n$ and preserves (resp. reverses) the orientation of $\mathbf{S}_{n-1}$ if n is even (resp. odd) (16.21.10). It follows therefore from (17.9.2.5) and (16.24.5.1) that $c_\nu = 0$ for all multi-indices ν of *odd* total degree. On the other hand, if $|\nu| = 2k$ is even, then the residue of $T_\zeta(f)$ at the point $-n-2k$ has the value

(17.9.2.6) $$\operatorname{res}_{-n-2k} T_\zeta(f) = \sum_{|\nu|=2k} c_\nu D_f^\nu(0) = \frac{1}{(2k)!} D^{2k} M_f(0)$$

as follows from substituting the expansion (17.9.2.3) in the expression for $M_f(\rho)$. To calculate this number, we introduce the *Laplacian*, which is the following differential operator on $\mathbf{R}^n$:

(17.9.2.7) $$\Delta = D_1^2 + D_2^2 + \cdots + D_n^2.$$

Using the formulas

$$D_i r^\zeta = \zeta r^{\zeta-2}\xi^i, \qquad D_i^2 r^\zeta = \zeta r^{\zeta-2} + \zeta(\zeta-2) r^{\zeta-4}(\xi^i)^2 \qquad (1 \leqq i \leqq n)$$

we obtain

(17.9.2.8) $$\Delta r^\xi = \zeta(\zeta + n - 2) r^{\zeta-2}$$

for all $\zeta \in \mathbf{C}$ and $x \in \mathbf{R}^n - \{0\}$. This implies, by (17.5.5.1), that

(17.9.2.9) $$\Delta T_{\zeta+2}(f) = (\zeta+2)(\zeta+n) T_\zeta(f)$$

for $\mathscr{R}\zeta > -n$ first of all; but since both sides of this relation are meromorphic functions of ζ, it follows that it remains valid for all ζ which are not poles of $T_\zeta(f)$, i.e., $\zeta \neq -n-2k$. By iteration we obtain, under the same conditions,

(17.9.2.10) $$\begin{aligned}\Delta^k T_{\zeta+2k}(f) &= T_{\zeta+2k}(\Delta^k f)\\ &= (\zeta+2)(\zeta+4)\cdots(\zeta+2k)(\zeta+n)(\zeta+n+2)\cdots(\zeta+n+2k-2) T_\zeta(f)\end{aligned}$$

and since none of the linear factors on the right-hand side vanishes when $\zeta = -n-2k$, the residue of $T_\zeta(f)$ at this pole is given by

$$\begin{aligned}&\operatorname{res}_{-n-2k} T_\zeta(f)\\ &\quad = ((-2)(-4)\cdots(-2k)(-n)(-n-2)\cdots(-n-2k+2))^{-1} \operatorname{res}_{-n} T_\zeta(\Delta^k f).\end{aligned}$$

Now apply the formula (17.9.2.6), replacing f by $\Delta^k f$ and k by 0; since

$$M_f(0) = \Omega_n f(0) = \frac{n\pi^{n/2}}{\Gamma(\frac{1}{2}n+1)} f(0)$$

by (17.9.2.2), we obtain finally

(17.9.2.11) $$\operatorname{res}_{-n-2k} T_\zeta(f) = \frac{1}{(2k)!} D^{2k} M_f(0) = \frac{\Omega_n}{2^k k!\, n(n+2)\cdots(n+2k-2)} \Delta^k f(0)$$

(when $k = 0$, the denominator is to be replaced by 1).

Let C_k denote the constant which appears in the right-hand side of this formula. We can now, following the method described in (17.9.1), define the *finite part* $\mathrm{Pf}(r^\zeta)$ for all $\zeta \in \mathbf{C}$. For values of ζ other than the poles of T_ζ, we define $\mathrm{Pf}(r^\zeta) = T_\zeta$, and for $\zeta = -n - 2k$

(17.9.2.12) $$\langle \mathrm{Pf}(r^{-n-2k}), f \rangle = \lim_{\zeta \to -n-2k} (T_\zeta(f) - C_k \Delta^k f(0)(\zeta + n + 2k)^{-1}).$$

The formula (17.9.2.9) gives the Laplacian of $\mathrm{Pf}(r^\zeta)$ for ζ not equal to a pole of T_ζ. To obtain its values at the poles, we may proceed as follows. For ζ not equal to a pole of T_ζ, write

$$T_\zeta(f) = \frac{C_{k-1}}{\zeta + n + 2k - 2} \Delta^{k-1} f(0) + B_{k-1,\zeta}(f)$$

so that the distribution $B_{k-1,\zeta}$ tends to $\mathrm{Pf}(r^{-n-2k+2})$ as $\zeta \to -n - 2k + 2$. Replacing f by Δf in this formula, and using (17.9.2.9), we obtain

$$B_{k-1,\zeta}(\Delta f) = \zeta(\zeta + n - 2) B_{k,\zeta-2}(f) + \frac{\zeta(\zeta + n - 2)C_k - C_{k-1}}{\zeta + n + 2k - 2} \Delta^k f(0)$$

and therefore, letting $\zeta \to -n - 2k + 2$, the formula

(17.9.2.13) $$\Delta(\mathrm{Pf}(r^{-n-2k+2})) = 2k(n + 2k - 2)\mathrm{Pf}(r^{-n-2k}) - \frac{(n + 4k - 2)\Omega_n}{2^k k!\, n(n+2)\cdots(n+2k-2)} \Delta^k \varepsilon_0$$

and, in particular, for $k = 0$,

(17.9.2.14) $$\Delta(\mathrm{Pf}(r^{2-n})) = -(n-2)\Omega_n \varepsilon_0 .$$

A variant of this method consists in remarking that, for $\mathscr{R}\zeta > -n$, $T_\zeta(f)$ is the limit as $\alpha \to 0$ of the integral

$$\int_{U_\alpha} r^\zeta f v_0 ,$$

where U_α is the exterior of the ball of radius $\alpha > 0$. For each $\alpha > 0$, this integral is defined for *all* $\zeta \in \mathbf{C}$ (assuming that f has compact support); for $\mathscr{R}\zeta > -n - 2m$, the right-hand side of (17.9.2.4) is therefore the limit of

$$\int_{U_\alpha} r^\zeta f v_0 + \sum_{0 \leqq k \leqq m} \frac{C_k \alpha^{\zeta+n+2k}}{\zeta + n + 2k} \Delta^k f(0) \tag{17.9.2.15}$$

when ζ is distinct from the poles of T_ζ, and for $\zeta = -n - 2m$ the finite part $\langle \mathrm{Pf}(r^{-n-2m}), f\rangle$ defined by (17.9.2.12) is the limit of the expression (17.9.2.15), provided that the last term in the sum is replaced by $(C_m \log \alpha)\, \Delta^m f(0)$.

(17.9.3) Next, take $n = 1$ and F to be the function x, so that $U_0 =]0, +\infty[$ and $P = \{0\}$; then we have

$$T_\zeta(f) = \int_0^{+\infty} x^\zeta f(x)\, dx.$$

The same method as before, but with much simpler calculations, gives us now an analytic continuation of $T_\zeta(f)$ to a meromorphic function on $\mathbf{C}$, with simple poles at the points $-k - 1$ ($k \in \mathbf{N}$), and residues

$$\mathrm{res}_{-k-1} T_\zeta(f) = \frac{1}{k!} D^k f(0).$$

We define $\mathrm{Pf}(x_+^\zeta)$ to be T_ζ when ζ is not a pole, and

$$\mathrm{Pf}(x_+^{-k-1}) = \lim_{\zeta \to -k-1} \left(T_\zeta - \frac{(-1)^k}{k!(\zeta + k + 1)} D^k \varepsilon_0 \right). \tag{17.9.3.1}$$

We find this time by the same method

$$D(\mathrm{Pf}(x_+^{-k-1})) = -(k+1)\mathrm{Pf}(x_+^{-k-2}) + \frac{(-1)^{k+1}}{(k+1)!} D^{k+1} \varepsilon_0 . \tag{17.9.3.2}$$

It is convenient to introduce at this point the distribution

$$Y_\zeta = \frac{1}{\Gamma(\zeta)} T_{\zeta-1} ; \tag{17.9.3.3}$$

by standard properties of the gamma-function, $Y_\zeta(f)$ is not only meromorphic on $\mathbf{C}$ but is an *entire* function of ζ, with values

$$Y_{-k} = D^k \varepsilon_0 \tag{17.9.3.4}$$

for $k \in \mathbf{N}$.

We observe that Y_1 is the Heaviside function already considered in (17.5.7). The formula (17.5.7.1) generalizes to

(17.9.3.5) $$DY_\zeta = Y_{\zeta-1}$$

for all $\zeta \in \mathbf{C}$. The support of Y_ζ is the half-line $[0, +\infty[$ if ζ is not equal to $-k$ $(k \in \mathbf{N})$, but the support of Y_{-k} is $\{0\}$.

(17.9.4) We shall sketch one last example, without entering into the details of the calculations. Take $X = \mathbf{R}^n$ and v_0 to be the canonical n-form. The coordinates in $\mathbf{R}^n$ will be denoted by $\xi^0, \xi^1, \ldots, \xi^{n-1}$, and F will be the function given by

$$s(x) = \left((\xi^0)^2 - \sum_{j=1}^{n-1} (\xi^j)^2\right)^{1/2}$$

when $\xi^0 \geqq 0$ and $\sum_{j=1}^{n-1} (\xi^j)^2 \leqq (\xi^0)^2$, and $s(x) = 0$ otherwise. We consider the integral

(17.9.4.1) $$T_\zeta(f) = \int_{U_0} s^{\zeta-n} f v_0$$

as a function of the complex variable ζ. Since f has compact support, the range of integration may be replaced by the subset of U_0 for which $\xi_0 \leqq a$ for some suitable $a > 0$. Put

$$x = \left(ar, ar\frac{1-t}{1+t}z\right) \in \mathbf{R} \times \mathbf{R}^{n-1},$$

where $z \in \mathbf{S}_{n-2}$, $0 \leqq r \leqq 1$, and $0 \leqq t \leqq 1$; then we obtain (first of all for $\mathscr{R}\zeta > n$)

(17.9.4.2) $$T_\zeta(f) = \int_0^1 r^{\zeta-1}\,dr \int_0^1 t^{(\zeta-n)/2}\, G(t, r, \zeta)\,dt,$$

where

(17.9.4.3) $$G(t, r, \zeta) = \frac{2^{\zeta-n+1}a^\zeta(1-t)^{n-2}}{(1+t)^\zeta}\int_{\mathbf{S}_{n-2}} f\left(ar, ar\frac{1-t}{1+t}z\right)\sigma(z)$$

(σ denoting as before the solid angle, but this time in $\mathbf{S}_{n-2}$). It is immediate that for fixed ζ the function $(t, r) \mapsto G(t, r, \zeta)$ is of class C^∞ for $r \in \mathbf{R}$ and

$t > -1$. By taking the Taylor series of this function in a neighborhood of $(0, 0)$, with an arbitrarily large number of terms, it is easily verified that the function of two complex variables (initially defined for $\mathscr{R}\alpha > 0$ and $\mathscr{R}\beta > 0$)

$$\mathrm{I}(\alpha, \beta) = \frac{1}{\Gamma(\alpha)\Gamma(\beta)} \int_0^1 \int_0^1 r^{\alpha-1} t^{\beta-1} \mathrm{G}(t, r, \alpha)\, dr\, dt$$

extends to an *entire* function. Finally, by taking the Taylor series of the function $r \mapsto \mathrm{G}(t, r, \zeta)$ in a neighborhood of $r = 0$, and using properties of the gamma-function, it can be shown that

(17.9.4.4) $$\lim_{\zeta \to 0} \frac{\mathrm{T}_\zeta(f)}{\Gamma(\zeta)\Gamma(\frac{1}{2}(\zeta + 2 - n))} = \pi^{(n/2)-1} f(0).$$

We now introduce the distribution

(17.9.4.5) $$\mathrm{Z}_\zeta(f) = \frac{\mathrm{T}_\zeta(f)}{\pi^{(n/2)-1} 2^{\zeta-1} \Gamma(\frac{1}{2}\zeta)\Gamma(\frac{1}{2}(\zeta + 2 - n))},$$

which is an analytic function of ζ in the half-plane $\mathscr{R}\zeta > 0$; the constants are chosen so that, by (17.9.4.4) and properties of the gamma-function, we have

(17.9.4.6) $$\lim_{\zeta \to 0,\, \zeta > 0} \mathrm{Z}_\zeta = \varepsilon_0 .$$

Next we introduce the *d'Alembertian*, which is the differential operator

(17.9.4.7) $$\Box = \mathrm{D}_0^2 - \mathrm{D}_1^2 - \cdots - \mathrm{D}_{n-1}^2 .$$

It is easily verified that

$$\Box\, s^\zeta = \zeta(\zeta + n - 2) s^{\zeta - 2}$$

for all $\zeta \in \mathbf{C}$ and x such that $s(x) \neq 0$. From this formula we obtain

(17.9.4.8) $$\Box(\mathrm{Z}_\zeta) = \mathrm{Z}_{\zeta-2}$$

at any rate for $\mathscr{R}\zeta > 2$. We shall deduce from this that Z_ζ is an *entire function* of ζ. From the properties of $\mathrm{I}(\alpha, \beta)$, and the Gauss–Legendre formula

$$\Gamma(\zeta) = 2^{\zeta-1}\pi^{-1/2}\Gamma(\tfrac{1}{2}\zeta)\Gamma(\tfrac{1}{2}(\zeta + 1)),$$

it follows that the only possible poles of Z_ζ are those of $\Gamma(\frac{1}{2}(\zeta + 1))$, that is to say the negative odd integers. By analytic continuation, the formula (17.9.4.8) is valid for all ζ except possibly the negative odd integers; but

$$\Box(\mathrm{Z}_\zeta)(f) = \mathrm{Z}_\zeta(\Box f)$$

has no pole at $\zeta = 1$, hence the same is true of $Z_{\zeta-2}(f)$; in other words, Z_ζ has no pole at $\zeta = -1$; applying the same argument repeatedly, we see that Z_ζ has no pole at any negative odd integer, which proves our assertion. By induction, it follows in particular that

(17.9.4.9)
$$\square^k Z_{2k} = \varepsilon_0 .$$

Remark

(17.9.5) By iterating the formula (17.9.2.8) and using (17.9.2.14), we obtain for *odd n* the analog of (17.9.4.9) for the iterated Laplacian:

(17.9.5.1)
$$\Delta^k(\mathrm{Pf}(r^{2k-n})) = 2^{k-1}(k-1)!(2k-n)(2k-2-n)\cdots(2-n)\Omega_n \varepsilon_0$$

(k a positive integer). For n even, say $n = 2p$, there is an analogous formula for the function $r^{2k-n} \log r$, with $k \geqq p$ (so that the function is integrable):

(17.9.5.2)
$$\Delta^k(r^{2k-n} \log r) = (-1)^{p-1} 2^{2k-2}(k-1)!\,(k-p)!\,(p-1)!\,\Omega_n \varepsilon_0 .$$

PROBLEMS

1. (a) For each function $f \in \mathscr{D}^{(1)}(\mathbf{R})$, show that as $\varepsilon > 0$ tends to 0, the sum

$$\int_{-\infty}^{-\varepsilon} \frac{f(t)}{t}\,dt + \int_{\varepsilon}^{\infty} \frac{f(t)}{t}\,dt$$

tends to a limit. This limit is called the *Cauchy principal value* of the (in general non-convergent) integral $\int_{-\infty}^{+\infty} f(t)/t\,dt$ and is denoted by $\mathrm{P.V.}\int_{-\infty}^{+\infty} f(t)/t\,dt$. Show that the mapping $f \mapsto \mathrm{P.V.}\int_{-\infty}^{+\infty} f(t)/t\,dt$ is a distribution of order 1 on $\mathbf{R}$. This distribution is denoted by P.V.$(1/x)$. Show that a primitive of P.V.$(1/x)$ is the integrable function $\log |x|$. Calculate the successive derivatives of P.V.$(1/x)$.

(b) More generally, for any function g of the form

$$g(x) = \frac{A}{x} + h(x),$$

where A is a constant and h is continuous, the distribution P.V.(g) (or P.V.$(g(x))$) is defined as in (a), by replacing the function $1/x$ by $g(x)$. Show that if π is an increasing diffeomorphism of $\mathbf{R}$ onto $\mathbf{R}$ such that $\pi(0) = 0$, then $\pi^{-1}(\mathrm{P.V.}(1/x)) = \mathrm{P.V.}(\pi'(x)/\pi(x))$.

2. Let m be an integer $\geqq 2$. Show that as $\varepsilon > 0$ tends to 0, the double integral

$$\iint_{|z| \geqq \varepsilon} \frac{f(x, y)}{z^m} \, dx \, dy$$

(where $z = x + iy$) tends to a limit, for each function $f \in \mathscr{D}(\mathbf{R}^2)$. (Reduce to the case where $f(x, y) = z^p \bar{z}^q$ and change to polar coordinates.) Show that the mapping

$$f \mapsto \lim_{\varepsilon \to 0} \left(\iint_{|z| \geqq \varepsilon} \frac{f(x, y)}{z^m} \, dx \, dy \right)$$

is a distribution on $\mathbf{R}^2$ of order $m - 1$. This distribution is denoted by P.V.$(1/z^m)$. Prove the formulas

$$\frac{\partial}{\partial z} \left(\text{P.V.} \left(\frac{1}{z^m} \right) \right) = -m \cdot \text{P.V.} \left(\frac{1}{z^{m+1}} \right),$$

$$\frac{\partial}{\partial \bar{z}} \left(\text{P.V.} \left(\frac{1}{z^m} \right) \right) = \frac{(-1)^{m+1} \pi}{(m-1)} \frac{\partial^{m-1} \varepsilon_0}{\partial z^{m-1}}.$$

3. Show that for $\zeta = n - 2 - 2k$ which is not of the form $-2m$ (where m is an integer $\geqq 0$), the support of Z_ζ (17.9.4.5) is the cone defined by $\xi^0 \geqq 0$, $s(x) = 0$.

4. Let f be a holomorphic function on an open set in $\mathbf{C}$ containing the closed unit disk $\mathrm{D} : |z| \leqq 1$, and let f_{D} be the function which is equal to f in D and vanishes outside D. Show that the distribution f_{D} on $\mathbf{R}^2$ has derivative $\partial f_{\mathrm{D}} / \partial \bar{z}$ equal to the distribution

$$g \mapsto \frac{i}{2} \int_\varepsilon fg \, dz,$$

where ε is the circuit $t \mapsto e^{it}$ $(0 \leqq t \leqq 2\pi)$. (Use the elementary Stokes' formula (16.24.11).)

5. Express in terms of Cauchy principal values (Problem 1) the distributions on $\mathbf{R}$ defined by the formulas

$$f \mapsto \lim_{y \to 0,\, y > 0} \int_{-\infty}^{+\infty} \frac{f(x) \, dx}{x - iy}$$

and

$$f \mapsto \lim_{y \to 0,\, y < 0} \int_{-\infty}^{+\infty} \frac{f(x) \, dx}{x - iy}.$$

6. For $\mathscr{R}\lambda > 0$ and $\mathscr{R}\mu > 0$, consider the distribution $\mathrm{T}_{\lambda, \mu}$ on $]0, 1[$ defined by

$$\mathrm{T}_{\lambda, \mu}(f) = \int_0^1 x^{\lambda - 1} (1 - x)^{\mu - 1} f(x) \, dx.$$

Show that $(\lambda, \mu) \mapsto \mathrm{T}_{\lambda, \mu}(f)$ extends to an analytic function except for $\lambda = -n$ or $\mu = -n$, where $n \in \mathbf{N}$, and determine the form of this function near these singular points. Outside the singular points, $\mathrm{T}_{\lambda, \mu}$ is a distribution on $]0, 1[$, denoted by $\mathrm{Pf}(x_+^{\lambda - 1} (1 - x)_+^{\mu - 1})$.

10. TENSOR PRODUCT OF DISTRIBUTIONS

(17.10.1) *Let* U *be an open subset of* $\mathbf{R}^n$, *let* T *be a distribution of order* $\leqq m$ *on* U, *let* E *be a metric space and* f *a mapping of* $U \times E$ *into* $\mathbf{C}$.

(i) *Suppose that there exists a compact set* $K \subset U$ *and a neighborhood* V *of a point* $z_0 \in E$ *such that*

(1) $f(\cdot, z) \in \mathscr{D}^{(m)}(U; K)$ *for all* $z \in V$;

(2) $(x, z) \mapsto D^\nu f(x, z)$ *is continuous on* $U \times V$, *for each multi-index* ν *such that* $|\nu| \leqq m$.

Then the function $z \mapsto F(z) = \langle T, f(\cdot, z)\rangle$ *is continuous on* V.

(ii) *Suppose that* E *is an open set in* $\mathbf{R}$ *or* $\mathbf{C}$, *and that the following condition is also satisfied:*

(3) *for each* $x \in U$ *and each multi-index* ν *such that* $|\nu| \leqq m$, *the function* $z \mapsto D^\nu f(x, z)$ *is differentiable on* V (*in the usual sense*) *and the function*

$$(x, z) \mapsto \frac{\partial}{\partial z}(D^\nu f(x, z))$$

is continuous on $U \times V$.

Then F *is differentiable on* V (*in the usual sense*) *and we have*

$$\frac{\partial F}{\partial z} = \left\langle T, \frac{\partial}{\partial z} f(\cdot, z)\right\rangle. \tag{17.10.1.1}$$

(i) Let (z_n) be a sequence in V converging to z_0. By virtue of (3.16.5) and the continuity hypothesis, each of the sequences of functions $(D^\nu f(x, z_n))$ converges *uniformly* to $D^\nu f(x, z_0)$ in K. Hence the result, having regard to the definition of distributions of order $\leqq m$ (17.3.2), and to (3.13.14).

(ii) By virtue of the hypothesis and (17.5.13), we have

$$\frac{\partial}{\partial z}(D^\nu f(x, z)) = D^\nu\left(\frac{\partial}{\partial z} f(x, z)\right).$$

The same reasoning shows that each of the sequences of functions

$$\left(D^\nu\left(\frac{f(x, z_n) - f(x, z_0)}{z_n - z_0} - \frac{\partial f}{\partial z}(x, z_0)\right)\right)$$

converges *uniformly* to 0 in K. Hence the sequence

$$\left(\frac{F(z_n) - F(z_0)}{z_n - z_0} - \left\langle T, \frac{\partial f}{\partial z}(\cdot, z_0)\right\rangle\right)$$

converges to 0, which proves (ii).

(17.10.2) *Let* X, Y *be two differential manifolds*, K (resp. L) *a compact subset of* X (resp. Y). *Let* K′ (resp. L′) *be a compact neighborhood of* K (resp. L). *Then as* u (resp. v) *runs through* $\mathscr{D}(X; K')$ (resp. $\mathscr{D}(Y; L')$), *the closure in* $\mathscr{D}(X \times Y; K' \times L')$ *of the set of linear combinations of functions of the form* $u \otimes v$ (13.21.14) *contains* $\mathscr{D}(X \times Y; K \times L)$.

The proof reduces immediately to the situation in which X (resp. Y) is an open subset of $\mathbf{R}^m$ (resp. $\mathbf{R}^n$). By a translation and a homothety, we may assume that $K' \times L'$ is contained in the cube $I = [-\frac{1}{2}, \frac{1}{2}]^{m+n}$ in $\mathbf{R}^{m+n}$. With the notation of (14.11.3), put

$$G_k(z) = a_k^{-(m+n)} \prod_{i=1}^{m+n} g_k(\zeta^i)$$

for $z = (\zeta^1, \dots, \zeta^{m+n}) \in \mathbf{R}^{m+n}$. If $h \in \mathscr{D}(X \times Y; K \times L)$, put $P_k = h * G_k$; then (13.8.6) we have $D^\nu P_k = D^\nu h * G_k = h * D^\nu G_k$. Hence (14.11.3) *in the cube* I, the function P_k coincides with a polynomial in the coordinates ζ^i, and for each multi-index ν, the $D^\nu P_k$ converge *uniformly* on I to $D^\nu h$. Let ρ (resp. σ) be a C^∞-function on X (resp. Y) with support contained in K′ (resp. L′) and equal to 1 on K (resp. L). Then the sequence of functions $(\rho \otimes \sigma)P_k$, which belong to $\mathscr{D}(X \times Y; K' \times L')$, converges in this space to h, by virtue of (17.1.4), and $(\rho \otimes \sigma)P_k$ is a linear combination of functions of the form $u \otimes v$ with $u \in \mathscr{D}(X; K')$ and $v \in \mathscr{D}(Y; L')$.

We may now follow the procedure of (13.21.1) for defining the product of two measures, to define the *product* $S \otimes T$ of a distribution $S \in \mathscr{D}'(X)$ and a distribution $T \in \mathscr{D}'(Y)$.

(17.10.3) *Let* X, Y *be two differential manifolds*, S *a distribution in* X, *and* T *a distribution on* Y. *Then there exists a unique distribution* R *on* X × Y *such that, for all* $f \in \mathscr{D}(X)$ *and* $g \in \mathscr{D}(Y)$,

(17.10.3.1) $$\langle R, f \otimes g \rangle = \langle S, f \rangle \langle T, g \rangle.$$

Furthermore, for each function $h \in \mathscr{D}(X \times Y)$, *the function* $x \mapsto H(x) = \langle T, h(x, \cdot) \rangle$ *belongs to* $\mathscr{D}(X)$, *and we have* $\langle R, h \rangle = \langle S, H \rangle$ *or equivalently (by abuse of notation)*

(17.10.3.2) $$\iint h(x, y)\, dR(x, y) = \int dS(x) \int h(x, y)\, dT(y).$$

The uniqueness of the distribution R clearly follows from (17.10.2). If $h = f \otimes g$, we may write $H(x) = f(x)\langle T, g \rangle$, and $\langle S, H \rangle = \langle S, f \rangle \langle T, g \rangle$. Hence we have only to show that the mapping $h \mapsto \langle S, H \rangle$, which is a linear form on each of the spaces $\mathscr{D}(X \times Y; M)$ (M a compact subset of $X \times Y$), is *continuous* on each of these spaces. Now, it follows already from (17.10.1) that $H \in \mathscr{D}(X; \mathrm{pr}_1(M))$, for each $h \in \mathscr{D}(X \times Y; M)$. Next, we need only prove that $h \mapsto \langle S, H \rangle$ is continuous when M is contained in a chart of $X \times Y$; in other words, we may assume that X and Y are open sets in $\mathbf{R}^m$ and $\mathbf{R}^n$, respectively. Let $D'^{\nu'}$ (resp. $D''^{\nu''}$) denote partial differentiations with respect to the coordinates in X (resp. Y). Let (h_j) be a sequence of functions in $\mathscr{D}(X \times Y; M)$, converging to 0 in this space. For each multi-index ν'', the derivatives $D''^{\nu''}h_j(x, y)$ converge to 0 *uniformly* in M; since $h_j(x, y) = 0$ for $y \notin \mathrm{pr}_2(M)$, it follows from (17.3.1.1) that the functions $H_j(x) = \langle T, h_j(x, \cdot) \rangle$ tend to 0 *uniformly* in $\mathrm{pr}_1(M)$. The same is true of the derivatives $D'^{\nu'}H_j(x)$ for each multi-index ν', because, by virtue of (17.10.1.1), $D'^{\nu'}H_j(x) = \langle T, D'^{\nu'}h_j(x, \cdot) \rangle$ and by hypothesis, for each multi-index ν'', $D''^{\nu''}(D'^{\nu'}h_j(x, y))$ tends to 0 uniformly in M, so that the argument above applies. The definition of distributions shows therefore that $\langle S, H_j \rangle \to 0$. Q.E.D.

This theorem therefore gives simultaneously a proof of existence and a method of calculation by successive application of the distributions S and T. Evidently we may invert the order of application and thus obtain the formula

$$\iint h(x, y)\, dR(x, y) = \int dT(y) \int h(x, y)\, dS(x).$$

The distribution R defined in (17.10.3) is called the *product* (or *tensor product*) of S and T, and is denoted by $S \otimes T$.

It is clear that when S and T are *measures* on X and Y, respectively, the distribution $S \otimes T$ is the product measure defined in (13.21), having regard to the uniqueness property of (17.10.3) and to (17.3.2).

Moreover, with the notation introduced above, the tensor product of distributions has the following properties:

(17.10.4) (i) $\mathrm{Supp}(S \otimes T) = \mathrm{Supp}(S) \times \mathrm{Supp}(T)$.

(ii) *If* S *has order* $\leqq r$ *and* T *has order* $\leqq s$, *then* $S \otimes T$ *has order* $\leqq r + s$. *Furthermore, if* $u \in \mathscr{E}(X)$ *and* $v \in \mathscr{E}(Y)$, *then*

(17.10.4.1) $$(u \otimes v) \cdot (S \otimes T) = (u \cdot S) \otimes (v \cdot T).$$

(iii) *If* (S_n) *is a sequence in* $\mathscr{D}'(X)$ *and* (T_n) *a sequence in* $\mathscr{D}'(Y)$, *and if one of these two sequences is weakly bounded, and the other converges weakly to* 0, *then the sequence* $(S_n \otimes T_n)$ *converges weakly to* 0.

(iv) *If* X *is an open subset of* $\mathbf{R}^m$ *and* Y *an open subset of* $\mathbf{R}^n$, *then*

(17.10.4.2) $$D'^{\nu'}D''^{\nu''}(S \otimes T) = (D'^{\nu'}S) \otimes (D''^{\nu''}T).$$

(i) It is clear that $\operatorname{Supp}(S \otimes T) \subset \operatorname{Supp}(S) \times \operatorname{Supp}(T)$. Conversely, if $(a, b) \in \operatorname{Supp}(S) \times \operatorname{Supp}(T)$ and if U (resp. V) is a compact neighborhood of a (resp. b), then there exists a function $f \in \mathscr{D}(X; U)$ and a function $g \in \mathscr{D}(Y; V)$ such that $\langle S, f\rangle \neq 0$ and $\langle T, g\rangle \neq 0$, so that $f \otimes g \in \mathscr{D}(X \times Y; U \times V)$ and $\langle S \otimes T, f \otimes g\rangle \neq 0$ by (17.10.3.1).

(ii) If T is of order s, then the derivatives $D'^{\nu'}H$ of order $|\nu'| \leqq r$ are majorized (by virtue of (17.10.1.1)) by seminorms involving only the derivatives of h of total order $\leqq r + s$, whence the first assertion follows. Also the values of the two sides of (17.10.4.1) for $h \in \mathscr{D}(X \times Y)$ are equal, by virtue of (17.10.3.1), when h is of the form $f \otimes g$, and therefore in general by (17.10.2). The same argument also proves (iv).

(iii) Suppose, to fix the ideas, that the sequence (T_n) is weakly bounded and that the sequence (S_n) converges weakly to 0, and let $h \in \mathscr{D}(X \times Y; M)$. Then, as x runs through $\operatorname{pr}_1(M)$, the functions $h(x, \cdot) : y \mapsto h(x, y)$ have their supports contained in $\operatorname{pr}_2(M)$, and for each multi-index (ν', ν'') the functions $y \mapsto D'^{\nu'}D''^{\nu''}h(x, y)$ are *uniformly bounded.* It follows therefore from (17.8.5(ii)) that for each multi-index ν' the sequence of functions $D'^{\nu'}H_n$, where $H_n(x) = \langle T_n, h(x, \cdot)\rangle$, is *uniformly bounded* in $\operatorname{pr}_1(M)$; but then this sequence is *relatively compact* in $\mathscr{D}(X; \operatorname{pr}_1(M))$ (17.2.2); and since the sequence (S_n) is *equicontinuous* in this space, by virtue of (17.8.5), the fact that it converges weakly implies that it converges *uniformly* on each compact subset of the metrizable space $\mathscr{D}(X; \operatorname{pr}_1(M))$ (7.5.6). Hence the sequence $\langle S_n, H_n\rangle$ converges to 0. Q.E.D.

Examples

(17.10.5) If ε'_a (resp. ε''_b) denotes the Dirac measure at the point $a \in X$ (resp. $b \in Y$), then it is immediately seen that the tensor product $\varepsilon'_a \otimes \varepsilon''_b$ is the Dirac measure $\varepsilon_{(a,b)}$. Hence, (17.10.4.2)

$$D'^{\nu'}D''^{\nu''}\varepsilon_{(a,b)} = (D'^{\nu'}\varepsilon'_a) \otimes (D''^{\nu''}\varepsilon''_b).$$

(17.10.6) Suppose that X is an open subset of $\mathbf{R}^n$ and that S is the distribution defined by a locally integrable function f (with respect to Lebesgue measure λ on X). Then, for each function $h \in \mathscr{D}(X \times Y)$, we have

(17.10.6.1) $$\langle f \otimes T, h\rangle = \int \langle T, h(x, \cdot)\rangle f(x)\, d\lambda(x) = \left\langle T, \int h(x, \cdot) f(x)\, d\lambda(x)\right\rangle.$$

(17.10.7) Let S be a distribution on X, let b be a point of Y, and let S_b denote the distribution on the submanifold $X \times \{b\}$ of $X \times Y$ which is the image of S under the diffeomorphism $x \mapsto (x, b)$. Then it is immediately verified that

(17.10.7.1) $$S \otimes \varepsilon''_b = j(S_b),$$

where $j : X \times \{b\} \to X \times Y$ is the canonical immersion. When $Y = \mathbf{R}^n$, this gives the derivatives $D''^{\nu''}(j(S_b)) = S \otimes D''^{\nu''}(\varepsilon''_b)$, by virtue of (17.10.4.2). When X is an open subset of $\mathbf{R}^m$ and S is a locally integrable function, the distribution $j(S_b)$ is sometimes said to be a *singlet or simple layer* on $X \times \{b\}$, and its derivatives $D''^{\nu''}(j(S_b))$ to be *multiplet layers* on $X \times \{b\}$.

(17.10.8) In the same way we define the tensor product

$$T = T_1 \otimes T_2 \otimes \cdots \otimes T_m = \bigotimes_{k=1}^{m} T_k,$$

where, for $1 \leq k \leq m$, T_k is a distribution on a manifold X_k; T is a distribution on the product manifold $X = \prod_{k=1}^{m} X_k$, and we leave to the reader the task of extending the results of this section to these multiple tensor products.

PROBLEMS

1. Let U be a bounded open subset of $\mathbf{R}^n$ and let $(g_n)_{n \geq 1}$ be a total orthonormal sequence in $\mathscr{L}^2_{\mathbf{R}}(U)$ (relative to the measure induced on U by Lebesgue measure) (Section 13.11, Problem 7).

 (a) Show that the functions $g_j \otimes g_k$ form a total orthonormal sequence in $\mathscr{L}^2_{\mathbf{R}}(U \times U)$. (Use the fact that the functions $u \otimes v$ with $u, v \in \mathscr{K}_{\mathbf{R}}(U)$ form a total set in $\mathscr{L}^2_{\mathbf{R}}(U \times U)$, cf. (13.21.1).)
 (b) Show that in the distribution space $\mathscr{D}'(U \times U)$ the series $\sum_{n=1}^{\infty}(g_n \otimes g_n)$ converges to the measure induced on $U \times U$ by the trace measure tr (Section 17.6, Problem 10) on $\mathbf{R}^n \times \mathbf{R}^n$. (Consider first the series $\sum_{n=1}^{\infty}(w \mid g_n \otimes g_n)$, where w is of the form $u \otimes v$, with $u, v \in \mathscr{D}(U)$; then use (17.10.2).)
 (c) Conversely, let $(g_n)_{n \geq 1}$ be an orthonormal sequence in $\mathscr{L}^2_{\mathbf{R}}(U)$ such that the series $\sum_{n=1}^{\infty}(g_n \otimes g_n)$ converges in $\mathscr{D}'(U \times U)$ to the measure induced by the trace measure tr. Show that the sequence (g_n) is total. (Use (6.5.2) and the fact that $\mathscr{D}(U)$ is dense in $\mathscr{L}^2_{\mathbf{R}}(U)$.)

2. Let U be an open subset of $\mathbf{R}^n$, let V be an open subset of $\mathbf{R}^m$, let u_j $(1 \leq j \leq r)$ be linearly independent continuous functions on U, and let v_k $(1 \leq k \leq s)$ be linearly

independent continuous functions on V. If T_j $(1 \leq j \leq r)$ are distributions belonging to $\mathscr{D}'(V)$, and S_k $(1 \leq k \leq s)$ distributions belonging to $\mathscr{D}'(U)$, such that the sum

$$\sum_j (u_j \otimes T_j) + \sum_k (S_k \otimes v_k)$$

is a continuous function on $U \times V$, show that the T_j are continuous functions on V and the S_k continuous functions on U. (Remark that there exist functions $\varphi_j \in \mathscr{D}(U)$ such that $\langle u_i, \varphi_j \rangle = \delta_{ij}$, and likewise for the v_k.)

3. Let T be a distribution belonging to $\mathscr{E}'(\mathbf{R}^m \times \mathbf{R}^n)$ whose (compact) support is contained in $\mathbf{R}^m \times \{0\}$. Show that there exists a unique decomposition

$$T = \sum_{|\alpha''| \leq p} S_{\alpha''} \otimes D^{\alpha''} \varepsilon_0,$$

where the $S_{\alpha''}$ are distributions belonging to $\mathscr{E}'(\mathbf{R}^m)$. (Consider the distributions

$$f \mapsto \langle T, f \otimes y^{\alpha''} \rangle$$

on $\mathbf{R}^m$ and use the Taylor series of an arbitrary function $g \in \mathscr{E}(\mathbf{R}^{m+n})$, considered as a function $y \mapsto g(x, y)$ on $\mathbf{R}^n$, with coefficients belonging to $\mathscr{E}(\mathbf{R}^m)$.)

11. CONVOLUTION OF DISTRIBUTIONS ON A LIE GROUP

(17.11.1) Let G be a Lie group, n an integer > 1 and let $m : G^n \to G$ be the mapping $(x_1, x_2, \ldots, x_n) \mapsto x_1 x_2 \cdots x_n$. A sequence $(T_1, \ldots, T_n)$ of n distributions on G is said to be *strictly convolvable* if the supports $A_k = \operatorname{Supp}(T_k)$ of these distributions have the following property: for every compact subset K of G, the set $m^{-1}(K) \cap \prod_{k=1}^{n} A_k$ is *compact* in G. Let $T = \bigotimes_{k=1}^{n} T_k$ be the product distribution on G^n, with support $A = \prod_{k=1}^{n} A_k$. For each function $f \in \mathscr{D}(G)$, it follows that $\langle T, f \circ m \rangle$ is defined; for if K is the support of f, then $f \circ m \in \mathscr{E}(G^n)$ and the support of $f \circ m$ is contained in $m^{-1}(K)$, hence our assertion follows from (17.4.5). Moreover, for each compact subset K of G, if a sequence of functions $f_p \in \mathscr{D}(G; K)$ converges to 0 in this space, then for each compact neighborhood V of $m^{-1}(K) \cap A$ the restrictions of the functions $f_p \circ m$ to V converge to 0 in $\mathscr{E}(V)$ (17.1.5), and hence we deduce from (17.4.5) that the sequence $(\langle T, f_p \circ m \rangle)$ tends to 0. Consequently the mapping $f \mapsto \langle T, f \circ m \rangle$ is a *distribution on* G, called the *convolution* (or *convolution product*) of the sequence $(T_1, \ldots, T_n)$ and denoted by $T_1 * T_2 * \cdots * T_n$. Equivalently, we may write (17.10.3)

(17.11.1.1) $$\langle T_1 * T_2 * \cdots * T_n, f \rangle = \iint \cdots \int f(x_1 x_2 \cdots x_n)\, dT_1(x_1) \cdots dT_n(x_n).$$

When the T_k are strictly convolvable *measures*, then by virtue of (17.3.4) and (17.10.4) their convolution product as distributions is identical with their convolution product as measures, in the sense defined in (14.5).

Examples

(17.11.2) If all the supports A_k with at most one exception are compact, then the sequence $(T_1, T_2, \dots, T_n)$ is strictly convolvable. The proof is the same as in (14.5.4).

Let P be the set of points $x = (\xi_j) \in \mathbf{R}^m$ such that $\xi_j \geqq 0$ for all j. If there exists $\mathbf{a} = (\alpha_j) \in \mathbf{R}^m$ such that the supports A_k all belong to the set $P + \mathbf{a}$, then the sequence (T_k) is strictly convolvable. For, by expressing the fact that a sequence of n points $\mathbf{x}_k = (\xi_{kj}) \in \mathbf{R}^m$ is such that the $\mathbf{x}_k$ belong to $P + \mathbf{a}$ and that $\mathbf{x}_1 + \mathbf{x}_2 + \cdots + \mathbf{x}_n$ belongs to a compact subset K of $\mathbf{R}^m$, one sees that there exists a constant C such that for $1 \leqq j \leqq m$,

$$\sum_{k=1}^{n} (\xi_{kj} - \alpha_j) \leqq C - n\alpha_j,$$

and since by hypothesis all the $\xi_{kj} - \alpha_j$ are $\geqq 0$, we have $0 \leqq \xi_{kj} - \alpha_j \leqq C - n\alpha_j$ for each index k, so that the point $(\mathbf{x}_1, \mathbf{x}_2, \dots, \mathbf{x}_n)$ belongs to a compact subset of $\mathbf{R}^{nm}$.

A finite sequence of *measures* on G may be convolvable (in the sense of (14.5)) without being strictly convolvable. In certain situations the convolution of distributions (other than measures) which are not strictly convolvable can be defined (Problem 1).

(17.11.3) *Suppose that the sequence* $(T_1, T_2, \dots, T_n)$ *is strictly convolvable. If* $A_k = \operatorname{Supp}(T_k)$ *for* $1 \leqq k \leqq n$, *then*

(17.11.3.1) $$\operatorname{Supp}(T_1 * T_2 * \cdots * T_n) \subset \overline{A_1 A_2 \cdots A_n}.$$

The proof is the same as (14.5.4), using the fact that the support of $T_1 \otimes T_2 \otimes \cdots \otimes T_n$ is $A_1 \times A_2 \times \cdots \times A_n$.

The following is a corollary of (17.11.3):

(17.11.4) *If the distributions* S *and* T *are strictly convolvable and if* $U \subset G$ *is an open set such that* $(\operatorname{Supp}(S))^{-1}U \cap \operatorname{Supp}(T) = \varnothing$, *then* $U \cap \operatorname{Supp}(S * T) = \varnothing$.

Let $A = \operatorname{Supp}(S)$, $B = \operatorname{Supp}(T)$. Since $A^{-1}U \cap B = \varnothing$, we have $AB \cap U = \varnothing$ and therefore (as U is open) $\overline{AB} \cap U = \varnothing$, whence the result.

Convolution of distributions has algebraic properties analogous to those of convolution of measures:

(17.11.5) *If a finite sequence of nonzero distributions* $(T_1, T_2, \dots, T_n)$ *is strictly convolvable, then for each* $h = 1, 2, \dots, n$ *the sequences* $(T_1, \dots, T_h)$, $(T_{h+1}, \dots, T_n)$ *are strictly convolvable; the distributions* $T_1 * \cdots * T_h$ *and* $T_{h+1} * \cdots * T_n$ *are strictly convolvable; and we have*

(17.11.5.1) $(T_1 * \cdots * T_h) * (T_{h+1} * \cdots * T_n) = T_1 * T_2 * \cdots * T_n$

("associativity of convolution").

Let $A_k = \text{Supp}(T_k)$ for $1 \leqq k \leqq n$; by hypothesis, these sets are nonempty. For $h + 1 \leqq k \leqq n$ let z_k be a point of A_k. Then the set of points $(x_1, \dots, x_h) \in G^h$ such that $x_1 x_2 \cdots x_h \in K$ and $x_k \in A_k$ for $1 \leqq k \leqq h$ is also the set of points $(x_1, \dots, x_h) \in G^h$ such that

$$x_1 x_2 \cdots x_h z_{h+1} \cdots z_n \in K z_{h+1} \cdots z_n ;$$

in other words, it is the section at the point $(z_{h+1}, \dots, z_n)$ of the set $m^{-1}(K z_{h+1} \cdots z_n) \cap \prod_{k=1}^{n} A_k$ considered as a subset of the product $G^h \times G^{n-h}$. Hence it is compact, which shows that the sequence $(T_1, \dots, T_h)$ is strictly convolvable. Similarly the sequence $(T_{h+1}, \dots, T_n)$ is strictly convolvable.

Now put $B = \overline{A_1 A_2 \cdots A_h}$, $C = \overline{A_{h+1} \cdots A_n}$. We shall show that for each compact subset K of G the set of pairs $(y, z) \in B \times C$ such that $yz \in K$ is compact; by virtue of (17.11.3), this will prove the second assertion. Let then V be a compact neighborhood of K in G; there exists in each A_k a sequence of points $(x_k^{(p)})$ such that the sequence of products $x_1^{(p)} x_2^{(p)} \cdots x_h^{(p)}$ tends to y, the sequence of products $x_{h+1}^{(p)} \cdots x_n^{(p)}$ tends to z, and such that

$$x_1^{(p)} \cdots x_h^{(p)} x_{h+1}^{(p)} \cdots x_n^{(p)} \in V$$

for all p. Since by hypothesis the set $L = m^{-1}(V) \cap \prod_{k=1}^{n} A_k$ is compact, the same is true of its projections M and N on G^h and G^{n-h}, respectively. The first part of the proof shows that y belongs to the (compact) image M′ of M under the mapping $(x_1, \dots, x_h) \mapsto x_1 x_2 \cdots x_h$, and that z belongs to the (compact) image N′ of N under the mapping $(x_{h+1}, \dots, x_n) \mapsto x_{h+1} \cdots x_n$. Hence $(y, z) \in M' \times N'$, which completes the proof of the second assertion.

Finally, put $R = T_1 * T_2 * \cdots * T_h$ and $S = T_{h+1} * \cdots * T_n$; then, by definition, for each function $f \in \mathscr{D}(G)$ we have

$$\begin{aligned}
\langle R * S, f \rangle &= \iint f(yz)\, dR(y)\, dS(z) = \int dR(y) \int f(yz)\, dS(z) \\
&= \int dR(y) \int \cdots \int f(y x_{h+1} \cdots x_n)\, dT_{h+1}(x_{h+1}) \cdots dT_n(x_n) \\
&= \iint \cdots \int f(x_1 \cdots x_h x_{h+1} \cdots x_n)\, dT_1(x_1) \cdots dT_h(x_h)\, dT_{h+1}(x_{h+1}) \cdots dT_n(x_n);
\end{aligned}$$

whence the formula (17.11.5.1) follows.

(17.11.6) *For each point $s \in \mathrm{G}$ and each distribution $\mathrm{T} \in \mathscr{D}'(\mathrm{G})$ we have*

(17.11.6.1) $$\varepsilon_s * \mathrm{T} = \gamma(s)\mathrm{T}, \qquad \mathrm{T} * \varepsilon_s = \delta(s^{-1})\mathrm{T}.$$

The proof is the same as in (14.6.1.1).

In particular, if e is the identity element of G, then $\varepsilon_e * \mathrm{T} = \mathrm{T} * \varepsilon_e = \mathrm{T}$.

(17.11.7) *For each distribution T on G, let $\check{\mathrm{T}}$ be the image of T under the diffeomorphism $x \mapsto x^{-1}$ of G onto G. If the sequence $(\mathrm{T}_1, \ldots, \mathrm{T}_n)$ is strictly convolvable, then so is the sequence $(\check{\mathrm{T}}_n, \ldots, \check{\mathrm{T}}_1)$ and we have*

(17.11.7.1) $$\check{\mathrm{T}}_n * \check{\mathrm{T}}_{n-1} * \cdots * \check{\mathrm{T}}_1 = (\mathrm{T}_1 * \cdots * \mathrm{T}_n)^{\vee}.$$

(17.11.8) *Suppose that G is commutative. If the distributions S and T on G are strictly convolvable (in that order), then so are T and S, and*

(17.11.8.1) $$\mathrm{T} * \mathrm{S} = \mathrm{S} * \mathrm{T}.$$

The proofs are immediate.

(17.11.9) *Let (S_n), (T_n) be two sequences of distributions on a Lie group G. Suppose that the supports of the S_n are contained in a fixed compact subset A of G. If one of the two sequences is weakly bounded and the other converges weakly to 0, then the sequence $(\mathrm{S}_n * \mathrm{T}_n)$ converges weakly to 0.*

Let $f \in \mathscr{D}(\mathrm{G})$ and let $\mathrm{K} = \mathrm{Supp}(f)$. If $m : \mathrm{G}^2 \to \mathrm{G}$ is the mapping $(y, z) \mapsto yz$, then the set $m^{-1}(\mathrm{K}) \cap (\mathrm{A} \times \mathrm{G})$ is compact, and if $h \in \mathscr{D}(\mathrm{G}^2)$ is equal to 1 on a compact neighborhood of this set, we have

$$\langle \mathrm{S}_n * \mathrm{T}_n, f \rangle = \langle \mathrm{S}_n \otimes \mathrm{T}_n, h(f \circ m) \rangle$$

for all n. Now apply (17.10.4(iii)).

It should be remarked that the conclusion of (17.11.9) may be false if the supports of the distribution S_n are not all contained in a fixed compact set. For an example we may take $\mathrm{G} = \mathbf{R}$ and S_n to be the Dirac measure ε_{-n} at the point $-n$, and for T the measure $\sum_{n=1}^{\infty} n\varepsilon_n$ defined by a mass n at each integer point $n > 0$. Then the sequence (S_n) converges weakly to 0, but the measure $\mathrm{S}_n * \mathrm{T}$ has mass n at the point 0, and therefore does not converge weakly to 0.

(17.11.10) *Let $\pi : G \to G'$ be a homomorphism of Lie groups, and let* S, T *be two distributions on* G. *Suppose that either* (1) *π is proper* (17.3.7) *and* S, T *are strictly convolvable, or* (2) *π is arbitrary and* S, T *have compact support. Then* $\pi(S)$ *and* $\pi(T)$ *are strictly convolvable, and we have*

(17.11.10.1) $$\pi(S * T) = \pi(S) * \pi(T).$$

In case (2), $\pi(S)$ and $\pi(T)$ are compactly supported (17.4.3), hence are strictly convolvable. In case (1) it is enough, by virtue of (17.4.3), to show that for each compact subset K' of G' the relations $x \in \mathrm{Supp}(S)$, $y \in \mathrm{Supp}(T)$, and $\pi(x)\pi(y) \in K'$ imply that the pair $(\pi(x), \pi(y))$ belongs to a compact subset of $G' \times G'$. However, since $\pi(x)\pi(y) = \pi(xy)$ and since $\pi^{-1}(K')$ is compact by hypothesis, the point (x, y) belongs to a compact subset of $G \times G$, where the result follows.

For each function $f \in \mathscr{D}(G')$, we have then

$$\begin{aligned}\int f(z')\, d(\pi(S * T))(z') &= \int f(\pi(z))\, d(S * T)(z) \\ &= \iint f(\pi(xy))\, dS(x)\, dT(y) \\ &= \iint f(\pi(x)\pi(y))\, dS(x)\, dT(y) \\ &= \int dT(y) \int f(x'\pi(y))\, d(\pi(S))(x') \\ &= \int d(\pi(S'))(x') \int f(x'y')\, d(\pi(T))(y') \\ &= \int f(z')\, d(\pi(S) * \pi(T))(z')\end{aligned}$$

which proves (17.11.10.1).

Remark

(17.11.10.2) If S and T are distributions *with support* $\{e\}$, then $\pi(S)$ and $\pi(T)$ have support $\{e'\}$, and the formula (17.11.10.1) remains valid, with the same proof, when π is a *local homomorphism* (16.9.9.4): for we need only consider functions f whose support is contained in a neighborhood V of e such that $\pi(xy)$ is defined and equal to $\pi(x)\pi(y)$ for all $x, y \in V$.

In the important case where $G = \mathbf{R}^n$, convolution of distributions behaves as follows relative to the operation of differentiation:

(17.11.11) *If* S *and* T *are strictly convolvable distributions on* $\mathbf{R}^n$, *then for* $1 \leqq k \leqq n$ *the distributions* $D_k S$ *and* T (resp. S *and* $D_k T$) *are strictly convolvable, and we have*

(17.11.11.1) $$D_k(S * T) = (D_k S) * T = S * (D_k T).$$

For each function $f \in \mathscr{D}(\mathbf{R}^n)$, put $g(x) = \int f(x+y)\,dT(y)$, so that by (17.10.1) the function g is indefinitely differentiable, and

$$D_k g(x) = \int D_k f(x+y)\,dT(y).$$

Since by definition $\langle D_k(S * T), f\rangle = -\langle S * T, D_k f\rangle$, we have

$$\langle D_k(S * T), f\rangle = -\langle S, D_k g\rangle = \langle D_k S, g\rangle = \langle (D_k S) * T, f\rangle,$$

from which the first of the equations (17.11.11.1) follows. The second is proved in the same way.

It follows by induction that, for any two multi-indices μ, ν, we have

(17.11.11.2) $$D^{\mu+\nu}(S * T) = (D^{\mu} S) * (D^{\nu} T).$$

In particular, the *derivatives* of a distribution on $\mathbf{R}^n$ can be expressed as *convolutions*:

(17.11.11.3) $$D^{\nu} T = (D^{\nu} \varepsilon_0) * T,$$

where ε_0 is the Dirac measure at the origin.

PROBLEMS

1. Let U be an open subset of $\mathbf{R}^n$. For each function $f \in \mathscr{E}(U)$ and each integer $r \geqq 0$, put $p_r(f) = \sup_{|\nu| \leqq r,\, x \in U} |D^{\nu} f(x)|$, which is a real number or $+\infty$. Let $\mathscr{F}(U)$ be the subspace of $\mathscr{E}(U)$ consisting of the functions f for which all the $p_r(f)$ are finite. The restrictions of the p_r to $\mathscr{F}(U)$ are norms on this vector space, with respect to which it is a Fréchet space.

(a) We have $\mathscr{D}(U) \subset \mathscr{F}(U)$. A distribution T on U is said to be *summable* if it is continuous relative to the topology on $\mathscr{D}(U)$ defined by the restrictions of the norms p_r. Such a distribution is necessarily of finite order. Let (K_m) be a fundamental sequence of compact subsets of U (17.1); for each distribution $T \in \mathscr{D}'(U)$ and each integer $r \geqq 0$, let $p_{m,r}(T)$ denote the least upper bound of the numbers $|T(f)|$ where $\operatorname{Supp}(f) \subset U - K_m$ and $(f) \leqq 1$. Show that a distribution T is summable if and only if, for some integer

$r \geq 0$, the sequence $(\rho_{m,r}(\mathrm{T}))_{m \geq 1}$ converges to 0. In particular, the summable distributions of order 0 are precisely the bounded measures (13.20). (Argue by contradiction to show that the condition is necessary.) Every derivative $\mathrm{D}^\nu \mathrm{T}$ of a summable distribution is summable.
(b) Suppose that $\mathrm{U} = \mathbf{R}^n$ and that K_m is the ball $\|x\| \leq m$. Let $h_1 : \mathbf{R}^n \to [0, 1]$ be a C^∞-mapping which is equal to 1 on K_1 and is 0 outside K_2, and put $h_m(x) = h_1(x/m)$ for all $m \geq 2$. If T is a summable distribution, show that for each function $f \in \mathscr{F}(\mathbf{R}^n)$ the sequence $(\mathrm{T}(h_m f))_{m \geq 1}$ tends to a limit, which we denote by $\mathrm{T}(f)$. In this way T is extended to a continuous linear form on the Fréchet space $\mathscr{F}(\mathbf{R}^n)$.
(c) Show with the help of (b) that the convolution of two summable distributions on $\mathbf{R}^n$ can be defined, and that this convolution is also a summable distribution.

2. (a) With the notation of (17.9.3) show that $\mathrm{Y}_\alpha * \mathrm{Y}_\beta = \mathrm{Y}_{\alpha+\beta}$ for any two complex numbers α, β. (Show that it is sufficient to prove the result for $\mathscr{R}\alpha > 0$ and $\mathscr{R}\beta > 0$.) If T is a distribution on $\mathbf{R}$ whose support is bounded below, then $\mathrm{Y}_{-k} * \mathrm{T} = \mathrm{D}^k\mathrm{T}$ for all integers $k > 0$, and $\mathrm{Y}_k * \mathrm{T}$ is the kth primitive of T whose support is bounded below. By extension, for each complex number ζ, the distribution $\mathrm{Y}_\zeta * \mathrm{T}$ is called the *primitive of order* ζ of T, and the distribution $\mathrm{Y}_{-\zeta} * \mathrm{T}$ is called the *derivative of order* ζ of T.
(b) Let α, β, γ be complex numbers such that $\mathscr{R}\beta > 0$ and $\mathscr{R}\gamma > 0$. The *hypergeometric function* $\mathrm{F}(\alpha, \beta, \gamma; x)$ is defined on the interval $]-1, 1[$ of $\mathbf{R}$ by the formula

$$\mathrm{F}(\alpha, \beta, \gamma; x) = \frac{\Gamma(\gamma)}{\Gamma(\beta)\Gamma(\gamma-\beta)} \int_0^1 t^{\beta-1}(1-t)^{\gamma-\beta-1}(1-tx)^{-\alpha}\, dt.$$

Show that, for γ complex and $\neq -n$, where $n \in \mathbf{N}$, the function can be extended to all values of $\beta \in \mathbf{C}$ in such a way that

$$\frac{x^{\gamma-1}}{\Gamma(\gamma)} \mathrm{F}(\alpha, \beta, \gamma; x) = \mathrm{Y}_{\gamma-\beta} * \frac{\mathrm{Pf}(x_+^{\beta-1}(1-x)_+^{-\alpha})}{\Gamma(\beta)}$$

(change the variable to $w = tx$). In particular, for $\beta = -k$ (where $k \in \mathbf{N}$) we have

$$\frac{x^{\gamma-1}}{\Gamma(\gamma)} \mathrm{F}(\alpha, -k, \gamma; x) = \mathrm{Y}_{k+\gamma} * ((1-x)^{-\alpha}\mathrm{D}^k\varepsilon_0).$$

Deduce that

$$\mathrm{F}(\alpha, -k, \gamma; x) = \sum_{r=0}^{k} (-1)^r \binom{k}{r} \frac{\Gamma(\alpha+r)}{\Gamma(\alpha)} \frac{\Gamma(\gamma)}{\Gamma(\gamma+r)} x^r$$

(*Jacobi polynomial*). (Expand the distribution $(1-x)^{-\alpha}\mathrm{D}^k\varepsilon_0$ as a sum of point-distributions with support $\{0\}$.)
(c) For each $p \in \mathbf{C}$ such that $\mathscr{R}p > -\frac{1}{2}$, the *Bessel function* of order p, defined for $x \in \mathbf{R}$, is given by the formula

$$\mathrm{J}_p(x) = \frac{2^{1-p}x^p}{\Gamma(p+\frac{1}{2})\pi^{1/2}} \int_0^1 (1-t^2)^{p-(1/2)} \cos xt\, dt.$$

Show that the function can be extended to all complex values of p in such a way that, for $u > 0$, we have

$$2^p\pi^{1/2}u^{p/2}\mathrm{J}_p(u^{1/2}) = \mathrm{Y}_{p+1/2} * (u^{-1/2}\cos u^{1/2}).$$

Deduce *Sonine's formula*

$$J_{p+q+1}(x) = \frac{x^{q+1}}{\Gamma(q+1)2^q} \int_0^{\pi/2} J_p(x \sin \theta) \sin^{p+1} \theta \cos^{2q+1} \theta \, d\theta.$$

(Convolve with Y_{q+1}.)

3. With the notation of (17.9.4), show that $Z_\alpha * Z_\beta = Z_{\alpha+\beta}$ for all complex numbers α, β. (Same method as in Problem 2.)

4. Consider $\mathscr{E}'(\mathbf{R}^n)$ as a vector space of linear forms *on* $\mathscr{E}(\mathbf{R}^n)$, and endow $\mathscr{E}'(\mathbf{R}^n)$ with the corresponding weak topology (12.15.2). For each distribution $S \in \mathscr{D}'(\mathbf{R}^n)$, show that the mapping $T \mapsto S * T$ of $\mathscr{E}'(\mathbf{R}^n)$ into $\mathscr{D}'(\mathbf{R}^n)$ is continuous.

5. Let u be a continuous linear mapping of $\mathscr{E}'(\mathbf{R}^n)$ into $\mathscr{D}'(\mathbf{R}^n)$.

   (a) Show that the following two properties are equivalent:
   (1) $\gamma(\mathbf{h})u(T) = u(\gamma(\mathbf{h})T)$ for all $T \in \mathscr{E}'(\mathbf{R}^n)$ and all $\mathbf{h} \in \mathbf{R}^n$;
   (2) $D_j u(T) = u(D_j T)$ for all $T \in \mathscr{E}'(\mathbf{R}^n)$ and $1 \leqq j \leqq n$.
   (Use formula (17.8.2.1) and consider, for each $f \in \mathscr{D}(\mathbf{R}^n)$, the function $\mathbf{h} \mapsto \langle u(\gamma(\mathbf{h})T), \gamma(\mathbf{h})f \rangle$; calculate its partial derivatives.)
   (b) If u satisfies the equivalent conditions of (a), show that u is necessarily of the form $T \mapsto S * T$, where $S \in \mathscr{D}'(\mathbf{R}^n)$. (Consider the linear mapping $R \mapsto u(R * T) - R * u(T)$ of $\mathscr{E}'(\mathbf{R}^n)$ into $\mathscr{D}'(\mathbf{R}^n)$, for a fixed distribution $T \in \mathscr{E}'(\mathbf{R}^n)$, and show that its kernel is the whole of $\mathscr{E}'(\mathbf{R}^n)$. For this purpose, observe that the Dirac measures ε_x ($x \in \mathbf{R}^n$) form a total set (12.13) in $\mathscr{E}'(\mathbf{R}^n)$, by using Problem 13 of Section 12.15; then remark that the ε_x belong to the kernel of the linear mapping in question.)

6. Let G, G′ be Lie groups and let S, T (resp. S′, T′) be strictly convolvable distributions on G (resp. G′). Show that $S \otimes S'$ and $T \otimes T'$ are strictly convolvable distribution on $G \times G'$ and that

$$(S \otimes S') * (T \otimes T') = (S * T) \otimes (S' * T').$$

12. REGULARIZATION OF DISTRIBUTIONS

(17.12.1) *Let p, m be two integers $\geqq 0$ such that $p \geqq m$. If $T \in \mathscr{D}'^{(m)}(\mathbf{R}^n)$ and $f \in \mathscr{D}^{(p)}(\mathbf{R}^n)$* (resp. *if $T \in \mathscr{E}'^{(m)}(\mathbf{R}^n)$ and $f \in \mathscr{E}^{(p)}(\mathbf{R}^n)$*), *then the distribution $T * f$ may be identified* (17.5.3) *with a function in $\mathscr{E}^{(p-m)}(\mathbf{R}^n)$ such that, for each $x \in \mathbf{R}^n$,*

(17.12.1.1) $$(T * f)(x) = \langle T, \check{f}(x)\gamma \rangle = \int f(x-y) \, dT(y).$$

The fact that the function $x \mapsto \int f(x-y)\, dT(y)$ belongs to $\mathscr{E}^{(p-m)}(\mathbf{R}^n)$ follows from the hypotheses and from (17.10.1). Next, if $S = T * f$, then we have by definition, for any $u \in \mathscr{D}(\mathbf{R}^n)$,

$$\langle S, u \rangle = \int dT(x) \int f(y)u(x + y)\, d\lambda(y) = \langle T \otimes \lambda, g \rangle,$$

where λ is Lebesgue measure on $\mathbf{R}^n$, and $g(x, y) = f(y)u(x + y)$. However,

$$\int f(y)u(x + y)\, d\lambda(y) = \int f(y - x)u(y)\, d\lambda(y);$$

hence $\langle S, u \rangle = \langle T \otimes (u \cdot \lambda), h \rangle$, where $h(x, y) = f(y - x)$; and therefore

$$\langle S, u \rangle = \int u(y)\, d\lambda(y) \int f(y - x)\, dT(x),$$

which proves (17.12.1.1).

In particular:

(17.12.2) *If* $T \in \mathscr{E}'(\mathbf{R}^n)$ *and* $f \in \mathscr{E}(\mathbf{R}^n)$, *or if* $T \in \mathscr{D}'(\mathbf{R}^n)$ *and* $f \in \mathscr{D}(\mathbf{R}^n)$, *then* $T * f \in \mathscr{E}(\mathbf{R}^n)$. *If* $T \in \mathscr{E}'(\mathbf{R}^n)$ *and if* (f_p) *is a sequence of functions in* $\mathscr{E}(\mathbf{R}^n)$ *converging to* 0 *in this space, then the sequence* $(T * f_p)$ *tends to* 0 *in* $\mathscr{E}(\mathbf{R}^n)$.

Let A be the (compact) support of T, let V be a compact neighborhood of A, and let K be any compact subset of $\mathbf{R}^n$. If we put $g_p(x, y) = f_p(x - y)$, then the sequence of elements $g_p(x, \cdot)$ of $\mathscr{E}(\mathbf{R}^n)$ converges to 0 in this space, *uniformly* with respect to $x \in K$. For the partial derivative with multi-index v of the function $y \mapsto g_p(x, y)$ is $(-1)^{|v|} D^v f_p(x - y)$. Since the sequence $(D^v f_p(z))$ converges uniformly to 0 in the compact set $K + (-V)$, the sequence of functions $(x, y) \mapsto D^v f_p(x - y)$ converges uniformly to 0 on $K \times V$; hence, by the definition of distributions, the sequence of functions $x \mapsto h_p(x) = \int f_p(x - y)\, dT(y)$ converges uniformly to 0 in K, and the same is true of the sequence of partial derivatives $x \mapsto D^v h_p(x)$ for any multi-index v, by virtue of (17.10.1).

(17.12.3) *For each open subset* U *of* $\mathbf{R}^n$, *the set* $\mathscr{D}(U)$ *of* C^∞-*functions on* U *with compact support, identified with a space of distributions on* U (17.5.3), *is weakly dense in* $\mathscr{D}'(U)$.

Let (K_m) be a fundamental sequence of compact subsets of U, and let h_m be a function belonging to $\mathscr{D}(U)$ which is equal to 1 on K_m. Since for each $u \in \mathscr{D}(U)$ the reexists an integer m such that K_m is a neighborhood of the support of u, it follows that for each distribution $T \in \mathscr{D}'(U)$ we have

$$\langle T, u \rangle = \langle T, h_m u \rangle = \langle h_m \cdot T, u \rangle$$

and therefore the sequence $(h_m \cdot \mathrm{T})$ converges weakly to T. We are therefore reduced to the case where T has *compact* support A. Let W be a neighborhood of 0 in $\mathbf{R}^n$ such that $\mathrm{A} + \mathrm{W} \subset \mathrm{U}$. If (g_k) is a regularizing sequence (17.1.2), then $\mathrm{Supp}(g_k) \subset \mathrm{W}$ for large k. Since the sequence of distributions (g_k) converges weakly to the Dirac measure ε_0, it follows from (17.11.9) that the sequence $(\mathrm{T} * g_k)$ converges weakly to $\mathrm{T} * \varepsilon_0 = \mathrm{T}$. This completes the proof.

This proof shows moreover that a distribution T with compact support can be approximated by functions belonging to $\mathscr{D}(\mathrm{U})$ whose support is contained in an *arbitrary neighborhood of the support of* T (cf. Problem 13).

(17.12.4) *Let* T *be a distribution on* $\mathbf{R}^n$, *with compact support* A. *Then for each neighborhood* V *of* A *there exists a finite number of continuous functions* f_k *on* $\mathbf{R}^n$, *with supports contained in* V, *such that* T *is equal to a sum of partial derivatives* $\sum_k \mathrm{D}^{\nu_k} f_k$ (*in which some of the* ν_k *may be zero*).

Let m be the order of T (17.7.1) and let p be an integer such that $2p - n \geqq m + 1$. Then it follows from (17.9.5) that there exists a function $\mathrm{E} \in \mathscr{E}^{(m)}(\mathbf{R}^n)$ such that $\Delta^p \mathrm{E} = \varepsilon_0$, or equivalently, such that $(\Delta^p \varepsilon_0) * \mathrm{E} = \varepsilon_0$. We then have

$$\mathrm{T} = \varepsilon_0 * \mathrm{T} = (\Delta^p \varepsilon_0) * (\mathrm{E} * \mathrm{T}) = \Delta^p(\mathrm{E} * \mathrm{T});$$

the distribution $\mathrm{E} * \mathrm{T}$ is in fact a *continuous function* on $\mathbf{R}^n$, by reason of the choice of p and (17.12.1), but its support is not in general contained in V. However, let W be a compact neighborhood of 0 in $\mathbf{R}^n$ such that $\mathrm{A} + \mathrm{W} \subset \mathrm{V}$, and let $g \in \mathscr{D}(\mathbf{R}^n)$ be a function with support contained in W, and equal to 1 in some neighborhood of 0. Since the function $1 - g$ vanishes on a neighborhood of 0, the function $(1 - g)\mathrm{E}$ belongs to $\mathscr{E}(\mathbf{R}^n)$ (17.9.5), and hence the same is true of $(\Delta^p((1 - g)\mathrm{E})) * \mathrm{T}$ (17.12.2). Moreover, the support of $g\mathrm{E}$ is contained in W, and hence so also is the support of $\Delta^p(g\mathrm{E})$, and therefore also the support of the function $u = \Delta^p((1 - g)\mathrm{E}) = \varepsilon_0 - \Delta^p(g\mathrm{E})$. Consequently $\mathrm{Supp}(u * \mathrm{T}) \subset \mathrm{V}$. If also we put $v = g\mathrm{E}$, then $\mathrm{Supp}(v) \subset \mathrm{W}$ and so $\mathrm{Supp}(v * \mathrm{T}) \subset \mathrm{V}$. Since $\mathrm{T} = \Delta^p(v * \mathrm{T}) + (u * \mathrm{T})$, the proof is complete.

Remark

(17.12.5) The formula (17.12.1.1) gives in particular

(17.12.5.1) $$(\mathrm{T} * f)(0) = \langle \mathrm{T}, \check{f} \rangle$$

for all $f \in \mathscr{D}^{(p)}(\mathbf{R}^n)$ and all $\mathrm{T} \in \mathscr{D}'^{(m)}(\mathbf{R}^n)$ with $p \geqq m$. In particular, this shows that if $\mathrm{T} * f = 0$ for *all* $f \in \mathscr{D}(\mathbf{R}^n)$, then $\mathrm{T} = 0$.

Since $\int f(x+y)\,dT(y) = \int f(x-y)\,d\check{T}(y)$, the definition of convolution (17.11.1) shows that if S, T are strictly convolvable distributions on $\mathbf{R}^n$, then for all $f \in \mathscr{D}(\mathbf{R}^n)$ we have

(17.12.5.2) $$\langle S * T, f\rangle = \langle S, \check{T} * f\rangle = \langle T, \check{S} * f\rangle.$$

PROBLEMS

1. Let S, T be two distributions belonging to $\mathscr{E}'(\mathbf{R}^n)$. Show that

$$\text{Supp. sing}(S * T) \subset \text{Supp. sing}(S) + \text{Supp. sing}(T).$$

(Decompose each of S, T into the sum of a distribution whose support is contained in an arbitrarily small neighborhood of the singular support and a function belonging to $\mathscr{D}(\mathbf{R}^n)$.)

2. What can be said about the convolution of an arbitrary distribution $T \in \mathscr{E}'(\mathbf{R}^n)$ with a polynomial? Deduce that every $T \in \mathscr{E}'(\mathbf{R}^n)$ is the limit of a sequence of polynomials, with respect to the weak topology of $\mathscr{D}'(\mathbf{R}^n)$.

3. What can be said about the convolution of an arbitrary distribution $T \in \mathscr{E}'(\mathbf{R}^n)$ with the product of a polynomial and an exponential $\exp(\langle x, x'\rangle)$, where x' is a linear form on $\mathbf{R}^n$?

4. (a) Let $T \in \mathscr{D}'(\mathbf{R}^n)$ be such that, for each $f \in \mathscr{K}(\mathbf{R}^n)$, the distribution $T * f$ is a locally bounded function. Let K be a compact subset of $\mathbf{R}^n$ and let H be a compact neighborhood of 0 in $\mathbf{R}^n$. Show that the mapping $f \mapsto (T * f)|H$ is a continuous mapping of the Banach space $\mathscr{K}(\mathbf{R}^n; K)$ into $\mathscr{L}^\infty(H, \lambda_H)$, where λ_H is Lebesgue measure on H. (Show first that for each function $g \in \mathscr{D}(\mathbf{R}^n)$ the mapping $f \mapsto ((T * f) * g)|H$ is continuous (14.10.6); then replace g by the functions belonging to a regularizing sequence, and use the Banach–Steinhaus theorem.)
(b) Show that, for each function $f \in \mathscr{K}(\mathbf{R}^n)$, the distribution $T * f$ is a continuous function. (Use (a) and the fact that $\mathscr{D}(\mathbf{R}^n; K)$ is dense in $\mathscr{K}(\mathbf{R}^n; K)$.)
(c) Deduce from (a) and (b) that T is a *measure* on $\mathbf{R}^n$. (Use the formula (17.12.5.1).)
(d) Show likewise that if $T * f$ is locally bounded for each $f \in \mathscr{L}^p(\lambda)$ $(1 \leq p < +\infty)$ with compact support (resp. $f \in \mathscr{D}^{(r)}(\mathbf{R}^n)$), then T is a function belonging to $\mathscr{L}^q(\lambda)$, where $(1/q) + (1/p) = 1$ (resp. a distribution belonging to $\mathscr{D}'^{(r)}(\mathbf{R}^n)$).
(e) Deduce from (c) that if $T * f$ is a measure for each $f \in \mathscr{L}^1(\lambda)$ with compact support, then T is a measure. (Consider the convolution $(T * f) * g$, where $g \in \mathscr{K}(\mathbf{R}^n)$.)

5. Let $T \in \mathscr{D}'(\mathbf{R}^n)$ be a distribution such that $T * f \in \mathscr{E}(\mathbf{R}^n)$ for all $f \in \mathscr{D}^{(r)}(\mathbf{R}^n)$. Show that $T \in \mathscr{E}(\mathbf{R}^n)$. (Argue as in (17.12.4).)

6. A distribution all of whose derivatives are of order $\leq m$ belongs to $\mathscr{E}(\mathbf{R}^n)$ (use (17.9.5)).

7. Let H be a subset of $\mathscr{D}'(\mathbf{R}^n)$. Show that the following conditions are equivalent:

(α) H is bounded in $\mathscr{D}'(\mathbf{R}^n)$ (12.15).
(β) For each $f \in \mathscr{D}(\mathbf{R}^n)$ and each compact $K \subset \mathbf{R}^n$, the set of restrictions to K of the functions $T * f$, where $T \in H$, is bounded in $\mathscr{C}(K)$.
(γ) For each $f \in \mathscr{D}(\mathbf{R}^n)$, the set of functions $T * f$, where $T \in H$, is bounded in $\mathscr{D}'(\mathbf{R}^n)$.
(δ) There exists an integer $m \geqq 0$ such that, for each relatively compact open subset U of $\mathbf{R}^n$, the restrictions to U of the $T \in H$ are sums of derivatives of order $\leqq m$ of continuous uniformly bounded functions on U.

(To show that (γ) implies (δ), remark first that if V is a relatively compact open neighborhood of 0 in $\mathbf{R}^n$ and if $f \in \mathscr{D}(\mathbf{R}^n; V)$, then the set of mappings $u \mapsto ((T * f) * u) | U$ of $\mathscr{D}(\mathbf{R}^n; V)$ into $\mathscr{L}^\infty(U, \lambda_U)$, where T runs through H, is equicontinuous. Apply Section 12.16, Problem 10 to deduce that there exists an integer m and a neighborhood W of 0 in $\mathscr{D}^{(m)}(\mathbf{R}^n; V)$ such that, for all $u, v \in W$ and $T \in H$, the restriction of $T * u * v$ to U is a function bounded above by 1 in absolute value. Using the fact that $\mathscr{D}(\mathbf{R}^n; V)$ is dense in $\mathscr{D}^{(m)}(V)$, show that for all $u, v \in \mathscr{D}^{(m)}(V)$, the restrictions to U of the distributions $T * u * v$ are continuous functions on U which are uniformly bounded as T runs through H. Finally, use the formula

$$T = \Delta^{2p}(gE * gE * T) - 2\Delta^p(gE * u * T) + u * u * T,$$

in the notation of the proof of (17.12.4).)

8. Let (T_k) be a sequence of distributions belonging to $\mathscr{D}'(\mathbf{R}^n)$. Show that the following conditions are equivalent:

(α) $T_k \to 0$ in $\mathscr{D}'(\mathbf{R}^n)$.
(β) For each $f \in \mathscr{D}(\mathbf{R}^n)$ the sequence of functions $(T_k * f)$ converges uniformly to 0 on each compact subset of $\mathbf{R}^n$.
(γ) For each $f \in \mathscr{D}(\mathbf{R}^n)$ the sequence of distributions $(T_k * f)$ converges to 0 in $\mathscr{D}'(\mathbf{R}^n)$.
(Argue as in Problem 7.)

9. Let $L : \mathscr{D}(\mathbf{R}^n) \to \mathscr{D}'(\mathbf{R}^n)$ be a linear mapping such that, for each compact subset K of $\mathbf{R}^n$, the restriction of L to $\mathscr{D}(\mathbf{R}^n; K)$ is continuous and $L(f * g) = L(f) * g$ for all $f, g \in \mathscr{D}(\mathbf{R}^n)$. Show that there exists a distribution $S \in \mathscr{D}'(\mathbf{R}^n)$ such that $L(f) = S * f$. (Observe that if (g_k) is a regularizing sequence, the sequence $(L(g_k) * f)$ converges to $L(f)$ in $\mathscr{D}'(\mathbf{R}^n)$, and use Problem 8.)

10. Let T be a distribution on $\mathbf{R}^n$. Show that, for T to be an analytic function on $\mathbf{R}^n$, it is necessary and sufficient that, for each $f \in \mathscr{D}(\mathbf{R}^n)$, $T * f$ should be an analytic function on $\mathbf{R}^n$. (To show that the condition is necessary, use Cauchy's inequalities to majorize the derivatives $D^\nu(T * f)$ on a compact set K. Conversely, if $T * f$ is analytic for each $f \in \mathscr{D}(\mathbf{R}^n)$, observe that for each relatively compact open set U and each relatively compact open neighborhood V of 0 in $\mathbf{R}^n$, the mapping

$$u_\nu : f \mapsto \sup |(D^\nu T * f)(x)/\nu!|^{1/|\nu|}$$

is finite and continuous on $\mathscr{D}(\mathbf{R}^n; \bar{V})$ for each multi-index ν; furthermore, the set of $u_\nu(f)$ is bounded above as ν runs through $\mathbf{N}^n$, for each $f \in \mathscr{D}(\mathbf{R}^n; \bar{V})$. Using Baire's theorem (12.16.2), deduce that there exists a constant $c > 0$ such that, for each $f \in \mathscr{D}(\mathbf{R}^n; \bar{V})$, the set of functions $D^\nu(T * f)/(\nu! c^{|\nu|})$ (for $\nu \in \mathbf{N}^n$) is uniformly bounded on U. Using Problem 7, show that there exists an integer $m \geqq 0$ such that the same property holds for each $f \in \mathscr{D}^{(m)}(V)$, and take f of the form gE, in the notation of the proof of (17.12.4).)

11. Let $T \in \mathscr{E}'(\mathbf{R}^n)$ be such that $S = \partial^n T/\partial x_1 \partial x_2 \cdots \partial x_n$ is a measure (resp. a locally integrable function). Show that T is then a bounded (resp. continuous) function. Convolve S with the function $Y(x_1)Y(x_2) \cdots Y(x_n)$, where Y is the Heaviside function, and use (14.10.6).)

12. (a) Show that every distribution $T \in \mathscr{E}'(\mathbf{R}^n)$ can be written in the form $D^\nu f$, where f is a continuous function on $\mathbf{R}^n$ (but not necessarily of compact support).
(b) Show that there is no compactly supported distribution T on $\mathbf{R}$ such that $\varepsilon_0 + D\varepsilon_0 = D^p T$ for some $p > 0$.
(c) Let B be the disk $\|x\| \leq 1$ in $\mathbf{R}^2$. Show that there exists no compactly supported distribution T on $\mathbf{R}^2$ such that $\varphi_B = D^\nu T$ for some multi-index $\nu \neq (0, 0)$.

13. Give an example of a distribution on $\mathbf{R}$ whose support is a compact interval I, not consisting of a single point, and which is not the limit in $\mathscr{D}'(\mathbf{R})$ of functions $f \in \mathscr{D}(\mathbf{R})$ with support contained in I.

14. Let S, T be two distributions belonging to $\mathscr{D}'(\mathbf{R})$, whose supports are *bounded below* in $\mathbf{R}$; then they are strictly convolvable (17.11.2). Show that $S * T = 0$ implies that either $S = 0$ or $T = 0$. (Remark first that for all $f, g \in \mathscr{D}(\mathbf{R})$ we have

$$(S * f) * (T * g) = 0,$$

and use Titchmarsh's theorem (Section 11.6, Problem 11) to deduce that for example $S * f = 0$ (with f not identically zero), and hence $(S * u) * f = 0$ for all $u \in \mathscr{D}(\mathbf{R})$; then use (17.12.5).)

15. Give an example of a sequence of functions (f_n) in $\mathscr{D}(\mathbf{R})$, with supports contained in a fixed compact set, which converges in $\mathscr{D}'(\mathbf{R})$ but which is such that the norms $\|f_n\|$ (in $\mathscr{C}^\infty(\mathbf{R})$) and $N_1(f_n)$ (in $L^1(\mathbf{R})$) are not bounded.

16. Let H be a bounded set of distributions belonging to $\mathscr{E}'(\mathbf{R})$, whose supports are all contained in a fixed compact set K. If I is a compact interval which is a neighborhood of K, then there exists an integer r such that the distributions $T \in H$ are of the form $D^r F$, where the functions F are continuous on I (Section 17.7, Problem 1 and Section 17.5, Problem 3). Show that the set H_r of functions $Y_r * T$, where $T \in H$, is bounded in $\mathscr{L}^1$ (use (17.8.5)). If H_{r+1} is the set of functions $Y_{r+1} * T$, then H_{r+1} is bounded in $\mathscr{C}(I)$. Deduce that if (T_n) is a sequence of distributions belonging to H which converges to $T \in \mathscr{E}'(\mathbf{R})$, then the sequence of primitives $(Y_{r+2} * T_n)$ is a sequence of continuous functions which converges *uniformly* on I to $Y_{r+2} * T$.

17. Let $T \in \mathscr{E}'(\mathbf{R})$. For each $\lambda > 0$, let T_λ denote the distribution defined by $T_\lambda(f) = \lambda^{-1} \int f(x/\lambda)\, dT(x)$ (in other words, $(T_\lambda)_{|\upsilon_0}$ is the image of $T_{|\upsilon_0}$ under the homothety $x \mapsto x/\lambda$).

Show that if T_λ converges to a distribution T_{0+} in $\mathscr{E}'(\mathbf{R})$ as $\lambda \to 0$, then this distribution must be a constant function on $\mathbf{R}$. (Observe that, for each function $f \in \mathscr{D}(\mathbf{R})$ and each $h \neq 0$, if $\langle T_\lambda, f\rangle$ tends to a limit, then $\langle T_\lambda, \gamma(h)f\rangle$ tends to the same limit, by expanding the difference $f(x + \lambda h) - f(x)$ by Taylor's formula, stopping at the order of T; then apply Problem 4 of Section 17.8.)

18. (a) Let f be a real-valued function defined on an open interval $I =]0, c[$. Suppose that there exists a compact interval $J = [a, b] \subset I$ such that $a < b$, n functions $a_0(\lambda), \ldots, a_{n-1}(\lambda)$ defined on a sufficiently small interval $]0, \gamma[$, and an increasing function $\varepsilon(\lambda)$ defined on this interval and tending to 0 as $\lambda \to 0$, such that, for all $x \in [a, b]$ and all $\lambda \in]0, \gamma[$, we have

$$|f(\lambda x) - a_0(\lambda) - a_1(\lambda)\lambda x - \cdots - a_{n-1}(\lambda)(\lambda x)^{n-1}| \leqq \varepsilon(\lambda)\lambda^n.$$

Then each of the functions $a_j(\lambda)$ tends to a limit b_j as $\lambda \to 0$, and we have

$$f(x) = b_0 + b_1 x + \cdots + b_{n-1}x^{n-1} + o(x^n)$$

as $x \to 0$ in I. (Let $a < x_0 < x_1 < \cdots < x_{n-1} < b$ be fixed points of J and let $\theta \in]0, 1[$ be such that $a < \theta x_0$. If t, t' are such that $0 < \theta t \leqq t' \leqq t < \gamma$, show that

$$|(a_0(t') - a_0(t)) + (a_1(t') - a_1(t))t'x_i + \cdots + (a_{n-1}(t') - a_{n-1}(t))(t'x_i)^{n-1}| \leqq 2\varepsilon(t)t^n$$

for $0 \leqq i \leqq n-1$, and deduce that there exists a constant A such that under these hypotheses

$$|a_j(t) - a_j(t')| \leqq A\theta^{-j}\varepsilon(t)^{n-j}$$

$(0 \leqq j \leqq n-1)$. Now consider the sequence (θ^p) tending to 0, and apply Cauchy's criterion.)

(b) With the notation of Problem 17, show that the limit T_{0+} exists if and only if there exists an integer $n \geqq 0$ and a continuous function F on an open interval $I =]0, c[$ such that $\lim_{x \to 0,\, x > 0} F(x)/x^n = A$ exists and such that the restriction of T to I is equal to $D^n F$; in which case $T_{0+} = n!A$. (To show that the condition is necessary, consider a compact interval $J = [\theta a, a]$, where $0 < \theta < 1$, contained in I; by the result of Problem 16, there exists an integer n, an open neighborhood V of J and, for each sufficiently small λ, a continuous function G_λ on V such that the restriction to V of T_λ is of the form $D^n G_\lambda$, and such that the functions G_λ converge *uniformly* to $n!Ax^n$ in V. Deduce that there exists a continuous function G on $]0, b[$ such that the restriction of T to $]0, b[$ is equal to $D^n G$. For all sufficiently small λ, we can write $G(\lambda x) - \lambda^n G_\lambda(x) = w_\lambda(x)$, where w_λ is a polynomial of degree $\leqq n-1$, with coefficients depending on λ. Now apply (a).)

(c) Deduce from (b) that if S is a distribution such that $DS = T$, and if the limit T_{0+} exists, then so does the limit S_{0+}. Also, for each function $f \in \mathscr{D}(\mathbf{R})$, the limit $(f \cdot T)_{0+}$ exists and is equal to $f(0)T_{0+}$.

(d) If $\alpha > 0$ and $\beta > 0$, show that the function $x^{-\alpha}\sin(x^{-\beta})$, which is defined and continuous for $x > 0$, extends to a distribution T on $\mathbf{R}$ for which the limit T_{0+} exists and is equal to 0.

13. DIFFERENTIAL OPERATORS AND FIELDS OF POINT-DISTRIBUTIONS

(17.13.1) Let X be a differential manifold and E, F two complex vector bundles over X. We have seen that the vector spaces $\Gamma(X, E)$ and $\Gamma(X, F)$ are canonically endowed with structures of separable complex Fréchet spaces (17.2.2). A C^∞ *linear differential operator* from E to F (or simply a *differential operator*, or even an *operator* if there is no risk of ambiguity) is by definition a *continuous linear mapping* $\mathbf{f} \mapsto P \cdot \mathbf{f}$ of the Fréchet space $\Gamma(X, E)$ into the Fréchet space $\Gamma(X, F)$ which satisfies the following condition:

(L) *For each open subset* U *of* X *and each section* $\mathbf{f} \in \Gamma(X, E)$ *such that* $\mathbf{f} | U = 0$, *we have* $P \cdot \mathbf{f} | U = 0$.

In other words, if two sections $\mathbf{f}, \mathbf{g}$ of E over X are equal on an open set U, then so are their images $P \cdot \mathbf{f}$ and $P \cdot \mathbf{g}$. An equivalent way of stating this is to say that P is an operator of *local* character.

Let (V, φ, n) be a chart of X such that E and F are trivializable over V. If π', π'' are the projections of the bundles E, F and if N′, N″ are their respective ranks over V, then there exist diffeomorphisms $z \mapsto (\varphi(\pi'(z)), v(z))$ of $\pi'^{-1}(V)$ onto $\varphi(V) \times \mathbf{C}^{N'}$ and $z \mapsto (\varphi(\pi''(z)), w(z))$ of $\pi''^{-1}(V)$ onto $\varphi(V) \times \mathbf{C}^{N''}$ such that v (resp. w) is a linear isomorphism of each fiber $\pi'^{-1}(x)$ (resp. $\pi''^{-1}(x)$) onto $\mathbf{C}^{N'}$ (resp. $\mathbf{C}^{N''}$). The mapping $\mathbf{f} \mapsto v \circ \mathbf{f} \circ \varphi^{-1}$ (resp. $\mathbf{f} \mapsto w \circ \mathbf{f} \circ \varphi^{-1}$) is then an isomorphism of $\Gamma(V, E)$ onto $(\mathscr{E}(\varphi(V)))^{N'}$ (resp. of $\Gamma(V, F)$ onto $(\mathscr{E}(\varphi(V)))^{N''}$). If P is a differential operator from E to F, then for each section $\mathbf{f} \in \Gamma(X, E)$ the value $P \cdot \mathbf{f} | V$ depends only on $\mathbf{f} | V$, and there is therefore a well-defined continuous linear mapping $\mathbf{g} \mapsto Q \cdot \mathbf{g}$ of $(\mathscr{E}(\varphi(V)))^{N'}$ into $(\mathscr{E}(\varphi(V)))^{N''}$ such that

$$w \circ (P \cdot \mathbf{f} | V) \circ \varphi^{-1} = Q \cdot (v \circ (\mathbf{f} | V) \circ \varphi^{-1}). \tag{17.13.2}$$

The linear mapping Q is said to be the *local expression* of the operator P corresponding to the chart (V, φ, n) and the mappings v and w.

(17.13.3) *In order that a linear mapping P of* $\Gamma(X, E)$ *into* $\Gamma(X, F)$ *should be a differential operator, it is necessary and sufficient that for each* $x \in X$ *there should exist a chart* (V, φ, n) *of* X *at the point x, such that* E *and* F *are trivializable over* V *and such that the corresponding local expression of P should be of the form*

$$\mathbf{g} \mapsto \sum_{|\nu| \leqq p} A_\nu \cdot D^\nu \mathbf{g}, \tag{17.13.3.1}$$

where, for each multi-index ν such that $|\nu| \leqq p$, *the mapping* $y \mapsto A_\nu(y)$ *is a* C^∞*-mapping of* $\varphi(V)$ *into the vector space* $\operatorname{Hom}_{\mathbf{C}}(\mathbf{C}^{N'}, \mathbf{C}^{N''})$ (which can be identified with the space of N″ × N′ matrices over **C**).

The condition is sufficient. First, it is clear that it implies the condition (L). Second, since each compact subset of X admits a finite covering by domains of definition of charts which satisfy the condition of the statement of the proposition, it is enough by virtue of (17.2) and (3.13.14) to verify that the mapping (17.13.3.1) is continuous, and this is a direct consequence of (17.1.3) and (17.1.4).

To show that the condition is necessary, we may clearly assume that X is an open subset of $\mathbf{R}^n$ and that $E = X \times \mathbf{C}^{N'}$, $F = X \times \mathbf{C}^{N''}$, so that $\Gamma(X, E) = (\mathscr{E}(X))^{N'}$ and $\Gamma(X, F) = (\mathscr{E}(X))^{N''}$. Replacing P by $p \circ P \circ j$, where j is a canonical injection of one of the factors of $(\mathscr{E}(X))^{N'}$ into this product, and p is a canonical projection of the product $(\mathscr{E}(X))^{N''}$ onto one of its factors, we reduce further to the case where $N' = N'' = 1$.

Replacing X if necessary by a relatively compact open set, we may suppose, by virtue of the definition of the topology of $\mathscr{E}(X)$ (17.1), that there exists a constant c and an integer p such that for all $x \in X$ and all $f \in \mathscr{E}(X)$ we have

(17.13.3.2) $$|(P \cdot f)(x)| \leqq c \cdot \sup_{y \in X,\, |\nu| \leqq p} |D^{\nu} f(y)|.$$

This shows that for each $x \in X$ the linear form $f \mapsto (P \cdot f)(x)$ is a *distribution* of order $\leqq p$ on X; furthermore, if $x \notin \operatorname{Supp}(f)$, then by hypothesis we have $(P \cdot f)(x) = 0$, so that the support of this distribution is $\{x\}$. Hence (17.7.3) it is of the form

$$f \mapsto \sum_{|\nu| \leqq p} a_{\nu}(x) D^{\nu} f(x),$$

where the $a_{\nu}(x)$ are scalars.

Replacing f successively by monomials x^{α}, we see that for $|\alpha| \leqq p$ the functions

$$x \mapsto \sum_{\nu \leqq \alpha} \nu! \binom{\alpha}{\nu} a_{\nu}(x) x^{\alpha - \nu}$$

are of class C^{∞}. It follows easily by induction on $|\nu|$ that all the a_{ν} are of class C^{∞}, and the proof is complete.

For each differential operator P from E to F, and each point $x \in X$, the *order of P at x* is defined to be the largest of the integers $|\nu|$ such that $A_{\nu}(x) \neq 0$ in a local expression of P in a neighborhood of x. It follows immediately from the rule for differentiating composite functions and from Leibniz's formula that this number cannot increase when we pass from one local expression to another, and hence it is *independent* of the particular local expression chosen.

By virtue of (17.13.3), a *differential operator of order* 0 may be written

(17.13.3.3) $$\mathbf{f} \mapsto \mathbf{A} \cdot \mathbf{f},$$

where, for each $x \in X$, $\boldsymbol{A}(x)$ is a linear mapping of the fiber E_x into the fiber F_x, in other words an element of $\mathrm{Hom}_{\mathbf{C}}(E_x, F_x)$, and $\boldsymbol{A} : x \mapsto \boldsymbol{A}(x)$ is a *section* (of class C^∞) of the vector bundle Hom(E, F) (which may be identified with the tensor product $E^* \otimes F$). Such a section may also be identified with a *linear* X-*morphism* $A : E \to F$; for each section $\boldsymbol{f}$ of E, $\boldsymbol{A} \cdot \boldsymbol{f}$ is the section $A \circ \boldsymbol{f}$ of F (16.16.4).

(17.13.4) If P is a differential operator from E to F, then for each open subset U of X we may define the *restriction* of P to U, which is a differential operator $P | U$ from $E | U$ to $F | U$, as follows: for each section $\boldsymbol{f}$ of E over U and each point $x \in U$, there exists a C^∞-function h on X with support contained in U, which is equal to 1 in a neighborhood of x; $h\boldsymbol{f}$ extended by 0 outside Supp(h) is a C^∞-section of E over X; hence $P \cdot (h\boldsymbol{f})$ is defined. The value $(P \cdot h\boldsymbol{f})(x)$ is independent of the function h chosen, and if we denote this value by $(P \cdot \boldsymbol{f})(x)$, it is immediate from (17.13.3) (or directly from the condition (L)) that $P \cdot \boldsymbol{f}$ is a section of F over U and that $\boldsymbol{f} \mapsto P \cdot \boldsymbol{f}$ is a differential operator from $E | U$ to $F | U$.

If now (U_λ) is an open covering of X and if, for each λ, we are given a differential operator P_λ from $E | U_\lambda$ to $F | U_\lambda$, so that for each pair of indices λ, μ the restrictions of P_λ and P_μ to $U_\lambda \cap U_\mu$ are equal, then it follows immediately from (17.13.3) that there exists one and only one differential operator P from E to F such that $P | U_\lambda = P_\lambda$ for each λ. For each section $\boldsymbol{f} \in \Gamma(X, E)$ and each $x \in X$, we define $(P \cdot \boldsymbol{f})(x)$ to be the common value of $(P_\lambda \cdot (\boldsymbol{f} | U_\lambda))(x)$ for all indices λ such that $x \in U_\lambda$.

(17.13.5) If P is a differential operator from E to F and if h is a C^∞-function on X, it is clear that the mapping $\boldsymbol{f} \mapsto h(P \cdot \boldsymbol{f})$ is also a differential operator from E to F, which we denote by hP (17.1.4). The set of differential operators from E to F is therefore an $\mathscr{E}(X)$-*module.*

Let E_1, E_2, E_3 be complex vector bundles on X, and let $P_1 : E_1 \to E_2$ and $P_2 : E_2 \to E_3$ be differential operators. Then it is clear that $P_2 \circ P_1$ is a differential operator from E_1 to E_3. Furthermore, from the local expressions of P_1 and P_2 it is immediately seen that if P_1 is of order p and P_2 of order q at a point $x \in X$, then $P_2 \circ P_1$ is of order $\leqq p + q$ at x.

(17.13.6) An important particular case is that in which E and F are both equal to the trivial complex line bundle $X \times \mathbf{C}$, so that $\Gamma(X, E) = \Gamma(X, F) = \mathscr{E}(X)$. The local expression of a differential operator P from $X \times \mathbf{C}$ to $X \times \mathbf{C}$ is then of the form

(17.13.6.1)
$$g \mapsto \sum_{|\nu| \leqq p} a_\nu D^\nu g,$$

where the mappings $y \mapsto a_\nu(y)$ are complex-valued C^∞-functions defined on an open subset $\varphi(V)$ of $\mathbf{R}^n$. The operator P is of order p at the point $x \in V$ if and only if at least one of the numbers $a_\nu(\varphi(x))$ for $|\nu| = p$ is nonzero. The differential operators from $X \times \mathbf{C}$ to $X \times \mathbf{C}$ clearly form a **C**-*algebra* with respect to the composition defined in (17.13.5); we denote this algebra by Diff(X). We have already seen (17.13.3) that for each $x \in X$ the mapping $f \mapsto (P \cdot f)(x)$ is a *distribution with support contained in* $\{x\}$; this distribution is denoted by $P(x)$, so that we have

(17.13.6.2) $$(P \cdot f)(x) = P(x) \cdot f.$$

Thus an operator $P \in \text{Diff}(X)$ is a C^∞-*field of point-distributions.*

(17.13.7) Let $u : X \to Y$ be a diffeomorphism. For each differential operator $P \in \text{Diff}(X)$ we *transport* P by means of u to a differential operator $u_*(P) \in \text{Diff}(Y)$ as follows: for each function $f \in \mathscr{E}(Y)$, we have

(17.13.7.1) $$u_*(P) \cdot f = (P \cdot (f \circ u)) \circ u^{-1}$$

or, in other words, for each $x \in X$,

(17.13.7.2) $$(u_*(P) \cdot f)(u(x)) = (P \cdot (f \circ u))(x),$$

which shows immediately, bearing in mind the definition of the image under u of a distribution on X (17.3.7), that for each $x \in X$ we have

(17.13.7.3) $$u(P(x)) = (u_*(P))(u(x)).$$

From (17.13.7.1) it follows immediately that if $P_1, P_2 \in \text{Diff}(X)$, then

(17.13.7.4) $$u_*(P_1 P_2) = u_*(P_1) u_*(P_2)$$

in the algebra Diff(Y). Further, if $v : Y \to Z$ is another diffeomorphism,

(17.13.7.5) $$(v \circ u)_* = v_* \circ u_*,$$

which shows that u_* is an *isomorphism* of the algebra Diff(X) onto the algebra Diff(Y).

With this notation, if (V, φ, n) is a chart of X, it is clear that the local expression of an operator $P \in \text{Diff}(X)$ is $\varphi_*(P|V)$.

(17.13.8) Let $P \in \mathrm{Diff}(\mathrm{X})$. For each compact subset K of X, the image under P of $\mathscr{D}(\mathrm{X}; \mathrm{K})$ is contained in $\mathscr{D}(\mathrm{X}; \mathrm{K})$. If T is a distribution on X, we may therefore consider, for each function $f \in \mathscr{D}(\mathrm{X}; \mathrm{K})$, the value $\langle \mathrm{T}, P \cdot f \rangle$, and it is clear that the linear mapping $f \mapsto \langle \mathrm{T}, P \cdot f \rangle$ of $\mathscr{D}(\mathrm{X}; \mathrm{K})$ into $\mathbf{C}$ is continuous. Hence we have a *distribution* ${}^tP \cdot \mathrm{T}$, defined by the relation

(17.13.8.1) $$\langle {}^tP \cdot \mathrm{T}, f \rangle = \langle \mathrm{T}, P \cdot f \rangle$$

for all $f \in \mathscr{D}(\mathrm{X})$ (12.15.3). It follows from this definition that $\mathrm{T} \mapsto {}^tP \cdot \mathrm{T}$ is a linear mapping of $\mathscr{D}'(\mathrm{X})$ into itself, and that if the restriction of T to an open set $\mathrm{U} \subset \mathrm{X}$ is zero, then also the restriction of ${}^tP \cdot \mathrm{T}$ to U is zero. In other words, we have

$$\mathrm{Supp}({}^tP \cdot \mathrm{T}) \subset \mathrm{Supp}(\mathrm{T}),$$

and for each open $\mathrm{U} \subset \mathrm{X}$ and each $\mathrm{T} \in \mathscr{D}'(\mathrm{X})$ we have

$${}^t(P \,|\, \mathrm{U}) \cdot (\mathrm{T} \,|\, \mathrm{U}) = ({}^tP \cdot \mathrm{T}) \,|\, \mathrm{U}.$$

If (V, φ, n) is a chart of X for which P has the local expression (17.13.6.1), then, for each function $f \in \mathscr{D}(\varphi(\mathrm{V}))$,

(17.13.8.2) $$\langle {}^tP \cdot \mathrm{T}, f \circ \varphi \rangle = \sum_{|\nu| \leqq p} (-1)^{|\nu|} \langle \mathrm{D}^\nu (a_\nu \varphi(\mathrm{T} \,|\, \mathrm{V})), f \rangle$$

having regard to the definition of the derivative of a distribution on an open subset of $\mathbf{R}^n$ (17.5.5).

Remark

(17.13.9) Let $P : \mathrm{E} \to \mathrm{F}$ be a differential operator. From the local expression (17.13.3.1) and from Leibniz's rule it follows that, for each section $\mathbf{f} \in \Gamma(\mathrm{X}, \mathrm{E})$, the mapping $\sigma \mapsto P(\sigma\mathbf{f})$, where σ runs through the set of C^∞ scalar-valued functions on X, is a differential operator from $\mathrm{X} \times \mathbf{C}$ to F; moreover, if P is of order p at a point x, and if $\mathbf{f}(x) \neq 0$, then $\sigma \mapsto P(\sigma\mathbf{f})$ is also of order p at x, by virtue of Leibniz's formula. In particular, in order to verify that a differential operator P is of order 0, it is enough to show that $P(\sigma\mathbf{f}) = \sigma P(\mathbf{f})$ for all $\sigma \in \mathscr{E}(\mathrm{X})$ and all $\mathbf{f} \in \Gamma(\mathrm{X}, \mathrm{E})$.

(17.13.10) The notion of a *real* differential operator is defined in exactly the same way, by replacing in (17.13.1) complex vector bundles by *real* vector bundles, so that $\Gamma(\mathrm{X}, \mathrm{E})$ and $\Gamma(\mathrm{X}, \mathrm{F})$ are *real* Fréchet spaces; we have only to replace $\mathbf{C}$ by $\mathbf{R}$ throughout in the developments of this section. In particular, to say that $P \in \mathrm{Diff}(\mathrm{X})$ is a real differential operator signifies that, for each

C^∞ *real*-valued function f on X, the function $P \cdot f$ is also *real*-valued. For each real distribution T on X, ${}^tP \cdot T$ is then also a real distribution. If E, F are real vector bundles on X and $E_{(\mathbf{C})}$, $F_{(\mathbf{C})}$ their complexifications, then every real differential operator P from E to F extends uniquely to a complex differential operator $P_{(\mathbf{C})}$ from $E_{(\mathbf{C})}$ to $F_{(\mathbf{C})}$, because $\Gamma(X, E_{(\mathbf{C})}) = \Gamma(X, E) \otimes_{\mathbf{R}} \mathbf{C}$.

(17.13.11) The results of this section are easily extended to continuous linear mappings of $\Gamma^{(r)}(X, E)$ into $\Gamma^{(s)}(X, F)$ (17.2), where r, s are integers $\geqq 0$. For such an operator one obtains a local expression (17.13.3.1), in which necessarily $p \leqq r - s$ and the A_ν are assumed only to be of class C^s. We leave it to the reader to modify appropriately the other results of this section.

PROBLEMS

1. With the notation of (17.13.1), let P be a linear mapping (*not assumed to be continuous*) of $\Gamma(X, E)$ into $\Gamma(X, F)$, satisfying the condition (L). Then P is *continuous*, and hence is a differential operator (Peetre's theorem). Begin by showing that, for each open subset U of X, the restriction of P to U may be defined as in (17.13.4). This allows us to reduce to the case in which X is an open set in $\mathbf{R}^n$ and $E = X \times \mathbf{C}^{N'}$, $F = X \times \mathbf{C}^{N''}$; we may then assume that $N' = N'' = 1$, so that $\Gamma(X, E) = \Gamma(X, F) = \mathscr{E}(X)$. Then proceed as follows:

   (1) For each $x \in X$, if $f \in \mathscr{E}(X)$ is such that $D^\nu f(x) = 0$ for each multi-index ν, then also $D^\nu(P \cdot f)(x) = 0$ for all ν. (Argue by contradiction, using Section 16.4, Problem 1: There exists a function φ such that φf is of class C^∞, equal to 0 for all $y \in X$ such that $\eta^j - \xi^j \leq 0$ for $1 \leq j \leq n$, and equal to f for all $y \in X$ such that $\eta^j - \xi^j \geq 0$ for $1 \leq j \leq n$. Derive a contradiction by considering $P \cdot (\varphi f)$.)
   (2) A point $x \in X$ is said to be *regular* for P if there exists an integer $k_x > 0$ with the following property: For each function $f \in \mathscr{E}(X)$ such that $D^\nu f(x) = 0$ for $|\nu| < k_x$, we have $(P \cdot f)(x) = 0$. Show that if U is an open set in X, all points of which are regular for P, then $P | U$ is a differential operator. (Prove that the k_x are bounded on each compact $K \subset U$; for this, argue by contradiction and use Problem 2 of Section 16.4. In each relatively compact open subset V of U we may then write $(P \cdot f)(x) = \sum_{|\nu| \leq p} a_\nu(x) D^\nu f(x)$; prove as in (17.13.3) that the a_ν are of class C^∞.) Deduce that the set S of nonregular points contains no *isolated* points.
   (3) Show that the set S is *empty*. (Prove that there cannot exist a sequence (x_k) of distinct points of S, converging to a point x, by using Section 16.4, Problem 2 again.)

2. If $P : \Gamma(X, E) \to \Gamma(X, F)$ (notation of (17.13.1)) is a linear mapping, show that the following conditions are equivalent:

   (a) P is a differential operator of order $\leq m$.
   (b) For each function $f \in \mathscr{E}(X)$, the linear mapping

$$\mathbf{s} \mapsto P \cdot (f\mathbf{s}) - fP \cdot \mathbf{s}$$

   is a differential operator of order $\leq m - 1$.

(c) For each family $(f_i)_{1 \leqq i \leqq m+1}$ of $m+1$ functions in $\mathscr{E}(\mathrm{X})$, and each $\mathbf{s} \in \Gamma(\mathrm{X}, \mathrm{E})$, we have

$$\sum_{\mathrm{H}} (-1)^{\mathrm{Card(H)}} (\prod_{i \in \mathrm{H}} f_i) P \cdot ((\prod_{i \notin \mathrm{H}} f_i)\mathbf{s}) = 0,$$

where H runs through all subsets of $\{1, 2, \ldots, m+1\}$. (Use induction on m.)

3. Let X be a differential manifold, E and F two complex vector bundles over X. Show that the differential operators of order $\leqq r$ from E to F may be identified with the linear X-morphisms of the bundle $\mathrm{P}^r(\mathrm{X}, \mathrm{E})$ of jets of global sections of E into the bundle F.

4. Let X be a differential manifold. For each point $x \in \mathrm{X}$ and each integer $r \geqq 0$ and let $\mathrm{T}_x^{(r)}(\mathrm{X})$ denote the vector space of real distributions of order $\leqq r$ with support contained in $\{x\}$. For each real-valued C^∞-function f on X and each real point-distribution $\mathrm{S}_x \in \mathrm{T}_x^{(r)}$, the value $\langle \mathrm{S}_x, f \rangle \in \mathbf{R}$ depends only on the jet $\mathrm{J}_x^{(r)}(f)$ of order r. If we denote the value by $\langle \mathrm{S}_x, \mathrm{J}_x^{(r)}(f) \rangle$, we have in this way defined a bilinear form on

$$\mathrm{T}_x^{(r)}(\mathrm{X}) \times \mathrm{P}_x^{(r)}(\mathrm{X}),$$

which identifies $\mathrm{T}_x^{(r)}(\mathrm{X})$ with the *dual* of $\mathrm{P}_x^r(\mathrm{X})$. The space $\mathrm{T}_x^{(r)}(\mathrm{X})$ is called the *tangent space of order r* of X at the point x. The disjoint union $\mathrm{T}^{(r)}(\mathrm{X})$ of the spaces $\mathrm{T}_x^{(r)}(\mathrm{X})$ as x runs through X is canonically endowed with a structure of a vector bundle over X; this bundle is canonically isormorphic to the *dual* of the bundle $\mathrm{P}^r(\mathrm{X})$ of jets of X into $\mathbf{R}$, and is called the *tangent bundle of order r* of X. Its rank is

$$n + \binom{n+1}{2} + \cdots + \binom{n+r-1}{r}$$

if $\dim(\mathrm{X}) = n$. The bundle $\mathrm{T}^{(1)}(\mathrm{X})$ may be canonically identified with the tangent bundle $\mathrm{T}(\mathrm{X})$ (whence the terminology).

The bundle $\mathrm{T}^{(r-1)}(\mathrm{X})$ is canonically isomorphic to a subbundle of $\mathrm{T}^{(r)}(\mathrm{X})$. Define a canonical isomorphism

$$\mathbf{S}_r(\mathrm{T}(\mathrm{X})) \to \mathrm{T}^{(r)}(\mathrm{X})/\mathrm{T}^{(r-1)}(\mathrm{X}),$$

where $\mathbf{S}_r(\mathrm{T}(\mathrm{X}))$ is the rth *symmetric power* (A.17.4) of the bundle $\mathrm{T}(\mathrm{X})$. (To a sequence $(Z_1, \ldots, Z_r)$ of r vector fields on X corresponds the image in $\mathrm{T}^{(r)}(\mathrm{X})/\mathrm{T}^{(r-1)}(\mathrm{X})$ of the section $\theta_{Z_1} \circ \theta_{Z_2} \circ \cdots \circ \theta_{Z_r}$ of $\mathrm{T}^{(r)}(\mathrm{X})$; use (17.14.3).)

For each C^∞-mapping $u : \mathrm{X} \to \mathrm{Y}$, define canonically a bundle morphism $\mathrm{T}^{(r)}(u)$: $\mathrm{T}^{(r)}(\mathrm{X}) \to \mathrm{T}^{(r)}(\mathrm{Y})$. The diagram

$$\begin{array}{ccccccccc}
0 & \longrightarrow & \mathrm{T}^{(r-1)}(\mathrm{X}) & \longrightarrow & \mathrm{T}^{(r)}(\mathrm{X}) & \longrightarrow & \mathbf{S}_r(\mathrm{T}(\mathrm{X})) & \longrightarrow & 0 \\
 & & \downarrow{\scriptstyle \mathrm{T}^{(r-1)}(u)} & & \downarrow{\scriptstyle \mathrm{T}^{(r)}(u)} & & \downarrow{\scriptstyle \mathbf{S}_r(\mathrm{T}(u))} & & \\
0 & \longrightarrow & \mathrm{T}^{(r-1)}(\mathrm{Y}) & \longrightarrow & \mathrm{T}^{(r)}(\mathrm{Y}) & \longrightarrow & \mathbf{S}_r(\mathrm{T}(\mathrm{Y})) & \longrightarrow & 0
\end{array}$$

is commutative.

14. VECTOR FIELDS AS DIFFERENTIAL OPERATORS

(17.14.1) Let M be a differential manifold, x a point of M, and $\mathbf{h}_x$ a tangent vector at the point x. The mapping $f \mapsto \langle d_x f, \mathbf{h}_x \rangle$ of $\mathscr{E}(\mathrm{M})$ into $\mathbf{R}$ is a *distribution* of order 1 and support $\{x\}$, if $\mathbf{h}_x \neq 0$. For if $c = (\mathrm{U}, \varphi, n)$ is a chart of M such that $x \in \mathrm{U}$, and if $\mathrm{F} = f \circ \varphi^{-1}$ is the local expression of f, and $\mathbf{v} = \boldsymbol{\theta}_c(\mathbf{h}_x)$ (16.5.1), then we have $\langle d_x f, \mathbf{h}_x \rangle = \mathrm{DF}(\varphi(x)) \cdot \mathbf{v}$, whence the assertion follows (17.3.1.1). This distribution is denoted by $\theta_{\mathbf{h}_x}$, and $\theta_{\mathbf{h}_x} \cdot f = \langle d_x f, \mathbf{h}_x \rangle$ is called the *derivative of f at x in the direction of the tangent vector* $\mathbf{h}_x$.

Now let X be a C^∞ *vector field* on M. With X we associate canonically a *differential operator* $\theta_X \in \mathrm{Diff(M)}$ by putting

(17.14.1.1) $$(\theta_X \cdot f)(x) = \theta_{X(x)} \cdot f = \langle d_x f, X(x) \rangle$$

for all $f \in \mathscr{E}(\mathrm{M})$.

By (17.13.3), to show that θ_X is a differential operator, we may assume that M is an open subset of $\mathbf{R}^n$, so that $\mathrm{T(M)} = \mathrm{M} \times \mathbf{R}^n$ and the vector field X is a mapping $x \mapsto (x, \mathbf{v}(x))$ of M into $\mathrm{M} \times \mathbf{R}^n$, with $\mathbf{v}$ of class C^∞; the cotangent bundle $\mathrm{T(M)}^*$ is also identified with $\mathrm{M} \times \mathbf{R}^n$, the covector $d_x f$ with $(x, \mathrm{D}f(x))$, and we have

(17.14.1.2) $$(\theta_X \cdot f)(x) = \langle \mathrm{D}f(x), \mathbf{v}(x) \rangle.$$

The criterion (17.13.3) therefore shows that θ_X is a *real differential operator of order* ≤ 1, *which annihilates the constants.*

(17.14.2) Conversely, we shall show that *every* differential operator $P \in \mathrm{Diff(M)}$ with these properties is of the form θ_X for some uniquely determined C^∞ vector field X on M. The hypothesis on P implies (17.13.3) that relative to a chart (V, φ, n) of M the local expression of P is of the type

$$g \mapsto \sum b_i \mathrm{D}_i g,$$

where the b_i are C^∞ real-valued functions on $\varphi(\mathrm{V})$. Clearly we may associate with this local expression the vector field X_φ on V defined by

$$\langle X_\varphi(x), d_x \varphi^i \rangle = b_i(\varphi(x))$$

for $1 \leq i \leq n$, and it has to be shown that, if (V, ψ, n) is another chart of M with the same domain of definition V, then the vector fields X_φ and X_ψ are the same. Now we have $\psi = \rho \circ \varphi$, where $\rho : \varphi(\mathrm{V}) \to \psi(\mathrm{V})$ is a diffeomorphism;

the assertion then follows from the formulas

$$d_x \psi^i = \sum_{j=1}^{n} (\mathrm{D}_j \rho^i(\varphi(x)))\, d_x \varphi^j,$$

$$(\mathrm{D}_j(h \circ \rho))(z) = \sum_{i=1}^{n} (\mathrm{D}_i h(\rho(z)))(\mathrm{D}_j \rho^i(z))$$

for $z \in \varphi(\mathrm{V})$ and h a C^∞-function on $\psi(\mathrm{V})$.

We shall often speak of the vector field X as a differential operator, and we shall write $X \cdot f$ in place of $\theta_X \cdot f$ when there is no risk of confusion.

(17.14.2.1) *For each* C^∞ *vector field* X *on* M, θ_X *is a real derivation of the algebra* $\mathscr{E}(\mathrm{M})$: *in other words, for* $f, g \in \mathscr{E}(\mathrm{M})$ *we have*

(17.14.2.2)
$$\theta_X \cdot (fg) = (\theta_X \cdot f)g + f(\theta_X \cdot g).$$

This is clear from the local expression (17.14.1.2).

Conversely, it can be shown that *every* real derivation of the algebra $\mathscr{E}(\mathrm{M})$ is of the form θ_X (Problem 1).

(17.14.3) *If* X, Y *are two* C^∞ *vector fields on* M, *there exists a unique* C^∞ *vector field on* M, *denoted by* $[X, Y]$, *such that*

(17.14.3.1)
$$\theta_{[X,Y]} = \theta_X \circ \theta_Y - \theta_Y \circ \theta_X .$$

It is enough to show that the differential operator on the right-hand side of (17.14.3.1) is of order $\leqq 1$, since clearly it annihilates constants. Since the question is local, we may assume that M is an open subset of $\mathbf{R}^n$ and that X, Y are the mappings $x \mapsto (x, \mathbf{u}(x))$ and $x \mapsto (x, \mathbf{v}(x))$. It follows then from (17.14.1.2) that

$$(\theta_Y \cdot (\theta_X \cdot f))(x) = \mathrm{D}^2 f(x) \cdot (\mathbf{v}(x), \mathbf{u}(x)) + \mathrm{D}f(x) \cdot (\mathrm{D}\mathbf{u}(x), \mathbf{v}(x)).$$

Hence the proposition follows from the symmetry of the bilinear form $\mathrm{D}^2 f(x)$ (8.12.2), and the vector field $[X, Y]$ is given by

(17.14.3.2)
$$x \mapsto (x, \mathrm{D}\mathbf{v}(x) \cdot \mathbf{u}(x) - \mathrm{D}\mathbf{u}(x) \cdot \mathbf{v}(x)).$$

The vector field $[X, Y]$ is called the *Lie bracket* of the fields X, Y. The following relations are immediately verified:

(17.14.3.3)
$$[X, X] = 0,$$

(17.14.3.4) $$[X, [Y, Z]] + [Y, [Z, X]] + [Z, [X, Y]] = 0.$$

These relations show that the real vector space $\mathscr{E}_1(\mathrm{M}) = \Gamma(\mathrm{T}(\mathrm{M}))$ of C^∞ vector fields on M becomes a *real Lie algebra* under the bracket operation.

(17.14.3.5) *Let* N *be a submanifold of* M *and let* X, Y *be* C^∞ *vector fields on* M. *If* X *and* Y *are tangent to* N *at all points of* N (i.e., if $X(x)$ and $Y(x)$ are in $\mathrm{T}_x(\mathrm{N})$ for all $x \in \mathrm{N}$), *then the same is true of* $[X, Y]$.

Since the question is local, we may assume that M is an open subset of $\mathbf{R}^n$ and $\mathrm{N} = \mathrm{M} \cap \mathbf{R}^m$, so that T(M) is identified with $\mathrm{M} \times \mathbf{R}^n$ and T(N) with $\mathrm{N} \times \mathbf{R}^m \subset \mathrm{N} \times \mathbf{R}^n$. With the notation of (17.14.3.2), the hypothesis signifies that for each $x \in \mathrm{M} \cap \mathbf{R}^m$ the last $n - m$ components $u_j(x)$, $v_j(x)$ of $\mathbf{u}(x)$, $\mathbf{v}(x)$ are zero ($m + 1 \leqq j \leqq n$). Since the jth component of $\mathrm{D}\mathbf{v}(x) \cdot \mathbf{u}(x)$ is $\langle \mathrm{D}v_j(x), \mathbf{u}(x)\rangle = \sum_{k=1}^{n} \mathrm{D}_k v_j(x) u_k(x)$, it is zero for $j \geqq m + 1$. Similarly the jth component of $\mathrm{D}\mathbf{u}(x) \cdot \mathbf{v}(x)$ is zero for $j \geqq m + 1$, and the proposition is proved.

(17.14.4) *For each* C^∞ *vector field* X *on* M, *the mapping* $Y \mapsto [X, Y]$ *is a real differential operator of order* $\leqq 1$ *from the tangent bundle* T(M) *to* T(M). *If we denote this operator by* θ_X, *then for any two* C^∞ *vector fields* Y, Z *on* M *and any real-valued* C^∞*-function* f *on* M *we have*

(17.14.4.1) $$\theta_X \cdot [Y, Z] = [\theta_X \cdot Y, Z] + [Y, \theta_X \cdot Z],$$

(17.14.4.2) $$\theta_X \cdot (fY) = (\theta_X \cdot f) Y + f(\theta_X \cdot Y).$$

The first assertion follows from (17.14.3.2), for we may assume that M is an open subset of $\mathbf{R}^n$. The formula (17.14.4.1) is simply another way of writing (17.14.3.4). To prove (17.14.4.2) it is enough to show that, for each function $g \in \mathscr{E}(\mathrm{M})$,

(17.14.4.3) $$\theta_{[X, fY]} \cdot g = (\theta_X \cdot f)(\theta_Y \cdot g) + f(\theta_{[X, Y]} \cdot g).$$

However, by definition, we have

$$\begin{aligned}\theta_{[X, fY]} \cdot g &= \theta_X \cdot (\theta_{fY} \cdot g) - \theta_{fY} \cdot (\theta_X \cdot g)\\ &= \theta_X \cdot (f(\theta_Y \cdot g)) - f(\theta_Y \cdot (\theta_X \cdot g))\\ &= (\theta_X \cdot f)(\theta_Y \cdot g) + f(\theta_X \cdot (\theta_Y \cdot g)) - f(\theta_Y \cdot (\theta_X \cdot g))\end{aligned}$$

bearing in mind (17.14.2.2), and this clearly proves (17.4.4.3).

(17.14.5) For each vector field X on M we have now defined two real differential operators of order $\leqq 1$, both denoted by θ_X: one from the trivial line bundle $\mathrm{M} \times \mathbf{R} = \mathbf{T}_0^0(\mathrm{M})$ to itself, and the other from the tangent bundle $\mathrm{T}(\mathrm{M}) = \mathbf{T}_0^1(\mathrm{M})$ to itself. We shall now show that for each pair (r, s) of integers $\geqq 0$ there is a unique canonically defined differential operator of order $\leqq 1$, again denoted by θ_X, of the bundle $\mathbf{T}_s^r(\mathrm{M})$ of tensors of type (r, s) to itself, such that for any two tensor fields $\mathbf{Z}' \in \mathscr{T}_{s'}^{r'}(\mathrm{M}) = \Gamma(\mathbf{T}_{s'}^{r'}(\mathrm{M}))$ and $\mathbf{Z}'' \in \mathscr{T}_{s''}^{r''}(\mathrm{M}) = \Gamma(\mathbf{T}_{s''}^{r''}(\mathrm{M}))$ we have

(17.14.5.1) $$\theta_X \cdot (\mathbf{Z}' \otimes \mathbf{Z}'') = (\theta_X \cdot \mathbf{Z}') \otimes \mathbf{Z}'' + \mathbf{Z}' \otimes \theta_X \cdot \mathbf{Z}''.$$

In view of (17.14.4.2) this assertion is a consequence of the following more general proposition:

(17.14.6) *Let X be a* C^∞ *vector field on* M *and let Q be a real differential operator of order $\leqq 1$ from $\mathbf{T}_0^1(\mathrm{M}) = \mathrm{T}(\mathrm{M})$ to itself, such that, for each function $f \in \mathscr{E}(\mathrm{M})$ and each* C^∞ *vector field Y on* M, *we have*

(17.14.6.1) $$Q \cdot (fY) = (\theta_X \cdot f)Y + f(Q \cdot Y).$$

Then for each pair of integers $r \geqq 0$, $s \geqq 0$, there exists a unique real differential operator D_s^r of order $\leqq 1$, from $\mathbf{T}_s^r(\mathrm{M})$ to itself, such that:

(i) *$D_0^0 = \theta_X$ and $D_0^1 = Q$;*
(ii) *for each* C^∞ *differential* 1*-form α and each* C^∞ *vector field Y on* M,

(17.14.6.2) $$\theta_X \cdot \langle Y, \alpha \rangle = \langle Q \cdot Y, \alpha \rangle + \langle Y, D_1^0 \cdot \alpha \rangle;$$

(iii) *for any two tensor fields $\mathbf{Z}' \in \mathscr{T}_{s'}^{r'}(\mathrm{M})$, $\mathbf{Z}'' \in \mathscr{T}_{s''}^{r''}(\mathrm{M})$,*

(17.14.6.3) $$D_{s'+s''}^{r'+r''} \cdot (\mathbf{Z}' \otimes \mathbf{Z}'') = (D_{s'}^{r'} \cdot \mathbf{Z}') \otimes \mathbf{Z}'' + \mathbf{Z}' \otimes (D_{s''}^{r''} \cdot \mathbf{Z}'').$$

Any tensor field $\mathbf{Z}$ on M belonging to $\mathscr{T}_s^r(\mathrm{M})$, where $r \geqq 1$ (resp. $s \geqq 1$) can always be expressed locally, in the domain of definition of a chart of M, as a sum of fields of the form $Y \otimes \mathbf{Z}'$ (resp. $\mathbf{Z}'' \otimes \alpha$), where $\mathbf{Z}' \in \mathscr{T}_s^{r-1}(\mathrm{M})$ and Y is a vector field (resp. $\mathbf{Z}'' \in \mathscr{T}_{s-1}^r(\mathrm{M})$ and α is a differential form). Hence the uniqueness of the D_s^r follows by induction from (17.14.6.3), once the uniqueness of D_1^0 has been established; but the latter follows immediately from (17.14.6.2), which determines $\langle Y, D_1^0 \cdot \alpha \rangle$ as a function of Y and α. Hence, in order to establish the existence of the D_s^r satisfying the stated conditions, we may assume that M is an open subset of $\mathbf{R}^n$ (17.13.4). Let X_i be the constant vector field equal to $\mathbf{e}_i$ at each point (where $\mathbf{e}_1, \ldots, \mathbf{e}_n$ is the canonical basis of $\mathbf{R}^n$) and let α_i denote the differential form $d\xi^i$ $(1 \leqq i \leqq n)$. Then $\mathbf{T}_s^r(M)$ is a free $\mathscr{E}(M)$-module having as basis the n^{r+s} tensor fields

(17.14.6.4) $$X_{i_1} \otimes \cdots \otimes X_{i_r} \otimes \alpha_{j_1} \otimes \cdots \otimes \alpha_{j_s}.$$

If $Q \cdot X_i = \sum_{j=1}^{n} a_{ij} X_j$, then the relations $\langle X_i, \alpha_j \rangle = \delta_{ij}$ show, by virtue of (17.14.6.2), that for a form $f\alpha_i$, where $f \in \mathscr{E}(M)$, we must have

$$\langle X_j, D_1^0 \cdot (f\alpha_i) \rangle + \sum_{k=1}^{n} f a_{jk} \langle X_k, \alpha_i \rangle = \delta_{ij} \theta_X \cdot f$$

for all j, and consequently

$$D_1^0 \cdot (f\alpha_i) = (\theta_X \cdot f)\alpha_i - f \sum_{j=1}^{n} a_{ji} \alpha_j.$$

It is clear that D_1^0, so defined, is a differential operator of order $\leqq 1$ from $T(M)^* = \mathbf{T}_1^0(M)$ to itself. To define the other D_s^r we may assume that either $r \geqq 1$, or that $r = 0$ and $s \geqq 2$. In the first case, by induction it is sufficient to define $D_s^r \cdot (fX_i \otimes \mathbf{Z})$, where $\mathbf{Z} \in \mathscr{T}_s^{r-1}(M)$. In conformity with (17.14.6.3) we put

(17.14.6.5) $$D_s^r \cdot (fX_i \otimes \mathbf{Z}) = Q \cdot (fX_i) \otimes \mathbf{Z} + fX_i \otimes D_s^{r-1} \cdot \mathbf{Z}.$$

In the second case, by induction it is sufficient to define $D_s^0 \cdot (\mathbf{Z} \otimes f\alpha_i)$, where $\mathbf{Z} \in \mathscr{T}_{s-1}^0(M)$, and this time we put

(17.14.6.6) $$D_s^0 \cdot (\mathbf{Z} \otimes f\alpha_i) = D_{s-1}^0 \cdot \mathbf{Z} \otimes (f\alpha_i) + \mathbf{Z} \otimes D_1^0 \cdot (f\alpha_i).$$

To prove (17.14.6.3), suppose first that $r' \geqq 1$, in which case we may assume that $\mathbf{Z}' = fX_i \otimes \mathbf{Z}$, with $\mathbf{Z}' \in \mathscr{T}_{s'}^{r'-1}(M)$, and the verification is then a trivial consequence of (17.14.6.5) and (17.14.6.1), using induction. If $r' = 0$ and $r'' \geqq 1$, we may assume that $\mathbf{Z}'' = fX_i \otimes \mathbf{Z}$ with $\mathbf{Z} \in \mathscr{T}_{s''}^{r''-1}(M)$; the multiplication law for mixed tensors (16.18.3.6) then gives $\mathbf{Z}' \otimes \mathbf{Z}'' = fX_i \otimes \mathbf{Z}' \otimes \mathbf{Z}$, and we use (17.14.6.5) again. Finally, if $r' = r'' = 0$, we use (17.14.6.6), the definition of D_1^0, and the fact that θ_X is a derivation.

(17.14.7) The operator θ_X defined on each tensor bundle $\mathbf{T}_s^r(M)$ by applying (17.14.6) (in view of (17.14.4.2)) is called the *Lie derivative* relative to the vector field X.

By induction on p, we obtain from (17.14.5.1) the formula

(17.14.7.1) $$\theta_X \cdot (Y_1 \otimes Y_2 \otimes \cdots \otimes Y_p) = \sum_{j=1}^{p} Y_1 \otimes \cdots \otimes Y_{j-1} \otimes (\theta_X \cdot Y_j) \otimes Y_{j+1} \otimes \cdots \otimes Y_p$$

for any C^∞ vector fields $Y_1, \ldots, Y_p$ on M, and the formula

(17.14.7.2) $$\theta_X \cdot (\alpha_1 \otimes \alpha_2 \otimes \cdots \otimes \alpha_p) = \sum_{j=1}^{p} \alpha_1 \otimes \cdots \otimes \alpha_{j-1} \otimes (\theta_X \cdot \alpha_j) \otimes \alpha_{j+1} \otimes \cdots \otimes \alpha_p$$

for any C^∞ differential 1-forms $\alpha_1, \ldots, \alpha_p$ on M.

Since locally a contravariant (resp. covariant) tensor field of order p is a sum of tensor products of p vector fields (resp. p differential forms), it follows that, for any permutation σ in the symmetric group $\mathfrak{S}_p$ and any tensor field $\mathbf{Z} \in \mathscr{T}_0^p(\mathrm{M})$ (resp. $\mathbf{Z} \in \mathscr{T}_p^0(\mathrm{M})$), we have

(17.14.7.3) $$\sigma(\theta_X \cdot \mathbf{Z}) = \theta_X \cdot (\sigma(\mathbf{Z})).$$

In particular, the operator θ_X commutes with the symmetrization and antisymmetrization operators on $\mathscr{T}_0^p(\mathrm{M})$ and $\mathscr{T}_p^0(\mathrm{M})$. Since locally a p-vector field (resp. a differential p-form) is obtained by antisymmetrizing a contravariant (resp. covariant) tensor field of order p, it follows that θ_X *acts on the p-vector fields* (resp. *the p-forms*) of class C^∞. Moreover, it follows from (17.14.7.3) and the definition of the exterior product by antisymmetrization (A.13.2) that if $\mathbf{Z}'$ is a p-vector field and $\mathbf{Z}''$ a q-vector field (resp. if α is a p-form and β a q-form) of class C^∞, then we have

(17.14.7.4) $$\theta_X \cdot (\mathbf{Z}' \wedge \mathbf{Z}'') = (\theta_X \cdot \mathbf{Z}') \wedge \mathbf{Z}'' + \mathbf{Z}' \wedge (\theta_X \cdot \mathbf{Z}''),$$

(resp.

(17.14.7.5) $$\theta_X \cdot (\alpha \wedge \beta) = (\theta_X \cdot \alpha) \wedge \beta + \alpha \wedge (\theta_X \cdot \beta)).$$

Next, let $\mathbf{Z} \in \mathscr{T}_0^p(\mathrm{M}) = \Gamma(\mathbf{T}_0^p(\mathrm{M}))$ be a contravariant tensor field of order p, and let $\mathbf{Z}^* \in \mathscr{T}_p^0(\mathrm{M}) = \Gamma(\mathbf{T}_p^0(\mathrm{M}))$ be a covariant tensor field *of the same order*. Then

(17.14.7.6) $$\theta_X \cdot \langle \mathbf{Z}, \mathbf{Z}^* \rangle = \langle \theta_X \cdot \mathbf{Z}, \mathbf{Z}^* \rangle + \langle \mathbf{Z}, \theta_X \cdot \mathbf{Z}^* \rangle.$$

It is enough to verify this relation locally, when we may assume that $\mathbf{Z} = X_1 \otimes X_2 \otimes \cdots \otimes X_p$ and $\mathbf{Z}^* = \alpha_1 \otimes \alpha_2 \otimes \cdots \otimes \alpha_p$, the X_j being vector fields and the α_j differential forms; application of (17.14.2.1) then gives the result. By antisymmetrization, it follows that for each differential p-form α and p vector fields $X_1, \ldots, X_p$,

(17.14.7.7) $$\theta_X \cdot \langle \alpha, X_1 \wedge X_2 \wedge \cdots \wedge X_p \rangle = \langle \theta_X \cdot \alpha, X_1 \wedge \cdots \wedge X_p \rangle + \sum_{j=1}^{p} \langle \alpha, X_1 \wedge \cdots \wedge X_{j-1} \wedge [X, X_j] \wedge X_{j+1} \wedge \cdots \wedge X_p \rangle.$$

More generally, the relations (17.14.5.1) and (17.14.6.2) show that the operator θ_X *commutes with contractions*: for $i \leqq r, j \leqq s$, and $\mathbf{Z} \in \mathscr{T}_s^r(\mathrm{M})$ we have

(17.14.7.8) $$\theta_X \cdot c_j^i(\mathbf{Z}) = c_j^i(\theta_X \cdot \mathbf{Z}).$$

In particular, we recall that when $r = s = 1$, a tensor field $\mathbf{A} \in \mathscr{T}_1^1(\mathrm{M})$ may be identified with an M-morphism of the tangent bundle T(M) into itself, and that if Y is any vector field on M, then $c_1^1(\mathbf{A} \otimes Y)$ is equal to the vector field $\mathbf{A} \cdot Y$. Hence, from (17.14.5.1) and (17.14.7.8), we deduce that

(17.14.7.9) $$\theta_X \cdot (\mathbf{A} \cdot Y) = (\theta_X \cdot \mathbf{A}) \cdot Y + \mathbf{A} \cdot (\theta_X \cdot Y).$$

Remarks

(17.14.8) With the notation of (17.14.1), let $\pi : \mathrm{M} \to \mathrm{N}$ be a C^∞-mapping of M into a differential manifold N. Then we have

(17.14.8.1) $$\pi(\theta_{\mathbf{h}_x}) = \theta_{\mathrm{T}(\pi) \cdot \mathbf{h}_x}$$

by virtue of the definition of the image of a point-distribution (17.7.1) and the formula (16.5.8.5) for the differential of a composite function.

(17.14.9) With the notation of (17.14.1), let $\mathbf{f}$ be a C^∞-function, defined in a neighborhood of x, with values in a finite-dimensional real vector space E. Then we may define $\theta_{\mathbf{h}_x} \cdot \mathbf{f}$, by replacing $\langle d_x f, \mathbf{h}_x \rangle$ in (17.14.1) by $d_x \mathbf{f} \cdot \mathbf{h}_x$ (16.5.7.1). If $(\mathbf{a}_j)_{1 \leqq j \leqq m}$ is a basis of E and if $\mathbf{f} = \sum_j f_j \mathbf{a}_j$, where the f_j are real-valued functions, then

$$\theta_{\mathbf{h}_x} \cdot \mathbf{f} = \sum_{j=1}^{m} (\theta_{\mathbf{h}_x} \cdot f_j) \mathbf{a}_j.$$

The definition of $\theta_X \cdot \mathbf{f}$ for a C^∞ vector field X on M follows immediately from this.

(17.14.10) The formula (17.14.7.5) shows that θ_X is a *derivation of degree* 0 on the anticommutative graded algebra $\mathscr{A}(\mathrm{M}) = \Gamma(\mathrm{M}, \bigwedge \mathrm{T}(\mathrm{M})^*)$ of C^∞ differential forms on M. Since locally a differential p-form α on M is expressible as a finite sum of forms of the type $f\, dg_1 \wedge dg_2 \wedge \cdots \wedge dg_p$, it follows from (17.14.7.5) that $\theta_X \cdot \alpha$ is determined by the values of $\theta_X \cdot f$ and $\theta_X \cdot df$ for functions $f \in \mathscr{E}(\mathrm{M})$. Now $\theta_X \cdot f$ is given by (17.14.1.1), and we shall show that

(17.14.10.1) $$\theta_X \cdot df = d(\theta_X \cdot f).$$

For by applying (17.14.7.7) with $p = 1$, we obtain

$$\langle \theta_X \cdot df, Y \rangle + \langle df, [X, Y] \rangle = \theta_X \cdot \langle df, Y \rangle = \theta_X \cdot (\theta_Y \cdot f)$$

for any C^∞ vector field Y on M; and since

$$\langle df, [X, Y] \rangle = \theta_{[X, Y]} \cdot f = \theta_X \cdot (\theta_Y \cdot f) - \theta_Y \cdot (\theta_X \cdot f)$$

by virtue of (17.14.3), we have

$$\langle \theta_X \cdot df, Y \rangle = \theta_Y \cdot (\theta_X \cdot f \rangle = \langle d(\theta_X \cdot f), Y \rangle,$$

which proves (17.14.10.1).

If X, Y are C^∞ vector fields on M, then

(17.14.10.2) $$\theta_X \circ i_Y - i_Y \circ \theta_X = i_{[X, Y]},$$

both sides being considered as operators on $\mathscr{A}(\mathrm{M})$.

For since i_Y is an *antiderivation* of degree -1 of $\mathscr{A}(\mathrm{M})$ (16.18.4), the same is true (A.18.7) of the left-hand side of (17.14.10.2). Since the question is local, it follows as above that it is enough to verify that the two sides of (17.14.10.2) take the same values for functions $f \in \mathscr{E}(\mathrm{M})$ and their differentials $df \in \mathscr{E}_1(\mathrm{M})$; but the operators i_Y vanish on $\mathscr{E}(\mathrm{M})$, and by (16.18.4.6) we have $i_Y \cdot df = \langle df, Y \rangle = \theta_Y \cdot f$. Hence, bearing in mind (17.14.10.1), the verification that the two sides of (17.14.10.2) take the same value for df reduces to (17.14.3).

(17.14.11) Although, for a function $f \in \mathscr{E}(\mathrm{M})$, the value of $\theta_X \cdot f$ at a point $x \in \mathrm{M}$ depends only on the value of $X(x)$, the same is *not* true for the value of $\theta_X \cdot Y$ if Y is a vector field. It may be shown that it is not possible to define "intrinsically" a vector in $\mathrm{T}_x(\mathrm{M})$ which should be the "derivative" of Y at the point x in the direction of a given tangent vector $\mathbf{h}_x$ (Problem 2; cf. (18.2.14)).

(17.14.12) The results of this section may be generalized without difficulty to vector and tensor fields of class C^r, where r is an integer $\geqq 0$. If X is a vector field of class C^r, then $\theta_X \cdot \boldsymbol{Z}$ is defined for tensor fields $\boldsymbol{Z}$ of class C^s, where $s \geqq 1$, and is a tensor field of class $C^{\inf(r, s-1)}$. All the formulas proved in (17.14.4)–(17.14.7) remain unchanged.

(17.14.13) We shall also leave to the reader the task of transposing the definitions and results of this section and the preceding one to the context of complex-analytic manifolds. Here we remark only that differential operators can no longer be defined by a local property; it is necessary to define them by means of their local expressions (relative of course to the charts of a complex-analytic atlas), and C^∞-functions and sections are replaced everywhere by holomorphic functions and sections.

PROBLEMS

1. Let D be a derivation of the ring $\mathscr{E}(M)$ of C^∞-functions on a differential manifold M. Show that there exists a unique C^∞ vector field X on M such that $D = \theta_X$. (First show that the condition (L) of (17.13.1) is satisfied. Then either use Problem 1 of Section 17.13, or else give a direct proof with the help of Section 8.14, Problem 7(b).)

2. Show that there exists no linear mapping $\mathbf{h}_x \mapsto D_{\mathbf{h}_x}$ of the tangent space $T_x(M)$ into $\operatorname{Hom}(\mathscr{T}_0^1(M), T_x(M))$ which is not identically zero and satisfies the following conditions:

   (1) for each vector field $Y \in \mathscr{T}_0^1(M)$ and each function $f \in \mathscr{E}(M)$,
   $$D_{\mathbf{h}_x} \cdot (fY) = f(x) D_{\mathbf{h}_x} \cdot Y + (\theta_{\mathbf{h}_x} \cdot f) Y(x);$$
   (2) for each diffeomorphism u of M onto M,
   $$D_{T_x(u) \cdot \mathbf{h}_x} \cdot (T(u) \cdot Y) = T_x(u) \cdot (D_{\mathbf{h}_x} \cdot Y).$$
   (Use Problem 11 of Section 16.26.)

3. Consider the n^2 vector fields $\xi^j D_k$ $(1 \leq j, k \leq n)$ on $\mathbf{R}^n$. Show that the vector space they generate is a Lie algebra $\mathfrak{g}$ and that $[\mathfrak{g}, \mathfrak{g}] \neq \mathfrak{g}$ (cf. (19.4.2.2)).

4. Let X be a C^∞ vector field on a differential manifold M. Show that if $X(x) \neq 0$, there exists a chart c of M at the point x such that X is equal to the vector field X_1 in the domain of definition of c, where the notation is that of (16.15.4.2).

5. Let M be a differential manifold and let $\mathscr{H}$ be a subset of the algebra $\mathscr{E}^{(r)}(M)$, where r is an integer > 0. Show that the subalgebra generated by $\mathscr{H}$ is dense in $\mathscr{E}^{(r)}(M)$ if and only if the following three conditions are satisfied:

   (1) for each $x \in M$ there exists $f \in \mathscr{H}$ such that $f(x) \neq 0$;
   (2) for each pair of distinct points $x, y \in M$ there exists $f \in \mathscr{H}$ such that $f(x) \neq f(y)$;
   (3) for each tangent vector $\mathbf{h}_x \neq 0$ in T(M), there exists $f \in \mathscr{H}$ such that $\theta_{\mathbf{h}_x} \cdot f \neq 0$.

(To show that the conditions are sufficient, consider a compact subset K of M and a relatively compact open neighborhood U of K. If $\dim_x(M) = n$, there exist n tangent vectors $\mathbf{h}_j$ at x and n functions $f_j \in \mathscr{H}$ such that $\theta_{\mathbf{h}_i} \cdot f_j = \delta_{ij}$. We can therefore cover $\bar{U}$ by a finite number of open neighborhoods V_i and define on each V_i functions g_{ij} $(1 \leq j \leq n)$ belonging to $\mathscr{H}$ and such that the mapping $x \mapsto (g_{i1}(x), \ldots, g_{in}(x))$ is a homeomorphism of V_i onto an open subset of $\mathbf{R}^n$, which together with its inverse is of class C^r. By considering the compact set L which is the complement of the union of the sets $(V_i \cap \bar{U}) \times (V_i \cap \bar{U})$ in $\bar{U} \times \bar{U}$, show that there exist a finite number of functions $h_k \in \mathscr{H}$ such that, for each $(x, y) \in L$, we have $h_k(x) \neq h_k(y)$ for some index k. Finally, there exist a finite number of functions $f_l \in \mathscr{H}$ such that, for each $x \in \bar{U}$, we have $f_l(x) \neq 0$ for some index l. Deduce that there exist N functions $F_i \in \mathscr{H}$ such that the mapping $\Phi : x \mapsto (F_1(x), \ldots, F_N(x))$ is a homeomorphism of U onto an open subset $\Phi(U)$ of $\mathbf{R}^n$ whose closure does not contain 0, and such that both Φ and Φ^{-1} are of class C^r. Hence, for each function $f \in \mathscr{E}^{(r)}(M)$, there exists a function $\varphi \in \mathscr{E}^{(r)}(\mathbf{R}^N)$ such that $\varphi(0) = 0$ and $f(x) = \varphi(F_1(x), \ldots, F_N(x))$ for all $x \in K$. Finally, use the Weierstrass approximation theorem.)

6. Let $\mathfrak{G}_c$ be the Lie algebra of compactly supported C^∞ vector fields on a differential manifold M.

(a) If $\mathfrak{a} \neq \mathfrak{G}_c$ is an ideal in $\mathfrak{G}_c$, show that there exists a point $x_0 \in M$ such that $X(x_0) = 0$ for all $X \in \mathfrak{a}$. (Argue by contradiction: If $X \in \mathfrak{a}$ is such that $X(x) \neq 0$, then for each field $Z \in \mathfrak{G}_c$ there exists a field $Y \in \mathfrak{G}_c$ such that Z and $[X, Y]$ coincide in a neighborhood of x, by using Problem 4. Then cover the support K of Z by a finite number of suitably chosen open neighborhoods V_i $(1 \leq i \leq m)$; take a partition of unity $(f_i)_{0 \leq i \leq m}$ of class C^∞ subordinate to the covering of M formed by $M - K$ and the V_i, and by considering successively the vector fields

$$Z_1 = f_1 Z, \quad Z_2 = f_2(Z - Z_1), \quad Z_3 = f_3(Z - Z_1 - Z_2), \quad \ldots,$$

prove that Z can be written in the form $\sum_i [X_i, Y_i]$, where $X_i \in \mathfrak{a}$ and $Y_i \in \mathfrak{G}_c$.)

(b) Let $x_0 \in M$ be such that $X(x_0) = 0$ for all $X \in \mathfrak{a}$, and let $c = (U, \varphi, n)$ be any chart of M at x_0. Show that the local expressions of the vector fields $X \in \mathfrak{a}$ relative to the chart c have all their derivatives of all orders zero at the point $\varphi(x_0)$. (Suppose that the result is false for some field $X \in \mathfrak{a}$, and consider the Lie brackets $[X_i, X]$, where X_i are the fields associated with the chart c (16.15.4.2)).

(c) Deduce from (a) and (b) that the maximal ideals of the Lie algebra $\mathfrak{G}_c$ are the $\mathfrak{J}_{x_0}$, where for each point $x_0 \in M$ the ideal $\mathfrak{J}_{x_0}$ consists of all $X \in \mathfrak{G}_c$ such that, for some chart $c = (U, \varphi, n)$ of M at x_0, all derivatives of X of all orders vanish at the point $\varphi(x_0)$ (in other words, such that X and the zero vector field have a contact of infinite order at the point x_0 : cf. Section 16.5, Problem 9).

(d) Let $X_0 \in \mathfrak{G}_c$ and $x_0 \in M$. Show that $[X_0, \mathfrak{G}_c] + \mathfrak{J}_{x_0} = \mathfrak{G}_c$ if and only if $X_0(x_0) \neq 0$. (Use Problem 3 to show that the condition is necessary.)

(e) Let M, N be two compact differential manifolds. Show that if there exists an isomorphism of the Lie algebra $\mathscr{T}_0^1(M)$ onto the Lie algebra $\mathscr{T}_0^1(N)$, then M and N are diffeomorphic and every isomorphism of $\mathscr{T}_0^1(M)$ onto $\mathscr{T}_0^1(N)$ is of the form

$$X \mapsto T(u) \cdot X,$$

where u is a diffeomorphism of M onto N. (First observe that there is a canonical bijective correspondence between the closed subsets of M and the ideals of the Lie algebra $\mathcal{T}_0^1(\mathrm{M})$: to each ideal $\mathfrak{a}$ corresponds the set of all $x \in \mathrm{M}$ such that $X(x) = 0$ for all $X \in \mathfrak{a}$. Deduce that an isomorphism v of $\mathcal{T}_0^1(\mathrm{M})$ onto $\mathcal{T}_0^1(\mathrm{N})$ defines a homeomorphism u of M onto N; then show with the help of (d) that if $x_0 \in \mathrm{M}$ is such that $X(x_0) \neq 0$, then we have $(v(X))(u(x_0)) \neq 0$, and deduce that if $f \in \mathscr{E}(\mathrm{M})$, we have $f \circ u^{-1} \in \mathscr{E}(\mathrm{N})$, by considering the vector field fX and its image under v.)

15. THE EXTERIOR DIFFERENTIAL OF A DIFFERENTIAL p-FORM

(17.15.1) Let M be a differential manifold. The mapping $f \mapsto df$ (16.20.2) is a real differential operator of order 1 from the trivial line bundle $\mathrm{M} \times \mathbf{R}$ to the cotangent bundle $\mathrm{T(M)}^*$, as follows immediately from its local expression. We shall now define, for each $p \geqq 1$, a differential operator of order 1 from $\bigwedge^p \mathrm{T(M)}^*$ to $\bigwedge^{p+1} \mathrm{T(M)}^*$.

(17.15.2) *Let* M *be a differential manifold. For each integer $p \geqq 0$, there exists a unique real differential operator d of order* 1 *from* $\bigwedge^p \mathrm{T(M)}^*$ *to* $\bigwedge^{p+1} \mathrm{T(M)}^*$, *satisfying the following conditions*:

(i) *If α is a* C^∞ *p-form and β a* C^∞ *q-form on* M, *then*

(17.15.2.1) $$d(\alpha \wedge \beta) = (d\alpha) \wedge \beta + (-1)^p \alpha \wedge d\beta$$

($p \geqq 0$ $q \geqq 0$) (in other words (A.18.4) d is an *antiderivation* of the algebra of differential forms on M).

(ii) *When $p = 0$, d is the differential $f \mapsto df$* (16.20.2).

(iii) *For each function $f \in \mathscr{E}(\mathrm{M})$, we have $d(df) = 0$.*

In the domain of definition V of a chart of M, every differential p-form may be written as the sum of a finite number of forms of the type

$$f\, dg_1 \wedge dg_2 \wedge \cdots \wedge dg_p,$$

where f and the g_k are real-valued functions of class C^∞ on V. Condition (i), applied by induction on p, together with conditions (ii) and (iii), shows that we must have

$$d(f\, dg_1 \wedge dg_2 \wedge \cdots \wedge dg_p) = df \wedge dg_1 \wedge dg_2 \wedge \cdots \wedge dg_p,$$

which proves the uniqueness of d. By virtue of this uniqueness, we need only establish the existence of d in the domain of definition U of a chart (U, φ, n) of M. Then a p-form α is uniquely expressible as

(17.15.2.2) $$\alpha = \sum_{i_1 < i_2 < \cdots < i_p} a_{i_1 i_2 \cdots i_p}\, d\varphi^{i_1} \wedge \cdots \wedge d\varphi^{i_p}$$

and $d\alpha$ is defined unambiguously by

$$d\alpha = \sum_{i_1 < i_2 < \cdots < i_p} da_{i_1 i_2 \cdots i_p} \wedge d\varphi^{i_1} \wedge \cdots \wedge d\varphi^{i_p}. \tag{17.15.2.3}$$

It remains to verify (i) and (iii). As to (iii), if F is the local expression of f relative to the chart (U, φ, n), we have

$$df = \sum_{i=1}^{n} \mathrm{D}_i \mathrm{F} \cdot d\varphi^i$$

and therefore

$$d(df) = \sum_{i=1}^{n} d(\mathrm{D}_i \mathrm{F}) \wedge d\varphi^i = \sum_{i=1}^{n} \sum_{j=1}^{n} \mathrm{D}_j(\mathrm{D}_i \mathrm{F})\, d\varphi^j \wedge d\varphi^i.$$

Since $\mathrm{D}_j(\mathrm{D}_i \mathrm{F}) = \mathrm{D}_i(\mathrm{D}_j \mathrm{F})$ and $d\varphi^i \wedge d\varphi^j = -d\varphi^j \wedge d\varphi^i$ for $i \neq j$, and $d\varphi^i \wedge d\varphi^i = 0$, (iii) is clear. As to (i), we may by linearity assume that $\alpha = f\, d\varphi^{i_1} \wedge \cdots d\varphi^{i_p}$, $\beta = g\, d\varphi^{j_1} \wedge \cdots \wedge d\varphi^{j_q}$, so that

$$\alpha \wedge \beta = fg\, d\varphi^{i_1} \wedge \cdots \wedge d\varphi^{i_p} \wedge d\varphi^{j_1} \wedge \cdots \wedge d\varphi^{j_q}$$

and consequently

$$\begin{aligned} d(\alpha \wedge \beta) &= (f \cdot dg + g \cdot df) \wedge d\varphi^{i_1} \wedge \cdots \wedge d\varphi^{i_p} \wedge d\varphi^{j_1} \wedge \cdots \wedge d\varphi^{j_q} \\ &= (df \wedge d\varphi^{i_1} \wedge \cdots \wedge d\varphi^{i_p}) \wedge (g\, d\varphi^{j_1} \wedge \cdots \wedge d\varphi^{j_q}) \\ &\quad + (-1)^p (f\, d\varphi^{i_1} \wedge \cdots \wedge d\varphi^{i_p}) \wedge (dg \wedge d\varphi^{j_1} \wedge \cdots \wedge d\varphi^{j_q}) \end{aligned}$$

having regard to the anticommutation relations in the exterior algebra (A.13.2.9). Hence (i) is verified and the proof is complete.

The differential operator d just defined is called the *exterior differential* in the bundle $\bigwedge \mathrm{T(M)}^*$, the direct sum of the $\bigwedge^p \mathrm{T(M)}^*$.

Example

(17.15.2.4) For any differential manifold M we have defined (16.20.6) the canonical differential 1-form κ_{M} on the manifold $\mathrm{T(M)}^*$. The 2-form $-d\kappa_{\mathrm{M}}$ is called the *canonical differential 2-form* on $\mathrm{T(M)}^*$. In the notation of (16.20.6), its local expression is $\sum_{i=1}^{n} d\xi^i \wedge d\eta_i$.

(17.15.3) (i) *For each* C^∞ *differential p-form* α *on* M *we have*

(17.15.3.1) $$d(d\alpha) = 0.$$

(ii) *For each* C^∞*-mapping* $u : M' \to M$ *and each p-form* α *on* M, *we have*

(17.15.3.2) $$d({}^t u(\alpha)) = {}^t u(d\alpha).$$

(iii) *For each* C^∞ *vector field* X *on* M *and each* C^∞ *differential p-form* α *on* M, *we have*

(17.15.3.3) $$\theta_X \cdot (d\alpha) = d(\theta_X \cdot \alpha),$$

(17.15.3.4) $$\theta_X \cdot \alpha = i_X \cdot d\alpha + d(i_X \cdot \alpha).$$

(iv) *If* α *is a* C^∞ *differential p-form on* M *and if* $X_0, X_1, \ldots, X_p$ *are* $p + 1$ C^∞ *vector fields on* M, *then*

(17.15.3.5)
$$\langle d\alpha, X_0 \wedge X_1 \wedge \cdots \wedge X_p\rangle = \sum_{j=0}^{p} (-1)^j \theta_{X_j} \cdot \langle \alpha, X_0 \wedge \cdots \wedge \hat{X}_j \wedge \cdots \wedge X_p\rangle$$
$$+ \sum_{0 \leq i < j \leq p} (-1)^{i+j} \langle \alpha, [X_i, X_j] \wedge X_0 \wedge \cdots \wedge \hat{X}_i \wedge \cdots \wedge \hat{X}_j \wedge \cdots \wedge X_p\rangle,$$

where as usual the circumflex over a symbol means that the symbol is to be omitted. In particular, for each C^∞ *differential* 1*-form* ω *on* M *and two vector fields* X, Y, *we have*

(17.15.3.6)
$$\langle d\omega, X \wedge Y\rangle = \theta_X \cdot \langle \omega, Y\rangle - \theta_Y \cdot \langle \omega, X\rangle - \langle \omega, [X, Y]\rangle.$$

(i) It is enough to verify this locally, in the domain of definition of a chart of M, and then it follows immediately from (17.1.2.3) and the fact that $d(df) = 0$.

(ii) Again the question is local, so that we may assume that

$$\alpha = f\, dg_1 \wedge dg_2 \wedge \cdots \wedge dg_p,$$

whence

$${}^t u(\alpha) = (f \circ u)\, {}^t u(dg_1) \wedge {}^t u(dg_2) \wedge \cdots \wedge {}^t u(dg_p)$$

and

$${}^t u(d\alpha) = {}^t u(df) \wedge {}^t u(dg_1) \wedge \cdots \wedge {}^t u(dg_p)$$

by virtue of (16.20.9.5). The result therefore follows from the relation $d(f \circ u) = {}^t u(df)$ (16.20.8.2).

(iii) To prove (17.15.3.3) we may again assume that α is given locally by the expression in (ii). By virtue of (17.14.7.5), we have

$$\theta_X \cdot (df \wedge dg_1 \wedge \cdots \wedge dg_p) = (\theta_X \cdot df) \wedge dg_1 \wedge \cdots \wedge dg_p + df \wedge (\theta_X \cdot dg_1) \wedge \cdots \wedge dg_p + \cdots + df \wedge dg_1 \wedge \cdots \wedge (\theta_X \cdot dg_p)$$

and

$$\theta_X \cdot (f\, dg_1 \wedge \cdots \wedge dg_p) = (\theta_X \cdot f)\, dg_1 \wedge \cdots \wedge dg_p + f(\theta_X \cdot dg_1) \wedge \cdots \wedge dg_p + \cdots + f\, dg_1 \wedge \cdots \wedge (\theta_X \cdot dg_p),$$

so that the result follows from (17.14.10.1).

To prove (17.15.3.4) we observe that, since i_X and d are antiderivations, $i_X \circ d + d \circ i_X$ is a *derivation* (A.18.7), and by virtue of (17.14.7.5) it is therefore enough to verify that θ_X and this derivation take the same values for functions f of class C^∞ and differential 1-forms $f\, dg$. Since $i_X \cdot f = 0$, the first property is nothing but the definition of $\theta_X \cdot f = \langle df, X\rangle = i_X \cdot df$. Also, by (17.14.10.1) and (17.14.7.5), we have

$$\theta_X \cdot (f\, dg) = (\theta_X \cdot f)\, dg + f(\theta_X \cdot dg) = (\theta_X \cdot f)\, dg + f\, d(\theta_X \cdot g)$$

and

$$i_X \cdot (df \wedge dg) = \langle df, X\rangle\, dg - \langle dg, X\rangle\, df = (\theta_X \cdot f)\, dg - (\theta_X \cdot g)\, df,$$
$$i_X \cdot (f\, dg) = f\langle dg, X\rangle = f(\theta_X \cdot g).$$

Consequently

$$d(i_X \cdot (f\, dg)) = (\theta_X \cdot g)\, df + f\, d(\theta_X \cdot g)$$

and (17.15.3.4) is verified for $\alpha = f\, dg$.

(iv) The proof is by induction on p. For $p = 0$, the formula (17.15.3.5) reduces to $\langle df, X\rangle = \theta_X \cdot f$ for $f \in \mathscr{E}(M)$, which is just the definition of the operator θ_X. If $p > 0$, by (17.14.7.7) we may write

$$\langle \theta_{X_0} \cdot \alpha, X_1 \wedge \cdots \wedge X_p\rangle = \theta_{X_0} \cdot \langle \alpha, X_1 \wedge \cdots \wedge X_p\rangle - \sum_{j=1}^{p} \langle \alpha, X_1 \wedge \cdots \wedge X_{j-1} \wedge [X_0, X_j] \wedge X_{j+1} \wedge \cdots \wedge X_p\rangle.$$

Also, by (17.15.3.4),

$$\langle \theta_{X_0} \cdot \alpha, X_1 \wedge \cdots \wedge X_p\rangle = \langle i_{X_0} \cdot d\alpha, X_1 \wedge \cdots \wedge X_p\rangle + \langle d(i_{X_0} \cdot \alpha), X_1 \wedge \cdots \wedge X_p\rangle.$$

Now by definition we have (16.18.4.5)

$$\langle i_{X_0} \cdot d\alpha, X_1 \wedge \cdots \wedge X_p \rangle = \langle d\alpha, X_0 \wedge X_1 \wedge \cdots \wedge X_p \rangle$$

and on the other hand, by the inductive hypothesis,

$$\begin{aligned}
&\langle d(i_{X_0} \cdot \alpha), X_1 \wedge \cdots \wedge X_p \rangle \\
&= \sum_{j=1}^{p} (-1)^{j-1} \theta_{X_j} \cdot \langle i_{X_0} \cdot \alpha, X_1 \wedge \cdots \wedge \hat{X}_j \wedge \cdots \wedge X_p \rangle \\
&+ \sum_{1 \leqq i < j \leqq p} (-1)^{i+j-1} \langle i_{X_0} \cdot \alpha, [X_i, X_j] \wedge X_1 \wedge \cdots \wedge \hat{X}_i \wedge \cdots \wedge \hat{X}_j \wedge \cdots \wedge X_p \rangle.
\end{aligned}$$

Using once again in this last expression the definition of the operator i_{X_0}, we obtain immediately the formula (17.15.3.5).

(17.15.4) For each $p \geqq 0$ and each compact subset K of M, the mapping d is a continuous linear mapping of the Fréchet space $\mathscr{D}_p(\mathrm{M}; \mathrm{K})$ into the Fréchet space $\mathscr{D}_{p+1}(\mathrm{M}; \mathrm{K})$, and therefore by transposition (12.15.4) defines a linear operator of the space $\mathscr{D}'_{p+1}(\mathrm{M}; \mathrm{K})$ of $(p+1)$-currents into the space $\mathscr{D}'_p(\mathrm{M}; \mathrm{K})$ of p-currents on M. We denote this operator by $\mathrm{T} \mapsto b\mathrm{T}$; the current $b\mathrm{T}$ is called the *boundary* of the current T. Hence, for each compactly supported differential p-form α on M and each $(p+1)$-current T on M, we have by definition

(17.15.4.1) $$\langle b\mathrm{T}, \alpha \rangle = \langle \mathrm{T}, d\alpha \rangle.$$

If $\pi : \mathrm{M} \to \mathrm{M}'$ is a proper mapping of class C^r ($r \geqq 1$), then

(17.15.4.2) $$\pi(b(\mathrm{T})) = b(\pi(\mathrm{T})).$$

For if α' is any differential p-form on M′ with compact support, we have

$$\begin{aligned}
\langle b(\pi(\mathrm{T})), \alpha' \rangle &= \langle \pi(\mathrm{T}), d\alpha' \rangle = \langle \mathrm{T}, {}^t\pi(d\alpha') \rangle \\
&= \langle \mathrm{T}, d({}^t\pi(\alpha')) \rangle = \langle b\mathrm{T}, {}^t\pi(\alpha') \rangle \\
&= \langle \pi(b\mathrm{T}), \alpha' \rangle.
\end{aligned}$$

A similar argument shows that

(17.15.4.3) $$b(b\mathrm{T}) = 0.$$

Examples

(17.15.5) Let X be an oriented pure manifold of dimension n. If $\pi : X \to X'$ is a proper mapping of class C^r ($r \geqq 1$), then $\mathfrak{b}(\pi(T_{\varphi_X})) = 0$. To prove this it is enough to show that, for each C^∞ differential $(n-1)$-form α on X *with compact support*, we have

$$\int_X d\alpha = 0. \tag{17.15.5.1}$$

Let K be a compact neighborhood of $\mathrm{Supp}(\alpha)$. Then K can be covered by a finite number of relatively compact open sets U_j ($1 \leqq j \leqq r$) such that each U_j is the domain of definition of a chart (U_j, φ_j, n) of X. There exists a family $(h_j)_{1 \leqq j \leqq r}$ of C^∞-mappings of X into $[0, 1]$ such that $\mathrm{Supp}(h_j) \subset U_j$ and $\sum_{j=1}^{r} h_j(x) = 1$ for all $x \in K$ (16.4.2). We then have $\alpha = \sum_{j=1}^{r} h_j \alpha$, so that it is enough to show that $\int_X d(h_j \alpha) = 0$ for each j; hence we may assume (by virtue of (17.15.3.2) and (16.24.5.1)) that X is an open subset of $\mathbf{R}^n$, and then we can write

$$\alpha = \sum_{j=1}^{n} f_j \, d\xi^1 \wedge \cdots \wedge \widehat{d\xi^j} \wedge \cdots \wedge d\xi^n$$

so that

$$d\alpha = \left(\sum_{j=1}^{n} (-1)^{j-1} D_j f_j \right) d\xi^1 \wedge \cdots \wedge d\xi^n.$$

Hence we are reduced to proving that if f is any C^∞-function on $\mathbf{R}^n$ *with compact support*, then

$$\int_{\mathbf{R}^n} D_j f(x) \, d\xi^1 \, d\xi^2 \cdots d\xi^n = 0,$$

and this follows immediately from the Lebesgue–Fubini theorem.

This result justifies the terminology "n-chain element without boundary" introduced in (17.5.2).

(17.15.6) With the hypotheses of the elementary version of Stokes' theorem (16.24.11), if we put $\alpha = d\xi^2 \wedge d\xi^3 \wedge \cdots \wedge d\xi^n$, then the formula (16.24.11.1) may be written in the form

$$\int_{E_b} f\alpha - \int_{E_a} f\alpha = \int_{U_{a,b}} d(f\alpha) \tag{17.15.6.1}$$

for any function $f \in \mathscr{E}(\mathrm{U})$. Now, given any two functions $f_1 \in \mathscr{D}(\mathrm{E}_a)$ and $f_2 \in \mathscr{D}(\mathrm{E}_b)$, there exists a function $f \in \mathscr{D}(\mathrm{U})$ whose restrictions to E_a and E_b are, respectively, f_1 and f_2. To see this, we may assume that $\mathrm{U} = \mathrm{I} \times \mathrm{V}$, where V is open in $\mathbf{R}^{n-1}$ and I is an open interval in $\mathbf{R}$. If g_1 (resp. g_2) is a C^∞-mapping of $\mathbf{R}$ into $[0, 1]$ which is equal to 1 at the point a (resp. b) and vanishes outside a neighborhood W_1 (resp. W_2) of a (resp. b) in I, where $\mathrm{W}_1 \cap \mathrm{W}_2 = \varnothing$, then we define $f(t, y) = g_1(t)f_1(y)$ for $t \in \mathrm{W}_1$; $f(t, y) = g_2(t)f_2(y)$ for $t \in \mathrm{W}_2$; and $f(t, y) = 0$ otherwise. The function f so defined clearly satisfies the required conditions. Since by hypothesis the form α is nonzero almost everywhere in E_a and in E_b, it follows therefore that if α_1, α_2 are compactly supported C^∞ differential $(n-1)$-forms on E_a, E_b, respectively, then there exists $f \in \mathscr{D}(\mathrm{U})$ such that α_1, α_2 are induced by $f\alpha$ on E_a, E_b, respectively. Hence from the formula (17.15.6.1) we deduce that (with the abuse of notation of (17.5.2))

(17.15.6.2) $$b\mathrm{U}_{a,b} = \mathrm{E}_b - \mathrm{E}_a.$$

In Chapter XXIV we shall generalize this result by defining a large class of "open n-chain elements" whose boundaries can be explicitly determined.

(17.15.7) Now let M be a *complex-analytic* manifold. Starting from the differential operator $f \mapsto df$ from the trivial bundle $\mathrm{M} \times \mathbf{C}$ to the cotangent bundle $\mathrm{T(M)}^*$, we can repeat the construction of the exterior differential in the bundle $\bigwedge \mathrm{T(M)}^*$, by replacing C^∞ real-valued functions by holomorphic functions throughout; but, furthermore, if $\mathrm{M}_0 = \mathrm{M}_{|\mathbf{R}}$ is the differential manifold underlying M, then the complex-analytic structure on M enables us to define canonically new differential operators on the bundle $\bigwedge (\mathrm{T}(\mathrm{M}_0)^*)_{(\mathbf{C})}$, the exterior algebra of the *complexification* $(\mathrm{T}(\mathrm{M}_0)^*)_{(\mathbf{C})}$ of the cotangent bundle of M_0.

For there is a canonically defined endomorphism ${}^tJ$ (16.20.16) of the *complex* vector bundle $(\mathrm{T}(\mathrm{M}_0)^*)_{(\mathbf{C})}$, with square $-I$, such that

$$p' = \tfrac{1}{2}(I - i \cdot {}^tJ), \qquad p'' = \tfrac{1}{2}(I + i \cdot {}^tJ)$$

are *projections* on $\mathrm{E} = (\mathrm{T}(\mathrm{M}_0)^*)_{(\mathbf{C})}$ whose sum is I and whose images may be canonically identified with $\mathrm{T(M)}^*$ and the "conjugate" bundle $\overline{\mathrm{T(M)}^*}$ (whose sections are, respectively, generated locally by the differentials df of holomorphic functions, and the differentials $d\bar{f}$ of complex conjugates of holomorphic functions). For each complex-valued C^∞-function f on M_0, we put

(17.15.7.1) $$d'f = p' \circ df, \qquad d''f = p'' \circ df,$$

so that d' is a differential operator of order 1 from $(\mathrm{T}(\mathrm{M}_0)^*)_{(\mathbf{C})}$ to $\mathrm{T}(\mathrm{M})^*$, and d'' a differential operator of order 1 from $(\mathrm{T}(\mathrm{M}_0)^*)_{(\mathbf{C})}$ to $\overline{\mathrm{T}(\mathrm{M})^*}$, such that $d = d' + d''$ and such that the relation $d''f = 0$ characterizes the *holomorphic* functions on M, and the relation $d'f = 0$ characterizes their complex conjugates. From the decomposition of $\mathrm{E} = (\mathrm{T}(\mathrm{M}_0)^*)_{(\mathbf{C})}$ as the direct sum of $\mathrm{T}(\mathrm{M})^*$ and $\overline{\mathrm{T}(\mathrm{M})^*}$, we have a canonical decomposition of $\bigwedge^{p} \mathrm{E}$ as the direct sum of $p + 1$ vector subbundles

(17.15.7.2) $$\bigwedge^{r,s} \mathrm{E} = \Big(\bigwedge^{r} \mathrm{T}(\mathrm{M})^*\Big) \otimes \Big(\bigwedge^{s} \overline{\mathrm{T}(\mathrm{M})^*}\Big)$$

$(0 \leqq r \leqq p,\ r + s = p)$. If (U, φ, n) is a chart of M, and $\varphi = (\varphi^j)_{1 \leqq j \leqq n}$, the sections of $\bigwedge^{r,s} \mathrm{E}$ over U form a free $\mathscr{E}(\mathrm{U})$-module with a basis consisting of the sections

(17.15.7.3) $$(d\varphi^{j_1} \wedge \cdots \wedge d\varphi^{j_r}) \wedge (d\bar{\varphi}^{k_1} \wedge \cdots \wedge d\bar{\varphi}^{k_s})$$

$(j_1 < \cdots < j_r,\ k_1 < \cdots < k_s)$.

For each pair (r, s) of positive integers there are then defined uniquely two (complex) differential operators of order 1:

(17.15.7.4) $$d' : \bigwedge^{r,s}\mathrm{E} \to \bigwedge^{r+1,s}\mathrm{E}, \qquad d'' : \bigwedge^{r,s}\mathrm{E} \to \bigwedge^{r,s+1}\mathrm{E},$$

which for $r = s = 0$ are the operators already denoted by d' and d'' and which have the following properties:

(i) $d' \circ d' = d'' \circ d'' = 0, \quad d' \circ d'' + d'' \circ d' = 0, \quad d = d' + d''$.

(ii) $d'(\alpha \wedge \beta) = (d'\alpha) \wedge \beta + (-1)^p \alpha \wedge (d'\beta)$,
$d''(\alpha \wedge \beta) = (d''\alpha) \wedge \beta + (-1)^p \alpha \wedge (d''\beta)$

for $\alpha \in \bigwedge^{r,s} \mathrm{E}$ and $\beta \in \bigwedge^{r',s'} \mathrm{E}$, with $r + s = p$.

(iii) $d''(d'f) = 0$ and $d'(d''\bar{f}) = 0$ for each holomorphic function f on M.

The proof is analogous to that of (17.15.2), by reducing to the case where the form $\alpha \in \bigwedge^{r,s} \mathrm{E}$ is the product of a function and a form of type (17.15.7.3); the only point to be noticed is that if f is a complex-valued C^∞-function on an open subset of $\mathbf{C}^n$, so that

$$d'f = \sum_{j=1}^{n} \frac{\partial f}{\partial z_j}\, d\zeta_j, \qquad d''f = \sum_{j=1}^{n} \frac{\partial f}{\partial \bar{z}_j}\, d\bar{\zeta}_j,$$

we have

$$\sum_{j=1}^{n} \left(d''\left(\frac{\partial f}{\partial z_j}\right) \wedge d\zeta_j + d'\left(\frac{\partial f}{\partial \bar{z}_j}\right) \wedge d\bar{\zeta}_j\right) = 0,$$

which follows from the fact that the operators $\partial/\partial z_j$ and $\partial/\partial \bar{z}_j$ commute.

(17.15.8) The definition of the exterior differential $d\alpha$ can be easily extended to the case of a *vector-valued differential p-form* $\boldsymbol{\alpha}$ (16.20.15), with values in a finite-dimensional vector space F. If $\boldsymbol{\alpha} = \sum_{i=1}^{q} \alpha_i \mathbf{e}_i$, where $(\mathbf{e}_i)_{1 \leqq i \leqq q}$ is a basis of F, then we define $d\boldsymbol{\alpha} = \sum_{i=1}^{q} (d\alpha_i)\mathbf{e}_i$, and it is immediate that this definition is independent of the choice of basis of F. Again we have $d(d\boldsymbol{\alpha}) = 0$. If $\boldsymbol{\omega}$ is a vector-valued C^∞ differential 2-form and X, Y are two C^∞ vector fields on M, the formula which replaces (17.15.3.6) is

(17.15.8.1) $$d\boldsymbol{\omega} \cdot (X \wedge Y) = \theta_X \cdot (\boldsymbol{\omega} \cdot Y) - \theta_Y \cdot (\boldsymbol{\omega} \cdot X) - \boldsymbol{\omega} \cdot [X, Y].$$

PROBLEMS

1. Let H be a vector subspace of $\mathscr{T}_q^p(\mathrm{M})$ which is stable under all diffeomorphisms of M onto itself. Let $L : \mathrm{H} \to \mathscr{T}_s^r(\mathrm{M})$ be a linear mapping such that $L(u(\mathbf{s})) = u(L(\mathbf{s}))$ for all diffeomorphisms $u : \mathrm{M} \to \mathrm{M}$ and all $\mathbf{s} \in \mathrm{H}$. Show that, if $r \neq s$, then L is necessarily a differential operator. (By virtue of Problem 1 of Section 17.13, the question reduces to verifying the condition (L) of (17.13.1). Observe that if $\mathbf{s} | \mathrm{U} = 0$ for some open neighborhood U of x, there exists a diffeomorphism u of M onto itself such that u is the identity on $\mathrm{M} - \mathrm{U}$, $u(x) = x$ and $\mathrm{T}_x(u)$ is a homothety whose ratio can be arbitrarily prescribed (Section 16.26, Problem 11).) Give an example in which $r = s = 0$ and L is not a differential operator (cf. (16.24.2)).

2. Let α be a C^∞ differential p-form on M. At a point $x_0 \in \mathrm{M}$ a system of local coordinates $(\varphi^j)_{1 \leqq j \leqq n}$ is said to be *privileged* relative to α if, in some neighborhood of x_0, the form α can be written either as $c\, d\varphi^1 \wedge d\varphi^2 \wedge \cdots \wedge d\varphi^p$, where c is a constant, or as $\varphi^{p+1}\, d\varphi^1 \wedge d\varphi^2 \wedge \cdots \wedge d\varphi^p$. Show that for each point $x_0 \in \mathrm{M}$ there exists a neighborhood U of x_0 in which α is the sum of a finite number of p-forms, for each of which there exists a privileged local coordinate system in U (depending on the p-form). (Observe that, for each C^∞-function g defined in a neighborhood of x_0 and each local coordinate system (φ^j) at x_0, there exists $\varepsilon > 0$ such that
$$(\varphi^1, \ldots, \varphi^p, \varphi^{p+1} + \varepsilon g, \varphi^{p+2}, \ldots, \varphi^n)$$
is a local coordinate system.)

3. Let M be a connected differential manifold of dimension n, and let P be a differential operator from $\bigwedge^p \mathrm{T}(\mathrm{M})^*$ to $\bigwedge^q \mathrm{T}(\mathrm{M})^*$ $(0 \leqq p, q \leqq n)$. If $P \cdot (u(\alpha)) = u(P \cdot \alpha)$ for all diffeomorphisms u of M onto M, show that (1) if $q \neq p, p+1$, then $P = 0$; (2) if $q = p$, then $P \cdot \alpha = c\alpha$, where c is a constant; (3) if $q = p + 1$, then $P \cdot \alpha = c \cdot d\alpha$, where c is a constant. (Use (17.13.3), Problem 11 of Section 16.26, and (16.26.8) to reduce to the case where $\mathrm{M} = \mathbf{R}^n$ and to prove that for $x = 0$ and all p-forms α on $\mathbf{R}^n$ we have $(P \cdot \alpha)(0) = 0$

if $q \neq p, p+1$; $(P \cdot \alpha)(0) = c\alpha(0)$ if $q = p$; and $(P \cdot \alpha)(0) = c \cdot d\alpha(0)$ if $q = p + 1$. Then using Problem 2, reduce to the case where α is one of the two forms

$$c \cdot d\xi^1 \wedge d\xi^2 \wedge \cdots \wedge d\xi^p \qquad \text{or} \qquad \xi^{p+1}\, d\xi^1 \wedge d\xi^2 \wedge \cdots \wedge d\xi^p,$$

and apply the condition on P by taking u to be a translation or a linear mapping given by a diagonal matrix.)

4. (a) Let $\alpha = f \cdot d\xi^1 \wedge d\xi^2 \wedge \cdots \wedge d\xi^p$ be a p-form on $\mathbf{R}^n$, with $p \leq n-1$ and f a C^∞-function with support contained in the ball $B : \|x\| < 1$, and such that $f(x) = \xi^{p+1}$ in the ball $rB : \|x\| < r < 1$. Let u be a diffeomorphism of $\mathbf{R}^n$ onto itself which is equal to a homothety in a neighborhood of the support of f, is the identity on $\mathbf{R}^n - B$ and is such that $u(\operatorname{Supp}(f)) \subset rB$ (Section **16.26**, Problem 11). Let $u(\alpha) = g \cdot d\xi^1 \wedge d\xi^2 \wedge \cdots \wedge d\xi^p$, where $\operatorname{Supp}(g) \subset rB$. Show that, for all sufficiently small $\varepsilon > 0$, the mapping

$$v : (\xi^1, \ldots, \xi^n) \mapsto (\xi^1, \ldots, \xi^p, \xi^{p+1} + \varepsilon g(x), \ldots, \xi^n)$$

is a diffeomorphism of $\mathbf{R}^n$ onto itself (Section **16.12**, Problem 1). We have then $u(\alpha) = \alpha + \varepsilon v(\alpha)$, so that $\alpha = v^{-1}u(\alpha/\varepsilon) - v^{-1}(\alpha/\varepsilon)$.

(b) Let M be a compact differential manifold of dimension n. Show that, for each $p \leq n-1$, every C^∞ differential p-form α on M is the sum of a finite number of p-forms β_j, such that for each β_j there exists a point $x_j \in M$ and a local coordinate system (φ^k) at x_j relative to which β_j is equal to $\varphi^{p+1}\, d\varphi^1 \wedge \cdots \wedge d\varphi^p$ in a neighborhood of x_j. (Use Problem 2 and remark that we may write $1 = A\xi^{p+1} + (1 - A\xi^{p+1})$.)

(c) If M is a compact differential manifold of dimension n and if $p < n$, show that if a linear form L on $\bigwedge^p T(M)^*$ satisfies $L(u(\alpha)) = u(L(\alpha))$ for all diffeomorphisms u of M onto itself, then $L = 0$. (Use (a) and (b).)

5. On a pure differential manifold M of dimension n, let X be a vector field, α a differential n-form, and f a real-valued function (all of class C^∞). Show that

$$df \wedge (i_X \cdot \alpha) = (\theta_X \cdot f)\alpha.$$

6. Let M_1, M_2 be two differential manifolds, f a real-valued function of class C^∞ on $M_1 \times M_2$. For each point $x = (x_1, x_2) \in M_1 \times M_3$, the tangent space $T_x(M_1 \times M_2)$ may be canonically identified with $T_{x_1}(M_1) \times T_{x_2}(M_2)$. Let $d_x^{(1)}f$ (resp. $d_x^{(2)}f$) denote the covector $(\mathbf{h}_1, \mathbf{h}_2) \mapsto \langle \mathbf{h}_1, d_{x_1}f(\cdot, x_2)\rangle$ (resp. $(\mathbf{h}_1, \mathbf{h}_2) \mapsto \langle \mathbf{h}_2, d_{x_2}f(x_1, \cdot)\rangle$). Then $d^{(1)}f : x \mapsto d_x^{(1)}f$ (resp. $d^{(2)}f : x \mapsto d_x^{(2)}f$) are two C^∞ differential forms on $M_1 \times M_2$, called the *partial differentials* of f. We have $df = d^{(1)}f + d^{(2)}f$. With the notation of Section **16.25**, Problem 14, show that one can define uniquely two differential operators $d^{(1)}$, $d^{(2)}$ of order 1:

$$d^{(1)} : \mathscr{E}_{r,s} \to \mathscr{E}_{r+1,s}, \qquad d^{(2)} : \mathscr{E}_{r,s} \to \mathscr{E}_{r,s+1},$$

which for $r = s = 0$ are the operators defined above, and which have the following properties:

$$d^{(1)} \circ d^{(1)} = d^{(2)} \circ d^{(2)} = 0, \qquad d^{(1)} \circ d^{(2)} + d^{(2)} \circ d^{(1)} = 0,$$

$$d = d^{(1)} + d^{(2)},$$

$$d^{(1)}(\alpha \wedge \beta) = (d^{(1)}\alpha) \wedge \beta + (-1)^p \alpha \wedge (d^{(1)}\beta),$$

$$d^{(2)}(\alpha \wedge \beta) = (d^{(2)}\alpha) \wedge \beta + (-1)^p \alpha \wedge (d^{(2)}\beta)$$

for $\alpha \in \mathscr{E}_{r,s}$ and $\beta \in \mathscr{E}_{r',s'}$, with $p = r + s$.

7. For the Dirac p-current ε_{zx}, calculate $\mathscr{E} \cdot \varepsilon_{zx}$, ${}^t i_X \cdot \varepsilon_{zx}$ and ${}^t\theta_X \cdot \varepsilon_{zx}$.

8. (a) Let $(\mathbf{e}_i)_{1 \leq i \leq n}$ be the canonical basis of $\mathbf{R}^n$, and let V′, V″ be any neighborhoods of $\mathbf{e}_1$ and $-\mathbf{e}_1$, respectively. Show that there exist two n-forms v', v'' of class C^∞ on $\mathbf{R}^n$ such that the supports of v', v'' are contained in V′, V″ respectively, and such that there exists an $(n-1)$-form σ of class C^∞, with compact support, satisfying $d\sigma = v' - v''$, and finally such that $\int v' = \int v'' > 0$. (Take v' to be

$$h(\xi^1 - 1)h((\xi^2)^2 + \cdots + (\xi^n)^2)\, d\xi^1 \wedge d\xi^2 \wedge \cdots \wedge d\xi^n,$$

where h is a nonnegative C^∞-function on $\mathbf{R}$ with arbitrarily small support, and such that $h(0) > 0$. Define v'' similarly.)
(b) Let M, N be oriented connected differential manifolds of the same dimension n, and let f be a *proper* (**17.3.7**) C^∞-mapping of M into N, such that the inverse image ${}^t f(v_0)$ of an n-form v_0 belonging to the orientation of N is ≥ 0 at all points of M (relative to the orientation of M). If f is not surjective, show that all points of M are *critical* points of f (**16.23**). (The set $f(\mathrm{M})$ is closed in N. Suppose that there exists a point $y_1 \in \mathrm{N} - f(\mathrm{M})$ and a point $x_2 \in \mathrm{M}$ which is not critical for f, and let $y_2 = f(x_2) \in f(\mathrm{M})$. Then there exists an open subset $\mathrm{U} \subset \mathrm{N}$, diffeomorphic to $\mathbf{R}^n$ and containing y_1 and y_2 (**16.26.9**). Use (a) to show that there exist two n-forms v', v'' on N and an $(n-1)$-form σ such that $d\sigma = v' - v''$, the forms v', v'' and σ being of class C^∞ and compactly supported, and such that $\int_{\mathrm{M}} {}^t f(v') = 0$ and $\int_{\mathrm{M}} {}^t f(v'') > 0$, contradicting (**17.15.5.1**).)

9. Let M be a pure differential manifold of dimension n. Let $\Omega = -d\kappa_{\mathrm{M}}$ be the canonical 2-form (**17.15.2.4**) on the cotangent bundle $\mathrm{N} = \mathrm{T}(\mathrm{M})^*$. For each $z \in \mathrm{N}$, the mapping $\mathbf{h}_z \mapsto i(\mathbf{h}_z) \cdot \Omega(z)$ is an N-isomorphism of the tangent bundle T(N) onto the cotangent bundle $\mathrm{T}(\mathrm{N})^*$. For each C^∞ differential 1-form α on N, there exists therefore a unique C^∞ vector field X_α such that $\alpha = i_{X_\alpha} \cdot \Omega$.
(a) If α, β are C^∞ differential 1-forms on N, their *Poisson bracket* $\{\alpha, \beta\}$ is defined to be the differential 1-form $-i_{[X_\alpha, X_\beta]} \cdot \Omega$. Show that

$$\{\alpha, \beta\} = -\theta_{X_\alpha} \cdot \beta + \theta_{X_\beta} \cdot \alpha + d(i_{X_\alpha} \cdot (i_{X_\beta} \cdot \Omega)).$$

For any three 1-forms α, β, γ on N, show that

$$\{\alpha, \{\beta, \gamma\}\} + \{\beta, \{\gamma, \alpha\}\} + \{\gamma, \{\alpha, \beta\}\} = 0.$$

If f, g are real-valued C^∞-functions on M, their *homogeneous Poisson bracket* is defined to be the function

$$\begin{aligned}\{f, g\} &= -i_{X_{df}} \cdot (i_{X_{dg}} \cdot \Omega)\\ &= -\theta_{X_{df}} \cdot g = \theta_{X_{dg}} \cdot f.\end{aligned}$$

Show that $d\{f, g\} = \{df, dg\}$ and that

$$\{f, \{g, h\}\} + \{g, \{h, f\}\} + \{h, \{f, g\}\} = 0,$$
$$\{f, gh\} = h\{f, g\} + g\{f, h\}$$

for any three functions f, g, h.

With the notation of (16.20.6), if the local expression of Ω is $\sum_{i=1}^{n} d\xi^i \wedge d\eta_i$, then

$$\{f, g\} = \sum_{i=1}^{n} \left(\frac{\partial f}{\partial \xi^i} \frac{\partial g}{\partial \eta_i} - \frac{\partial f}{\partial \eta_i} \frac{\partial g}{\partial \xi^i} \right)$$

(b) Replace n by $n+1$ and denote a point of $N = T(M)^*$ by $((\xi^i), (\eta_i))$ $(0 \leq i \leq n)$, where M is an open set in $\mathbf{R}^{n+1}$. Given any function $F(z, x^1, \ldots, x^n, y_1, \ldots, y_n)$ defined on $M \times \mathbf{R}^n$, construct the function

$$f(\xi^0, \xi^1, \ldots, \xi^n, \eta_0, \eta_1, \ldots, \eta_n) = F\left(-\xi^0, \xi^1, \ldots, \xi^n, \frac{\eta_1}{\eta_0}, \ldots, \frac{\eta_n}{\eta_0}\right)$$

defined on the open subset $U : \eta_0 \neq 0$ in N. If G is another function on $M \times \mathbf{R}^n$ and g the corresponding function on U, and if F, G are both of class C^∞, then the function $\eta_0\{f, g\}$ corresponds to the function

$$\sum_{i=1}^{n} \left(\left(\frac{\partial F}{\partial x^i} + y_i \frac{\partial F}{\partial z} \right) \frac{\partial G}{\partial y_i} - \left(\frac{\partial G}{\partial x^i} + y_i \frac{\partial G}{\partial z} \right) \frac{\partial F}{\partial y_i} \right),$$

which is denoted by $\{F, G\}$ if there is no risk of ambiguity, and is called the *nonhomogeneous Poisson bracket* of F and G. If ω is the differential 1-form

$$\omega = dz - \sum_{i=1}^{n} y_i \, dx^i,$$

then we have

$$\{F, G\} \, dz \wedge dx^1 \wedge dy_1 \wedge \cdots \wedge dx^n \wedge dy_n = \omega \wedge (d\omega)^{\wedge(n-1)} \wedge dF \wedge dG.$$

16. CONNECTIONS IN A VECTOR BUNDLE

(17.16.1) Let G be a Lie group. We have seen (16.15.6) that for each pair of points x, y of G there is a well-determined isomorphism $\mathbf{h}_x \mapsto yx^{-1} \cdot \mathbf{h}_x$ of the vector space $T_x(G)$ onto $T_y(G)$, which depends only on the points x, y; but the existence of such an isomorphism depends essentially on the supplementary *Lie group* structure on the differentiable manifold G. In the absence of such a structure, there is no *canonical* isomorphism of the tangent space at a point x of a pure manifold M onto the tangent space at another point y, i.e., no isomorphism which is determined uniquely by the manifold structure of M and the two points x, y.

The notion of a *connection* in a vector bundle E over M is the mathematical expression of the idea of defining, for each point $x \in M$, a procedure for providing an isomorphism of E_x onto E_y for points y "infinitely near" to x.

Since the question is local, we shall consider first the case of an open subset U of $\mathbf{R}^n$ and a trivial bundle $E = U \times \mathbf{R}^p$. Let x be a point of U, and suppose that for each vector $\mathbf{h} \in \mathbf{R}^n$ such that $x + \mathbf{h} \in \mathrm{U}$ we have a linear isomorphism

(17.16.1.1) $$(x, \mathbf{u}) \mapsto (x + \mathbf{h}, F(\mathbf{h})^{-1} \cdot \mathbf{u})$$

of the fiber $\mathrm{E}_x = \{x\} \times \mathbf{R}^p$ onto the fiber $\mathrm{E}_{x+h} = \{x + \mathbf{h}\} \times \mathbf{R}^p$, so that $\mathbf{h} \mapsto F(\mathbf{h})$ is a mapping of a neighborhood V of 0 in $\mathbf{R}^n$ such that $x + \mathrm{V} \subset \mathrm{U}$, into the vector space $\mathscr{L}(\mathbf{R}^p)$ of all endomorphisms of $\mathbf{R}^p$ (a space which is isomorphic to $\mathbf{R}^{p^2}$). Suppose that $F(0) = 1_{\mathbf{R}^p}$ and that F is of class C^∞ in V, so that the mapping

(17.16.1.2) $$(\mathbf{h}, \mathbf{u}) \mapsto (x + \mathbf{h}, F(\mathbf{h})^{-1} \cdot \mathbf{u})$$

of $\mathrm{V} \times \mathbf{R}^p$ into $\mathrm{U} \times \mathbf{R}^p$ is indefinitely differentiable. Its *derivative* at the point $(0, \mathbf{u})$ is the value at the point $(x, \mathbf{u}) \in \mathrm{E}_x$ of the linear isomorphism "infinitely near" to the identity which we wish to consider; this value is, by virtue of (8.1.5), (8.9.1), and (8.3.2),

(17.16.1.3) $$(\mathbf{k}, \mathbf{v}) \mapsto (\mathbf{k}, \mathbf{v} - (\mathrm{D}F(0) \cdot \mathbf{k}) \cdot \mathbf{u}),$$

a linear mapping of $\mathbf{R}^n \times \mathbf{R}^p$ into itself; $\mathrm{D}F(0)$ belongs to $\mathscr{L}(\mathbf{R}^n; \mathscr{L}(\mathbf{R}^p))$, and therefore (5.7.8) the mapping $(\mathbf{k}, \mathbf{u}) \mapsto (\mathrm{D}F(0) \cdot \mathbf{k}) \cdot \mathbf{u}$ is a *bilinear* mapping of $\mathbf{R}^n \times \mathbf{R}^p$ into $\mathbf{R}^p$, which we shall denote by

(17.16.1.4) $$(\mathbf{k}, \mathbf{u}) \mapsto \Gamma_x(\mathbf{k}, \mathbf{u}).$$

Conversely, if we prescribe arbitrarily such a bilinear mapping and put $F(\mathbf{h}) = 1_{\mathbf{R}^p} + \Gamma_x(\mathbf{h}, \cdot)$, then $F(\mathbf{h})$ is an automorphism of $\mathbf{R}^p$ for all sufficiently small $\mathbf{h}$ (8.3.2), such that $(\mathrm{D}F(0) \cdot \mathbf{k}) \cdot \mathbf{u} = \Gamma_x(\mathbf{k}, \mathbf{u})$.

(17.16.2) Let us now interpret these remarks in terms which are *independent of the trivialization* of E chosen. Since $\mathrm{E} = \mathrm{U} \times \mathbf{R}^p$ and since $(x, \mathbf{u})$ is a point of the fiber E_x, the tangent space $\mathrm{T}_{(x, \mathbf{u})}(\mathrm{E})$ may be identified with

$$\mathrm{T}_x(\mathrm{U}) \times \mathrm{T}_{\mathbf{u}}(\mathbf{R}^p),$$

and therefore with $(\{x\} \times \mathbf{R}^n) \times (\{\mathbf{u}\} \times \mathbf{R}^p)$. To a pair of vectors $(x, \mathbf{k}) \in \mathrm{T}_x(\mathrm{M})$ and $(x, \mathbf{u}) \in \mathrm{E}_x$ we associate, by means of (17.16.1.4), the vector

(17.16.2.1) $$\mathbf{C}_x((x, \mathbf{k}), (x, \mathbf{u})) = ((x, \mathbf{u}), (\mathbf{k}, -\Gamma_x(\mathbf{k}, \mathbf{u})))$$

belonging to $\mathrm{T}_{(x, \mathbf{u})}(\mathrm{E})$. If $\pi : \mathrm{E} \to \mathrm{U}$ is the projection of the bundle E onto its base, then we have

$$\mathrm{T}_{(x, \mathbf{u})}(\pi) \cdot ((x, \mathbf{u}), (\mathbf{k}, \mathbf{v})) = (x, \mathbf{k}) \qquad \text{for all} \quad ((x, \mathbf{u}), (\mathbf{k}, \mathbf{v})) \in \mathrm{T}_{(x, \mathbf{u})}(\mathrm{E}).$$

Moreover, T(E) is a vector bundle over E (16.15.4); if $o_E : T(E) \to E$ is its projection, then we have $o_E((x, \mathbf{u}), (\mathbf{k}, \mathbf{v})) = (x, \mathbf{u})$. Finally, $((x, \mathbf{k}), (x, \mathbf{u}))$ may be identified with the vector $(x, (\mathbf{k}, \mathbf{u}))$ of the fiber over x of the vector bundle $T(U) \oplus E$ over U (16.16.1). Hence the mapping $\mathbf{C}_x$ defined by (17.16.2.1) satisfies the following conditions:

(17.16.2.2) $$T_{(x,\mathbf{u})}(\pi) \cdot \mathbf{C}_x((x, \mathbf{k}), (x, \mathbf{u})) = (x, \mathbf{k}),$$

(17.16.2.3) $$o_E(\mathbf{C}_x((x, \mathbf{k}), (x, \mathbf{u}))) = (x, \mathbf{u});$$

the mapping

(17.16.2.4) $$(x, \mathbf{k}) \mapsto \mathbf{C}_x((x, \mathbf{k}), (x, \mathbf{u}))$$

is a *linear* mapping of $T_x(U)$ into $T_{(x,\mathbf{u})}(E)$; and the mapping

(17.16.2.5) $$(x, \mathbf{u}) \mapsto \mathbf{C}_x((x, \mathbf{k}), (x, \mathbf{u}))$$

is a *linear* mapping of E_x into $T(\pi)^{-1}(x, \mathbf{k})$.

(17.16.3) It is now easy to give the definition of a *connection* (or *linear connection*) in an arbitrary *vector bundle* E over a differential manifold M. We have merely to replace in the definitions and conditions of (17.16.2) the vector $(x, \mathbf{u})$ by a vector $\mathbf{u}_x \in E_x$, and the tangent vector $(x, \mathbf{k})$ by a vector $\mathbf{k}_x \in T_x(M)$. We denote again by $\pi : E \to M$, $o_M : T(M) \to M$, and $o_E : T(E) \to E$ the canonical projections, and we observe that T(E) is a vector bundle over T(M) with projection $T(\pi)$ (16.15.7). Finally, the composite mappings $\pi \circ o_E$ and $o_M \circ T(\pi)$ of T(E) into M are equal, and if ϖ denotes this mapping, then the triple $(T(E), M, \varpi)$ is again a *fibration*, but *not* a vector bundle (16.15.7).

Having said this, a *connection* (or *linear connection*) *in the vector bundle* E is defined to be an M-morphism

(17.16.3.1) $$\mathbf{C} : T(M) \oplus E \to T(E)$$

of fiber bundles over M (16.12.1), having the following properties:

(17.16.3.2) $$T(\pi) \cdot \mathbf{C}_x(\mathbf{k}_x, \mathbf{u}_x) = \mathbf{k}_x, \qquad o_E(\mathbf{C}_x(\mathbf{k}_x, \mathbf{u}_x)) = \mathbf{u}_x;$$

the mapping

(17.16.3.3) $$\mathbf{k}_x \mapsto \mathbf{C}_x(\mathbf{k}_x, \mathbf{u}_x)$$

of $T_x(M)$ into $T_{\mathbf{u}_x}(E)$ is *linear*; the mapping

(17.16.3.4) $$\mathbf{u}_x \mapsto \mathbf{C}_x(\mathbf{k}_x, \mathbf{u}_x)$$

is a *linear* mapping of E_x into $(T(E))_{\mathbf{k}_x}$, the fiber over $\mathbf{k}_x$ of T(E) considered as a vector bundle over T(M) with projection $T(\pi)$ (16.15.7). In particular, for each scalar $c \in \mathbf{R}$, we have

(17.16.3.5) $$T(m_c) \cdot \mathbf{C}_x(\mathbf{k}_x, \mathbf{u}_x) = \mathbf{C}_x(\mathbf{k}_x, c \cdot \mathbf{u}_x),$$

where m_c is the mapping $\mathbf{u}_x \mapsto c \cdot \mathbf{u}_x$ of E into itself.

We remark that the conditions (17.16.3.2) imply that the linear mappings $\mathbf{k}_x \mapsto \mathbf{C}_x(\mathbf{k}_x, \mathbf{u}_x)$ and $\mathbf{u}_x \mapsto \mathbf{C}_x(\mathbf{k}_x, \mathbf{u}_x)$ are *injective*. For each $\mathbf{u}_x \in E_x$, the subspace of $T_{\mathbf{u}_x}(E)$ which is the image of $T_x(M)$ under the mapping

$$\mathbf{k}_x \mapsto \mathbf{C}_x(\mathbf{k}_x, \mathbf{u}_x)$$

is *supplementary* to the subspace $T_{\mathbf{u}_x}(\pi)^{-1}(\mathbf{0}_x)$ formed by the *vertical* tangent vectors to E at the point $\mathbf{u}_x$ (16.12.1). This supplementary subspace is sometimes called the space of *horizontal* tangent vectors to E at the point $\mathbf{u}_x$, *relative to the connection* **C**. For each vector field X on M, the *horizontal lifting* (relative to **C**) of X is a vector field on E, denoted by $\mathrm{rel}_{\mathbf{C}}(X)$, and defined by

(17.16.3.6) $$\mathrm{rel}_{\mathbf{C}}(X)(\mathbf{u}_x) = \mathbf{C}_x(X(x), \mathbf{u}_x).$$

(17.16.4) Let (U, φ, n) be a chart of M such that E is trivializable over U, and let $(\pi^{-1}(U), \theta, n+p)$ be a corresponding *fibered chart* of E (16.15.1), with $\theta(\pi^{-1}(U)) = \varphi(U) \times \mathbf{R}^p$. Then, for each point $x \in \varphi(U)$ and each vector $(\mathbf{k}, \mathbf{u}) \in \mathbf{R}^n \times \mathbf{R}^p$, we have

(17.16.4.1) $$\begin{aligned}\mathbf{C}_{\varphi^{-1}(x)}(T_x(\varphi^{-1}) \cdot (x, \mathbf{k}), \theta^{-1}(x, \mathbf{u}))\\ = T_{(x, \mathbf{u})}(\theta) \cdot ((x, \mathbf{u}), (\mathbf{k}, -\Gamma_x(\mathbf{k}, \mathbf{u})),\end{aligned}$$

where $(\mathbf{k}, \mathbf{u}) \mapsto \Gamma_x(\mathbf{k}, \mathbf{u})$ is a *bilinear* mapping of $\mathbf{R}^n \times \mathbf{R}^p$ into $\mathbf{R}^p$ (17.16.1); the mapping $(\mathbf{k}, \mathbf{u}) \mapsto (\mathbf{k}, -\Gamma_x(\mathbf{k}, \mathbf{u}))$ is called the *local expression* of **C** corresponding to the fibered chart. More explicitly, this mapping may be written as

$$(k^1, \ldots, k^n, u^1, \ldots, u^p) \mapsto (k^1, \ldots, k^n, -\Gamma_x^1(\mathbf{k}, \mathbf{u}), \ldots, -\Gamma_x^p(\mathbf{k}, \mathbf{u}))$$

$(x \in \varphi(U))$, where each Γ_x^i is a *bilinear* form

(17.16.4.2) $$\Gamma_x^i(\mathbf{k}, \mathbf{u}) = \sum_{h, l} \Gamma_{hl}^i(x) k^h u^l,$$

the Γ_{hl}^i being C^∞-functions on $\varphi(U)$.

Now consider another fibered chart, corresponding to a chart $(\mathrm{U}, \varphi', n)$ defined on the same open set U. The transition diffeomorphism

$$\psi : \varphi(\mathrm{U}) \times \mathbf{R}^p \to \varphi'(\mathrm{U}) \times \mathbf{R}^p$$

is of the form

$$(x^1, \ldots, x^n, u^1, \ldots, u^p) \mapsto (\Phi(x), A(x) \cdot \mathbf{u}), \tag{17.16.4.3}$$

where $\Phi(x) = (\Phi_1(x), \ldots, \Phi_n(x))$ and $A(x) = (a_{ij}(x))$ is a square matrix of order p, so that the derivative $\mathrm{D}\psi$ is the linear mapping

$$(k^1, \ldots, k^n, v^1, \ldots, v^p) \mapsto (k'^1, \ldots, k'^n, v'^1, \ldots, v'^p)$$

given by

$$\begin{cases} k'^\beta = \sum_h \mathrm{D}_h \Phi_\beta \cdot k^h, \\ v'^\gamma = \sum_l a_{\gamma l} v^l + \sum_{h,l} \mathrm{D}_h a_{\gamma l} \cdot k^h u^l \end{cases} \tag{17.16.4.4}$$

((8.9.1) and (8.9.2)). Now let

$$(k'^1, \ldots, k'^n, u'^1, \ldots, u'^p) \mapsto (k'^1, \ldots, k'^n, -\Gamma'^1_{x'}(\mathbf{k}', \mathbf{u}'), \ldots, -\Gamma'^p_{x'}(\mathbf{k}', \mathbf{u}'))$$

$(x' \in \varphi'(\mathrm{U}))$ be the local expression of the connection **C** relative to the second chart, and let

$$\Gamma'^\alpha_{x'}(\mathbf{k}', \mathbf{u}') = \sum_{\beta, \gamma} \Gamma'^\alpha_{\beta\gamma}(x') k'^\beta u'^\gamma. \tag{17.16.4.5}$$

Also put

$$\Phi^{-1}(x') = (\tilde{\Phi}_1(x'), \ldots, \tilde{\Phi}_n(x')), \qquad A^{-1}(\Phi^{-1}(x')) = (\tilde{a}_{\alpha\gamma}(x')).$$

Then we have from above, for $x' = \Phi(x)$ and $\mathbf{u}' = A(x) \cdot \mathbf{u}$,

$$\Gamma'^\alpha_{x'}(\mathrm{D}\Phi(x) \cdot \mathbf{k}, A(x) \cdot \mathbf{u}) = \sum_i a_{\alpha i}(x) \Gamma^i_x(\mathbf{k}, \mathbf{u}) + \sum_{h,l} \mathrm{D}_h a_{\alpha l}(x) \cdot k^h u^l;$$

also the first of the formulas (17.16.4.4) gives

$$k^h = \sum_\beta \mathrm{D}_\beta \tilde{\Phi}_h(x') \cdot k'^\beta$$

and we have

$$u^l = \sum_\gamma \tilde{a}_{l\gamma}(x') u'^\gamma,$$

so that finally we obtain the following expression for the $\Gamma'^\alpha_{\beta\gamma}(x')$:

(17.16.4.6)

$$\Gamma'^{\alpha}_{\beta\gamma}(x') = \sum_{h,l} (D_\beta \tilde{\Phi}_h(x'))\tilde{a}_{l\gamma}(x')(D_h a_{\alpha l}(\Phi^{-1}(x')) + \sum_i a_{\alpha i}(\Phi^{-1}(x'))\Gamma^i_{hl}(\Phi^{-1}(x')).$$

We see from this that, contrary to what one might have expected from (17.16.1), the presence of a connection on M *does not enable us* to define, for all $x \in M$, a *bilinear* mapping Γ_x of $T_x(M) \times E_x$ into E_x; the mapping Γ_x corresponding to a trivialization of E *depends* on this trivialization. In particular, to say that a connection is "zero" has no meaning, because all the Γ^i_{hl} can vanish without the $\Gamma'^{\alpha}_{\beta\gamma}$ vanishing.

(17.16.5) Since, for each vector $(\mathbf{k}_x, \mathbf{u}_x)$ in $T(M) \oplus E$, the value $\mathbf{C}_x(\mathbf{k}_x, \mathbf{u}_x)$ of a connection belongs to the tangent space $T_{\mathbf{u}_x}(E)$, it follows that the *sum* of the values of two connections at the point $(\mathbf{k}_x, \mathbf{u}_x)$ is defined, as is also the product of $\mathbf{C}_x(\mathbf{k}_x, \mathbf{u}_x)$ by a scalar. We may therefore form *linear combinations* $\sum_j f_j \mathbf{C}_j$ of connections in the bundle E, the coefficients f_j being real-valued C^∞-functions on M; but in general the M-morphisms so obtained *are not connections*, by reason of the first condition (17.16.3.2). Nevertheless, there are two important particular cases.

(17.16.6) In the first place, if $\mathbf{C}$ and $\mathbf{C}'$ are two connections in E, their *difference* $\mathbf{C}' - \mathbf{C}$ is no longer a connection, but is an M-morphism

(17.16.6.1) $$A : T(M) \oplus E \to T(E)$$

such that

(17.16.6.2) $$T(\pi) \cdot A_x(\mathbf{k}_x, \mathbf{u}_x) = \mathbf{0}_x, \quad o_E(A_x(\mathbf{k}_x, \mathbf{u}_x)) = \mathbf{u}_x.$$

The first of these relations shows that $A_x(\mathbf{k}_x, \mathbf{u}_x)$ is an element of $T_{\mathbf{u}_x}(E_x)$, the tangent space to the *fiber* E_x at the point $\mathbf{u}_x$, identified with the subspace of "vertical" tangent vectors in $T_{\mathbf{u}_x}(E)$. Furthermore, $\mathbf{k}_x \mapsto A_x(\mathbf{k}_x, \mathbf{u}_x)$ is a *linear* mapping of $T_x(M)$ into $T_{\mathbf{u}_x}(E_x)$, and $\mathbf{u}_x \mapsto A_x(\mathbf{k}_x, \mathbf{u}_x)$ a *linear* mapping of E_x into $(T(E))_{\mathbf{0}_x}$, the fiber of T(E) over the point $\mathbf{0}_x \in T_x(M)$ relative to the fibration $(T(E), T(M), T(\pi))$, which is isomorphic to the tangent bundle $T(E_x)$, identified with $E_x \times E_x$. Also we have a canonical isomorphism $\boldsymbol{\tau}_{\mathbf{u}_x} : T_{\mathbf{u}_x}(E_x) \to E_x$ (16.5.2) which, modulo the preceding identifications, is the restriction to $T_{\mathbf{u}_x}(E_x)$ of the second projection. It follows that, if we put

(17.16.6.3) $$B_x(\mathbf{k}_x, \mathbf{u}_x) = \boldsymbol{\tau}_{\mathbf{u}_x}(A_x(\mathbf{k}_x, \mathbf{u}_x)),$$

B_x is a *bilinear* mapping of $T_x(M) \times E_x$ into E_x. In other words, we have defined a *bilinear* M-morphism $B : T(M) \oplus E \to E$, which, by abuse of language, is called the *difference* of the connections $\mathbf{C}'$ and $\mathbf{C}$. Relative to a local trivialization of E in which $\mathbf{C}$ and $\mathbf{C}'$ correspond, respectively, to the bilinear functions Γ and Γ' (17.16.1), the morphism B corresponds to $-\Gamma' + \Gamma$. Conversely, if $B : T(M) \oplus E \to E$ is any bilinear M-morphism, then if we put

$$A_x(\mathbf{k}_x, \mathbf{u}_x) = \tau_{\mathbf{u}_x}^{-1}(B_x(\mathbf{k}_x, \mathbf{u}_x)),$$

$\mathbf{C} + A$ is a connection, for every connection $\mathbf{C}$ in E.

(17.16.7) *Let* $(U_\alpha, \varphi_\alpha, n)$ *be a family of charts of* M *such that* E *is trivializable over each* U_α, *and such that the family* (U_α) *is locally finite. For each* α, *let* $\mathbf{C}_\alpha$ *be a connection in the vector bundle* $\pi^{-1}(U_\alpha)$, *and let* (f_α) *be a* C^∞-*partition of unity subordinate to* (U_α) (16.4.1). *Then* $\sum_\alpha f_\alpha \mathbf{C}_\alpha$ *is a connection in* E (with the understanding that we replace $f_\alpha(x)(\mathbf{C}_\alpha)_x(\mathbf{k}_x, \mathbf{u}_x)$ by the origin of the tangent space $T_{\mathbf{u}_x}(E)$ for each point $x \notin U_\alpha$).

Since for each point $x \in M$ there is a neighborhood of x which meets only a finite number of the U_α, and since $\sum_\alpha f_\alpha(x) = 1$, the only point that needs to be checked is that the functions $f_\alpha \mathbf{C}_\alpha$ are C^∞-mappings of the *whole* of $T(M) \oplus E$ into E. If x is a frontier point of some U_α, there is a neighborhood V of x in M which does not meet the support of f_α, and therefore, for each point $(\mathbf{k}_y, \mathbf{u}_y)$ of $T(M) \oplus E$ lying over a point $y \in V$, by definition $f_\alpha(y)(\mathbf{C}_\alpha)_y(\mathbf{k}_y, \mathbf{u}_y)$ is the origin of the vector space $T_{\mathbf{u}_y}(E)$. The assertion now follows by taking a trivialization of T(E) (considered as a bundle over M) in a neighborhood of x.

In particular:

(17.16.8) *There exists a connection in any vector bundle.*

Choose a family of charts of M having the properties stated in (17.6.7), and define each connection $\mathbf{C}_\alpha$ by taking a particular trivialization of $\pi^{-1}(U_\alpha)$ and taking the corresponding mapping (17.16.1.4) to be, e.g., zero for all $x \in \varphi_\alpha(U_\alpha)$.

(17.16.9) If (f, g) is an isomorphism of a vector bundle E over M onto a vector bundle E′ over M′, then every connection $\mathbf{C}$ in E is transported by (f, g) to a connection $\mathbf{C}'$ in E′, such that

$$\mathbf{C}'_{x'}(\mathbf{k}'_{x'}, \mathbf{u}'_{x'}) = T(T(f)) \cdot \mathbf{C}_{f^{-1}(x')}(T(f^{-1}) \cdot \mathbf{k}'_{x'}, g^{-1}(\mathbf{u}'_{x'}))).$$

17. DIFFERENTIAL OPERATORS ASSOCIATED WITH A CONNECTION

(17.17.1) We take up again the situation considered in (17.16.1). Let Z be an open subset of some $\mathbf{R}^q$, and let G be a C^∞-mapping of Z into the bundle E, so that we may write $G(z) = (f(z), \mathbf{g}(z))$, where f (resp. $\mathbf{g}$) is a C^∞-mapping of Z into U (resp. into $\mathbf{R}^p$). Let W be a neighborhood of 0 in $\mathbf{R}^q$ such that $z + W \subset Z$, and for each $\mathbf{w} \in W$ put $\mathbf{h}(\mathbf{w}) = f(z + \mathbf{w}) - f(z)$. Since the point $G(z + \mathbf{w})$ lies in the fiber $E_{f(z)+\mathbf{h}(\mathbf{w})}$, we may consider the point

$$(f(z), F(\mathbf{h}(\mathbf{w})) \cdot \mathbf{g}(z + \mathbf{w})),$$

which lies in the fiber $E_{f(z)}$, and we are thus led to take as "derivative" of G the derivative of the mapping

$$\mathbf{w} \mapsto F(\mathbf{h}(\mathbf{w})) \cdot \mathbf{g}(z + \mathbf{w})$$

at the point $\mathbf{w} = 0$; bearing in mind that $F(0) = 1_{\mathbf{R}^p}$, and using (8.1.4) and (8.2.1), this gives us the following linear mapping of $\mathbf{R}^q$ into $\mathbf{R}^p$:

(17.17.1.1) $$\mathbf{w} \mapsto D\mathbf{g}(z) \cdot \mathbf{w} + (DF(0) \cdot (Df(z) \cdot \mathbf{w})) \cdot \mathbf{g}(z).$$

Since $DG(z) \cdot \mathbf{w} = (Df(z) \cdot \mathbf{w}, D\mathbf{g}(z) \cdot \mathbf{w})$, the right-hand side of (17.17.1.1) can also be written as

$$DG(z) \cdot \mathbf{w} - (Df(z) \cdot \mathbf{w}, -\Gamma_{f(z)}(Df(z) \cdot \mathbf{w}, \mathbf{g}(z)))$$

in which appears the value $\mathbf{C}_{f(z)}(Df(z) \cdot \mathbf{w}, \mathbf{g}(z))$ of the *connection* $\mathbf{C}$.

(17.17.2) We shall now give an intrinsic definition of the *covariant derivative* relative to a connection $\mathbf{C}$ in a vector bundle E over M. Let N be a differential manifold and let $\mathbf{G}$ be a C^r-mapping ($r \geqq 1$) of N into E. It follows from (17.16.3.2) that, for each $z \in N$ and each tangent vector $\mathbf{h}_z \in T_z(N)$, the vector

$$T_z(\mathbf{G}) \cdot \mathbf{h}_z - \mathbf{C}_{\pi(\mathbf{G}(z))}(T_z(\pi \circ \mathbf{G}) \cdot \mathbf{h}_z, \mathbf{G}(z))$$

belongs to the tangent space $T_{\mathbf{G}(z)}(E_{\pi(\mathbf{G}(z))})$ to the *fiber* through the point $\mathbf{G}(z)$ of the bundle E. Since this fiber is a vector space, we may apply the canonical translation $\tau_{\mathbf{G}(z)}$ (16.5.2) to the above vector: the vector so obtained *in the fiber* $E_{\pi(\mathbf{G}(z))}$

(17.17.2.1) $$\nabla_{\mathbf{h}_z} \cdot \mathbf{G} = \tau_{\mathbf{G}(z)}(T_z(\mathbf{G}) \cdot \mathbf{h}_z - \mathbf{C}_{\pi(\mathbf{G}(z))}(T_z(\pi \circ \mathbf{G}) \cdot \mathbf{h}_z, \mathbf{G}(z)))$$

is called the *covariant derivative of* $\mathbf{G}$ *at the point* z (relative to the connection $\mathbf{C}$) *in the direction of the tangent vector* $\mathbf{h}_z$.

When M and N are open sets in $\mathbf{R}^m$ and $\mathbf{R}^n$ (a situation to which we can always reduce by a suitable choice of charts), $\mathbf{G}$ is a function of the form $z \mapsto (f(z), \mathbf{g}(z))$ and the connection $\mathbf{C}$ is given by (17.16.2.1). We have then, with $\mathbf{h}_z = (z, \mathbf{h})$,

(17.17.2.2) $$\nabla_{\mathbf{h}_z} \cdot \mathbf{G} = (f(z), \mathrm{D}\mathbf{g}(z) \cdot \mathbf{h} + \Gamma_{f(z)}(\mathrm{D}f(z) \cdot \mathbf{h}, \mathbf{g}(z)))$$

or, expressing everything in terms of coordinates as in (17.16.4),

(17.17.2.3) $$(\nabla_{\mathbf{h}_z} \cdot \mathbf{G})^i = \sum_j \frac{\partial g^i}{\partial z^j} h^j + \sum_{j,k,l} \Gamma^i_{jk}(f(z)) \frac{\partial f^j}{\partial z^l} g^k(z) h^l$$

for the ith component of $\nabla_{\mathbf{h}_z} \cdot \mathbf{G}$ in $\mathbf{R}^p$.

It is clear that $\mathbf{h}_z \mapsto \nabla_{\mathbf{h}_z} \cdot \mathbf{G}$ is a *linear* mapping of $\mathrm{T}_z(\mathrm{N})$ into $\mathrm{E}_{\pi(\mathbf{G}(z))}$. If $\mathbf{G}_1$, $\mathbf{G}_2$ are two mappings of class C^r ($r \geqq 1$) of N into E, such that $\pi \circ \mathbf{G}_1 = \pi \circ \mathbf{G}_2$ (in other words, two *liftings* to E of the *same* C^r-mapping of N into M), then we have

(17.17.2.4) $$\nabla_{\mathbf{h}_z} \cdot (\mathbf{G}_1 + \mathbf{G}_2) = \nabla_{\mathbf{h}_z} \cdot \mathbf{G}_1 + \nabla_{\mathbf{h}_z} \cdot \mathbf{G}_2 .$$

For every *scalar* function σ of class C^r on N, the mapping $z \mapsto \sigma(z)\mathbf{G}(z)$ is a C^r-mapping of N into E which we denote by $\sigma\mathbf{G}$; clearly $\pi \circ (\sigma\mathbf{G}) = \pi \circ \mathbf{G}$. Bearing in mind the definition of (17.14.9), it is easily seen that

(17.17.2.5) $$\nabla_{\mathbf{h}_z} \cdot (\sigma\mathbf{G}) = (\theta_{\mathbf{h}_z} \cdot \sigma) \cdot \mathbf{G}(z) + \sigma(z)(\nabla_{\mathbf{h}_z} \cdot \mathbf{G}),$$

by reducing to the case where the covariant derivative is given by (17.17.2.2).

Finally, if $u : \mathrm{N}_1 \mapsto \mathrm{N}$ is a mapping of class C^r, then for each point $z_1 \in \mathrm{N}_1$ and each tangent vector $\mathbf{h}_{z_1} \in \mathrm{T}_{z_1}(\mathrm{N}_1)$, it follows immediately from the definitions that

(17.17.2.6) $$\nabla_{\mathbf{h}_{z_1}} \cdot (\mathbf{G} \circ u) = \nabla_{\mathbf{h}_z} \cdot \mathbf{G},$$

where $z = u(z_1)$ and $\mathbf{h}_z = \mathrm{T}_{z_1}(u) \cdot \mathbf{h}_{z_1}$.

Remark

(17.17.2.7) By virtue of (17.17.2.1), the relation $\nabla_{\mathbf{h}_z} \cdot \mathbf{G} = 0$ signifies that the tangent vector $\mathrm{T}_z(\mathbf{G}) \cdot \mathbf{h}_z$ is horizontal (17.16.3) at the point $\mathbf{G}(z)$.

(17.17.3) Having defined the covariant derivative of the mapping $\mathbf{G} : \mathrm{N} \to \mathrm{E}$ at a point $z \in \mathrm{N}$ in the direction of a tangent vector $\mathbf{h}_z \in \mathrm{T}_z(\mathrm{N})$, it is now easy to define the *covariant derivative of* $\mathbf{G}$ (relative to $\mathbf{C}$) *in the direction of a vector field* $Z \in \mathcal{T}_0^1(\mathrm{N})$; this is the mapping $\nabla_Z \cdot \mathbf{G}$ of N into E which at each point $z \in \mathrm{N}$ has the value $\nabla_{Z(z)} \cdot \mathbf{G}$. The results of (17.17.2) give rise to the formulas

(17.17.3.1) $$\nabla_{Z_1+Z_2} \cdot \mathbf{G} = \nabla_{Z_1} \cdot \mathbf{G} + \nabla_{Z_2} \cdot \mathbf{G}$$

for two vector fields Z_1, Z_2 on N;

(17.17.3.2) $$\nabla_{\sigma Z} \cdot \mathbf{G} = \sigma(\nabla_Z \cdot \mathbf{G})$$

for every scalar function σ of class C^r on N;

(17.17.3.3) $$\nabla_Z \cdot (\mathbf{G}_1 + \mathbf{G}_2) = \nabla_Z \cdot \mathbf{G}_1 + \nabla_Z \cdot \mathbf{G}_2$$

for two liftings $\mathbf{G}_1$, $\mathbf{G}_2$ to E of the *same* mapping $f: \mathrm{N} \to \mathrm{M}$ of class C^r;

(17.17.3.4) $$\nabla_Z \cdot (\sigma \mathbf{G}) = (\theta_Z \cdot \sigma)\mathbf{G} + \sigma(\nabla_Z \cdot \mathbf{G})$$

for every scalar function σ of class C^r on N.

In the particular case where $\mathrm{N} = \mathrm{M}$ and $\pi \circ \mathbf{G} = 1_{\mathrm{M}}$ (that is, where $\mathbf{G}$ is a *section* of E), we see that for each C^∞ vector field X on M, the operator ∇_X is a real *differential operator* of order $\leqq 1$ from E to E.

18. CONNECTIONS ON A DIFFERENTIAL MANIFOLD

(17.18.1) Given a differential manifold M, a *connection* (or *linear connection*) *on* M is (by abuse of language) a connection *in the tangent vector bundle* T(M), that is to say an M-morphism of $\mathrm{T(M)} \oplus \mathrm{T(M)}$ into T(T(M)) satisfying the conditions of (17.16.3), with E replaced by T(M). Given such a connection $\mathbf{C}$ we define, for each vector $\mathbf{h}_x \in \mathrm{T}_x(\mathrm{M})$ and each vector field Y on M, the *covariant derivative* (relative to $\mathbf{C}$) $\nabla_{\mathbf{h}_x} \cdot Y$ of Y at the point x in the direction of $\mathbf{h}_x$ (17.17.2.1).

Hence, for each vector field X on M, we have a differential operator $Y \mapsto \nabla_X \cdot Y$ from T(M) to T(M) (17.17.3), with the following properties:

(17.18.1.1) $$\nabla_{X_1+X_2} \cdot Y = \nabla_{X_1} \cdot Y + \nabla_{X_2} \cdot Y,$$

(17.18.1.2) $$\nabla_{\sigma X} \cdot Y = \sigma(\nabla_X \cdot Y),$$

(17.18.1.3) $$\nabla_X \cdot (Y_1 + Y_2) = \nabla_X \cdot Y_1 + \nabla_X \cdot Y_2,$$

(17.18.1.4) $$\nabla_X \cdot (\sigma Y) = (\theta_X \cdot \sigma) Y + \sigma(\nabla_X \cdot Y),$$

where σ is any C^∞ scalar function on M. In particular, if U is an open subset of M over which T(M) is trivializable, so that the $\mathscr{E}(U)$-module $\mathscr{T}_0^1(U)$ of C^∞ vector fields on U admits a *basis* $Y_1, \ldots, Y_n$, then the above formulas show that a knowledge of the vector fields $\nabla_{Y_i} \cdot Y_j$ on U for $1 \leqq i, j \leqq n$ determines $\nabla_X \cdot Y$ completely for any two vector fields X, $Y \in \mathscr{T}_0^1(U)$.

In particular, if (U, φ, n) is a chart of M, and if $(X_i)_{1 \leqq i \leqq n}$ is the basis of $\mathscr{T}_0^1(U)$ over $\mathscr{E}(U)$ associated with this chart (16.15.4.2), then it follows from (17.17.2.3) that

(17.18.1.5) $$\nabla_{X_j} \cdot X_k = \sum_i \Gamma_{jk}^i X_i,$$

which shows that if two connections give rise to the *same* covariant derivative, they are *identical* (cf. Problem 2).

(17.18.2) *Let* **C** *be a connection on a differential manifold* M, *and let* $f : N \to M$ *be a mapping of class* C^m $(m \geqq 1)$. *For each pair of integers* $r \geqq 0$, $s \geqq 0$, *let* $\mathscr{R}_s^r(f)$ *denote the vector space of all* C^m *liftings of* f *to the bundle* $\mathbf{T}_s^r(M)$. *Then, for each* $z \in N$ *and each tangent vector* $\mathbf{h}_z \in T_z(N)$, *there exists a unique linear mapping* $\mathbf{G} \mapsto \nabla_{\mathbf{h}_z} \cdot \mathbf{G}$ *of* $\mathscr{R}_s^r(f)$ *into the fiber* $(\mathbf{T}_s^r(M))_{f(z)}$ *which*: (1) *for* $r = s = 0$ *coincides with* $\theta_{\mathbf{h}_z}$ (17.14.1); (2) *for* $r = 1$ *and* $s = 0$ *coincides with the covariant derivative defined in* (17.17.2.1); (3) *satisfies the following conditions*:

(17.18.2.1) $$\nabla_{\mathbf{h}_z} \cdot (\mathbf{G}' \otimes \mathbf{G}'') = (\nabla_{\mathbf{h}_z} \cdot \mathbf{G}') \otimes \mathbf{G}''(z) + \mathbf{G}'(z) \otimes (\nabla_{\mathbf{h}_z} \cdot \mathbf{G}'')$$

for $\mathbf{G}' \in \mathscr{R}_{s'}^{r'}(f)$ *and* $\mathbf{G}'' \in \mathscr{R}_{s''}^{r''}(f)$;

(17.18.2.2) $$\theta_{\mathbf{h}_z} \cdot \langle \mathbf{G}, \mathbf{G}^* \rangle = \langle \nabla_{\mathbf{h}_z} \cdot \mathbf{G}, \mathbf{G}^*(z) \rangle + \langle \mathbf{G}(z), \nabla_{\mathbf{h}_z} \cdot \mathbf{G}^* \rangle$$

for $\mathbf{G} \in \mathscr{R}_s^r(f)$ *and* $\mathbf{G}^* \in \mathscr{R}_r^s(f)$.

The proof follows that of (17.14.6) step by step, replacing the vector field X by the vector $\mathbf{h}_z$, and tensor fields on M by liftings of f. (It is also possible to obtain (17.14.6) and (17.18.2) simultaneously as corollaries of the same algebraic lemma: see Problem 1.)

(17.18.3) If now Z is a C^m vector field on N, we define $\nabla_Z \cdot \mathbf{G}$ as in (17.17.3) for a lifting $\mathbf{G} \in \mathscr{R}_s^r(f)$, by putting $(\nabla_Z \cdot \mathbf{G})(z) = \nabla_{Z(z)} \cdot \mathbf{G}$ for all $z \in N$; it is a lifting of f to $\mathbf{T}_s^r(M)$, of class C^{m-1}. The formulas of (17.17.3) remain valid without any change.

(17.18.4) Consider the particular case where $M = N$ and f is the identity mapping. The liftings of f are then *tensor fields* on M. If $\mathbf{U} \in \mathscr{T}^r_s(M)$ is a tensor field, the mapping

$$(\mathbf{V}^*, X) \mapsto \langle \nabla_X \cdot \mathbf{U}, \mathbf{V}^* \rangle$$

of $\mathscr{T}^s_r(M) \times \mathscr{T}^1_0(M)$ into $\mathscr{E}(M)$ is $\mathscr{E}(M)$-*bilinear* by virtue of (17.17.3.2). Consequently this morphism defines a *tensor field on* M, belonging to $\mathscr{T}^r_{s+1}(M)$, which is called the *covariant derivative of* $\mathbf{U}$ (relative to the connection $\mathbf{C}$) and is denoted by $\nabla \mathbf{U}$ or $\nabla_{\mathbf{C}} \mathbf{U}$. Thus we have

(17.18.4.1) $$\langle \nabla \mathbf{U}, \mathbf{V}^* \otimes X \rangle = \langle \nabla_X \cdot \mathbf{U}, \mathbf{V}^* \rangle$$

and it is clear that $\mathbf{U} \mapsto \nabla \mathbf{U}$ is a *differential operator* of order $\leqq 1$ from $\mathbf{T}^r_s(M)$ to $\mathbf{T}^r_{s+1}(M)$. If $\sigma \in \mathscr{E}(M)$ is any scalar function, then

(17.18.4.2) $$\nabla \sigma = d\sigma$$

by virtue of (17.14.1.1) and (17.18.2). Furthermore, we have

(17.18.4.3) $$\nabla(\sigma \mathbf{U}) = \sigma(\nabla \mathbf{U}) + \mathbf{U} \otimes d\sigma.$$

For with the notation introduced above, we have

$$\langle \nabla_X \cdot (\sigma \mathbf{U}), \mathbf{V}^* \rangle = \sigma \langle \nabla_X \cdot \mathbf{U}, \mathbf{V}^* \rangle + (\theta_X \cdot \sigma) \langle \mathbf{U}, \mathbf{V}^* \rangle$$

by use of (17.17.3.4); but $\theta_X \cdot \sigma = \langle d\sigma, X \rangle$, and

$$\langle d\sigma, X \rangle \langle \mathbf{U}, \mathbf{V}^* \rangle = \langle \mathbf{U} \otimes d\sigma, \mathbf{V}^* \otimes X \rangle$$

by the definition of duality in tensor products (A.11.1.4); the formula (17.18.4.3) now follows.

(17.18.5) *Let* M *be a differential manifold and* N *a closed submanifold of* M. *Then every connection on* N *is the restriction of a connection on* M.

There exists a locally finite denumerable covering of M by open sets which are domains of definition of charts of M for which the conditions of (16.8.1) are satisfied for the submanifold N. Using a partition of unity subordinate to this covering and (17.16.7), we reduce to the situation in which N is an open subset of $\mathbf{R}^m$ and $M = N \times P$, where P is an open subset of $\mathbf{R}^{n-m}$, so that T(N) may be identified with $N \times \mathbf{R}^m$, T(P) with $P \times \mathbf{R}^{n-m}$, and T(M) with $T(N) \times T(P)$. Then if $y \in N$ and $\mathbf{h}', \mathbf{k}' \in \mathbf{R}^m$, the connection $\mathbf{C}'$ given on N may be written (17.16.2.1) in the form

$$\mathbf{C}'_y(\mathbf{h}', \mathbf{k}') = (\mathbf{h}', -\Gamma'_y(\mathbf{h}', \mathbf{k}')),$$

where Γ'_y is a bilinear mapping of $\mathbf{R}^m \times \mathbf{R}^m$ into $\mathbf{R}^m$, and $y \mapsto \Gamma'_y$ is of class C^∞ on N. We then take for $\mathbf{C}$ the connection on M given by

$$\mathbf{C}_x(\mathbf{h}, \mathbf{k}) = (\mathbf{h}, -\Gamma'_{\mathrm{pr}_1 x}(\mathrm{pr}_1 \mathbf{h}, \mathrm{pr}_1 \mathbf{k})),$$

where $x \in$ M and $\mathbf{h}, \mathbf{k} \in \mathbf{R}^n$ (considered as the product $\mathbf{R}^m \times \mathbf{R}^{n-m}$).

(17.18.6) Let f: M′ → M be a *local diffeomorphism* (16.5.6) and let $\mathbf{C}$ be a connection on M. Then there exists a unique connection $\mathbf{C}'$ on M′ such that, for each open subset U′ of M′ for which $f \mid$ U′ is a diffeomorphism of U′ onto the open subset f(U′) of M, the restriction of $\mathbf{C}$ to f(U′) is the image under f of the restriction of $\mathbf{C}'$ to U′ (17.16.9). For this requirement determines $\mathbf{C}'_{x'}$ at every point x' of M, because if U′, V′ are two open neighborhoods of x' such that $f \mid$ U′ and $f \mid$ V′ are diffeomorphisms onto f(U′) and f(V′), respectively, then $f \mid$(U′ ∩ V′) is a diffeomorphism onto f(U′ ∩ V′). The connection $\mathbf{C}'$ is said to be the *inverse image* of $\mathbf{C}$ under f.

This remark may be applied in particular to the situation in which (M′, M, f) is a *covering* of M. If x'_1, x'_2 are two distinct points of M′ such that $f(x'_1) = f(x'_2) = x \in$ M, then there exist disjoint open neighborhoods U'_1 of x'_1 and U'_2 of x'_2 in M′ such that $f(U'_1) = f(U'_2)$, and a diffeomorphism $g : U'_1 \to U'_2$ such that $f \mid U'_1 = g \circ (f \mid U'_2)$. With the notation used above, if $\mathbf{C}'_1$ and $\mathbf{C}'_2$ are the restrictions of $\mathbf{C}'$ to U'_1 and U'_2, respectively, then $\mathbf{C}'_2$ is the image of $\mathbf{C}'_1$ by g. Conversely, if a connection $\mathbf{C}'$ on the covering space M′ has this property, it is immediately seen that we may define a connection $\mathbf{C}$ on M by taking (with the notation of the beginning of this subsection) the restriction of $\mathbf{C}$ to f(U′) to be the image under $f \mid$ U′ of the restriction of $\mathbf{C}'$ to U′. The above condition on $\mathbf{C}'$ then ensures that $\mathbf{C}$ is unambiguously defined.

PROBLEMS

1. Let A, A′ be two commutative rings; ρ : A → A′ a surjective ring homomorphism; E, F two free A-modules with bases $(e_i)_{1 \le i \le n}$ and $(f_i)_{1 \le i \le n}$; Φ a bilinear form on E × F such that $\Phi(e_i, f_j) = \delta_{ij}$. Also let E′, F′ be two free A′-modules with bases $(e'_i)_{1 \le i \le n}$ and $(f'_i)_{1 \le i \le n}$, and Φ' a bilinear form on E′ × F′ such that $\Phi'(e'_i, f'_j) = \delta_{ij}$. Let $\rho_0^1 : \mathrm{E} \to \mathrm{E}'$ be the A-module homomorphism defined by $\rho_0^1(e_i) = e'_i$ $(1 \le i \le n)$, and $\rho_1^0 : \mathrm{F} \to \mathrm{F}'$ the A-module homomorphism defined by $\rho_1^0(f_i) = f'_i$ $(1 \le i \le n)$. Suppose also that we are given a mapping L : A → A′ such that $\mathrm{L}(ab) = \mathrm{L}(a)\rho(b) + \rho(a)\mathrm{L}(b)$, and a linear mapping $\mathrm{L}_0^1 : \mathrm{E} \to \mathrm{E}'$ such that $\mathrm{L}_0^1(ax) = \mathrm{L}(a)\rho_0^1(x) + \rho(a)\mathrm{L}_0^1(x)$ for all $a, b \in$ A and $x \in$ E.

   (a) Show that there exists a unique mapping $\mathrm{L}_1^0 : \mathrm{F} \to \mathrm{F}'$ such that

$$\mathrm{L}_1^0(ay) = \mathrm{L}(a)\rho_1^0(y) + \rho(a)\mathrm{L}_1^0(y)$$

for all $a \in A$ and $y \in F$, and

$$L(\Phi(x, y)) = \Phi'(L_0^1(x), \rho_1^0(y)) + \Phi'(\rho_0^1(x), L_1^0(y))$$

for all $x \in E$ and $y \in F$.

(b) Show that for each pair (p, q) of integers $\geqq 0$ there exists a unique mapping

$$L_q^p : E^{\otimes p} \otimes F^{\otimes q} \to E'^{\otimes p} \otimes F'^{\otimes q}$$

such that

$$L_q^p(az) = L(a)\rho_q^p(z) + \rho(a)L_q^p(z)$$

for $a \in A$ and $z \in E^{\otimes p} \otimes F^{\otimes q}$, where ρ_q^p is the canonical extension of ρ_0^1 and ρ_1^0 to the tensor product);

$$L_{q+s}^{p+r}(u \otimes v) = L_q^p(u) \otimes \rho_s^r(v) + \rho_q^p(u) \otimes L_s^r(v)$$

for $u \in E^{\otimes p} \otimes F^{\otimes q}$ and $v \in E^{\otimes r} \otimes F^{\otimes s}$; and

$$L(\Phi(u, v^*)) = \Phi'(L_q^p(u), \rho_p^q(v^*)) + \Phi'(\rho_q^p(u), L_p^q(v^*))$$

for $u \in E^{\otimes p} \otimes F^{\otimes q}$ and $v^* \in E^{\otimes q} \otimes F^{\otimes p}$, where Φ and Φ' are the canonical extensions of the bilinear forms to the tensor product. (Follow the proof of (17.14.6).)

2. Let E be a vector bundle over M, and let $\text{Diff}_1(E)$ denote the $\mathscr{E}(M)$-module of differential operators of order $\leqq 1$ from E to E. Show that every $\mathscr{E}(M)$-linear mapping $X \mapsto P_X$ of $\mathscr{T}_0^1(M)$ into $\text{Diff}_1(E)$ such that $P_X \cdot (\sigma \mathbf{Y}) = (\theta_X \cdot \sigma)\mathbf{Y} + \sigma(P_X \cdot \mathbf{Y})$ for all $\sigma \in \mathscr{E}(M)$ and all C^∞-sections $\mathbf{Y}$ of E, is of the form $X \mapsto \nabla_X$ relative to a unique connection in E. (Show first that if X vanishes in an open set U, then $P_X | U = 0$.)

3. Generalize the result of (17.18.1) to linear connections in an arbitrary vector bundle E over M. Consider in particular a C^∞-section $\mathbf{G}^*$ of the dual E^* of E, and associate with it the scalar function on E given by $\mathbf{u}_x \mapsto \Omega(\mathbf{u}_x) = \langle \mathbf{u}_x, \mathbf{G}^*(x) \rangle$. Show that

$$\langle \mathbf{u}_x, \nabla_{\mathbf{h}_x} \cdot \mathbf{G}^* \rangle = \theta_{\mathbf{C}_x(\mathbf{h}_x, \mathbf{u}_x)} \cdot \Omega.$$

4. With the notation of Section 16.19, Problem 11s, how that a linear connection in E is a mapping $\mathbf{C} : E \times_B T(B) \to T(E)$ such that $\mu \circ \mathbf{C} = 1_{E \times_B T(B)}$, and such that $\mathbf{C}$ is a bundle morphism of $E \times_B T(B)$ into T(E) both as bundles over E and as bundles over T(B).

5. With the notation of (17.18.4), show that $\nabla(c_j^i \mathbf{U}) = c_j^i(\nabla \mathbf{U})$ for any contraction c_j^i.

19. THE COVARIANT EXTERIOR DIFFERENTIAL

(17.19.1) The formula (17.15.3.6) enables us to calculate at each point $x \in M$ the value $\langle d(\omega(x), \mathbf{h}_x \wedge \mathbf{k}_x \rangle$ of the exterior differential of a 1-form ω by considering two C^∞ vector fields X, Y on M, such that $X(x) = \mathbf{h}_x$ and $Y(x) = \mathbf{k}_x$, and calculating the value of the right-hand side of (17.15.3.6) at the point x. It is remarkable that although each of the terms on the right-hand side depends on the values of the fields X, Y *in a neighborhood of* x and not merely *at the point* x, yet the left-hand side does not. We shall see that an analogous phenomenon presents itself when the Lie derivative is replaced by the covariant derivative relative to a connection.

(17.19.2) In detail, let E be a vector bundle over M, and let f be a C^∞-mapping of a differential manifold N into M. If $\boldsymbol{\omega} : \bigwedge^p T(N) \to E$ is a C^∞-mapping such that $(f, \boldsymbol{\omega})$ is a *vector bundle morphism* (16.15.2), then $\boldsymbol{\omega}$ is said to be a C^∞ *differential p-form on* N *with values in* E (relative to the mapping f) (cf. Section 20.6, Problem 2). If $Z_1, Z_2, \ldots, Z_p$ are C^∞ vector fields on N, then for each point $z \in N$ the element $\boldsymbol{\omega}(Z_1(z) \wedge Z_2(z) \wedge \cdots \wedge Z_p(z))$ belongs by definition to $E_{f(z)}$; moreover, it is immediate that the mapping

$$z \mapsto \boldsymbol{\omega}(Z_1(z) \wedge Z_2(z) \wedge \cdots \wedge Z_p(z))$$

is a C^∞-*lifting* of f to E, which we denote by

$$\boldsymbol{\omega} \cdot (Z_1 \wedge Z_2 \wedge \cdots \wedge Z_p).$$

(17.19.3). Now suppose that we are given a connection **C** on E, and consider first the case $p = 1$. Let $\boldsymbol{\omega}$ be a differential 1-form of class C^∞ on N, with values in E, and let X, Y be two C^∞-vector fields on N. By analogy with (17.15.8.1) we form the C^∞-lifting of f to E

(17.19.3.1) $$\nabla_X \cdot (\boldsymbol{\omega} \cdot Y) - \nabla_Y \cdot (\boldsymbol{\omega} \cdot X) - \boldsymbol{\omega} \cdot [X, Y].$$

We shall show that the value of this mapping at a point $z \in N$ *depends only on the values* $X(z)$ *and* $Y(z)$ *of the fields* X, Y *at the point* z. For this it is enough to show that if we replace X (resp. Y) by σX (resp. σY), where σ is a *scalar* function of class C^∞ on N, the value of (17.19.3.1) is obtained by *multiplying by* $\sigma(z)$ the value for X and Y. For, reducing to the case where M, N are open subsets of $\mathbf{R}^p$, $\mathbf{R}^q$ and $E = M \times \mathbf{R}^n$, we have $X(z) = (z, \mathbf{g}(z))$ and $Y(z) = (z, \mathbf{h}(z))$, and the formula (17.17.2.2) shows that (17.19.3.1) is a *bilinear* function of the vectors $(\mathbf{g}(z), D\mathbf{g}(z))$ and $(\mathbf{h}(z), D\mathbf{h}(z))$; hence, by virtue of (8.1.4), the condition stated above is necessary and sufficient for this function not to depend on $D\mathbf{g}(z)$ (resp. $D\mathbf{h}(z)$).

Now we have

$$\nabla_{\sigma X} \cdot (\boldsymbol{\omega} \cdot Y) = \sigma \nabla_X \cdot (\boldsymbol{\omega} \cdot Y)$$

by (17.17.3.2), and

$$\boldsymbol{\omega} \cdot (\sigma X) = \sigma(\boldsymbol{\omega} \cdot X),$$

$$\nabla_Y(\boldsymbol{\omega} \cdot (\sigma X)) = (\theta_Y \cdot \sigma)(\boldsymbol{\omega} \cdot X) + \sigma(\nabla_Y \cdot (\boldsymbol{\omega} \cdot X))$$

by (17.17.3.4), and finally

$$[\sigma X, Y] = -(\theta_Y \cdot \sigma)X + \sigma[X, Y],$$
$$\boldsymbol{\omega} \cdot [\sigma X, Y] = -(\theta_Y \cdot \sigma)(\boldsymbol{\omega} \cdot X) + \sigma(\boldsymbol{\omega} \cdot [X, Y])$$

by (17.14.4.2). Hence the result.

Since (17.19.3.1) is an alternating bilinear function of (X, Y), there exists a unique C^∞ *differential* 2-*form on* N *with values in* E, called the *covariant exterior differential* of $\boldsymbol{\omega}$ (relative to **C**) and denoted by $\mathbf{d}\boldsymbol{\omega}$, such that for any two C^∞ vector fields X, Y on N we have

(17.19.3.2) $$\mathbf{d}\boldsymbol{\omega} \cdot (X \wedge Y) = \nabla_X \cdot (\boldsymbol{\omega} \cdot Y) - \nabla_Y \cdot (\boldsymbol{\omega} \cdot X) - \boldsymbol{\omega} \cdot [X, Y].$$

(17.19.4) This result generalizes easily to differential p-forms on N with values in E, where $p > 1$. We take the analog of the formula (17.15.3.5) by proving that if $\boldsymbol{\omega}$ is a C^∞ differential p-form on N with values in E and if $X_0, X_1, \dots, X_p$ are $p + 1$ C^∞ vector fields on N, then the C^∞-lifting of f to E

(17.19.4.1)

$$\sum_{j=0}^{p} (-1)^j \nabla_{X_j} \cdot (\boldsymbol{\omega} \cdot (X_0 \wedge \cdots \wedge \hat{X}_j \wedge \cdots \wedge X_p)$$
$$+ \sum_{0 \leq i < j \leq p} (-1)^{i+j} \boldsymbol{\omega} \cdot ([X_i, X_j] \wedge X_0 \wedge \cdots \wedge \hat{X}_i \wedge \cdots \wedge \hat{X}_j \wedge \cdots \wedge X_p)$$

has at each point $z \in$ N a value *which depends only on the values $X_j(z)$ of the vector fields X_j at z* $(0 \leq j \leq p)$. The method of proof is exactly the same: we replace each X_j successively by σX_j. In this way we establish the existence and uniqueness of a C^∞ differential $(p + 1)$-form $\mathbf{d}\boldsymbol{\omega}$ on N, with values in E, such that $\mathbf{d}\boldsymbol{\omega} \cdot (X_0 \wedge \cdots \wedge X_p)$ has as its value (17.19.4.1); $\mathbf{d}\boldsymbol{\omega}$ is said to be the *covariant exterior differential of* $\boldsymbol{\omega}$ (relative to **C**).

Finally, since by convention $\bigwedge^0 T(N) = N \times \mathbf{R}$ (16.16.2), a C^∞ differential 0-form on N with values in E is identified with a C^∞ *lifting* **G** of f to E. For each C^∞ vector field X on N, the value of $\nabla_X \cdot \mathbf{G}$ at a point $z \in$ N depends only on $X(z)$ by definition (17.17.3). Hence there exists a unique differential 1-form of class C^∞ on N, with values in E, which is called the *covariant exterior differential* of **G** and is denoted by $\mathbf{dG}$, such that

(17.19.4.2) $$\mathbf{dG} \cdot X = \nabla_X \cdot \mathbf{G}.$$

(17.19.5) With the notation and hypotheses of (17.19.2), let $u : \mathrm{N}_1 \to \mathrm{N}$ be a C^∞-mapping, and consider the composite mapping $f_1 = f \circ u : \mathrm{N}_1 \to \mathrm{M}$. It is clear that the pair $(f_1, \boldsymbol{\omega}_1)$, where $\boldsymbol{\omega}_1 = \boldsymbol{\omega} \circ \bigwedge^p \mathrm{T}(u)$, is a morphism of vector bundles. The differential p-form $\boldsymbol{\omega}_1$ on N_1 with values in E is called the *inverse image* of $\boldsymbol{\omega}$ by u.

Now suppose that we are given a connection **C** on E. Then the covariant exterior differentials relative to f and to $f_1 = f \circ u$ satisfy the relation

(17.19.5.1) $$\mathbf{d}(\boldsymbol{\omega} \wedge \bigwedge^p \mathrm{T}(u)) = (\mathbf{d}\boldsymbol{\omega}) \circ \bigwedge^{p+1} \mathrm{T}(u)$$

for any C^∞ exterior differential p-form $\boldsymbol{\omega}$ on N with values in E. We shall give the proof only for $p = 0$ and $p = 1$ (cf. Problem 1). For $p = 0$, in view of the definition (17.19.4.2), the formula (17.19.5.1) reduces to (17.17.2.6). For $p = 1$, since the question is local with respect to N, N_1, and M, we may assume that M, N, N_1 are open subsets of $\mathbf{R}^m$, $\mathbf{R}^n$, $\mathbf{R}^{n_1}$, respectively, and that $\mathrm{E} = \mathrm{M} \times \mathbf{R}^q$. Then $\boldsymbol{\omega}$ may be written as $(z, \mathbf{h}) \mapsto (f(z), A(z) \cdot \mathbf{h})$, where $z \mapsto A(z)$ is a C^∞-mapping of N into $\mathscr{L}(\mathbf{R}^n; \mathbf{R}^q)$; this implies that $\mathrm{D}A(z)$ is an element of $\mathscr{L}_2(\mathbf{R}^n; \mathbf{R}^q)$ ((5.7.8) and (8.12)). It now follows from the formula (17.17.2.2), the definition (17.19.3.2), and the rules for calculating derivatives in vector spaces (Chapter VIII) that $\mathbf{d}\boldsymbol{\omega}$ is the mapping

(17.19.5.2)
$$\begin{aligned}(z, \mathbf{h} \wedge \mathbf{k}) \mapsto (f(z), \mathrm{D}A(z) \cdot ((\mathbf{h}, \mathbf{k}) - (\mathbf{k}, \mathbf{h})) \\ + \Gamma_{f(z)}(\mathrm{D}f(z) \cdot \mathbf{h}, A(z) \cdot \mathbf{k}) - \Gamma_{f(z)}(\mathrm{D}f(z) \cdot \mathbf{k}, A(z) \cdot \mathbf{h})).\end{aligned}$$

(Incidentally, this calculation provides another proof of the fact that (17.19.3.1) depends only on the values $X(z)$ and $Y(z)$.) Likewise, $\boldsymbol{\omega}_1$ may be written as $(z_1, \mathbf{h}_1) \mapsto (f_1(z_1), A_1(z_1) \cdot \mathbf{h}_1)$, where $f_1 = f \circ u$ and

$$A_1(z_1) = A(u(z_1)) \circ \mathrm{D}u(z_1).$$

We have then $\mathrm{D}f_1(z_1) \cdot \mathbf{h}_1 = \mathrm{D}f(u(z_1)) \cdot (\mathrm{D}u(z_1) \cdot \mathbf{h}_1)$; also the mapping $\mathbf{h}_1 \mapsto \mathrm{D}A_1(z_1) \cdot (\mathbf{h}_1, \mathbf{k}_1)$ is the derivative of $z_1 \mapsto A_1(z_1) \cdot \mathbf{k}_1$; consequently, in view of the definition of A_1, we have

$$\begin{aligned}\mathrm{D}A_1(z_1) \cdot (\mathbf{h}_1, \mathbf{k}_1) = \mathrm{D}A(u(z_1)) \cdot (\mathrm{D}u(z_1) \cdot \mathbf{h}_1, \mathrm{D}u(z_1) \cdot \mathbf{k}_1) \\ + A(u(z_1)) \cdot (\mathrm{D}^2 u(z_1) \cdot (\mathbf{h}_1, \mathbf{k}_1)).\end{aligned}$$

Inserting these values of $\mathrm{D}f_1(z_1)$ and $\mathrm{D}A_1(z_1)$ into the expression for $\mathbf{d}\boldsymbol{\omega}_1$ analogous to (17.19.5.2), we obtain, using the symmetry of $\mathrm{D}^2 u(z_1)$ (8.12.2),

$$(\mathbf{d}\boldsymbol{\omega}_1)(z_1, \mathbf{h}_1 \wedge \mathbf{k}_1) = (\mathbf{d}\boldsymbol{\omega})(z, \mathbf{h} \wedge \mathbf{k}),$$

where $z = u(z_1)$, $\mathbf{h} = \mathrm{D}u(z_1) \cdot \mathbf{h}_1$, $\mathbf{k} = \mathrm{D}u(z_1) \cdot \mathbf{k}_1$. This proves (17.19.5.1) in the case $p = 1$.

PROBLEMS

1. Prove the formula (17.19.5.1) for arbitrary p, as follows. To calculate the value of the left-hand side at a $(p+1)$-vector $\mathbf{k}_0 \wedge \mathbf{k}_1 \wedge \cdots \wedge \mathbf{k}_p \in \bigwedge^{p+1} \mathrm{T}_{z_1}(\mathrm{N}_1)$, consider separately the cases where the vectors $\mathrm{T}_{z_1}(u) \cdot \mathbf{k}_j$ $(0 \leqq j \leqq p)$ are linearly independent or linearly dependent. In the first case, reduce to the situation where N_1 is a submanifold of N of dimension $p+1$ by use of a chart, and use the fact that in calculating the value of (17.19.4.1) we may assume that the X_j are fields of tangent vectors to N_1. In the second case, we may assume that $\mathrm{T}_{z_1}(u) \cdot \mathbf{k}_0 = 0$ and that the fields X_j such that $X_j(z_1) = \mathbf{k}_j$ for $0 \leqq j \leqq p$ are such that $[X_j, X_h] = 0$; use the formula (17.17.2.6).

2. Let M be a differential manifold. Given two integers $p \geqq 1$, $q \geqq 0$, and M-morphisms $\boldsymbol{P}: \bigwedge^{p} \mathrm{T(M)} \to \mathrm{T(M)}$, $\boldsymbol{Q}: \bigwedge^{q} \mathrm{T(M)} \to \mathrm{T(M)}$, we define an M-morphism

$$\boldsymbol{P} \barwedge \boldsymbol{Q}: \bigwedge^{p+q-1} \mathrm{T(M)} \to \mathrm{T(M)}$$

by the formula

$$(\boldsymbol{P} \barwedge \boldsymbol{Q}) \cdot (X_1 \wedge X_2 \wedge \cdots \wedge X_{p+q-1})$$
$$= \frac{1}{(p-1)!q!} \sum_{\sigma \in \mathfrak{S}_{p+q-1}} \varepsilon_\sigma \boldsymbol{P} \cdot (\boldsymbol{Q} \cdot (X_{\sigma(1)} \wedge \cdots \wedge X_{\sigma(q)}) \wedge X_{\sigma(q+1)} \wedge \cdots \wedge X_{\sigma(p+q-1)}).$$

(When $q = 0$, $\boldsymbol{Q}$ is identified with a vector field X, and $\boldsymbol{Q} \cdot (X_{\sigma(1)} \wedge \cdots \wedge X_{\sigma(q)})$ has to be replaced by X.)

Likewise, for each scalar-valued differential p-form α on M, we define a $(p+q-1)$-form $\alpha \barwedge \boldsymbol{Q}$ by the formula

$$\langle \alpha \barwedge \boldsymbol{Q}, X_1 \wedge X_2 \wedge \cdots \wedge X_{p+q-1} \rangle$$
$$= \frac{1}{(p-1)!q!} \sum_{\sigma \in \mathfrak{S}_{p+q-1}} \varepsilon_\sigma \langle \alpha, \boldsymbol{Q} \cdot (X_{\sigma(1)} \wedge \cdots \wedge X_{\sigma(q)}) \wedge X_{\sigma(q+1)} \wedge \cdots \wedge X_{\sigma(p+q-1)} \rangle$$

(with the same convention for $q = 0$). Extend this definition to the case $p = 0$ by putting $f \barwedge \boldsymbol{Q} = 0$ for all functions $f \in \mathscr{E}(\mathrm{M})$. Show that:

(a) If $p \geqq 1$, $q \geqq 1$ and if α is a scalar-valued r-form, we have

$$(\alpha \barwedge \boldsymbol{P}) \barwedge \boldsymbol{Q} - \alpha \barwedge (\boldsymbol{P} \barwedge \boldsymbol{Q}) = (-1)^{(p-1)(q-1)}((\alpha \barwedge \boldsymbol{Q}) \barwedge \boldsymbol{P} - \alpha \barwedge (\boldsymbol{Q} \barwedge \boldsymbol{P})).$$

(b) $1_{\mathrm{T(M)}} \barwedge \boldsymbol{Q} = \boldsymbol{Q}$; $\boldsymbol{P} \barwedge 1_{\mathrm{T(M)}} = p\boldsymbol{P}$.

(c) If $\boldsymbol{u}, \boldsymbol{v}$ are endomorphisms of T(M) (1-forms with values in T(M)), or tensor fields of type (1, 1), then

$$\boldsymbol{u} \barwedge \boldsymbol{v} = \boldsymbol{u} \circ \boldsymbol{v}, \qquad \boldsymbol{u} \barwedge X = \boldsymbol{u} \cdot X$$

for any vector field X.

3. The notation is as in Problem 2.

(a) The mapping $\alpha \mapsto \alpha \barwedge Q$ of the exterior algebra $\mathscr{A}$ of scalar-valued differential forms on M into itself is an *antiderivation* of degree $q-1$ which is zero on $\mathscr{E}_0$. This antiderivation is denoted by i_Q. Conversely, every antiderivation of $\mathscr{A}$ of degree $q-1$ which vanishes on $\mathscr{E}_0$ is of this form. (Observe that such an antiderivation is a differential operator from T(M) to $\bigwedge^q$ T(M).)
(b) The mapping

$$\mathrm{D}: \quad \alpha \mapsto (d\alpha) \barwedge Q + (-1)^q d(\alpha \barwedge Q)$$

of $\mathscr{A}$ into itself is an *antiderivation* of degree q such that

$$\mathrm{D} \circ d = (-1)^q d \circ \mathrm{D}. \tag{1}$$

(We have $\mathrm{D} = i_Q \circ d - (-1)^{q-1} d \circ i_Q$.) This antiderivation is denoted by d_Q. Conversely, every antiderivation D of $\mathscr{A}$ of degree q which satisfies condition (1) is of the form d_Q. (Observe that D is a differential operator from $\mathrm{M} \times \mathbf{R}$ into $\bigwedge^q$ T(M).)
(c) Every antiderivation D of $\mathscr{A}$ of degree r can be written uniquely as $\mathrm{D} = i_P + d_Q$, where P (resp. Q) is an M-morphism of $\bigwedge^{r+1}$ T(M) (resp. $\bigwedge^r$ T(M)) into T(M). (Determine Q by the condition that d_Q coincides with D on $\mathscr{E}_0$.) (*Fröhlicher–Nijenhuis theorem.*)
(d) If Q is a vector field X (i.e., if $q = 0$), then i_Q coincides with the interior product i_X. If Q is an endomorphism u of T(M) (i.e., if $q = 1$), then

$$\langle i_u \cdot \alpha, X_1 \wedge \cdots \wedge X_p \rangle = \sum_{j=1}^{p} \langle \alpha, X_1 \wedge \cdots \wedge (u \cdot X_j) \wedge \cdots \wedge X_p \rangle$$

and, in particular, $i_{1_{\mathrm{T(M)}}} \cdot \alpha = p\alpha$.

If Q is a vector field X, then d_Q coincides with the Lie derivative θ_X. We have $d_{1_{\mathrm{T(M)}}} = d$ (exterior differentiation in $\mathscr{A}$).
(e) With the same notation, there exists a unique M-morphism $[P, Q]$ of $\bigwedge^{p+q}$ T(M) into T(M) such that

$$[d_P, d_Q] = d_{[P,Q]}. \tag{2}$$

We have

$$[Q, P] = (-1)^{pq+1}[P, Q],$$
$$[1_{\mathrm{T(M)}}, Q] = 0,$$
$$(-1)^{pr}[P, [Q, R]] + (-1)^{qp}[Q, [R, P]] + (-1)^{rq}[R, [P, Q]] = 0.$$

(f) If $P = X$ and $Q = Y$ are vector fields, then $[P, Q]$ is the usual Lie bracket $[X, Y]$. For each Q, $[X, Q]$ is the Lie derivative $\theta_X \cdot Q$ (where Q is identified with a tensor field of type $(1, q)$). If $P = u$ and $Q = v$ are endomorphisms of T(M), then

$$[u, v] \cdot (X \wedge Y) = [u \cdot X, v \cdot Y] + [v \cdot X, u \cdot Y] + u \cdot (v \cdot [X, Y]) + v \cdot (u \cdot [X, Y])$$
$$- u \cdot [v \cdot X, Y] - v \cdot [u \cdot X, Y] - u \cdot [X, v \cdot Y] - v \cdot [X, u \cdot Y]$$

and, in particular,

$$\tfrac{1}{2}[u, u] \cdot (X \wedge Y) = [u \cdot X, u \cdot Y] + u \cdot (u \cdot [X, Y]) - u \cdot [u \cdot X, Y] - u \cdot [X, u \cdot Y]$$

(the *Nijenhuis torsion* of u).
(g) Show that

$$[i_Q, d_P] = d \barwedge Q + (-1)^p i_{[P,Q]}.$$

4. Let M be a pure differential manifold of dimension n. Consider the canonical exact sequence (Section **16.19**, Problem 11)

$$0 \to \mathrm{T(M)} \times_{\mathrm{M}} \mathrm{T(M)} \xrightarrow{\lambda} \mathrm{T(T(M))} \xrightarrow{\mu} \mathrm{T(M)} \times_{\mathrm{M}} \mathrm{T(M)} \to 0.$$

The endomorphism $\boldsymbol{J} = \lambda \circ \mu$ is called the *vertical endomorphism* of T(T(M)). Its local expression relative to a chart of M at a point x is

$$(x, \mathbf{h}_x, \mathbf{u}_x, \mathbf{k}_{\mathbf{h}_x}) \mapsto (x, \mathbf{h}_x, 0, \mathbf{u}_x).$$

It is a T(M)-morphism for the vector bundle structure on T(T(M)) with projection $o_{\mathrm{T(M)}}$, but not for the bundle structure with projection $\mathrm{T}(o_{\mathrm{M}})$. It is of rank n at every point, and we have $\boldsymbol{J} \circ \boldsymbol{J} = 0$.

(a) With the notation of Problems 2 and 3, show that

$$\boldsymbol{J} \barwedge \boldsymbol{J} = 0, \qquad [\boldsymbol{J}, \boldsymbol{J}] = 0,$$
$$[i_{\boldsymbol{J}}, d_{\boldsymbol{J}}] = 0, \qquad [d, d_{\boldsymbol{J}}] = 0, \qquad d_{\boldsymbol{J}} \circ d_{\boldsymbol{J}} = 0.$$

(b) Show that, for all vector fields Z, Z' on T(M), we have

$$[\boldsymbol{J} \cdot Z, \boldsymbol{J} \cdot Z'] = \boldsymbol{J} \cdot [\boldsymbol{J} \cdot Z, Z'] + \boldsymbol{J} \cdot [Z, \boldsymbol{J} \cdot Z'],$$
$$[i_{\boldsymbol{J}}, i_Z] = -i_{\boldsymbol{J} \cdot Z}.$$

20. CURVATURE AND TORSION OF A CONNECTION

(17.20.1) Let E be a vector bundle over M, let **C** be a connection in E, let $f: \mathrm{N} \to \mathrm{M}$ be a C^∞-mapping and $\mathbf{G}: \mathrm{N} \to \mathrm{E}$ a C^∞-lifting of f. Since $\mathbf{dG}$ is a C^∞ differential 1-form on N with values in E, we may consider the differential 2-form $\mathbf{d}(\mathbf{dG})$ on N with values in E. By contrast with the case of the exterior differential (17.15.3.1), however, $\mathbf{d}(\mathbf{dG})$ *is not identically zero* in general; but, for all $\mathbf{h}_z$, $\mathbf{k}_z$ in $\mathrm{T}_z(\mathrm{N})$, the vector $\mathbf{d}(\mathbf{dG}) \cdot (\mathbf{h}_z \wedge \mathbf{k}_z) \in \mathrm{E}_{f(z)}$ *depends only on the value* $\mathbf{G}(z) \in \mathrm{E}_{f(z)}$ of $\mathbf{G}$ at the point z (and not on its values in a neighborhood of z). To see this, let X, Y be two C^∞ vector fields on N such that $X(z) = \mathbf{h}_z$ and $Y(z) = \mathbf{k}_z$. Then the vector $\mathbf{d}(\mathbf{dG}) \cdot (\mathbf{h}_z \wedge \mathbf{k}_z)$ is by definition ((17.19.3) and (17.9.4)) the value at z of the following lifting of f to E:

(17.20.1.1) $$\nabla_X \cdot (\nabla_Y \cdot \mathbf{G}) - \nabla_Y \cdot (\nabla_X \cdot \mathbf{G}) - \nabla_{[X, Y]} \cdot \mathbf{G}.$$

By the same argument as in (17.19.3), it is enough to show that if σ is any C^∞ scalar-valued function on N, the value of (17.20.1.1) for $\sigma\mathbf{G}$ is obtained by multiplying the corresponding value for $\mathbf{G}$ by $\sigma(z)$. Now, by virtue of (17.17.3.4), we have

$$\begin{aligned}\nabla_X \cdot (\nabla_Y \cdot (\sigma\mathbf{G})) &= \nabla_X \cdot ((\theta_Y \cdot \sigma)\mathbf{G} + \sigma(\nabla_Y \cdot \mathbf{G})) \\ &= (\theta_X \cdot (\theta_Y \cdot \sigma))\mathbf{G} + (\theta_Y \cdot \sigma)(\nabla_X \cdot \mathbf{G}) \\ &\quad + (\theta_X \cdot \sigma)(\nabla_Y \cdot \mathbf{G}) + \sigma(\nabla_X \cdot (\nabla_Y \cdot \mathbf{G})).\end{aligned}$$

Interchanging X and Y in this formula, and remembering that

$$\nabla_{[X,Y]} \cdot (\sigma \mathbf{G}) = (\theta_{[X,Y]} \cdot \sigma)\mathbf{G} + \sigma(\nabla_{[X,Y]} \cdot \mathbf{G}),$$

it follows that our assertion is a consequence of the definition of $[X, Y]$ (17.14.3).

(17.20.2) The expression (17.20.1.1) is clearly a linear function of $\mathbf{G}$. Hence there is an *endomorphism* $R_z\,(\mathbf{h}_z \wedge \mathbf{k}_z)$ of $E_{f(z)}$ such that

$$\mathbf{d}(\mathbf{dG}) \cdot (\mathbf{h}_z \wedge \mathbf{k}_z) = R_z(\mathbf{h}_z \wedge \mathbf{k}_z) \cdot \mathbf{G}(z).$$

Furthermore, it is immediately seen that the mapping

(17.20.2.1) $$\boldsymbol{r}_f : \mathbf{h}_z \wedge \mathbf{k}_z \mapsto R_z(\mathbf{h}_z \wedge \mathbf{k}_z)$$

is such that $(f, \boldsymbol{r}_f)$ is a *vector bundle morphism* of $\bigwedge^2 T(N)$ into $\mathrm{Hom}(E, E) = E^* \otimes E$ (in other words, $\boldsymbol{r}_f$ is a *differential* 2-*form* on N with values in $E^* \otimes E$).

(17.20.3) Now let $u : N_1 \to N$ be a C^∞-mapping. By applying the formula (17.19.5.1) for $p = 0$ and $p = 1$ we obtain

(17.20.3.1) $$\mathbf{d}(\mathbf{d}(\mathbf{G} \circ u)) = (\mathbf{d}(\mathbf{dG})) \circ \bigwedge^2 T(u).$$

In the notation introduced in (17.20.2), this takes the form

(17.20.3.2) $$\boldsymbol{r}_{f \circ u} = \boldsymbol{r}_f \circ \bigwedge^2 T(u);$$

for if $\mathbf{G}_1 : N_1 \to E$ is any lifting of $f \circ u$, and if $z_1 \in N_1$, then there exists a lifting $\mathbf{G}$ of f to E such that $\mathbf{G}(u(z_1)) = \mathbf{G}_1(z_1)$ ((16.15.1.2) and (16.19.1)).

(17.20.4) Consider in particular the case where $N = M$ and $f = 1_M$. The M-morphism $\boldsymbol{r}_{1_M}$ is denoted simply by

$$\boldsymbol{r} : \bigwedge^2 T(M) \to E^* \otimes E$$

and is called the *curvature* M-*morphism* (or simply the *curvature*) of the connection $\mathbf{C}$ in E. Knowledge of this morphism determines all the differentials $\mathbf{d}(\mathbf{dG})$, by virtue of (17.20.3.2); in other words, with the notation of (17.20.1), if $\mathbf{G}$ is any C^∞-lifting of f, we have

(17.20.4.1) $$\nabla_X \cdot (\nabla_Y \cdot \mathbf{G}) - \nabla_Y \cdot (\nabla_X \cdot \mathbf{G}) - \nabla_{[X,Y]} \cdot \mathbf{G} = (\boldsymbol{r} \cdot ((T(f) \cdot X) \wedge (T(f) \cdot Y))) \cdot \mathbf{G}.$$

When M is an open set in $\mathbf{R}^m$ and $E = M \times \mathbf{R}^q$, then we have $\mathbf{h}_x = (x, \mathbf{h})$ and $\mathbf{k}_x = (x, \mathbf{k})$ with $\mathbf{h}, \mathbf{k} \in \mathbf{R}^m$ for all $x \in M$; if $\mathbf{u}_x = (x, \mathbf{u})$ is any element of E_x, an easy calculation using (17.17.2.2) and (17.19.5.2) gives

(17.20.4.2)
$$(\boldsymbol{r} \cdot (\mathbf{h}_x \wedge \mathbf{k}_x)) \cdot \mathbf{u}_x = (x, D\Gamma_x \cdot (\mathbf{h}, \mathbf{k}, \mathbf{u}) - D\Gamma_x \cdot (\mathbf{k}, \mathbf{h}, \mathbf{u}) + \Gamma_x(\mathbf{h}, \Gamma_x(\mathbf{k}, \mathbf{u})) - \Gamma_x(\mathbf{k}, \Gamma_x(\mathbf{h}, \mathbf{u}))).$$

(Since $x \mapsto \Gamma_x$ is a mapping of M into $\mathscr{L}_2(\mathbf{R}^m, \mathbf{R}^q; \mathbf{R}^q)$, it follows that $D\Gamma_x$ belongs to $\mathscr{L}(\mathbf{R}^m; \mathscr{L}_2(\mathbf{R}^m, \mathbf{R}^q; \mathbf{R}^q))$, identified with the space of trilinear mappings $\mathscr{L}_3(\mathbf{R}^m, \mathbf{R}^m, \mathbf{R}^q; \mathbf{R}^q)$.)

(17.20.5) Consider in particular the case where $E = T(M)$, so that $\mathbf{C}$ is a *linear connection on* M. Then the curvature morphism $\boldsymbol{r}$ of $\mathbf{C}$ defines a bilinear M-morphism $(\mathbf{h}_x, \mathbf{k}_x) \mapsto \boldsymbol{r} \cdot (\mathbf{h}_x \wedge \mathbf{k}_x)$ of $T(M) \oplus T(M)$ into $\mathbf{T}_1^1(M)$ and hence may be identified with a *tensor field* $\mathbf{r}$ *of type* (1, 3) (16.18.3), called the *curvature tensor field* (or, by abuse of language, the *curvature tensor*) of the connection $\mathbf{C}$. If (U, φ, m) is a chart of M, and $(X_i)_{1 \leq i \leq m}$ the basis of $\mathscr{T}_0^1(U)$ over $\mathscr{E}(U)$ associated with this chart (16.15.4.2), then from (17.20.4.2) we have

(17.20.5.1)
$$(\boldsymbol{r} \cdot (X_j \wedge X_k)) \cdot X_i = \sum_l \left(\frac{\partial \Gamma_{ki}^l}{\partial x^j} - \frac{\partial \Gamma_{ji}^l}{\partial x^k} + \sum_h (\Gamma_{ki}^h \Gamma_{jh}^l - \Gamma_{ji}^h \Gamma_{kh}^l) \right) X_l,$$

from which we obtain the corresponding components of the curvature tensor $\boldsymbol{r}$:

(17.20.5.2)
$$r_{ijk}^l = \frac{\partial \Gamma_{ki}^l}{\partial x^j} - \frac{\partial \Gamma_{ji}^l}{\partial x^k} + \sum_h (\Gamma_{ki}^h \Gamma_{jh}^l - \Gamma_{ji}^h \Gamma_{kh}^l).$$

(17.20.6) Again assume that $E = T(M)$. The identity mapping $1_{T(M)}$ of T(M) can be considered as a *differential* 1-*form on* M *with values in* T(M). Its covariant exterior differential $\boldsymbol{t} = \mathbf{d}(1_{T(M)})$ is therefore an M-morphism of $\bigwedge^2 T(M)$ into T(M), which is called the *torsion* M-*morphism* (or simply the *torsion*) of the linear connection $\mathbf{C}$ on M. So, by definition (17.19.3), we have

(17.20.6.1)
$$\boldsymbol{t} \cdot (X \wedge Y) = \nabla_X \cdot Y - \nabla_Y \cdot X - [X, Y]$$

for any two C^∞ vector fields X, Y on M. This morphism defines a bilinear M-morphism $(\mathbf{h}_x, \mathbf{k}_x) \mapsto \boldsymbol{t} \cdot (\mathbf{h}_x \wedge \mathbf{k}_x)$ of $T(M) \oplus T(M)$ into T(M), and hence may be identified with a *tensor field* $\mathbf{t}$ *of type* (1, 2), called the *torsion tensor field* (or, by abuse of language, the *torsion tensor*) of the connection **C**. If (U, φ, m) is a chart of M and $(X_i)_{1 \leq i \leq m}$ the basis of $\mathscr{T}_0^1(U)$ over $\mathscr{E}(U)$ associated with this chart (16.15.4.2), then from (17.18.1.5) we have

(17.20.6.2) $$\mathbf{t} \cdot (X_j \wedge X_k) = \sum_i (\Gamma^i_{jk} - \Gamma^i_{kj}) X_i,$$

from which we obtain the corresponding components of the torsion tensor **t**:

(17.20.6.3) $$t^i_{jk} = \Gamma^i_{jk} - \Gamma^i_{kj}.$$

If $f: N \to M$ is any C^∞-mapping, $T(f)$ is a lifting of f to T(M), and we may write it as $T(f) \circ 1_{T(M)}$. The formula (17.19.5.1) consequently shows that, for any two vector fields X, Y of class C^∞ on N, we have

(17.20.6.4)
$$\begin{aligned}\nabla_X \cdot (T(f) \cdot Y) - \nabla_Y \cdot (T(f) \cdot X) - T(f) \cdot [X, Y] \\ = \mathbf{t} \cdot ((T(f) \cdot X) \wedge (T(f) \cdot Y)).\end{aligned}$$

(17.20.7) *Let* **C**′, **C**″ *be two linear connections on* M, **t**′ *and* **t**″ *their respective torsions. If* B *is the bilinear* M*-morphism of* T(M) ⊕ T(M) *into* T(M) *which is the difference of* **C**′ *and* **C**″ (17.16.6), *then we have*

(17.20.7.1) $$\mathbf{t}'' \cdot (X \wedge Y) - \mathbf{t}' \cdot (X \wedge Y) = B(X, Y) - B(Y, X).$$

If we denote by $\nabla'_{\mathbf{h}_x}$ and $\nabla''_{\mathbf{h}_x}$ the covariant derivatives relative to **C**′ and **C**″, respectively, then it follows immediately from (17.17.2.1) that

$$\begin{aligned}\nabla'_{\mathbf{h}_x} \cdot Y - \nabla''_{\mathbf{h}_x} \cdot Y &= \tau_{Y(x)}(-\mathbf{C}'_x(\mathbf{h}_x, Y(x)) + \mathbf{C}''_x(\mathbf{h}_x, Y(x))) \\ &= -B_x(\mathbf{h}_x, Y(x))\end{aligned}$$

from the definition of B. Hence we have

(17.20.7.2) $$\nabla'_X \cdot Y - \nabla''_X \cdot Y = -B(X, Y)$$

and the formula (17.20.7.1) follows immediately from this and the definition (17.20.6.1) of the torsion.

APPENDIX

MULTILINEAR ALGEBRA

(The numbering of the sections in this Appendix continues that of the Appendix to Volume I.)

8. MODULES. FREE MODULES

(A.8.1) None of the results of (A.1.1)–(A.3.5) inclusive involves the *field* structure of K, and therefore all these results remain valid without modification when K is replaced by an arbitrary *commutative ring* A (with identity element). In place of K-*vector spaces* we speak of A-*modules* (in (A.2.3), h_λ is bijective if and only if λ is invertible in A). By abuse of language, the elements of an A-module are sometimes called *vectors* and the elements of A are called *scalars*.

(A.8.2) The definitions of a *free family* and of a *basis* of a vector space, given in (A.4.1) and (A.4.4), require no modification for an A-module, and the same is true of (A.4.3). On the other hand, the condition given in (A.4.2) for a family to be free is no longer valid in general (in the **Z**-module **Z**, for example, the number 1 does not belong to the **Z**-module 2**Z** generated by 2, but (1, 2) is not a free family). An A-module possessing a basis is said to be *free*.

(A.8.3) Everything in (A.5.1)–(A.6.6) inclusive, on determinants and matrices, also remains valid when the field K is replaced by an arbitrary commutative ring A. In (A.6.4) it is merely necessary to choose the form f_0 so that it takes the value $f_0(b_1, \ldots, b_n) = 1$ for a basis (b_i) of E. We remark that these results prove that any two bases of the same A-module E (assumed to have a finite basis) necessarily have the *same* number of elements (which one might call the "dimension" of E, but it should be realized that most of the results concerning dimensions of vector spaces *do not generalize* to free modules). The determinant calculations in (A.6.8) and (A.7.4) are also valid for an arbitrary commutative ring A.

9. DUALITY FOR FREE MODULES

(A.9.1) If E is an A-module, the A-module Hom(E, A) of *linear forms* on E (A.2.4) is called the *dual* of the module E and is often written E^*. If $x \in E$ and $x^* \in E^*$, we shall often write $\langle x, x^* \rangle$ or $\langle x^*, x \rangle$ in place of $x^*(x)$. The mapping $(x, x^*) \mapsto \langle x, x^* \rangle$ is a *bilinear form*, called the *canonical* bilinear form, on $E \times E^*$. For each $x \in E$, the mapping $x^* \mapsto \langle x, x^* \rangle$ of E^* into A is a linear form on E^*, in other words an element $c_E(x)$ of the *bidual* $E^{**} = (E^*)^*$ of E, and the mapping c_E (called the *canonical* mapping) of E into E^{**} is linear.

(A.9.2) Suppose that E is a free module having a *finite basis* $(e_i)_{1 \leqq i \leqq n}$ (also called a *finitely-generated* free module). For each index i, let e_i^* be the linear form on E (called the ith *coordinate function*) such that $\langle e_i, e_j^* \rangle = \delta_{ij}$ (Kronecker delta). Then for each $x = \sum_{i=1}^{n} \xi_i e_i \in E$, where $\xi_i \in A$ $(1 \leqq i \leqq n)$, we have $\langle x, e_i^* \rangle = \xi_i$. Hence (A.5.1) $(e_i^*)_{1 \leqq i \leqq n}$ is a *basis* of the dual A-module E^*, called the basis *dual* to the basis (e_i). From this definition it follows immediately that if $(e_i^{**})_{1 \leqq i \leqq n}$ is the basis of E^{**} dual to (e_i^*), then we have $c_E(e_i) = e_i^{**}$ for all i, and hence c_E is an *isomorphism*, by means of which we shall *identify* E^{**} with E, so that each element $x \in E$ is regarded as a *linear form* on E^*, namely the form $x^* \mapsto \langle x, x^* \rangle$.

(A.9.3) Let E and F be two A-modules and u a linear mapping of E into F. Then for each linear form $y^* \in F^*$, the function $y^* \circ u$ is a linear form on E, hence an element of E^*, and it is immediately verified that the mapping $y^* \mapsto y^* \circ u$ of F^* into E^* is linear. This mapping is called the *transpose* of u and is denoted by ${}^t u$. Clearly we have

(A.9.3.1) $${}^t(u_1 + u_2) = {}^t u_1 + {}^t u_2, \qquad {}^t(\lambda u) = \lambda \cdot {}^t u$$

for all $u_1, u_2, u \in \mathrm{Hom}(E, F)$ and $\lambda \in A$. If G is another A-module and $v : F \to G$ a linear mapping, then

(A.9.3.2) $${}^t(v \circ u) = {}^t u \circ {}^t v.$$

The definition of the transpose is contained in the *fundamental duality formula:*

(A.9.3.3) $$\langle u(x), y^* \rangle = \langle x, {}^t u(y^*) \rangle$$

for all $x \in \mathrm{E}$ and $y^* \in \mathrm{F}^*$. If u is an *isomorphism* of E onto F, then ${}^t u$ is an isomorphism of F^* onto E^*. Its inverse ${}^t u^{-1}$ (which is also the transpose of u^{-1}) is called the isomorphism *contragredient* to u. It satisfies the relation

(A.9.3.4) $$\langle u(x), {}^t u^{-1}(x^*) \rangle = \langle x, x^* \rangle$$

for all $x \in \mathrm{E}$ and $x^* \in \mathrm{E}^*$.

(A.9.4) Suppose now that E and F are free modules with finite bases $(a_i)_{1 \leqq i \leqq n}$ and $(b_j)_{1 \leqq j \leqq m}$, respectively. If $U = (\alpha_{ji})$ is the matrix of u with respect to these bases (A.5.2), then $\alpha_{ji} = \langle u(a_i), b_j^* \rangle$, and the formula (A.9.3.3) therefore shows that $\langle a_i, {}^t u(b_j^*) \rangle = \alpha_{ji}$. Since (a_i) is the basis dual to (a_i^*), the matrix of ${}^t u$ with respect to the bases (b_j^*) and (a_i^*) is obtained by interchanging the rows and columns of the matrix U of u (it is called the *transpose* of U and is denoted by ${}^t U$). Further, it is immediately clear that

(A.9.4.1) $${}^t({}^t u) = u.$$

(A.9.5) Suppose that the finitely-generated free A-module E is the *direct sum* $\mathrm{M} \oplus \mathrm{N}$ of two finitely-generated free A-modules (A.3.1). To the canonical projections $p : \mathrm{E} \to \mathrm{M}$, $q : \mathrm{E} \to \mathrm{N}$ (A.2.3) there correspond by transposition canonical injections ${}^t p : \mathrm{M}^* \to \mathrm{E}^*$, ${}^t q : \mathrm{N}^* \to \mathrm{E}^*$, such that E^* is the *direct sum* of the submodules $\mathrm{N}^0 = {}^t p(\mathrm{M}^*)$ and $\mathrm{M}^0 = {}^t q(\mathrm{N}^*)$. To see this, we take a basis of E consisting of the elements of a basis $(a_i)_{1 \leqq i \leqq m}$ of M and a basis $(b_j)_{1 \leqq j \leqq n}$ of N. It is then immediate that if (a_i^*) and (b_j^*) are the bases *dual* to (a_i) and (b_j), respectively (A.9.2), the elements ${}^t p(a_i^*)$ and ${}^t q(b_j^*)$ form the basis of E^* dual to the chosen basis of E. The submodule M^0 (resp. N^0) can also be defined as the set of linear forms $x^* \in \mathrm{E}^*$ such that $\langle x, x^* \rangle = 0$ *for all* $x \in \mathrm{M}$ (resp. *all* $x \in \mathrm{N}$), and is called the *annihilator* of M (resp. N) in E^*. By reason of the identification of a finitely generated free module with its bidual, it is clear that $(\mathrm{M}^0)^0 = \mathrm{M}$ and $(\mathrm{N}^0)^0 = \mathrm{N}$. Finally, note that if $i : \mathrm{M} \to \mathrm{E}$ and $j : \mathrm{N} \to \mathrm{E}$ are the canonical injections (A.2.3), their transposes ${}^t i : \mathrm{E}^* \to \mathrm{M}^*$ and ${}^t j : \mathrm{E}^* \to \mathrm{N}^*$ are the canonical projections, when we identify E^* with the direct sum $\mathrm{M}^* \oplus \mathrm{N}^*$.

(A.9.6) If E is a finite-dimensional *vector space* over a (commutative) field K, *every* vector subspace M of E admits a supplement N in E, and the results of

(A.9.5) can be applied to the direct sum decomposition $M \oplus N$ of E. The annihilator M^0 of M in E^* does not depend on the choice of the supplementary subspace N. It is clear that

(A.9.6.1) $$\operatorname{codim} M^0 = \dim M,$$

(A.9.6.2) $$(M^0)^0 = M.$$

Also, if M_1, M_2 are two subspaces of E,

(A.9.6.3) $$(M_1 + M_2)^0 = M_1^0 \cap M_2^0, \qquad (M_1 \cap M_2)^0 = M_1^0 + M_2^0;$$

this is easily seen by taking a basis of E as in (A.4.12).

Let F be another finite-dimensional vector space over K, and let $u : E \to F$ be a linear mapping. Then, with the same notation, we have

(A.9.6.4) $$(\operatorname{Ker} u)^0 = \operatorname{Im}({}^t u), \qquad (\operatorname{Im} u)^0 = \operatorname{Ker}({}^t u)$$

as is easily shown by decomposing E into the direct sum of $\operatorname{Ker}(u)$ and a supplementary subspace, and F into the direct sum of $\operatorname{Im}(u)$ and a supplementary subspace. It follows that

(A.9.6.5) $$\operatorname{rk}({}^t u) = \operatorname{rk}(u).$$

10. TENSOR PRODUCTS OF FREE MODULES

(A.10.1) Let $E_1, \ldots, E_r$ be A-modules, and for each index j let $x_j^* \in E_j^*$ be a linear form on E_j. Then the mapping

$$(x_1, x_2, \ldots, x_r) \mapsto \prod_{j=1}^{r} \langle x_j, x_j^* \rangle$$

of $E_1 \times E_2 \times \cdots \times E_r$ into A is an *r-linear form* (A.6.1), which is called the *tensor product* of the forms $x_1^*, \ldots, x_r^*$ and is denoted by

(A.10.1.1) $$x_1^* \otimes x_2^* \otimes \cdots \otimes x_r^*.$$

The set $\mathscr{L}_r(E_1, \ldots, E_r; A)$ of all r-linear forms is an A-module, and it follows directly from the definition that the mapping

(A.10.1.2) $$(x_1^*, x_2^*, \ldots, x_r^*) \mapsto x_1^* \otimes x_2^* \otimes \cdots \otimes x_r^*$$

of $E_1^* \times E_2^* \times \cdots \times E_r^*$ into $\mathscr{L}_r(E_1, \ldots, E_r; A)$ is *r-linear.*

(A.10.2) Suppose now that each E_j is a free A-module with a finite basis $(e_{jk})_{1 \leqq k \leqq n_j}$. Then it follows from (A.6.2) that the mapping (A.10.1.2) is *bijective* and that the elements

(A.10.2.1) $$e^*_{1,k_1} \otimes e^*_{2,k_2} \otimes \cdots \otimes e^*_{r,k_r}$$

form a *basis* consisting of $n_1 n_2 \cdots n_r$ elements of the A-module $\mathscr{L}_r(E_1, \ldots, E_r; A)$. This A-module is called the *tensor product* of $E^*_1, \ldots, E^*_r$, and is denoted by $E^*_1 \otimes_A E^*_2 \otimes_A \cdots \otimes_A E^*_r$, or simply $E^*_1 \otimes E^*_2 \otimes \cdots \otimes E^*_r$.

(A.10.3) Since, under the hypotheses of (A.10.2), E_j is identified with E^{**}_j (A.9.2), it follows that we may define in the same way the *tensor product* $E_1 \otimes_A E_2 \otimes_A \cdots \otimes_A E_r$ (or $E_1 \otimes E_2 \otimes \cdots \otimes E_r$) which has a basis consisting of the $n_1 n_2 \cdots n_r$ elements

(A.10.3.1) $$e_{1,k_1} \otimes e_{2,k_2} \otimes \cdots \otimes e_{r,k_r}$$

$(1 \leqq k_j \leqq n_j,\ 1 \leqq j \leqq r)$. The fundamental property of this A-module is the following:

For every r-linear mapping u of $E_1 \times E_2 \times \cdots \times E_r$ *into an arbitrary* A-*module* F, *there exists a unique* A-*linear mapping v of* $E_1 \otimes E_2 \otimes \cdots \otimes E_r$ *into* F *such that*

(A.10.3.2) $$v(x_1 \otimes x_2 \otimes \cdots \otimes x_r) = u(x_1, x_2, \ldots, x_r)$$

for all $x_j \in E_j$ $(1 \leqq j \leqq r)$.

For if $u(e_{1,k_1}, e_{2,k_2}, \ldots, e_{r,k_r}) = c_{k_1 k_2 \ldots k_r} \in F$, we can *define* v by the conditions

$$v(e_{1,k_1} \otimes e_{2,k_2} \otimes \cdots \otimes e_{r,k_r}) = c_{k_1 k_2 \cdots k_r},$$

and the mapping v so defined clearly has the required properties.

If E and F are two finitely-generated free A-modules, and if $(e_i)_{1 \leqq i \leqq m}$ and $(f_j)_{1 \leqq j \leqq n}$ are bases of E and F, respectively, then the $e_i \otimes f_j$ form a basis of $E \otimes F$, and hence every element of $E \otimes F$ is therefore uniquely expressible in the form $z = \sum_{i,j} \xi_{ij} e_i \otimes f_j$. This expression can also be written as

(A.10.3.3). $$z = \sum_{j=1}^{n} x_j \otimes f_j = \sum_{i=1}^{m} e_i \otimes y_i,$$

where the x_j (resp. y_i) are elements of E (resp. F) uniquely determined by z.

(A.10.4) The above definition implies the existence of *canonical isomorphisms* between tensor products of finitely-generated free A-modules. If E_1, E_2, E_3 are three finitely-generated free A-modules, there is a unique isomorphism (the *associativity isomorphism*)

(A.10.4.1) $$(E_1 \otimes E_2) \otimes E_3 \to E_1 \otimes E_2 \otimes E_3$$

which maps $(x_1 \otimes x_2) \otimes x_3$ to $x_1 \otimes x_2 \otimes x_3$. It can be *defined* by this property for the basis elements $(e_{1,k_1} \otimes e_{2,k_2}) \otimes e_{3,k_3}$ of the left-hand side. Likewise, there is a unique isomorphism (the *distributivity isomorphism*)

(A.10.4.2) $$(E_1 \oplus E_2) \otimes E_3 \to (E_1 \otimes E_3) \oplus (E_2 \otimes E_3)$$

which maps $(x_1 \oplus x_2) \otimes x_3$ to $(x_1 \otimes x_3) \oplus (x_2 \otimes x_3)$; it is defined in the same way as before.

(A.10.5) We have already seen (A.10.3) that there is a canonical isomorphism of $\mathrm{Hom}(E_1 \otimes E_2, F)$ onto the A-module $\mathscr{L}_2(E_1, E_2; F)$ of bilinear mappings of $E_1 \times E_2$ into F. Moreover, there is also a canonical isomorphism

(A.10.5.1) $$\mathrm{Hom}(E_1, \mathrm{Hom}(E_2, F)) \to \mathrm{Hom}(E_1 \otimes E_2, F).$$

For if $x_1 \mapsto v_{x_1}$ is a linear mapping of E_1 into $\mathrm{Hom}(E_2, F)$, then $(x_1, x_2) \mapsto v_{x_1}(x_2)$ is a bilinear mapping of $E_1 \times E_2$ into F, and we apply (A.10.3).

Now let E_1, E_2, F_1, F_2 be four finitely-generated free A-modules. To each pair of A-linear mappings $u_1 : E_1 \to F_1$, $u_2 : E_2 \to F_2$ we associate the bilinear mapping $(x_1, x_2) \mapsto u_1(x_1) \otimes u_2(x_2)$ of $E_1 \times E_2$ into $F_1 \otimes F_2$; to this bilinear mapping there corresponds (by (A.10.3)) a *linear* mapping

$$u_1 \otimes u_2 : E_1 \otimes E_2 \to F_1 \otimes F_2$$

such that

(A.10.5.2) $$(u_1 \otimes u_2)(x_1 \otimes x_2) = u_1(x_1) \otimes u_2(x_2).$$

Furthermore, the mapping $(u_1, u_2) \mapsto u_1 \otimes u_2$ is bilinear; hence (A.10.3) there corresponds to it a linear mapping

(A.10.5.3) $$\mathrm{Hom}(E_1, F_1) \otimes \mathrm{Hom}(E_2, F_2) \to \mathrm{Hom}(E_1 \otimes E_2, F_1 \otimes F_2),$$

which is in fact an *isomorphism* (and therefore the fact that the symbol $u_1 \otimes u_2$ has different meanings in the two sides of (A.10.5.3) is unimportant). If (a_i), (b_j), (c_h), (d_k) are bases of E_1, E_2, F_1, F_2, respectively, if $v_{ih} \in \mathrm{Hom}(E_1, F_1)$ is defined by the conditions $v_{ih}(a_i) = c_h$, and $v_{ih}(a_m) = 0$ for $m \neq i$, and if likewise

$w_{jk} \in \mathrm{Hom}(E_2, F_2)$ is defined by $w_{jk}(b_j) = d_k$, $w_{jk}(b_n) = 0$ for $n \neq j$, then it is immediate that the linear mappings $v_{ih} \otimes w_{jk}$ form a basis of $\mathrm{Hom}(E_1 \otimes E_2, F_1 \otimes F_2)$.

In particular, taking $F_1 = F_2 = A$, since $A \otimes A$ is canonically isomorphic to A (considered as the free A-module with basis consisting of the element 1), we obtain a canonical isomorphism

(A.10.5.4) $$E_1^* \otimes E_2^* \to (E_1 \otimes E_2)^*.$$

If on the other hand we take $F_1 = E_2 = A$, we obtain a canonical isomorphism

(A.10.5.5) $$E^* \otimes F \to \mathrm{Hom}(E, F)$$

under which $x^* \otimes y$ (where $x^* \in E^*$ and $y \in F$) corresponds to the linear mapping $x \mapsto \langle x, x^* \rangle y$ of E into F.

(A.10.6) Let B be a commutative ring containing A which is a finitely-generated free A-module and has the same identity element as A. If E is any finitely-generated free A-module, there is a unique B-module structure on $E \otimes_A B$ such that $(x \otimes \beta)\lambda = x \otimes (\beta\lambda)$ for all $x \in E$ and β, $\lambda \in B$. For if $(e_i)_{1 \leqq i \leqq n}$ is a basis of E, every element of $E \otimes_A B$ is uniquely of the form $\sum_i e_i \otimes \beta_i$ with $\beta_i \in B$, and the B-module structure required may be defined by $\left(\sum_i e_i \otimes \beta_i\right)\lambda = \sum_i e_i \otimes (\beta_i \lambda)$. This B-module is said to be obtained from E by *extension of the ring of scalars to* B and is denoted by $E_{(B)}$. The elements $e_i \otimes 1$ $(1 \leqq j \leqq n)$ form a basis of $E_{(B)}$, and E may be identified with the sub-A-module of $E_{(B)}$ generated by these basis elements, by means of the canonical injection $x \mapsto x \otimes 1$. Every A-*linear* mapping f of E into a B-*module* G extends uniquely to a B-*linear* mapping $\bar{f}$ of $E_{(B)}$ into G, such that $\bar{f}(x \otimes \beta) = f(x)\beta$. We may define $\bar{f}$ by the conditions $\bar{f}(e_i \otimes 1) = f(e_i)$ for $1 \leqq i \leqq n$.

In particular, if F is another finitely generated free A-module and if $j : F \to F_{(B)}$ is the canonical injection, then to each A-homomorphism $u : E \to F$ there corresponds the extension of $j \circ u$ to $E_{(B)}$, which is a B-linear mapping $u_{(B)} : E_{(B)} \to F_{(B)}$ such that $u_{(B)}(x \otimes \beta) = u(x) \otimes \beta$. In this way we define a canonical isomorphism

(A.10.6.1) $$(\mathrm{Hom}_A(E, F))_{(B)} \to \mathrm{Hom}_B(E_{(B)}, F_{(B)}),$$

which, in particular, gives an isomorphism

(A.10.6.2) $$(E^*)_{(B)} \to (E_{(B)})^*.$$

Finally, there is also a canonical isomorphism

(A.10.6.3) $$E_{(B)} \otimes_B F_{(B)} \to (E \otimes_A F)_{(B)},$$

where E, F are finitely generated free A-modules; the element

$$(x \otimes \beta) \otimes (y \otimes \beta') \qquad (x \in E,\ y \in F,\ \beta, \beta' \in B)$$

is mapped to $(x \otimes y) \otimes (\beta\beta')$.

11. TENSORS

(A.11.1) If E is a finitely generated free A-module, we denote by $E^{\otimes n}$ or $\mathbf{T}^n(E)$ or $\mathbf{T}_0^n(E)$ the tensor product of n copies of E, for $n \geqq 2$. This A-module is called the nth *tensor power* of E. We also define $\mathbf{T}_0^1(E)$ to be E itself and $\mathbf{T}_0^0(E)$ to be the ring A, considered as an A-module. Likewise we denote by $\mathbf{T}_n^0(E)$ the tensor product $(E^*)^{\otimes n}$, with the convention that $\mathbf{T}_1^0(E) = E^*$. Finally, if p and q are two integers >0, we denote by $\mathbf{T}_q^p(E)$ the tensor product $(E^*)^{\otimes q} \otimes (E^{\otimes p})$. The elements of $\mathbf{T}_0^n(E)$ (resp. $\mathbf{T}_n^0(E)$) are called *n-fold contravariant* (resp. *n-fold covariant*) *tensors*; the elements of $\mathbf{T}_q^p(E)$, for $p > 0$ and $q > 0$, are called *mixed tensors of type* (p, q), and p (resp. q) is the *contravariant* (resp. *covariant*) index.

It follows from (A.10.5.4) and (A.10.5.5) that $\mathbf{T}_q^p(E)$ may be canonically identified with $\mathrm{Hom}(E^{\otimes q}, E^{\otimes p})$. Hence by (A.10.5.3) we have a canonical isomorphism

(A.11.1.1) $$\mathbf{T}_q^p(E) \otimes \mathbf{T}_s^r(E) \to \mathbf{T}_{q+s}^{p+r}(E)$$

in which the product $u \otimes v$ of a tensor $u = x_1^* \otimes \cdots \otimes x_q^* \otimes x_1 \otimes \cdots \otimes x_p$ and a tensor $v = y_1^* \otimes \cdots \otimes y_2^* \otimes y_1 \otimes \cdots \otimes y_r$ corresponds to the tensor

$$x_1^* \otimes \cdots \otimes x_q^* \otimes y_1^* \otimes \cdots \otimes y_s^* \otimes x_1 \otimes \cdots \otimes x_p \otimes y_1 \otimes \cdots \otimes y_r.$$

When $p = q = 0$ or $r = s = 0$, the isomorphisms (A.11.1.1) are the linear mappings corresponding to the bilinear mappings $(\lambda, z) \mapsto \lambda z$ and $(z, \lambda) \mapsto \lambda z$, respectively. With these definitions it is immediate that for any three tensors u, v, w we have

(A.11.1.2) $$(u \otimes v) \otimes w = u \otimes (v \otimes w).$$

Again, by reason of the identification of a finitely generated free module with its bidual, and the canonical isomorphism (A.10.5.4), we have a canonical isomorphism

(A.11.1.3) $$(\mathbf{T}_q^p(E))^* \to \mathbf{T}_p^q(E)$$

such that, if the dual of $\mathbf{T}_q^p(\mathrm{E})$ is identified with $\mathbf{T}_p^q(\mathrm{E})$ by means of this isomorphism, we have

(A.11.1.4) $$\langle x_1^* \otimes \cdots \otimes x_q^* \otimes x_1 \otimes \cdots \otimes x_p, y_1^* \otimes \cdots \otimes y_p^* \otimes y_1 \otimes \cdots \otimes y_q \rangle = \left(\prod_{i=1}^{p} \langle x_i, y_i^* \rangle\right)\left(\prod_{j=1}^{q} \langle y_j, x_j^* \rangle\right).$$

(A.11.2) If $(e_i)_{1 \leq i \leq m}$ is a basis of E and (e_i^*) the dual basis of E^* (A.9.2), the elements

(A.11.2.1) $$e_{j_1}^* \otimes e_{j_2}^* \otimes \cdots \otimes e_{j_q}^* \otimes e_{i_1} \otimes e_{i_2} \otimes \cdots \otimes e_{i_p}$$

of $\mathbf{T}_q^p(\mathrm{E})$, where the indices i_h and j_k run independently through the set $\{1, 2, \ldots, m\}$, form a *basis* of $\mathbf{T}_q^p(\mathrm{E})$ which is called the basis *associated* to (e_i). A tensor belonging to $\mathbf{T}_q^p(\mathrm{E})$ then has a unique expression of the form

$$\sum \alpha_{i_1 i_2 \cdots i_p}^{j_1 j_2 \cdots j_q} e_{j_1}^* \otimes \cdots \otimes e_{j_q}^* \otimes e_{i_1} \otimes \cdots \otimes e_{i_p}$$

the sum being over all m^{p+q} families of indices $(j_1, \ldots, j_q, i_1, \ldots, i_p)$.

(A.11.3) Given two indices i, j such that $1 \leq i \leq p$ and $1 \leq j \leq q$, there exists a unique *linear* mapping

$$c_j^i : \mathbf{T}_q^p(\mathrm{E}) \to \mathbf{T}_{q-1}^{p-1}(\mathrm{E}),$$

called the *contraction* of the contravariant index i and the covariant index j such that, for $x_1, \ldots, x_p \in \mathrm{E}$ and $x_1^*, \ldots, x_q^* \in \mathrm{E}^*$,

(A.11.3.1)
$$c_j^i(x_1^* \otimes \cdots \otimes x_q^* \otimes x_1 \otimes \cdots \otimes x_p) = \langle x_i, x_j^* \rangle x_1^* \otimes \cdots \otimes \widehat{x_j^*} \otimes \cdots \otimes x_q^* \otimes x_1 \otimes \cdots \otimes \widehat{x_i} \otimes \cdots \otimes x_p,$$

wherein the circumflex accent signifies that the term underneath it is to be omitted from the tensor product. The mapping c_j^i may indeed be *defined* by this formula for the basis elements (A.11.2.1).

In particular, taking $p = q = 1$, we have $c_1^1(x^* \otimes x) = \langle x, x^* \rangle \in \mathrm{A}$. The elements of $\mathrm{E}^* \otimes \mathrm{E}$ correspond canonically to the endomorphisms of E (A.10.5.5); the value of the contraction c_1^1 for the tensor corresponding to an endomorphism u is called the *trace* of u and is denoted by $\mathrm{Tr}(u)$. If (with the above notation) u corresponds to the tensor $e_j^* \otimes e_i$, i.e., if u is the endomorphism $x \mapsto \langle x, e_j^* \rangle e_i$, then its trace is δ_{ij} (Kronecker delta). It follows easily that if $U = (\alpha_{ij})$ is the matrix of u with respect to the basis (e_i), then

(A.11.3.2)
$$\mathrm{Tr}(u) = \sum_i \alpha_{ii},$$

the sum of the *diagonal* elements of the matrix U; this is also called the *trace* of the matrix U and is denoted by $\mathrm{Tr}(U)$. It is immediately verified that, for any two endomorphisms u, v of E, we have

(A.11.3.3)
$$\mathrm{Tr}(u \circ v) = \mathrm{Tr}(v \circ u)$$

(it is enough to consider the case where u, v correspond to "decomposed" tensors $a^* \otimes a$ and $b^* \otimes b$).

12. SYMMETRIC AND ANTISYMMETRIC TENSORS

From now on we shall assume that the ring A contains the field $\mathbf{Q}$ of rational numbers, so that for each $\alpha \in \mathrm{A}$ and each integer $m \neq 0$ the element $m^{-1}\alpha$ belongs to A and is the only element $\xi \in \mathrm{A}$ such that $m\xi = \alpha$.

(A.12.1) Consider the A-module $\mathbf{T}^n(\mathrm{E})$ of n-fold contravariant tensors over a finitely-generated free A-module E. We define an action of the symmetric group $\mathfrak{S}_n$ on $\mathbf{T}^n(\mathrm{E})$ as follows. For each permutation $\sigma \in \mathfrak{S}_n$, the mapping

$$(x_1, x_2, \ldots, x_n) \mapsto x_{\sigma^{-1}(1)} \otimes x_{\sigma^{-1}(2)} \otimes \cdots \otimes x_{\sigma^{-1}(n)}$$

of E^n into $\mathbf{T}^n(\mathrm{E})$ is n-linear, hence factorizes as

$$(x_1, x_2, \ldots, x_n) \mapsto x_1 \otimes x_2 \otimes \cdots \otimes x_n \mapsto \sigma \cdot (x_1 \otimes x_2 \otimes \cdots \otimes x_n),$$

where σ is an *endomorphism* of the A-module $\mathbf{T}^n(\mathrm{E})$, defined by

(A.12.1.1)
$$\sigma \cdot (x_1 \otimes x_2 \otimes \cdots \otimes x_n) = x_{\sigma^{-1}(1)} \otimes x_{\sigma^{-1}(2)} \otimes \cdots \otimes x_{\sigma^{-1}(n)}.$$

From this definition it follows immediately that, if σ, τ are any two permutations in $\mathfrak{S}_n$, we have

(A.12.1.2)
$$\tau \cdot (\sigma \cdot z) = (\tau\sigma) \cdot z$$

for all $z \in \mathbf{T}^n(\mathrm{E})$.

A tensor $z \in \mathbf{T}^n(\mathrm{E})$ is said to be *symmetric* (resp. *antisymmetric*) if $\sigma \cdot z = z$ (resp. $\sigma \cdot z = \varepsilon_\sigma z$, where ε_σ is the *signature* of the permutation σ) for all

$\sigma \in \mathfrak{S}_n$. If we take the basis of $\mathbf{T}^n(\mathrm{E})$ associated with a basis (e_i) of E (A.11.2), a tensor is symmetric if and only if its components satisfy the conditions

(A.12.1.3) $$\alpha_{i_{\sigma(1)}i_{\sigma(2)}\cdots i_{\sigma(n)}} = \alpha_{i_1 i_2 \cdots i_n}$$

and antisymmetric if and only if

(A.12.1.4) $$\alpha_{i_{\sigma(1)}i_{\sigma(2)}\cdots i_{\sigma(n)}} = \varepsilon_\sigma \cdot \alpha_{i_1 i_2 \cdots i_n}$$

for all indices $i_1, i_2, \ldots, i_n$ and all $\sigma \in \mathfrak{S}_n$. It is sufficient that these relations should be satisfied for all transpositions $\tau \in \mathfrak{S}_n$.

(A.12.2) If z is any tensor belonging to $\mathbf{T}^n(\mathrm{E})$, we obtain from z a *symmetric* tensor (called the *symmetrization* of z)

(A.12.2.1) $$\boldsymbol{s} \cdot z = \sum_{\sigma \in \mathfrak{S}_n} \sigma \cdot z$$

and an *antisymmetric* tensor (called the *antisymmetrization* of z)

(A.12.2.2) $$\boldsymbol{a} \cdot z = \sum_{\sigma \in \mathfrak{S}_n} \varepsilon_\sigma(\sigma \cdot z).$$

It is evident that $\boldsymbol{s} \cdot z$ is symmetric; to show that $\boldsymbol{a} \cdot z$ is antisymmetric, we observe that, for any $\rho \in \mathfrak{S}_n$,

$$\rho \cdot (\boldsymbol{a} \cdot z) = \sum_{\sigma \in \mathfrak{S}_n} \varepsilon_\sigma(\rho \circ (\sigma \cdot z)) = \varepsilon_\rho \cdot \sum_{\sigma \in \mathfrak{S}_n} \varepsilon_{\rho\sigma}((\rho\sigma) \cdot z)$$
$$= \varepsilon_\rho \cdot (\boldsymbol{a} \cdot z).$$

If z is already symmetric, then

(A.12.2.3) $$\boldsymbol{s} \cdot z = n!\, z,$$

and if z is already antisymmetric,

(A.12.2.4) $$\boldsymbol{a} \cdot z = n!\, z.$$

The n-linear mapping $(x_1, \ldots, x_n) \mapsto \boldsymbol{s}(x_1 \otimes \cdots \otimes x_n)$ of E^n into $\mathbf{T}^n(\mathrm{E})$ is *symmetric*, and the n-linear mapping $(x_1, \ldots, x_n) \mapsto \boldsymbol{a}(x_1 \otimes \cdots \otimes x_n)$ of E^n into $\mathbf{T}^n(\mathrm{E})$ is *alternating* (or *antisymmetric*). In particular, if $x_i = x_j$ for some pair of distinct indices i, j, then $\boldsymbol{a}(x_1 \otimes \cdots \otimes x_n) = 0$.

(A.12.3) If $z \in \mathbf{T}^n(\mathrm{E})$ and $z^* \in \mathbf{T}^n(\mathrm{E}^*)$, then for each permutation $\sigma \in \mathfrak{S}_n$ we have

$$\langle \sigma \cdot z, z^* \rangle = \langle z, \sigma^{-1} \cdot z^* \rangle \tag{A.12.3.1}$$

by virtue of the formula (A.11.1.2), because

$$\prod_{i=1}^{n} \langle x_{\sigma^{-1}(i)}, x_i^* \rangle = \prod_{i=1}^{n} \langle x_i, x_{\sigma(i)}^* \rangle.$$

If we identify covariant tensors with multilinear forms on E (A.10.3), we have therefore

$$(\sigma \cdot z^*)(x_1, \ldots, x_n) = z^*(x_{\sigma(1)}, \ldots, x_{\sigma(n)});$$

consequently, symmetric (resp. antisymmetric) covariant tensors may be identified with symmetric (resp. antisymmetric or alternating) multilinear forms (A.6.3).

13. THE EXTERIOR ALGEBRA

All the tensors considered in this section are *contravariant*.

(A.13.1) Let E be a finitely generated free A-module, $(e_i)_{1 \leqq i \leqq m}$ a basis of E. The antisymmetrization $\boldsymbol{a}(e_{i_1} \otimes e_{i_2} \otimes \cdots \otimes e_{i_n})$ is zero whenever two of the indices i_k are equal (A.12.2). On the other hand, if the indices i_k are all distinct (which requires that $n \leqq m$), there is a unique permutation $\sigma \in \mathfrak{S}_n$ such that $i_{\sigma(1)} < i_{\sigma(2)} < \cdots < i_{\sigma(n)}$. For each subset $\mathrm{H} = \{i_1, i_2, \ldots, i_n\}$ of n elements of the set $\{1, 2, \ldots, m\}$ such that $i_1 < i_2 < \cdots < i_n$, the elements $e_{\mathrm{H}} = \boldsymbol{a} \cdot (e_{i_1} \otimes e_{i_2} \otimes \cdots \otimes e_{i_n})$ therefore form a *basis* (consisting of $\binom{m}{n}$ elements) of the A-module $\mathbf{A}_n(\mathrm{E})$ of antisymmetric tensors of order n over E.

(A.13.2) Given two antisymmetric tensors $z_p \in \mathbf{A}_p(\mathrm{E})$, $z_q \in \mathbf{A}_q(\mathrm{E})$, their *exterior product*, denoted by $z_p \wedge z_q$, is defined to be the antisymmetric tensor of order $p + q$ given by the formula

$$z_p \wedge z_q = \frac{1}{p!\,q!}\, \boldsymbol{a}(z_p \otimes z_q). \tag{A.13.2.1}$$

We shall prove the following two fundamental properties:

(A.13.2.2) (Anticommutativity) *If* $z_p \in \mathbf{A}_p(\mathrm{E})$ *and* $z_q \in \mathbf{A}_q(\mathrm{E})$, *then*

$$z_q \wedge z_p = (-1)^{pq} z_p \wedge z_q.$$

(A.13.2.3) (Associativity) *If* $z_p \in \mathbf{A}_p(\mathrm{E})$, $z_q \in \mathbf{A}_q(\mathrm{E})$, $z_r \in \mathbf{A}_r(\mathrm{E})$, *then*

$$z_p \wedge (z_q \wedge z_r) = (z_p \wedge z_q) \wedge z_r.$$

To begin with, we shall establish two preliminary results:

(A.13.2.4) *If* t_p (resp. t_q) *is a tensor of order* p (resp. q), *then*

$$\boldsymbol{a}(\boldsymbol{a}(t_p) \otimes t_q) = p!\,\boldsymbol{a}(t_p \otimes t_q),$$
$$\boldsymbol{a}(t_p \otimes \boldsymbol{a}(t_q)) = q!\,\boldsymbol{a}(t_p \otimes t_q).$$

We have

$$\boldsymbol{a}(\boldsymbol{a}(t_p) \otimes t_q) = \sum_{\sigma \in \mathfrak{S}_{p+q}} \varepsilon_\sigma \sum_{\tau \in \mathfrak{S}_p} \varepsilon_\tau \sigma \cdot ((\tau \cdot t_p) \otimes t_q);$$

but we can identify $\mathfrak{S}_p$ with the subgroup of $\mathfrak{S}_{p+q}$ which fixes the integers $>p$ in $\{1, 2, \ldots, p+q\}$. We have then

$$\begin{aligned}\boldsymbol{a}(\boldsymbol{a}(t_p) \otimes t_q) &= \sum_{\sigma \in \mathfrak{S}_{p+q}} \left(\sum_{\tau \in \mathfrak{S}_p} \varepsilon_{\sigma\tau} (\sigma\tau) \cdot (t_p \otimes t_q) \right) \\ &= p! \sum_{\rho \in \mathfrak{S}_{p+q}} \varepsilon_\rho \rho \cdot (t_p \otimes t_q) = p!\,\boldsymbol{a}(t_p \otimes t_q).\end{aligned}$$

The other formula is proved in the same way.

(A.13.2.5) *If* t_p (resp. t_q) *is a tensor of order* p (resp. q) *then*

$$\boldsymbol{a}(t_p \otimes t_q) = (-1)^{pq} \boldsymbol{a}(t_q \otimes t_p).$$

Let τ be the permutation in $\mathfrak{S}_{p+q}$ defined by

$$\tau(i) = p + i \quad \text{for} \quad 1 \leqq i \leqq q, \qquad \tau(q+i) = i \quad \text{for} \quad 1 \leqq i \leqq p.$$

Then by definition we have $\varepsilon_\tau = (-1)^{pq}$, and in the particular case where $t_p = x_1 \otimes x_2 \otimes \cdots \otimes x_p$, $t_q = x_{p+1} \otimes x_{p+2} \otimes \cdots \otimes x_{p+q}$, with $x_j \in \mathrm{E}$ $(1 \leqq j \leqq p+q)$,

$$\begin{aligned}\boldsymbol{a}(t_p \otimes t_q) &= \sum_{\sigma \in \mathfrak{S}_{p+q}} \varepsilon_{\sigma\tau} x_{\sigma\tau(1)} \otimes \cdots \otimes x_{\sigma\tau(q)} \otimes x_{\sigma\tau(q+1)} \otimes \cdots \otimes x_{\sigma\tau(p+q)} \\ &= (-1)^{pq} \sum_{\sigma \in \mathfrak{S}_{p+q}} \varepsilon_\sigma x_{\sigma(p+1)} \otimes \cdots \otimes x_{\sigma(p+q)} \otimes x_{\sigma(1)} \otimes \cdots \otimes x_{\sigma(p)}.\end{aligned}$$

On the other hand, if we put $y_i = x_{\tau(i)}$, then the permutation σ' such that $y_{\sigma'(i)} = x_{\sigma\tau(i)}$ is equal to $\tau^{-1}\sigma\tau$, so that $\varepsilon_{\sigma'} = \varepsilon_\sigma$. Consequently

$$\begin{aligned}\sum_{\sigma \in \mathfrak{S}_{p+q}} \varepsilon_\sigma x_{\sigma(p+1)} \otimes \cdots \otimes x_{\sigma(p+q)} \otimes x_{\sigma(1)} \otimes \cdots \otimes x_{\sigma(p)} &= \sum_{\sigma' \in \mathfrak{S}_{p+q}} \varepsilon_{\sigma'} y_{\sigma'(1)} \otimes \cdots \otimes y_{\sigma'(q)} \otimes y_{\sigma'(q+1)} \otimes \cdots \otimes y_{\sigma'(p+q)} \\ &= \boldsymbol{a}(t_q \otimes t_p).\end{aligned}$$

The general case now follows by linearity.

(A.13.2.6) To prove (A.13.2.2) and (A.13.2.3) it is now enough to observe that

$$z_p = \frac{1}{p!}\boldsymbol{a}(z_p), \qquad z_q = \frac{1}{q!}\boldsymbol{a}(z_q), \qquad z_r = \frac{1}{r!}\boldsymbol{a}(z_r).$$

The relation (A.13.2.2) is then an immediate consequence of (A.13.2.5). As to (A.13.2.3), we have

$$\begin{aligned} z_p \wedge (z_q \wedge z_r) &= \frac{1}{p!(q+r)!}\boldsymbol{a}(z_p \otimes (z_q \wedge z_r)) \\ &= \frac{1}{p!(q+r)!q!r!}\boldsymbol{a}(z_p \otimes \boldsymbol{a}(z_q \otimes z_r)) \\ &= \frac{1}{p!q!r!}\boldsymbol{a}(z_p \otimes z_q \otimes z_r) \end{aligned}$$

by the second formula of (A.13.2.4). Similarly the first formula of (A.13.2.4) gives us

$$(z_p \wedge z_q) \wedge z_r = \frac{1}{p!q!r!}\boldsymbol{a}(z_p \otimes z_q \otimes z_r)$$

and (A.13.2.3) is therefore proved.

This proof shows, by induction, that for any family of h antisymmetric tensors $z_{p_k} \in \mathbf{A}_{p_k}(\mathrm{E})$ $(1 \leqq k \leqq h)$,

(A.13.2.7) $$z_{p_1} \wedge z_{p_2} \wedge \cdots \wedge z_{p_h} = \frac{1}{p_1!\,p_2!\cdots p_h!}\boldsymbol{a}(z_{p_1} \otimes z_{p_2} \otimes \cdots \otimes z_{p_h}).$$

In particular, for n *vectors* $x_j \in \mathrm{E}$ $(1 \leqq j \leqq n)$, we have

(A.13.2.8) $$x_1 \wedge x_2 \wedge \cdots \wedge x_n = \boldsymbol{a}(x_1 \otimes x_2 \otimes \cdots \otimes x_n).$$

Consequently

(A.13.2.9) $$x_{\sigma(1)} \wedge x_{\sigma(2)} \wedge \cdots \wedge x_{\sigma(n)} = \varepsilon_\sigma x_1 \wedge x_2 \wedge \cdots \wedge x_n$$

for all permutations $\sigma \in \mathfrak{S}_n$; and if $x_i = x_j$ for two distinct indices i, j, then $x_1 \wedge x_2 \wedge \cdots \wedge x_n = 0$.

(A.13.3) By reason of the last formula, the module $\mathbf{A}_n(\mathrm{E})$ is called the nth exterior power of the finitely-generated free A-module E, and is denoted by $\bigwedge^n \mathrm{E}$. The basis of $\bigwedge^n \mathrm{E}$ associated with the basis (e_i) of E consists of the $\binom{m}{n}$ elements

(A.13.3.1) $$e_\mathrm{H} = e_{i_1} \wedge e_{i_2} \wedge \cdots \wedge e_{i_n},$$

where H runs through the set of subsets $\{i_1, i_2, \ldots, i_n\}$ of $\{1, 2, \ldots, m\}$, and $i_1 < i_2 < \cdots < i_n$.

The fundamental property of this module is the following:

For each alternating n-linear mapping u (A.6.3) *of* E^n *into an arbitrary* A-*module* F, *there exists a unique* A-*linear mapping* $v : \bigwedge^n \mathrm{E} \to \mathrm{F}$ *such that*

(A.13.3.2) $$v(x_1 \wedge x_2 \wedge \cdots \wedge x_n) = u(x_1, x_2, \ldots, x_n)$$

for all $x_j \in \mathrm{E}$.

For if $u(e_{i_1}, e_{i_2}, \ldots, e_{i_n}) = c_\mathrm{H} \in \mathrm{F}$ for $i_1 < i_2 < \cdots < i_n$ in $\{1, 2, \ldots, m\}$, we *define* v by the conditions $v(e_\mathrm{H}) = c_\mathrm{H}$; clearly v has the required properties, in view of the hypothesis on u.

(A.13.4) Consider two finitely generated free A-modules E, F, and let $u : \mathrm{E} \to \mathrm{F}$ be an A-linear mapping. Then the mapping

$$(x_1, x_2, \ldots, x_n) \mapsto u(x_1) \wedge u(x_2) \wedge \cdots \wedge u(x_n)$$

of E^n into $\bigwedge^n \mathrm{F}$ is clearly alternating and n-linear. By virtue of (A.13.3), there exists therefore a unique *linear* mapping $v : \bigwedge^n \mathrm{E} \to \bigwedge^n \mathrm{F}$ such that

$$v(x_1 \wedge x_2 \wedge \cdots \wedge x_n) = u(x_1) \wedge u(x_2) \wedge \cdots \wedge u(x_n)$$

for all $x_j \in \mathrm{E}$ $(1 \leq j \leq n)$. The mapping v is called the nth *exterior power* of u and is denoted by $\bigwedge^n u$.

Let $(e_j)_{1 \leqq j \leqq m}$ be a basis of E, and let $(f_i)_{1 \leqq i \leqq p}$ be a basis of F. Suppose that $n \leqq \inf(m, p)$ (otherwise $\bigwedge^n u = 0$) and let $X = (\alpha_{ij})$ be the matrix of u (with p rows and m columns) relative to these two bases. We shall calculate the matrix of $\bigwedge^n u$ relative to the associated bases (e_{K}) of $\bigwedge^n \mathrm{E}$ and (f_{H}) of $\bigwedge^n \mathrm{F}$, where K (resp. H) runs through the set of all subsets of n elements of $\{1, 2, \ldots, m\}$ (resp. $\{1, 2, \ldots, p\}$), these $\binom{m}{n}$ $\left(\text{resp. } \binom{p}{n}\right)$ sets being arranged in an arbitrary order. We have by definition

$$u(e_j) = \sum_{i=1}^{p} \alpha_{ij} f_i \qquad (1 \leqq j \leqq m)$$

so that, if $\mathrm{K} = \{j_1, j_2, \ldots, j_n\}$ with $j_1 < j_2 < \cdots < j_n$,

$$\begin{aligned}\left(\bigwedge^n u\right)(e_{\mathrm{K}}) &= \left(\sum_i \alpha_{ij_1} f_i\right) \wedge \left(\sum_i \alpha_{ij_2} f_i\right) \wedge \cdots \wedge \left(\sum_i \alpha_{ij_n} f_i\right) \\ &= \sum_{(i_1, i_2, \ldots, i_n)} \alpha_{i_1 j_1} \alpha_{i_2 j_2} \cdots \alpha_{i_n j_n} f_{i_1} \wedge f_{i_2} \wedge \cdots \wedge f_{i_n},\end{aligned}$$

the summation being over all sequences $(i_1, i_2, \ldots, i_n)$ of n *distinct* elements of the index set $\{1, 2, \ldots, p\}$. Now group together all the terms in the sum for which the set $\mathrm{H} = \{i_1, i_2, \ldots, i_n\}$ is the same; among the $n!$ sequences having H as underlying set there will be one for which $i_1 < i_2 < \cdots < i_n$, and the others will all be of the form $(i_{\sigma(1)}, i_{\sigma(2)}, \ldots, i_{\sigma(n)})$, where σ runs through $\mathfrak{S}_n$; hence, by virtue of (A.13.2.9), we have

$$\left(\bigwedge^n u\right)(e_{\mathrm{K}}) = \sum_{\mathrm{H}} \left(\sum_{\sigma \in \mathfrak{S}_n} \varepsilon_\sigma \alpha_{i_{\sigma(1)} j_1} \alpha_{i_{\sigma(2)} j_2} \cdots \alpha_{i_{\sigma(n)} j_n}\right) f_{\mathrm{H}},$$

where H runs through the set of subsets of n elements of $\{1, 2, \ldots, p\}$. The scalar factor multiplying f_{H} is the *determinant* of the $n \times n$ matrix $X^{\mathrm{HK}} = (\beta_{hk})_{1 \leqq h, k \leqq n}$, where $\beta_{hk} = \alpha_{i_h j_k}$ (A.6.8.1). Hence, with this notation, we may write

(A.13.4.1)
$$\left(\bigwedge^n u\right)(e_{\mathrm{K}}) = \sum_{\mathrm{H}} \det(X^{\mathrm{HK}}) f_{\mathrm{H}}$$

and therefore the matrix of $\bigwedge^n u$ relative to the bases (e_{K}) and (f_{H}) is the matrix $(\det(X^{\mathrm{HK}}))$ formed by the $n \times n$ minors of X.

In particular, if $x_i = \sum_{j=1}^{m} \xi_{ij} e_j$ are n elements of the A-module E, these considerations can be applied to the mapping $u : \mathrm{A}^n \to \mathrm{E}$ which maps the elements $a_1, a_2, \ldots, a_n$ of the canonical basis of A^n respectively to $x_1, x_2, \ldots, x_n$; we obtain

(A.13.4.2) $$x_1 \wedge x_2 \wedge \cdots \wedge x_n = \sum_{\mathrm{H}} \det(X^{\mathrm{H}}) e_{\mathrm{H}},$$

where H runs through the set of all subsets of n elements $j_1 < j_2 < \cdots < j_n$ of $\{1, 2, \ldots, m\}$, and X^{H} is the $n \times n$ matrix (η_{hk}), where $\eta_{hk} = \xi_{h, j_k}$ for $1 \leqq h, k \leqq n$.

In particular, if $n = m$,

(A.13.4.3) $$x_1 \wedge x_2 \wedge \cdots \wedge x_m = \det(X) e_1 \wedge e_2 \wedge \cdots \wedge e_m,$$

where X is the square matrix whose jth column consists of the components ξ_{ij} $(1 \leqq i \leqq m)$ of x_j.

(A.13.5) The definition of the exterior product (A.13.2.1) enables us to define a structure of an (associative) A-*algebra* on the A-module which is the direct sum of the exterior powers of E,

(A.13.5.1) $$\bigwedge \mathrm{E} = \overset{0}{\bigwedge} \mathrm{E} \oplus \overset{1}{\bigwedge} \mathrm{E} \oplus \overset{2}{\bigwedge} \mathrm{E} \oplus \cdots \oplus \overset{m}{\bigwedge} \mathrm{E}$$

(where conventionally $\overset{0}{\bigwedge} \mathrm{E} = \mathrm{A}$ and $\overset{1}{\bigwedge} \mathrm{E} = \mathrm{E}$), the multiplication being defined by the formula

(A.13.5.2) $$\Big(\sum_j z_j\Big) \wedge \sum_j z_j' = \sum_{j, k} (z_j \wedge z_k')$$

for all z_j, z_j' in $\overset{j}{\bigwedge} \mathrm{E}$. (When $j = 0$, z_j is an element λ of A, and the product $z_j \wedge z_k'$ is taken by definition to be $\lambda z_k'$ in the module $\overset{k}{\bigwedge} \mathrm{E}$.) Associativity follows directly from (A.13.2.3), and the identity element of A is also the identity element of $\bigwedge \mathrm{E}$. This algebra is called the *exterior algebra* of E; as A-module it admits a basis consisting of the 2^m elements e_{H}, where H runs through the set of all 2^m subsets of $\{1, 2, \ldots, m\}$ (we define $e_{\varnothing}$ to be $1 \in \mathrm{A}$). The multiplication table for this basis is given by

(A.13.5.3) $$\begin{cases} e_{\mathrm{H}} \wedge e_{\mathrm{K}} = 0 & \text{if } \mathrm{H} \cap \mathrm{K} \neq \varnothing, \\ e_{\mathrm{H}} \wedge e_{\mathrm{K}} = \rho_{\mathrm{H,K}} e_{\mathrm{H} \cup \mathrm{K}} & \text{if } \mathrm{H} \cap \mathrm{K} = \varnothing, \end{cases}$$

with $\rho_{\mathrm{H,K}} = (-1)^{\nu}$, where ν is the number of pairs $(i, j) \in \mathrm{H} \times \mathrm{K}$ such that $i > j$. This follows immediately from (A.13.2.9), by considering the permutation which is the product of the transpositions τ_{ij} (where $\tau_{ij}(i) = j$, $\tau_{ij}(j) = i$, and $\tau_{ij}(h) = h$ for $h \neq i, j$ in $\mathrm{H} \cup \mathrm{K}$) for $i \in \mathrm{H}$, $j \in \mathrm{K}$, and $i > j$.

It follows immediately from above that if E is the direct sum of two submodules M, N, then $\bigwedge^n M$ and $\bigwedge^n N$ may be identified canonically with submodules of $\bigwedge^n E$.

(A.13.6) Let B be a commutative ring which contains A, has the same identity element as A, and is a finitely-generated free A-module. Then the isomorphism (A.10.6.3) generalizes to an arbitrary finite number of factors, and in particular gives rise to a canonical A-module isomorphism

$$\mathbf{T}^n(E_{(B)}) \to (\mathbf{T}^n(E))_{(B)}. \tag{A.13.6.1}$$

It is immediately clear that this isomorphism transforms antisymmetric tensors into antisymmetric tensors, and therefore induces a canonical isomorphism

$$\bigwedge(E_{(B)}) \to (\bigwedge E)_{(B)} \tag{A.13.6.2}$$

which is a B-*algebra* isomorphism, as is easily verified.

(A.13.7) If E is a *vector space* of finite dimension m over a field K, the notion of exterior product enables us to express in a simple form the *linear independence* of n vectors $x_1, \dots, x_n$ in E: a necessary and sufficient condition for this is that

$$x_1 \wedge x_2 \wedge \cdots \wedge x_n \neq 0. \tag{A.13.7.1}$$

For it is clear that the exterior product $x_1 \wedge x_2 \wedge \cdots \wedge x_n$ will vanish if one of the x_j is a linear combination of the others. Conversely, if the x_j are linearly independent, then there exists a basis of E in which $x_1, \dots, x_n$ are the first n vectors (A.4.8); hence $x_1 \wedge x_2 \wedge \cdots \wedge x_n$ is an element of the basis of $\bigwedge^n E$ associated with this basis of E, and therefore is $\neq 0$.

The elements of $\bigwedge^p E$ are often called *p-vectors* (even when E is a module over a *ring* A).

14. DUALITY IN THE EXTERIOR ALGEBRA

(A.14.1) Let E be a free A-module with a finite basis $(e_i)_{1 \leq i \leq m}$. It follows from the definitions (A.13.1) that the A-module $\mathbf{T}^n(E)$ of n-fold contravariant tensors over E splits up into the *direct sum* of the submodule $\bigwedge^n E = \mathbf{A}_n(E)$

and the submodule spanned by the basis elements $e_{i_1} \otimes e_{i_2} \otimes \cdots \otimes e_{i_n}$ for which the sequence $(i_1, i_2, \ldots, i_n)$ either has two terms equal, or all terms distinct but not in increasing order. Hence (A.9.5) every *linear form* on the A-module $\mathbf{A}_n(\mathrm{E}) = \bigwedge^n(\mathrm{E})$ is the *restriction* to this submodule of a linear form $z \mapsto \langle z, t^* \rangle$ on $\mathbf{T}^n(\mathrm{E})$, where $t^* \in \mathbf{T}^n(\mathrm{E}^*)$ (A.11.1.4). Now, by virtue of (A.12.3.1), we have $\langle z, t^* \rangle = \varepsilon_\sigma \langle \sigma \cdot z, t^* \rangle = \langle z, \varepsilon_\sigma \sigma^{-1} \cdot t^* \rangle$ for all $z \in \mathbf{A}_n(\mathrm{E})$ and $\sigma \in \mathfrak{S}_n$; summing over all permutations σ, we obtain

(A.14.1.1) $$\langle z, t^* \rangle = \frac{1}{n!} \langle z, \boldsymbol{a}(t^*) \rangle.$$

In other words, if for each $z^* \in \mathbf{A}_n(\mathrm{E}^*)$ we denote by $\delta(z^*)$ the restriction of the linear form $z \mapsto (1/n!)\langle z, z^* \rangle$ to $\mathbf{A}_n(\mathrm{E})$, then δ is a *surjective* linear mapping of $\mathbf{A}_n(\mathrm{E}^*)$ onto the dual $(\mathbf{A}_n(\mathrm{E}))^*$ of $\mathbf{A}_n(\mathrm{E})$; but δ is also *injective*, because if $z^* \in \mathbf{A}_n(\mathrm{E}^*)$ is such that $\langle \boldsymbol{a}(t), z^* \rangle = 0$ for all $t \in \mathbf{T}^n(\mathrm{E})$, the same argument together with the fact that $\boldsymbol{a}(z^*) = n!\, z^*$ gives $\langle t, z^* \rangle = 0$ for all $t \in \mathbf{T}^n(\mathrm{E})$, that is to say, $z^* = 0$.

The mapping δ therefore *identifies* the exterior power $\bigwedge^n(\mathrm{E}^*)$ with the dual $\left(\bigwedge^n \mathrm{E}\right)^*$. If x_j $(1 \leqq j \leqq n)$ are elements of E and x_j^* $(1 \leqq j \leqq n)$ elements of E^*, then by virtue of (A.14.1.1) we have

$$\begin{aligned} &\langle x_1 \wedge x_2 \wedge \cdots \wedge x_n, x_1^* \wedge x_2^* \wedge \cdots \wedge x_n^* \rangle \\ &\quad = \langle \boldsymbol{a}(x_1 \otimes x_2 \otimes \cdots \otimes x_n), x_1^* \otimes x_2^* \otimes \cdots \otimes x_n^* \rangle \\ &\quad = \sum_{\sigma \in \mathfrak{S}_n} \varepsilon_\sigma \langle x_{\sigma(1)}, x_1^* \rangle \langle x_{\sigma(2)}, x_2^* \rangle \cdots \langle x_{\sigma(n)}, x_n^* \rangle, \end{aligned}$$

that is to say (A.6.8.1),

(A.14.1.2) $$\langle x_1 \wedge x_2 \wedge \cdots \wedge x_n, x_1^* \wedge x_2^* \wedge \cdots \wedge x_n^* \rangle = \det(\langle x_i, x_j^* \rangle).$$

If (e_i^*) is the basis of E^* dual to (e_i), and if (e_{H}) and (e_{H}^*) are the bases of $\bigwedge^n \mathrm{E}$ and $\bigwedge^n \mathrm{E}^*$ associated with (e_i) and (e_i^*), respectively, then it follows that

(A.14.1.3) $$\langle e_{\mathrm{H}}, e_{\mathrm{K}}^* \rangle = \delta_{\mathrm{HK}} \qquad \text{(Kronecker delta)}$$

for any two subsets H, K of n elements of $\{1, 2, \ldots, m\}$. In other words, (e_{H}^*) is the basis *dual* to (e_{H}) when $\bigwedge^n \mathrm{E}^*$ is identified with the dual of $\bigwedge^n \mathrm{E}$.

The elements of $\bigwedge^n E^*$ may be canonically identified with the *alternating n-linear forms* on E^* (A.13.3); these are also called *exterior forms of degree n* on E, or *n-forms* (or *n-covectors* when E is a vector space).

(A.14.2) Suppose that E is the direct sum $M \oplus N$ of two finitely generated free modules M, N. We have seen (A.13.5) that $\bigwedge^n M$ and $\bigwedge^n N$ may be identified with submodules of $\bigwedge^n E$. For each $z^* \in \bigwedge^n E^*$, considered as a linear form on $\bigwedge^n E$, we may therefore speak of the *restriction* of z^* to $\bigwedge^n M$ (or, as is also said, the restriction of z^* to M, considered as an alternating n-linear form on M).

Let p, q be two integers >0, let $u \in \bigwedge^p M$, $v \in \bigwedge^q N$, and let u^* be an element of $\bigwedge^p E^*$ and $v^* = x_1^* \wedge x_2^* \wedge \cdots \wedge x_q^*$ an element of $\bigwedge^q E^*$ such that the linear forms x_j^* $(1 \leqq j \leqq q)$ are *zero on* M. Then we have

(A.14.2.1) $$\langle u \wedge v, u^* \wedge v^* \rangle = \langle u, u^* \rangle \langle v, v^* \rangle.$$

It is enough to verify this relation when $u = a_1 \wedge a_2 \wedge \cdots \wedge a_p$ and $v = b_1 \wedge b_2 \wedge \cdots \wedge b_q$ and $u^* = c_1^* \wedge c_2^* \wedge \cdots \wedge c_p^*$, where the a_i belong to M, the b_j to N and the c_i^* are arbitrary elements of E^*. The formula then follows from (A.14.1.2) and the rule for calculating the determinant of a matrix by blocks (A.7.4.1), since $\langle a_i, x_k^* \rangle = 0$ for $1 \leqq i \leqq p$ and $1 \leqq k \leqq q$.

15. INTERIOR PRODUCTS

(A.15.1) We retain the hypotheses and notation of Section 14. Let p, q be two integers $\geqq 0$ and let $z_q \in \mathbf{T}^q(E)$ be a contravariant tensor; then the mapping $v_p \mapsto v_p \otimes z_q$ of $\mathbf{T}^p(E)$ into $\mathbf{T}^{p+q}(E)$ is *linear*, and its *transpose* (A.9.3) may therefore be identified with a linear mapping of $\mathbf{T}^{p+q}(E^*)$ into $\mathbf{T}^p(E^*)$, which is denoted by

$$u_{p+q}^* \mapsto z_q \lrcorner\, u_{p+q}^*$$

and is called the *interior product* of z_q and u_{p+q}^*. From this definition we have

(A.15.1.1) $$\langle v_p, z_q \lrcorner\, u_{p+q}^* \rangle = \langle v_p \otimes z_q, u_{p+q}^* \rangle$$

for all $v_p \in \mathbf{T}^p(E)$ and $u_{p+q}^* \in \mathbf{T}^{p+q}(E^*)$. Since, by virtue of (A.11.1.4),

$$\begin{aligned} &\langle e_{i_1} \otimes \cdots \otimes e_{i_p}, (e_{j_1} \otimes \cdots \otimes e_{j_q}) \lrcorner\, (e_{k_1}^* \otimes \cdots \otimes e_{k_{p+q}}^*) \rangle \\ &\quad = \langle e_{i_1}, e_{k_1}^* \rangle \cdots \langle e_{i_p}, e_{k_p}^* \rangle \langle e_{j_1}, e_{k_{p+1}}^* \rangle \cdots \langle e_{j_q}, e_{k_{p+q}}^* \rangle, \end{aligned}$$

it follows immediately that

(A.15.1.2) $$(e_{j_1} \otimes \cdots \otimes e_{j_q}) \lrcorner (e^*_{k_1} \otimes \cdots \otimes e^*_{k_{p+q}}) = 0$$

unless $j_h = k_{p+h}$ for $1 \leqq h \leqq q$; and that

(A.15.1.3) $$(e_{j_1} \otimes \cdots \otimes e_{j_q}) \lrcorner (e^*_{k_1} \otimes \cdots \otimes e^*_{k_{p+q}}) = e^*_{k_1} \otimes \cdots \otimes e^*_{k_q}$$

if $j_h = k_{p+h}$ for $1 \leqq h \leqq q$.

(A.15.2) It follows directly from the definition (A.15.1) and the associativity of the tensor product (A.11.1.2) that, for any three integers $p, q, r \geqq 0$,

(A.15.2.1) $$z'_q \lrcorner (z''_r \lrcorner u^*_{p+q+r}) = (z'_q \otimes z''_r) \lrcorner u^*_{p+q+r}.$$

In particular, the interior product $u^*_{p+1} \mapsto x \lrcorner u^*_{p+1}$ by an element $x \in \mathrm{E}$ is sometimes denoted by $i(x)$. The interior product by a tensor

$$x_1 \otimes x_2 \otimes \cdots \otimes x_p \in \mathbf{T}^p(\mathrm{E})$$

may therefore be written as $i(x_1) \circ i(x_2) \circ \cdots \circ i(x_p)$.

(A.15.3) Now consider the linear mapping $v_p \mapsto v_p \wedge z_q$ of $\bigwedge^p \mathrm{E}$ into $\bigwedge^{p+q} \mathrm{E}$, where z_q is an element of $\bigwedge^q \mathrm{E}$; its transpose, which is a linear mapping of $\bigwedge^{p+q} \mathrm{E}^*$ into $\bigwedge^p \mathrm{E}^*$, is denoted by

$$u^*_{p+q} \mapsto z_q \lrcorner u^*_{p+q}$$

and is called the *interior product* of z_q and u^*_{p+q}. However, it should be remarked that this interior product is *not* the same as the restriction to antisymmetric tensors of the interior product defined in (A.15.1) (this double use of the same notation does not in practice cause any confusion). Hence we have now

(A.15.3.1) $$\langle v_p, z_q \lrcorner u^*_{p+q} \rangle = \langle v_p \wedge z_q, u^*_{p+q} \rangle$$

for all $v_p \in \bigwedge^p \mathrm{E}$ and $u^*_{p+q} \in \bigwedge^{p+q} \mathrm{E}^*$. With the notation of (A.13.5), we have

$$\langle e_\mathrm{H}, e_\mathrm{K} \lrcorner e^*_\mathrm{L} \rangle = \langle e_\mathrm{H} \wedge e_\mathrm{K}, e^*_\mathrm{L} \rangle$$

and therefore, by virtue of (A.13.5.3),

(A.15.3.2) $$\begin{cases} e_\mathrm{K} \lrcorner e^*_\mathrm{L} = 0 & \text{if } \mathrm{K} \not\subset \mathrm{L}, \\ e_\mathrm{K} \lrcorner e^*_\mathrm{L} = \rho_{\mathrm{L-K,K}} e^*_{\mathrm{L-K}} & \text{if } \mathrm{K} \subset \mathrm{L}. \end{cases}$$

(A.15.4) From the associativity of the exterior product (A.13.2.3) we deduce immediately

(A.15.4.1) $$z'_q \rfloor (z''_r \rfloor u^*_{p+q+r}) = (z'_q \wedge z''_r) \rfloor u^*_{p+q+r}.$$

The interior product $u^*_{p+1} \mapsto x \rfloor u^*_{p+1}$ by an element $x \in \mathrm{E}$ is denoted by $i(x)$, so that the interior product by $x_1 \wedge x_2 \wedge \cdots \wedge x_p$ may be written as $i(x_1) \circ i(x_2) \circ \cdots \circ i(x_p)$. In particular,

(A.15.4.2) $$i(x) \circ i(x) = 0.$$

Explicitly, $i(x)$ is given by the formula

(A.15.4.3)

$$i(x)(x_1^* \wedge x_2^* \wedge \cdots \wedge x_{p+1}^*) = \sum_{i=1}^{p+1} (-1)^{i+1} \langle x, x_i^* \rangle x_1^* \wedge \cdots \wedge \widehat{x_i^*} \wedge \cdots \wedge x_{p+1}^*$$

with the usual convention that the symbol below the circumflex is to be omitted. By linearity, it is enough to consider the case where $x = e_j$ and $x_k^* = e_{i_k}^*$, where $i_1 < i_2 < \cdots < i_{p+1}$, and then the above formula follows from (A.15.3.2).

16. NONDEGENERATE ALTERNATING BILINEAR FORMS. SYMPLECTIC GROUPS

(A.16.1) Let E be a *vector space* of dimension m over a (commutative) field K, and let B be an *alternating bilinear form* on E × E, so that B may be identified with an element of $\bigwedge^2 \mathrm{E}^*$. We shall show, by induction on m, that there exists a basis $(e_i)_{1 \leqq i \leqq m}$ of E and an integer $r \geqq 0$ such that, if (e_i^*) is the dual basis of E*, then we have

(A.16.1.1) $$\mathrm{B} = e_1^* \wedge e_2^* + e_3^* \wedge e_4^* + \cdots + e_{2r-1}^* \wedge e_{2r}^*.$$

We may assume that $\mathrm{B} \neq 0$, or there is nothing to prove. Then there exists a bivector $e_1 \wedge e_2$ such that $\mathrm{B}(e_1, e_2) \neq 0$. The vectors e_1, e_2 are therefore linearly independent, and by multiplying one of them by a nonzero scalar we may assume that $\mathrm{B}(e_1, e_2) = 1$. Let F be the subspace of E consisting of the vectors x such that $\mathrm{B}(e_1, x) = \mathrm{B}(e_2, x) = 0$. Then F is of codimension $\leqq 2$ in E (A.9.6); and if P is the plane spanned by e_1 and e_2, then $\mathrm{P} \cap \mathrm{F} = \{0\}$ (for if $\mathrm{B}(e_1, \alpha e_1 + \beta e_2) = \mathrm{B}(e_2, \alpha e_1 + \beta e_2) = 0$ we obtain $\beta = 0$ and $-\alpha = 0$).

Hence the subspaces F and P are supplementary. If B_1 is the restriction of B to $F \times F$, then by the inductive hypothesis there exists a basis $(e_j)_{3 \leqq j \leqq m}$ of F such that

$$B_1 = e_3^* \wedge e_4^* + \cdots + e_{2r-1}^* \wedge e_{2r}^*.$$

In view of the definition of F, we conclude that

$$\text{(A.16.1.2)} \quad \begin{cases} B(e_{2j-1}, e_{2j}) = -B(e_{2j}, e_{2j-1}) = 1 & \text{for } 1 \leqq j \leqq r, \\ B(e_h, e_k) = 0 & \text{for all other pairs } (h, k), \end{cases}$$

which is equivalent to (A.16.1.1).

The number r is independent of the basis (e_i) satisfying (A.16.1.1), for it follows from this relation that if $B^{\wedge s}$ denotes the product of s bivectors equal to B in $\bigwedge E^*$, then $B^{\wedge(r+1)} = 0$, because this product is a sum of products of $2r + 2$ factors taken from the set $\{e_1^*, \ldots, e_{2r}^*\}$, and each such product must contain a repeated factor and therefore vanishes. On the other hand, since the bivectors $e_{2i-1}^* \wedge e_{2i}^*$ and $e_{2j-1}^* \wedge e_{2j}^*$ commute, we have

$$\text{(A.16.1.3)} \quad B^{\wedge r} = r!\, e_1^* \wedge e_2^* \wedge e_3^* \wedge e_4^* \wedge \cdots \wedge e_{2r-1}^* \wedge e_{2r}^* \neq 0.$$

The integer $2r$ is called the *rank* of B, and if $2r = m$, then B is said to be *nondegenerate*. Equivalently, B is nondegenerate if and only if there exists no vector $x \neq 0$ in E such that $B(x, y) = 0$ for all $y \in E$.

When $m = 2r$ and B is nondegenerate, a basis $(e_i)_{1 \leqq i \leqq m}$ of E for which (A.16.1.1) holds is called a *symplectic basis* of E (relative to B).

(A.16.2) For each $x \in E$, $i(x)(B) = \Phi(x)$ is a vector belonging to the dual space E^*. If B is given by (A.16.1.1), it is immediately seen that $\Phi(e_{2j}) = -e_{2j-1}^*$ and $\Phi(e_{2j-1}) = e_{2j}^*$ for $1 \leqq j \leqq r$, and that $\Phi(e_k) = 0$ for $k > 2r$ (A.15.4.3), so that Φ is a linear mapping of rank $2r$ (A.4.16) of E into E^*. Consequently Φ is bijective if and only if B is nondegenerate.

(A.16.3) Suppose for the rest of this section that B is nondegenerate (so that $m = \dim(E)$ is *even*). Two vectors $x, y \in E$ are said to be *orthogonal* (relative to B) if $B(x, y) = 0$; by (A.15.3.1), this is equivalent to $\langle y, \Phi(x)\rangle = 0$. If V is any vector subspace of E, the set of vectors $y \in E$ orthogonal to *all* $x \in E$ is a vector subspace $V^\perp$ of E, which is equal to $\Phi^{-1}(V^0)$ in the notation of (A.9.6), and is called the *orthogonal supplement* of V (relative to B). Since Φ is bijective, it follows from (A.9.6) that:

$$\text{(A.16.3.1)} \quad \operatorname{codim} V^\perp = \dim V,$$

(A.16.3.2) $$(V^{\perp})^{\perp} = V,$$

(A.16.3.3) $$(V_1 + V_2)^{\perp} = V_1^{\perp} \cap V_2^{\perp}, \qquad (V_1 \cap V_2)^{\perp} = V_1^{\perp} + V_2^{\perp}.$$

A subspace V of E is said to be *isotropic* if $V \cap V^{\perp} \neq \{0\}$, and *totally isotropic* if $V \subset V^{\perp}$. Every one-dimensional subspace of E is totally isotropic. If V is isotropic, then $V \cap V^{\perp}$ is totally isotropic by virtue of (A.16.3.2) and (A.16.3.3). It follows from (A.16.3.1) that if V is totally isotropic, then $\dim V \leqq \frac{1}{2}m$, and this upper bound is attained by the subspace generated by the vectors $e_1, \ldots, e_r$ (where $2r = m$) of a symplectic basis of E.

(A.16.4) The bijective linear mappings $u : E \to E$ such that $B(u(x), u(y)) = B(x, y)$ for all $x, y \in E$ (or, equivalently, such that $\overset{2}{\bigwedge}({}^t u)(B) = B$) are called *symplectic automorphisms* of E (relative to B) and form a group called the *symplectic group* of E (relative to B), which is denoted by **Sp**(E, B). The relation $\overset{2}{\bigwedge}({}^t u)(B) = B$ implies immediately (A.13.3.2) that

$$\overset{m}{\bigwedge}({}^t u)(B^{\wedge r}) = B^{\wedge r},$$

where $r = \frac{1}{2}m$; hence, from (A.16.1.3) and (A.13.4.3), we have $\det(u) = 1$ for all $u \in \mathbf{Sp}(E, B)$. In other words, **Sp**(E, B) is a subgroup of the special linear group $\mathbf{SL}(E) \subset \mathbf{GL}(E)$ (the group of automorphisms of the vector space E which have determinant equal to 1).

Relative to a *symplectic basis* of E, the matrices of symplectic automorphisms are the matrices U which satisfy the relation

(A.16.4.1) $${}^t U \cdot J \cdot U = J,$$

where J is the square matrix of order $2r = m$:

$$J = \begin{pmatrix} 0 & 1 & 0 & 0 & \cdots & 0 & 0 \\ -1 & 0 & 0 & 0 & \cdots & 0 & 0 \\ 0 & 0 & 0 & 1 & \cdots & 0 & 0 \\ 0 & 0 & -1 & 0 & \cdots & 0 & 0 \\ \cdots & \cdots & \cdots & \cdots & \cdots & \cdots & \cdots \\ 0 & 0 & 0 & 0 & \cdots & 0 & 1 \\ 0 & 0 & 0 & 0 & \cdots & -1 & 0 \end{pmatrix}.$$

All the symplectic groups **Sp**(E, B) relative to different nondegenerate alternating forms B on E × E are therefore *isomorphic*. The group of matrices U satisfying (A.16.4.1) is denoted by **Sp**(m, K) (it is therefore defined only for even m).

17. THE SYMMETRIC ALGEBRA

The developments of (A.13) can be repeated by replacing antisymmetric tensors throughout by symmetric tensors, and the antisymmetrization operator $\boldsymbol{a}$ by the symmetrization operator $\boldsymbol{s}$. The A-module $\mathbf{S}_n(\mathrm{E})$ of symmetric tensors of order n has a basis obtained as follows: For each integer $p \geqq 1$, let $e_i^{\otimes p}$ denote the tensor product $e_i \otimes \cdots \otimes e_i$ with p factors, and for each multi-index $\alpha = (\alpha_1, \ldots, \alpha_m) \in \mathbf{N}^m$, put

$$e^\alpha = \boldsymbol{s}(e_1^{\otimes \alpha_1} \otimes e_2^{\otimes \alpha_2} \otimes \cdots \otimes e_m^{\otimes \alpha_m}).$$

Then the e^α such that $|\alpha| = n$ form a basis of $\mathbf{S}_n(\mathrm{E})$.

Next we define the *symmetric product* $z_p z_q$ of an element $z_p \in \mathbf{S}_p(\mathrm{E})$ and an element $z_q \in \mathbf{S}_q(\mathrm{E})$ by the formula

(A.17.1) $$z_p z_q = \frac{1}{p!q!} \boldsymbol{s}(z_p \otimes z_q),$$

and just as in (A.13) one proves that

(A.17.2) $$z_p z_q = z_q z_p \qquad \text{(commutativity)},$$

(A.17.3) $$(z_p z_q) z_r = z_p (z_q z_r) \qquad \text{(associativity)}.$$

In particular, for n elements $x_1, \ldots, x_n$ of E we have

(A.17.4) $$x_1 x_2 \cdots x_n = \boldsymbol{s}(x_1 \otimes x_2 \otimes \cdots \otimes x_n).$$

For this reason the module $\mathbf{S}_n(\mathrm{E})$ is called the nth *symmetric power* of E.

The (infinite) *direct sum* of the symmetric powers of E:

(A.17.5) $$\mathbf{S}(\mathrm{E}) = \mathbf{S}_0(\mathrm{E}) \oplus \mathbf{S}_1(\mathrm{E}) \oplus \cdots \oplus \mathbf{S}_n(\mathrm{E}) \oplus \cdots,$$

where $\mathbf{S}_0(\mathrm{E}) = \mathrm{A}$ and $\mathbf{S}_1(\mathrm{E}) = \mathrm{E}$, becomes a commutative and associative A-algebra with the multiplication defined by

(A.17.6) $$\left(\sum_n z_n\right)\left(\sum_n z'_n\right) = \sum_{p,\,q} (z_p z'_q)$$

(when $p = 0$, so that $z_p = \lambda \in \mathrm{A}$, the product $z_p z'_q$ is taken to be the product $\lambda z'_q$ in the A-module $\mathbf{S}_q(\mathrm{E})$). This algebra $\mathbf{S}(\mathrm{E})$ is called the *symmetric algebra* of the A-module E; it has a basis consisting of the e^α, where $\alpha \in \mathbf{N}^m$ (and e^0 is taken to be the identity element 1 of A), and the multiplication of basis elements is given by

(A.17.7)
$$e^\alpha e^\beta = e^{\alpha+\beta}.$$

This proves that the symmetric algebra $\mathbf{S}(E)$ is *isomorphic to the polynomial algebra in m indeterminates* $A[X_1, \ldots, X_m]$.

Finally, as in (A.14), we can put $\mathbf{S}_n(E)$ and $\mathbf{S}_n(E^*)$ in duality, in such a way that

(A.17.8)
$$\langle x_1 x_2 \cdots x_n, x_1^* x_2^* \cdots x_n^* \rangle = \sum_{\sigma \in \mathfrak{S}_n} \langle x_{\sigma(1)}, x_1^* \rangle \langle x_{\sigma(2)}, x_2^* \rangle \cdots \langle x_{\sigma(n)}, x_n^* \rangle$$

but it should be remarked that $\mathbf{S}(E^*)$ is only a *submodule* of the dual of $\mathbf{S}(E)$.

18. DERIVATIONS AND ANTIDERIVATIONS OF GRADED ALGEBRAS

(A.18.1) In this and the following section, the algebras under consideration will *not* be assumed necessarily to be associative. An algebra E over a commutative ring A with identity element is therefore an A-module E together with an A-*bilinear mapping* $E \times E \to E$, denoted by $(x, y) \mapsto xy$.

The algebra E is said to be *graded* if E is the *direct sum* of a sequence $(E_n)_{n \geqq 0}$ of submodules, such that

(A.18.1.1)
$$E_m E_n \subset E_{m+n}$$

for all pairs of integers $m, n \geqq 0$. If E admits an identity element e (so that $ex = xe = x$ for all $x \in E$), we assume also that $e \in E_0$.

The elements of E_n are said to be *homogeneous of degree n*. The zero element 0 is therefore homogeneous of all degrees, but a homogeneous element $x \neq 0$ belongs to only one E_n, and the integer n is called its *degree*.

The exterior algebra (A.13.5) and the symmetric algebra (A.17) of a finitely-generated free A-module M are graded algebras, graded, respectively, by the submodules $\bigwedge^n M$ and $\mathbf{S}_n(M)$.

(A.18.2) Let E be an A-algebra. A mapping $d : E \to E$ is called a *derivation* of E if it is A-linear and satisfies

(A.18.2.1) $$d(xy) = (dx)\,y + x\,(dy)$$

for all $x, y \in E$.

For example, if E is associative and $a \in E$, the mapping

(A.18.2.2) $$\mathrm{ad}(a): \quad x \mapsto ax - xa = [a, x]$$

is a derivation (called an *inner* derivation).

If E is a graded algebra, define a linear mapping $d: E \to E$ by putting $d(x_p) = px_p$ for all $p \geqq 0$ and all $x_p \in E_p$. Then d is a derivation, for if $x_p \in E_p$ and $x_q \in E_q$, we have

$$d(x_p x_q) = (p + q)x_p x_q = (dx_p)\,x_q + x_p\,(dx_q).$$

If E is associative, then by induction on n we obtain from (A.18.2.1)

(A.18.2.3) $$d(x_1 x_2 \cdots x_n) = \sum_{i=1}^{n} x_1 \cdots x_{i-1}\,(dx_i)x_{i+1} \cdots x_n$$

for all $x_1, \ldots, x_n \in E$ and any derivation $d: E \to E$.

If E has an identity element $e \neq 0$, we have

(A.18.2.4) $$d(e) = 0$$

because from $e^2 = e$ we obtain $d(e) = d(e^2) = e \cdot d(e) + d(e) \cdot e = d(e) + d(e)$.

(A.18.3) Let d_1, d_2 be two derivations of an algebra E. Then the linear mapping

(A.18.3.1) $$[d_1, d_2]: \quad x \mapsto d_1\,(d_2\,x) - d_2\,(d_1 x)$$

is a derivation, since we have

$$\begin{aligned} d_1\,(d_2(xy)) &= d_1\,((d_2\,x)y + x\,(d_2\,y)) \\ &= (d_1\,(d_2\,x))y + (d_2\,x)\,(d_1 y) + (d_1 x)\,(d_2\,y) + x\,(d_1(d_2\,y)) \end{aligned}$$

and our assertion follows by interchanging the indices 1 and 2 and subtracting.

(A.18.4) When E is a *graded* A-algebra, a derivation d of E is said to be of *degree r* (where r is an integer, positive, negative, or zero) if

(A.18.4.1) $$d(E_n) \subset E_{n+r}$$

for all $n \geqq 0$ (with the convention that $E_m = \{0\}$ for $m < 0$).

With the same convention, an *antiderivation of* E *of degree r* is defined to be an A-linear mapping $d : \mathrm{E} \to \mathrm{E}$ which satisfies (A.18.4.1) and the relations

(A.18.4.2) $$d(x_m x_n) = (dx_m)\, x_n + (-1)^{mr} x_m\, (dx_n)$$

for all $m, n \geqq 0$ and $x_m \in \mathrm{E}_m$, $x_n \in \mathrm{E}_n$. An antiderivation of *even* degree r is therefore a derivation of degree r.

(A.18.5) Let M be a finitely generated free A-module and let $\mathrm{E} = \bigwedge(\mathrm{M}^*)$ be the exterior algebra of the dual module M*. For each element $x \in \mathrm{M}$, the mapping $i(x)$ is defined on $\bigwedge^n \mathrm{M}^*$ for all integers $n \geqq 1$ (A.15.4), and we extend it to the whole of E by linearity and by taking it to be the zero mapping on $\mathrm{A} = \bigwedge^0 \mathrm{M}^*$. This mapping is an *antiderivation of* E *of degree* -1. To prove this assertion we must show that

(A.18.5.1) $$i(x)(u_p^* \wedge u_q^*) = (i(x)(u_p^*)) \wedge u_q^* + (-1)^p u_p^* \wedge (i(x)(u_q^*))$$

for all $u_p^* \in \bigwedge^p \mathrm{M}^*$ and $u_q^* \in \bigwedge^q \mathrm{M}^*$. By linearity, we may assume that $u_p^* = x_1^* \wedge x_2^* \wedge \cdots \wedge x_p^*$ and $u_q^* = x_{p+1}^* \wedge \cdots \wedge x_{p+q}^*$, where the x_j^* are arbitrary elements of M*, and then the result follows immediately from (A.15.4.3).

(A.18.6) If the graded algebra E has an identity element e (necessarily of degree 0), then we have $d(e) = 0$ for every antiderivation d of E, for the same reason as in (A.18.2.4).

Suppose that E is associative and let $x_j \in \mathrm{E}_{m_j}$ for $1 \leqq j \leqq n$; then if d is any antiderivation of E of degree r, we have

(A.18.6.1)
$$d(x_1 x_2 \cdots x_n) = \sum_{i=1}^{n} (-1)^{r(m_1 + \cdots + m_{i-1})} x_1 \cdots x_{i-1} (dx_i) x_{i+1} \cdots x_n$$

by induction on n.

From this formula and from (A.18.2.3) it follows that if two derivations (resp. antiderivations) of E *of the same degree* coincide on a set of homogeneous *generators* of E, then they are identical.

(A.18.7) (i) *If d is an antiderivation of odd degree r, its square $d \circ d$ is a derivation of degree $2r$.*

(ii) *If d_1* (resp. d_2) *is an antiderivation of degree r* (resp. s), *then the linear mapping*

(A.18.7.1) $$x \mapsto d_1(d_2 x) - (-1)^{rs} d_2(d_1 x)$$

is an antiderivation of degree $r + s$.

Let x be a homogeneous element of E of degree n. Then, for all $y \in \mathrm{E}$, we have

$$d_1(d_2(xy)) = (d_1(d_2 x))y + (-1)^{r(s+n)}(d_2 x)(d_1 y) + (-1)^{sn}(d_1 x)(d_2 y) + (-1)^{(r+s)n} x(d_1(d_2 y)).$$

If $d_1 = d_2 = d$ and $r = s$ is odd, this shows that $d \circ d$ is a derivation. Next, if we interchange d_1 and d_2 in this equation, we obtain

$$\begin{aligned} d_1(d_2(xy)) &- (-1)^{rs} d_2(d_1(xy)) \\ &= (d_1(d_2 x))y - (-1)^{rs}(d_2(d_1 x))y \\ &\quad + (-1)^{(r+s)n} x(d_1(d_2 y)) - (-1)^{rs+(r+s)n} x(d_2(d_1 y)), \end{aligned}$$

which proves (ii).

19. LIE ALGEBRAS

A *Lie algebra* over a commutative ring A (with identity element) is an A-module $\mathfrak{g}$ endowed with an A-bilinear mapping of $\mathfrak{g} \times \mathfrak{g}$ into $\mathfrak{g}$, usually denoted by $(x, y) \mapsto [x, y]$, which satisfies the identities

(A.19.1) $$[x, x] = 0,$$

(A.19.2) $$[x, [y, z]] + [y, [z, x]] + [z, [x, y]] = 0$$

for all $x, y, z \in \mathfrak{g}$. The identity (A.19.2) is called the *Jacobi identity*. From (A.19.1) we deduce immediately that

(A.19.3) $$[x, y] = -[y, x].$$

If E is an associative A-algebra, then the A-module E endowed with the A-bilinear mapping $(x, y) \mapsto xy - yx$ is a Lie algebra.

A *subalgebra* $\mathfrak{h}$ of a Lie algebra $\mathfrak{g}$ (i.e., a submodule $\mathfrak{h}$ of $\mathfrak{g}$ such that $[x, y] \in \mathfrak{h}$ for all $x, y \in \mathfrak{h}$) is clearly a Lie algebra. If E is an arbitrary A-algebra, the subalgebra of the algebra $\mathrm{End}_{\mathrm{A}}(\mathrm{E})$ of A-module endomorphisms of E, formed by the *derivations* of E, is a Lie algebra with respect to the bracket (A.18.3.1).

A submodule $\mathfrak{a}$ of a Lie algebra $\mathfrak{g}$ is an *ideal* of $\mathfrak{g}$ if the relations $x \in \mathfrak{g}$ and $y \in \mathfrak{a}$ imply $[x, y] \in \mathfrak{a}$ (or equivalently $[y, x] \in \mathfrak{a}$). If $\dot{x}$, $\dot{y}$ are two elements of the quotient A-module $\mathfrak{g}/\mathfrak{a}$ (i.e., cosets of $\mathfrak{a}$ in $\mathfrak{g}$), then the values of $[x, y]$ for all $x \in \dot{x}$ and $y \in \dot{y}$ belong to the same coset of $\mathfrak{a}$ in $\mathfrak{g}$, because if $x' - x \in \mathfrak{a}$ and $y' - y \in \mathfrak{a}$, we have $[x', y'] - [x, y] = [x', y' - y] + [x' - x, y]$. This coset is denoted by $[\dot{x}, \dot{y}]$, and it is immediately verified that the mapping $(\dot{x}, \dot{y}) \mapsto [\dot{x}, \dot{y}]$ defines a Lie algebra structure on $\mathfrak{g}/\mathfrak{a}$. The A-module $\mathfrak{g}/\mathfrak{a}$, endowed with this structure, is called the *quotient Lie algebra* (of $\mathfrak{g}$ by $\mathfrak{a}$).

If $\mathfrak{g}$, $\mathfrak{g}'$ are two Lie algebras, a *homomorphism* f of $\mathfrak{g}$ into $\mathfrak{g}'$ is an A-linear mapping such that $f([x, y]) = [f(x), f(y)]$ for all $x, y \in \mathfrak{g}$. The kernel of f is an ideal $\mathfrak{a}$ of $\mathfrak{g}$ and the image of f a Lie subalgebra of $\mathfrak{g}'$, canonically isomorphic to $\mathfrak{g}/\mathfrak{a}$.

If $\mathfrak{g}_1$, $\mathfrak{g}_2$ are two Lie algebras over A, the product A-module $\mathfrak{g}_1 \times \mathfrak{g}_2$ is a Lie algebra with respect to the law of composition

$$((x_1, x_2), (y_1, y_2)) \mapsto ([x_1, y_1], [x_2, y_2]).$$

The verification of the axioms (A.19.1) and (A.19.2) is immediate. This Lie algebra is called the *product* of $\mathfrak{g}_1$ and $\mathfrak{g}_2$. The mapping $x_1 \mapsto (x_1, 0)$ (resp. $x_2 \mapsto (0, x_2)$) is an isomorphism of $\mathfrak{g}_1$ (resp. $\mathfrak{g}_2$) onto an ideal of $\mathfrak{g}_1 \times \mathfrak{g}_2$; usually we shall identify $\mathfrak{g}_1$ and $\mathfrak{g}_2$ with these ideals of $\mathfrak{g} \times \mathfrak{g}_2$. The quotient algebra $(\mathfrak{g}_1 \times \mathfrak{g}_2)/\mathfrak{g}_1$ (resp. $(\mathfrak{g}_1 \times \mathfrak{g}_2)/\mathfrak{g}_2$) is then identified with $\mathfrak{g}_2$ (resp. $\mathfrak{g}_1$). If f is a homomorphism of $\mathfrak{g}_1$ into $\mathfrak{g}_2$, its *graph* in $\mathfrak{g}_1 \times \mathfrak{g}_2$ is a Lie subalgebra of $\mathfrak{g}_1 \times \mathfrak{g}_2$, and the mapping $x_1 \mapsto (x_1, f(x_1))$ is an isomorphism of $\mathfrak{g}_1$ onto this subalgebra.

(A.19.4) *Let $\mathfrak{g}$ be a Lie algebra. For each $x \in \mathfrak{g}$, the linear mapping $y \mapsto [x, y]$ is a derivation of $\mathfrak{g}$, denoted by $\operatorname{ad}_{\mathfrak{g}}(x)$ or $\operatorname{ad}(x)$. The mapping $x \mapsto \operatorname{ad}(x)$ is a homomorphism of $\mathfrak{g}$ into the Lie algebra $\operatorname{Der}(\mathfrak{g})$ of derivations of $\mathfrak{g}$. For each derivation $\mathrm{D} \in \operatorname{Der}(\mathfrak{g})$ we have $[\mathrm{D}, \operatorname{ad}(x)] = \operatorname{ad}(\mathrm{D}x)$.*

In view of (A.19.3), the Jacobi identity may be written in the form

$$\operatorname{ad}(x) \cdot [y, z] = [\operatorname{ad}(x) \cdot y, z] + [y, \operatorname{ad}(x) \cdot z],$$

which proves the first assertion. It can also be written in the form

$$\operatorname{ad}([x, y]) \cdot z = \operatorname{ad}(x) \cdot (\operatorname{ad}(y) \cdot z) - \operatorname{ad}(y) \cdot (\operatorname{ad}(x) \cdot z),$$

which proves the second. Finally, we have from the definitions

$$\begin{aligned}[\mathrm{D}, \operatorname{ad}(x)] \cdot y &= \mathrm{D}([x, y]) - [x, \mathrm{D}y] \\ &= [\mathrm{D}x, y] = \operatorname{ad}(\mathrm{D}x) \cdot y,\end{aligned}$$

which proves the last assertion.

A derivation of $\mathfrak{g}$ of the form $\operatorname{ad}(x)$ is called an *inner derivation.* A submodule $\mathfrak{a}$ of $\mathfrak{g}$ is an ideal if and only if it is stable under all inner derivations.

If $\mathfrak{a}$ and $\mathfrak{b}$ are ideals in a Lie algebra $\mathfrak{g}$, then $\mathfrak{a} + \mathfrak{b}$ and $\mathfrak{a} \cap \mathfrak{b}$ are ideals in $\mathfrak{g}$. If $\mathfrak{h}$ is a Lie subalgebra of $\mathfrak{g}$ and $\mathfrak{a}$ is an ideal of $\mathfrak{g}$, then $\mathfrak{h} + \mathfrak{a}$ is a Lie subalgebra of $\mathfrak{g}$, and $(\mathfrak{h} + \mathfrak{a})/\mathfrak{a}$ is the image of $\mathfrak{h}$ under the canonical homomorphism of $\mathfrak{g}$ onto $\mathfrak{g}/\mathfrak{a}$, which is canonically isomorphic to $\mathfrak{h}/(\mathfrak{h} \cap \mathfrak{a})$.

If $\mathfrak{a}$, $\mathfrak{b}$ are two submodules of a Lie algebra, we denote by $[\mathfrak{a}, \mathfrak{b}]$ (by abuse of notation) the submodule of $\mathfrak{g}$ *generated* by all $[x, y]$ with $x \in \mathfrak{a}$ and $y \in \mathfrak{b}$. Clearly $[\mathfrak{a}, \mathfrak{b}] = [\mathfrak{b}, \mathfrak{a}]$. If $\mathfrak{a}$, $\mathfrak{b}$ are *ideals* in $\mathfrak{g}$, so is $[\mathfrak{a}, \mathfrak{b}]$; this follows immediately from the Jacobi identity.

A Lie algebra $\mathfrak{g}$ is said to be *commutative* if $[x, y] = 0$ for all $x, y \in \mathfrak{g}$. In that case every submodule of $\mathfrak{g}$ is an ideal, and every quotient algebra of $\mathfrak{g}$ is commutative.

The ideal $[\mathfrak{g}, \mathfrak{g}]$ is the smallest ideal $\mathfrak{a}$ such that $\mathfrak{g}/\mathfrak{a}$ is commutative. It is called the *derived ideal* of $\mathfrak{g}$ and is denoted by $\mathscr{D}\mathfrak{g}$. The *derived series* of $\mathfrak{g}$ is the decreasing sequence of ideals defined by induction in the following way:

$$\mathscr{D}^0\mathfrak{g} = \mathfrak{g}, \qquad \mathscr{D}^{p+1}\mathfrak{g} = [\mathscr{D}^p\mathfrak{g}, \mathscr{D}^p\mathfrak{g}].$$

A Lie algebra $\mathfrak{g}$ is called *solvable* if there is an integer $p \geqq 0$ such that $\mathscr{D}^p\mathfrak{g} = \{0\}$.

REFERENCES

VOLUME I

[1] Ahlfors, L., "Complex Analysis," McGraw-Hill, New York, 1953.

[2] Bachmann, H., "Transfinite Zahlen" (Ergebnisse der Math., Neue Folge, Heft 1). Springer, Berlin, 1955.

[3] Bourbaki, N., "Eléments de Mathématique," Livre I, "Théorie des ensembles" (Actual. Scient. Ind., Chaps. I, II, No. 1212; Chap III, No. 1243). Hermann, Paris, 1954–1956.

[4] Bourbaki, N., "Eléments de Mathématique," Livre II, "Algèbre" (Actual Scient. Ind., Chap. II, Nos. 1032, 1236, 3rd ed.). Hermann, Paris, 1962.

[5] Bourbaki, N., "Eléments de Mathématique," Livre III, "Topologie générale" (Actual. Scient. Ind., Chaps. I, II, Nos. 858, 1142, 4th ed.; Chap. IX, No. 1045, 2nd ed.; Chap. X, No. 1084, 2nd ed.). Hermann, Paris, 1958–1961.

[6] Bourbaki, N., "Eléments de Mathématique," Livre V, "Espaces vectoriels topologiques" (Actual. Scient. Ind., Chap. I, II, No. 1189, 2nd ed.; Chaps. III–V, No. 1229). Hermann, Paris, 1953–1955.

[7] Cartan, H., Séminaire de l'Ecole Normale Supérieure, 1951–1952: "Fonctions analytiques et faisceaux analytiques."

[8] Cartan, H., "Théorie Élémentaire des Fonctions Analytiques." Hermann, Paris, 1961.

[9] Coddington, E., and Levinson, N., "Theory of Ordinary Differential Equations." McGraw-Hill, New York, 1955.

[10] Courant, R., and Hilbert, D., "Methoden der mathematischen Physik," Vol. I, 2nd ed. Springer, Berlin, 1931.

[11] Halmos, P., "Finite Dimensional Vector Spaces," 2nd ed. Van Nostrand-Reinhold, Princeton, New Jersey, 1958.

[12] Ince, E., "Ordinary Differential Equations," Dover, New York, 1949.

[13] Jacobson, N., "Lectures in Abstract Algebra," Vol. II, "Linear algebra." Van Nostrand-Reinhold, Princeton, New Jersey, 1953.

[14] Kamke, E., "Differentialgleichungen reeller Funktionen." Akad. Verlag, Leipzig, 1930.

[15] Kelley, J., "General Topology." Van Nostrand-Reinhold, Princeton, New Jersey, 1955.

[16] Landau, E., "Foundations of Analysis." Chelsea, New York, 1951.

[17] Springer, G., "Introduction to Riemann Surfaces." Addison-Wesley, Reading, Massachusetts, 1957.
[18] Weil, A., "Introduction à l'Étude des Variétés Kählériennes" (Actual. Scient. Ind., No. 1267). Hermann, Paris, 1958.
[19] Weyl, H., "Die Idee der Riemannschen Fläche," 3rd ed. Teubner, Stuttgart, 1955.

VOLUME II

[20] Akhiezer, N., "The Classical Moment Problem." Oliver and Boyd, Edinburgh–London, 1965.
[21] Arnold, V. and Avez, A., "Théorie Ergodique des Systèmes Dynamiques." Gauthier-Villars, Paris, 1967.
[22] Bourbaki, N., "Eléments de Mathématique," Livre VI, "Intégration" (Actual. Scient. Ind., Chap. I–IV, No. 1175, 2nd ed., Chap. V, No. 1244, 2nd ed., Chap VII–VIII, No. 1306). Hermann, Paris, 1963–67.
[23] Bourbaki, N., "Eléments de Mathématique: Théories Spectrales" (Actual. Scient. Ind., Chap I, II, No. 1332). Hermann, Paris, 1967.
[24] Dixmier, J., "Les Algèbres d'Opérateurs dans l'Espace Hilbertien." Gauthier–Villars, Paris, 1957.
[25] Dixmier, J., "Les C*-Algèbres et leurs Représentations." Gauthier–Villars, Paris, 1964.
[26] Dunford, N. and Schwartz, J., "Linear Operators. Part II: Spectral Theory." Wiley (Interscience), New York, 1963.
[27] Hadwiger, H., "Vorlesungen über Inhalt, Oberfläche und Isoperimetrie." Springer, Berlin, 1957.
[28] Halmos, P., "Lectures on Ergodic Theory." Math. Soc. of Japan, 1956.
[29] Hoffman, K., "Banach Spaces of Analytic Functions." New York, 1962.
[30] Jacobs, K., "Neuere Methoden und Ergebnisse der Ergodentheorie" (Ergebnisse der Math., Neue Folge, Heft 29). Springer, Berlin, 1960.
[31] Kaczmarz, S. and Steinhaus, H., "Theorie der Orthogonalreihen." New York, 1951.
[32] Kato, T., "Perturbation Theory for Linear Operators." Springer, Berlin, 1966.
[33] Montgomery, D. and Zippin, L., "Topological Transformation Groups." Wiley (Interscience), New York, 1955.
[34] Naimark, M., "Normal Rings." P. Nordhoff, Groningen, 1959.
[35] Rickart, C., "General Theory of Banach Algebras." Van Nostrand-Reinhold, New York, 1960.
[36] Weil, A., "Adeles and Algebraic Groups." The Institute for Advanced Study, Princeton, New Jersey, 1961.

VOLUME III

[37] Abraham, R., "Foundations of Mechanics." Benjamin, New York, 1967.
[38] Cartan, H., Séminaire de l'École Normale Supérieure, 1949–50: "Homotopie: espaces fibrés."
[39] Chern, S. S., "Complex Manifolds" (Textos de matematica, No. 5). Univ. do Recife Brazil, 1959.
[40] Gelfand, I. M. and Shilov, G. E., "Les Distributions," Vols. 1 and 2. Dunod, Paris, 1962.

[41] Gunning, R., "Lectures on Riemann Surfaces." Princeton Univ. Press, Princeton, New Jersey, 1966.
[42] Gunning, R., "Lectures on Vector Bundles over Riemann Surfaces." Princeton Univ. Press, Princeton, New Jersey, 1967.
[43] Hu, S. T., "Homotopy Theory." Academic Press, New York, 1969.
[44] Husemoller, D., "Fiber Bundles." McGraw-Hill, New York, 1966.
[45] Kobayashi, S., and Nomizu, K., "Foundations of Differential Geometry," Vols. 1 and 2. Wiley (Interscience), New York, 1963 and 1969.
[46] Lang, S., "Introduction to Differentiable Manifolds." Wiley (Interscience), New York, 1962.
[47] Porteous, I. R., "Topological Geometry." Van Nostrand-Reinhold, Princeton, New Jersey, 1969.
[48] Schwartz, L., "Théorie des Distributions," New ed. Hermann, Paris, 1966.
[49] Steenrod, N., "The Topology of Fiber Bundles." Princeton Univ. Press, Princeton, New Jersey, 1951.
[50] Sternberg, S., "Lectures on Differential Geometry." Prentice-Hall, Englewood Cliffs, New Jersey, 1964.

INDEX

In the following index the first reference number refers to the chapter in which the subject may be found and the second to the section within the chapter.

A

Analytic manifold defined by a holomorphic function : 16.8, prob. 12
Analytic mapping : 16.3
Analytically compatible charts : 16.1
Antiderivation of degree r : A.18.4
Antisymmetric tensor : A.12.1
Antisymmetrization : A.12.2
Apsidal transformation : 16.20, prob. 4
Arcwise-connected : 16.27, prob. 1
Arcwise-connected component : 16.27, prob. 1
Area of a face of a polyhedron : 16.24, prob. 3
Area of the frontier of a convex body : 16.24, prob. 4
Atlas : 16.1
Automorphism of a Lie group : 16.9

B

Base of a fibration : 16.12
Basis of a module : A.8.2
Basis of a tangent (cotangent) space associated with a chart : 16.5
Bernstein's theorem : 17.8, prob. 3
Bessel function : 17.11, prob. 2
Bidual of a module : A.9.1
Bieberbach's theorem on injective functions: 16.22, prob. 5
Bilinear morphism of vector bundles : 16.16
Blowing up a manifold at a point : 16.11, prob. 3
Boundary of a current : 17.15
Bundle : 16.12
Bundle associated with a principal bundle: 16.14
Bundle of antisymmetric (symmetric) tensors of order m : 16.17

C

C^0-fibration : 16.25, prob. 8
C^0-vector bundle : 16.25, prob. 8
Canonical basis of the module of differential p-forms on an open set in $\mathbf{R}^n$: 16.20
Canonical bijection of $T_x(E)$ onto E (E a vector space) : 16.5
Canonical bilinear form on $E \times E^*$: A.9.1
Canonical chart on an open set in $\mathbf{R}^n$: 16.1
Canonical Lie group structure on a vector space : 16.9
Canonical manifold structure on a vector space : 16.2
Canonical mapping of a module into its bidual : A.9.1
Canonical morphisms : 16.18
$E' \oplus E'' \to E' \otimes E''$
$E_1 \otimes (E_2 \otimes E_3) \to (E_1 \otimes E_2) \otimes E_3$

$E_1 \otimes (E_2 \oplus E_3) \to (E_1 \otimes E_2) \oplus (E_1 \otimes E_3)$
$\mathrm{Hom}(E \otimes F, G) \to \mathrm{Hom}(E, \mathrm{Hom}(F, G))$
$\mathrm{Hom}(E' \oplus E'', F) \to \mathrm{Hom}(E', F) \oplus \mathrm{Hom}(E'', F)$
$\mathrm{Hom}(E, F' \oplus F'') \to \mathrm{Hom}(E, F') \oplus \mathrm{Hom}(E, F'')$
$\mathrm{Hom}(E', F') \otimes \mathrm{Hom}(E'', F'') \to \mathrm{Hom}(E' \otimes E'', F' \otimes F'')$
$E \to E^{**}$
$\mathrm{Hom}(E, F) \to \mathrm{Hom}(F^*, E^*)$
$(E \otimes F)^* \to E^* \otimes F^*$
$E^* \otimes F \to \mathrm{Hom}(E, F)$
$E^* \otimes E \to B \times \mathbf{R}$
$\mathbf{T}^p_q(E) \otimes \mathbf{T}^r_s(E) \to \mathbf{T}^{p+r}_{q+s}(E)$
$\mathbf{T}^p_q(E) \otimes \mathbf{T}^r_s(E) \to \mathrm{Hom}(\mathbf{T}^q_p(E), \mathbf{T}^r_s(E))$
$(\mathbf{T}^p_q(E))^* \to \mathbf{T}^q_p(E)$
$E^{\otimes m} \to \bigwedge^m E$
$(\bigwedge^p E) \otimes (\bigwedge^q E) \to \bigwedge^{p+q} E$
$(\bigwedge^p E)^* \to \bigwedge^p (E^*)$
$E \otimes (\bigwedge^p E^*) \to \bigwedge^{p-1} E^*$
$E_{(\mathbf{C})} \otimes E'_{(\mathbf{C})} \to (E \otimes E')_{(\mathbf{C})}$
$(\mathrm{Hom}(E, E'))_{(\mathbf{C})} \to \mathrm{Hom}(E_{(\mathbf{C})}, E'_{(\mathbf{C})})$
$(E^*)_{(\mathbf{C})} \to (E_{(\mathbf{C})})^*$
$(\bigwedge^m E)_{(\mathbf{C})} \to \bigwedge^m (E_{(\mathbf{C})})$
Canonical orientation of a complex manifold : 16.21
Canonical orientation of $\mathbf{R}^n$: 16.21
Canonical trivialization of T(M) (M open in $\mathbf{R}^n$) : 16.15
Canonical 2-form on T(M)* : 17.15
Canonical vector bundle over a Grassmannian : 16.16
Cauchy principal value of an integral : 17.9, prob. 1
Cauchy's formula for convex bodies : 16.24, prob. 4
Cauchy's formula for convex polyhedra: 16.24, prob. 3
Chart, chart at a point : 16.1
Class C^r, C^∞ (functions) : 16.3
Coarse C^r-topology : 17.1, prob. 1
Compatible atlases, charts : 16.1
Compatible (group structure and manifold structure) : 16.9
Completely monotone function : 17.8, prob. 3
Complex-analytic atlas : 16.1
Complex-analytic fibration : 16.12
Complex-analytic Lie group : 16.9
Complex-analytic manifold : 16.1
Complex vector bundle : 16.15
Composition of two jets : 16.9, prob. 1
Cone of revolution : 16.1, prob. 1
Conjugate of a complex current : 17.8
Connected sum of two manifolds : 16.26, prob. 15
Connection in a vector bundle : 17.16
Connection on a differential manifold : 17.18
Contact of order $\geq k$ (of mappings) : 16.5, prob. 9
Contact transformation : 16.20
Contingent : 16.8, prob. 5
Contractible manifold : 16.27
Contraction of a tensor : 16.18 and A.11.3
Contragredient of an isomorphism : A.9.3
Convergent integral : 17.5, prob. 4
Convex body : 16.5, prob. 6
Convex polyhedron : 16.5, prob. 6
Convolution of a finite sequence of distributions : 17.11
Coordinate functions : A.9.2
Coordinates relative to a chart : 16.1
Cotangent bundle : 16.20
Covariant derivative : 17.17
Covariant differential of a tensorfield : 17.18
Covariant exterior differential of a differential p-form with values in a vector bundle : 17.19
Covector (tangent) : 16.5
Covering homology theorem : 16.28, prob. 10
Covering space : 16.12
Critical point : 16.5
Critical value of a function : 16.5
Current : 17.3
Curvature of a connection, curvature morphism, curvature tensor field, curvature tensor : 17.20
Curve : 16.1

D

D'Alembertian : 17.9
Degree of a homogeneous element of a graded algebra : A.18.1
Derivation in an algebra : A.18.2

Derivation of degree r in a graded algebra : A.18.4
Derivative of a distribution : 17.5
Derivative of a function in the direction of a tangent vector : 17.14
Derived ideal of a Lie algebra, derived series : A.19
Diffeomorphic manifolds : 16.2
Diffeomorphism : 16.2
Difference of two linear connections : 17.16
Differentiable action (of a Lie group on a manifold) : 16.10
Differentiable mapping : 16.3
Differential fibration : 16.12
Differential manifold : 16.1
Differential manifold underlying an analytic manifold : 16.1
Differential of a function : 16.20
Differential of a function at a point : 16.5
Differential operator : 17.13
Differential p-form : 16.20
Differential p-form with values in a vector bundle : 17.19
Dilatation : 16.20, prob. 4
Dimension of a chart : 16.1
Dimension of a convex body : 16.5, prob. 6
Dimension of a manifold at a point : 16.1
Dimension of a pure manifold : 16.1
Dirac p-current : 17.3
Direct sequence of vector fields : 16.21
Direct sum of vector bundles : 16.16
Distribution : 17.3
Divisor, divisor of a function : 16.14, prob. 3
Domain of definition of a chart : 16.1
Dual basis : A.9.2
Dual of a module : A.9.1

E

Embedding of a manifold : 16.8
Equivariant actions : 16.10
Étale mapping : 16.5
Euclidean sphere : 16.2
Exact homotopy sequence : 16.30, prob. 5
Exact homotopy sequence of fiber bundles : 16.30, prob. 6
Exact sequence of vector bundles : 16.17
Extensions of sections : 16.25, prob. 9
Exterior algebra : A.13.5
Exterior differential of a differential p-form : 17.15
Exterior powers of a finitely-generated free module : A.13.3
Exterior powers of a linear mapping : A.13.3
Exterior powers of a vector bundle : 16.16
Exterior product of antisymmetric tensors, p-vectors : A.13.2

F

Face of a convex polyhedron : 16.5, prob. 6
Fiber, fiber at a point : 16.12
Fiber bundle : 16.12
Fiber product : 16.12
Fiber-type : 16.12
Fibered chart : 16.15
Fibered chart of T(M)* associated with a chart of M : 16.20
Fibered manifold : 16.12
Fibration : 16.12
Fibration underlying an analytic fibration : 16.12
Field of point-distributions : 17.13
Fine C^r-topology : 17.1, prob. 2
Finite part of an integral : 17.9
Frame, frame at a point, framing : 16.15
Frame dual to a frame of E : 16.16
Frame of T(M) associated with a chart : 16.15
Frame of T(M)* associated with a chart : 16.20
Frame of $\mathbf{T}^p_q$(E) induced by a frame of E : 16.16
Free family of elements of a module : A.8.2.
Fröhlicher–Nijenhuis theorem : 17.19, prob. 3
Function of class C^r, C^∞ : 16.3
Function of support : 16.5, prob. 7
Fundamental divisor on $\mathbf{P}_1(\mathbf{C})$: 16.14, prob. 3
Fundamental group of a connected manifold : 16.27
Fundamental groups of a manifold at a point : 16.27
Fundamental 1-form on T(M)* : 16.20

G

Galois covering : 16.28, prob. 2
Gaussian mapping : 16.19, prob. 8
Global section : 16.12

Graded algebra : A.18.1
Grassmannian (real, complex, quaternionic) : 16.11

H

Heaviside function : 17.5
Hessian matrix : 16.5
Hessian of a function at a point : 16.5
Holomorphic mapping : 16.3
Holomorphic vector bundle : 16.15
Homogeneous contact transformation : 16.20
Homogeneous element : A.18.1
Homomorphism of Lie algebras : A.19
Homomorphism of Lie groups : 16.9
Homotopic mappings : 16.26
Homotopy, C^p-homotopy : 16.26
Homotopy equivalence : 16.26, prob. 2
Homotopy groups : 16.30, prob. 3
Homotopy type : 16.26, prob. 2
Hopf fibration : 16.14
Horizontal lifting of a vector field : 17.16
Hypergeometric function : 17.11, prob. 2
Hypersurface : 16.8

I

Ideal of a Lie algebra : A.19
Image of a current : 17.3 and 17.7
Imaginary part of a current : 17.6
Immersion : 16.7
Implicit function theorem : 16.6
Induced covariant tensor field on a submanifold : 16.20
Induced current on an open set : 17.4
Induced differential form on a submanifold : 16.20
Induced fibration on a submanifold : 16.12
Induced orientation on an open set : 16.21
Induced principal bundle on a submanifold : 16.14
Induced structure of differential manifold on an open set : 16.2
Induced vector bundle on a submanifold : 16.19
Inner derivation in a Lie algebra : A.19.4
Inner derivation in an associative algebra : A.18.2
Integral of a differential form along the fibers : 16.24, prob. 11
Integral of an n-form : 16.24
Interior product of a p-vector and a $(p+q)$-form : A.15.3
Invariant distribution : 17.8, prob. 4
Inverse image of a covariant tensor field : 16.20
Inverse image of a current under a local diffeomorphism : 17.4
Inverse image of a differential form : 16.20
Inverse image of a differential form with values in a vector bundle : 17.19
Inverse image of a distribution : 17.5, probs. 8 and 9
Inverse image of a fibration : 16.12
Inverse image of a section : 16.12
Inverse image of a vector bundle : 16.19
Invertible jet : 16.9, prob. 1
Isomorphism of differential manifolds : 16.2
Isomorphism of fibrations, B-isomorphism of fibrations : 16.12
Isomorphism of Lie groups : 16.9
Isomorphism of principal bundles, B-isomorphism of principal bundles : 16.14
Isomorphism of vector bundles, B-isomorphism of vector bundles : 16.15
Isotopy : 16.26
Isotropic subspace : A.16.3

J

Jacobi identity : A.19
Jacobi polynomial : 17.11, prob. 2
James embedding: 16.25, prob. 16
Jet of order k from X to Y : 16.5, prob. 9
Jet of order k of a mapping at a point : 16.5, prob. 9

K

Klein bottle : 16.14
Kneser–Glaeser theorem : 16.23, prob. 1
Kronecker tensor field : 16.17

L

Lagrange's method of undetermined multipliers : 16.20, prob. 5
Laplacian : 17.9

Lebesgue measure : 16.22
Legendre transformation : 16.20
Lie algebra : A.19
Lie bracket of two vector fields : 17.14
Lie group, Lie group homomorphism : 16.9
Lie subgroup : 16.9
Lie's transformation : 16.20, prob. 4
Lifting of a mapping to a fiber bundle : 16.12
Lifting of a path-homotopy : 16.28
Linear connection : 17.16
Linear differential operator : 17.13
Linear representation of a Lie group : 16.9
Local coordinates at a point : 16.1
Local diffeomorphism : 16.5
Local expression of a B-morphism of fiber bundles : 16.13
Local expression of a Hessian $\mathrm{Hess}_x(f)$: 16.5
Local expression of a mapping : 16.3
Local expression of a morphism of vector bundles : 16.15
Local expression of a p-form : 16.5
Local expression of a section of a vector bundle : 16.15
Local expression of a tangent linear mapping $\mathrm{T}_x(f)$: 16.5
Local expression of a tangent vector $\mathbf{h}_x$: 16.5
Local expression of a vector field: 16.15
Local homomorphism : 16.9
Local triviality : 16.12
Locally arcwise-connected : 16.27, prob. 1
Locally integrable n-form : 16.24
Locally isomorphic Lie groups : 16.9
Lorentz group : 16.11

M

Manifold : 16.1
Manifold obtained by patching : 16.2
Manifold of class C^r : 16.1, prob. 2
Measurable n-form : 16.24
Meromorphic function : 16.14, prob. 2
Möbius strip : 16.14
Module : A.8.1
Module obtained by extending the ring of scalars : A.10.6
Monodromy principle : 16.28
Morphism of differential manifolds : 16.3
Morphism of fibrations, B-morphism of fibrations : 16.12
Morphism of principal bundles : 16.14
Morphism of vector bundles, B-morphism of vector bundles : 16.15
Morse index of a function at a point : 16.5, prob. 3
Morton Brown's theorem : 16.2, prob. 5
Multiple of a vector bundle : 16.16
Multiplet layer : 17.10

N

Natural continuation of a holomorphic function : 16.8, prob. 12
n-chain element without boundary : 17.5
Negative differential form (relative to an orientation) : 16.21
Negative sequence of vector fields : 16.21
Negligible n-form : 16.24
Nijenhuis torsion : 17.19, prob. 3
Nondegenerate alternating bilinear form : A.16.1
Nondegenerate critical point : 16.4
Nonhomogeneous contact transformation: 16.20, prob. 3
Normal bundle of a submanifold : 16.19
n-sheeted covering : 16.12

O

Open n-chain, open n-chain element : 17.5
Operator of local character : 17.13
Opposite (of a Lie group) : 16.9
Orbit manifold : 16.10
Order of a current : 17.3
Order of a differential operator : 17.13
Orientable manifold: 16.21
Orientation of a manifold : 16.21
Orientation of $f^{-1}(y)$ induced from those of X and Y (f: X → Y a submersion) : 16.21
Orientation of X induced from an orientation of Y by an étale morphism f: X → Y : 16.21
Orientation preserving, reversing : 16.21
Oriented manifold : 16.21
Orthogonal supplement (of a vector subspace relative to an alternating bilinear form) : A.16.3

Orthogonal vectors (relative to an alternating bilinear form) : A.16.3

P

Parallelizable manifold : 16.15, prob. 1
Paratingent : 16.8, prob. 5
Patching condition for fiber bundles : 16.13
p-covector : A.14.2
Periodic current : 17.9
p-form : A.14.2
Poincaré–Volterra theorem : 16.8, prob. 11
Point-distribution : 17.7
Poisson bracket of differential 1-forms : 17.15, prob. 9
Positive differential form (relative to an orientation) : 16.21
Positive sequence of vector fields : 16.21
Predivisor : 16.14, prob. 3
Principal bundle : 16.14
Principal C^0-bundle : 16.25, prob. 8
Principal divisor : 16.14, prob. 3
Privileged local coordinate system (relative to a p-form) : 17.15, prob. 2
Product manifold : 16.6
Product of fibrations : 16.12
Product of Lie algebras : A.19
Product of manifolds : 16.6
Product of orientations : 16.21
Product of two manifolds over a manifold : 16.8, prob. 10
Projective bundle : 16.19, prob. 9
Projective space (real, complex, quaternionic) : 16.11
Proper mapping : 17.3
Pure differential manifold : 16.1
p-vector : A.13.6

Q

Quotient bundle : 16.17
Quotient Lie algebra : A.19

R

Rank of a C^1-mapping at a point : 16.5
Rank of a vector bundle at a point : 16.15
Rank of an alternating bilinear form : A.16.1
Real-analytic atlas : 16.1
Real-analytic fibration : 16.12
Real-analytic group : 16.9
Real-analytic manifold : 16.1
Real-analytic manifold underlying a complex manifold : 16.1
Real part of a current : 17.6
Real p-current : 17.6
Real vector bundle underlying a complex vector bundle : 16.15
Reciprocal polars : 16.20, prob. 4
Regular value : 16.23
Regularizing sequence : 17.1
Relative homotopy groups : 16.30, prob. 4
Restriction of a chart to an open set : 16.1
Restriction of a differential operator to an open set : 17.13
Restriction of an atlas to an open set : 16.2
Retrograde sequence of vector fields : 16.21
ρ-extension of a principal bundle : 16.14, prob. 17
Riemann sphere : 16.11
Riemann surface : 16.1
Riemann surface defined by a function of two variables : 16.8
Riemann surface defined by a holomorphic function : 16.8, prob. 12
Riemann surface of the logarithmic function : 16.8
Rotation group : 16.11

S

Sard's theorem : 16.23
Saturated atlas : 16.1
Section of a fibration, fiber bundle : 16.12
Simple layer : 17.10
Simply-connected manifold : 16.27
Singular support of a current : 17.5
Solid angle : 16.24
Sonine's formula : 17.11, prob. 2
Source of a jet : 16.5, prob. 9
Space of a fibration : 16.12
Special linear group : 16.9
Special orthogonal group : 16.9
Special unitary group : 16.11
Sphere oriented toward the outside (inside) : 16.21
Stationary at a point : 16.5
Steiner–Minkowski formula : 16.24, prob. 7

Stereographic projection : 16.2
Stiefel manifolds : 16.11
Stokes' theorem : 16.24
Strictly convolvable distributions : 17.11
Subalgebra of a Lie algebra : A.19
Subbundle : 16.17
Subimmersion, subimmersion at a point : 16.7
Submanifold : 16.8
Submersion, submersion at a point : 16.7
Summable distribution : 17.11, prob. 1
Supplement (of a vector subbundle) : 16.17
Support of a current : 17.4
Surface : 16.1
Surface measure on S_n : 16.24
Symmetric algebra : A.17
Symmetric powers of a finitely-generated free module : A.17
Symmetric product of symmetric tensors : A.17
Symmetric tensor : A.12.1
Symmetrization of a tensor : A.12.2
Symplectic automorphism : A.16.4
Symplectic basis : A.16.4
Symplectic group : A.16.4
System of local coordinates : 16.1

T

Tangent affine-linear variety : 16.8
Tangent bundle : 16.15
Tangent covector : 16.5
Tangent (functions) : 16.5
Tangent hyperplane : 16.8
Tangent linear mapping : 16.5
Tangent plane : 16.8
Tangent to a curve : 16.8
Tangent vector at a point of a differential manifold : 16.5
Tangent vector field : 16.15
Tangent vector space at a point of a differential manifold : 16.5
Target of a jet : 16.5, prob. 9
Tautological vector bundle over a Grassmannian : 16.16
Tensor : A.11.1
Tensor bundle of type (p, q) : 16.16
Tensor (by abuse of language) : 16.20
Tensor field of type (p, q) : 16.18 and 16.20
Tensor multiplication : 16.18
Tensor powers of a finitely-generated free module : A.11.1
Tensor powers of a vector bundle : 16.16
Tensor product of distributions : 17.10
Tensor product of elements of free A-modules : A.10.1
Tensor product of finitely-generated free A-modules : A.10.3
Tensor product of linear forms : A.10.1
Tensor product of linear mappings : A.10.5
Tensor product of vector bundles : 16.16
Thom's transversality theorem : 16.25, prob. 17
Topological manifold : 16.1
Topological space underlying a manifold : 16.1
Torsion, torsion morphism, torsion tensor, torsion tensor field : 17.20
Totally isotropic subspace : A.16.2
Trace measure on $G \times G$: 17.5, prob. 10
Trace of a matrix : A.11.3
Trace of an endomorphism of a finitely-generated free module : A.11.3
Trace of an endomorphism of a vector bundle : 16.18
Transition diffeomorphism : 16.13
Transition homeomorphism : 16.1
Transpose of a homomorphism of finitely-generated free modules : A.9.4
Transpose of a matrix : A.9.4
Transpose of a morphism of vector bundles : 16.16
Transversal at a point, transversal over a submanifold (mappings) : 16.8, prob. 9
Transversal mappings : 16.8, prob. 10
Transversal submanifolds : 16.8, prob. 9
Trivial fiber bundle : 16.12
Trivial principal bundle : 16.14
Trivial vector bundle : 16.15
Trivializable, trivializable over an open set (bundle or fibration) : 16.12
Trivializable principal bundle : 16.14
Trivializable vector bundle : 16.15
Trivialization of a fiber bundle, fibration : 16.12
Trivialization of a principal bundle : 16.14
Trivialization of a vector bundle : 16.15
Twisted torus : 16.14

U

Unending path : 16.27
Unimodular group : 16.9
Unitary group : 16.11
Universal covering : 16.29
Universal covering group of a Lie group : 16.30

V

Vector bundle (real, complex) : 16.15
Vector field : 16.15
Vector part of a section : 16.15
Vector-valued differential p-form : 16.20
Vertical tangent vector : 16.12
Volume, volume form : 16.24

W

Weak integral of a variable current : 17.8
Whitney sum of vector bundles : 16.16
Whitney's embedding theorem : 16.25, probs. 2 and 13
Whitney's extension theorem : 16.4, prob. 6

Z

Zero section of a vector bundle : 16.15

Pure and Applied Mathematics

A Series of Monographs and Textbooks

Editors **Paul A. Smith and Samuel Eilenberg**

Columbia University, New York

1: Arnold Sommerfeld. Partial Differential Equations in Physics. 1949 (Lectures on Theoretical Physics, Volume VI)

2: Reinhold Baer. Linear Algebra and Projective Geometry. 1952

3: Herbert Busemann and Paul Kelly. Projective Geometry and Projective Metrics. 1953

4: Stefan Bergman and M. Schiffer. Kernel Functions and Elliptic Differential Equations in Mathematical Physics. 1953

5: Ralph Philip Boas, Jr. Entire Functions. 1954

6: Herbert Busemann. The Geometry of Geodesics. 1955

7: Claude Chevalley. Fundamental Concepts of Algebra. 1956

8: Sze-Tsen Hu. Homotopy Theory. 1959

9: A. M. Ostrowski. Solution of Equations and Systems of Equations. Third Edition, in preparation

10: J. Dieudonné. Treatise on Analysis: Volume I, Foundations of Modern Analysis, enlarged and corrected printing, 1969. Volume II, 1970. Volume III, 1972

11: S. I. Goldberg. Curvature and Homology. 1962

12: Sigurdur Helgason. Differential Geometry and Symmetric Spaces. 1962

13: T. H. Hildebrandt. Introduction to the Theory of Integration. 1963

14: Shreeram Abhyankar. Local Analytic Geometry. 1964

15: Richard L. Bishop and Richard J. Crittenden. Geometry of Manifolds. 1964

16: Steven A. Gaal. Point Set Topology. 1964

17: Barry Mitchell. Theory of Categories. 1965

18: Anthony P. Morse. A Theory of Sets. 1965

19: Gustave Choquet. Topology. 1966

20: Z. I. Borevich and I. R. Shafarevich. Number Theory. 1966

21: José Luis Massera and Juan Jorge Schaffer. Linear Differential Equations and Function Spaces. 1966

22: Richard D. Schafer. An Introduction to Nonassociative Alegbras. 1966

23: Martin Eichler. Introduction to the Theory of Algebraic Numbers and Functions. 1966

24: Shreeram Abhyankar. Resolution of Singularities of Embedded Algebraic Surfaces. 1966

25: François Treves. Topological Vector Spaces, Distributions, and Kernels. 1967

26: Peter D. Lax and Ralph S. Phillips. Scattering Theory. 1967

27: Oystein Ore. The Four Color Problem. 1967

28: Maurice Heins. Complex Function Theory. 1968

29: R. M. Blumenthal and R. K. Getoor. Markov Processes and Potential Theory. 1968

30: L. J. Mordell. Diophantine Equations. 1969

31: J. Barkley Rosser. Simplified Independence Proofs: Boolean Valued Models of Set Theory. 1969

32: William F. Donoghue, Jr. Distributions and Fourier Transforms. 1969

33: Marston Morse and Stewart S. Cairns. Critical Point Theory in Global Analysis and Differential Topology. 1969

34: Edwin Weiss. Cohomology of Groups. 1969

35: Hans Freudenthal and H. De Vries. Linear Lie Groups. 1969

36: Laszlo Fuchs. Infinite Abelian Groups: Volume I, 1970. Volume II, 1973

37: Keio Nagami. Dimension Theory. 1970

38: Peter L. Duren. Theory of H^p Spaces. 1970

39: Bodo Pareigis. Categories and Functors. 1970

40: Paul L. Butzer and Rolf J. Nessel. Fourier Analysis and Approximation: Volume 1, One-Dimensional Theory. 1971

41: Eduard Prugovečki. Quantum Mechanics in Hilbert Space. 1971

42: D. V. Widder: An Introduction to Transform Theory. 1971

43: Max D. Larsen and Paul J. McCarthy. Multiplicative Theory of Ideals. 1971

44: Ernst-August Behrens. Ring Theory. 1972

45: Morris Newman. Integral Matrices. 1972

46: Glen E. Bredon. Introduction to Compact Transformation Groups. 1972

47: Werner Greub, Stephen Halperin, and Ray Vanstone. Connections, Curvature, and Cohomology: Volume I, De Rham Cohomology of Manifolds and Vector Bundles, 1972. Volume II, Lie Groups, Principal Bundles, and Characteristic Classes, in preparation

48: Xia Dao-xing. Measure and Integration Theory of Infinite-Dimensional Spaces: Abstract Harmonic Analysis. 1972

49: Ronald G. Douglas. Banach Algebra Techniques in Operator Theory. 1972

50: Willard Miller, Jr. Symmetry Groups and Their Applications. 1972

In preparation

Arthur A. Sagle and Ralph E. Walde. Introduction to Lie Groups and Lie Algebras

Gerald J. Janusz. Algebraic Number Fields

T. Benny Rushing. Topological Embeddings

Samuel Eilenberg. Automata, Languages, and Machines: Volume A

James W. Vick. Homology Theory: An Introduction to Algebraic Topology

E. R. Kolchin. Differential Algebra and Algebraic Groups